Progress in Nonlinear Differential Equations and Their Applications

Volume 7

Editor
Haim Brezis
Université Pierre et Marie Curie
Paris
and
Rutgers University
New Brunswick, N.J.

Nonlinear Diffusion Equations and Their Equilibrium States, 3

*Proceedings from a Conference held
August 20-29, 1989 in Gregynog, Wales*

N. G. Lloyd
W. M. Ni
L. A. Peletier
J. Serrin

Editors

1992

Birkhäuser

Boston · Basel · Berlin

N.G. Lloyd
Department of Mathematics
University College of Wales
Aberystwyth DYFED SY3BZ
United Kingdom

W.M. Ni
School of Mathematics
University of Minnesota
Minneapolis, MN 55455
U.S.A.

L. A. Peletier
Mathematical Institute
Leiden University
PB 9512, 2300 RA Leiden
The Netherlands

J. Serrin
School of Mathematics
University of Minnesota
Minneapolis, MN 55455
U.S.A.

Library of Congress Cataloging-in-Publication Data
Nonlinear diffusion equations and their equilibrium states, 3 and 4 :
   proceedings of a conference at Gregynog, Wales, August 20-30, 1989 /
   edited by L.A. Peletier et al.
      p.   cm. -- (Progress in nonlinear differential equations and
their applications)
  Includes bibliographical references.
  ISBN 0-8176-3531-9
  1. Differential equations, Partial -- Congresses. 2. Differential
equations, Nonlinear--Congresses. 3. Diffusion--Mathematical
models--Congresses. I. Peletier, L. A. (Lambertus A.) II. Series.
QA377.N644   1992   91-44360
515 ' .353--dc20   CIP

ISBN 0-8176-3531-9
ISBN 3-7643-3531-9

Camera-ready copy prepared in TeX.
Printed and bound by Quinn Woodbine, Woodbine, NJ.
Printed in the USA.

9 8 7 6 5 4 3 2 1

Contents

Preface

Nonlinear diffusion equations have held a prominent place in the theory of partial differential equations, both for the challenging and deep mathematical questions posed by such equations and the important role they play in many areas of science and technology. Examples of current interest are biological and chemical pattern formation, semiconductor design, environmental problems such as solute transport in groundwater flow, phase transitions and combustion theory.

Central to the theory is the equation

$$u_t = \Delta\varphi(u) + f(u).$$

Here Δ denotes the n-dimensional Laplacian, φ and f are given functions and the solution is defined on some domain $\Omega \times [0, T]$ in space-time. Fundamental questions concern the existence, uniqueness and regularity of solutions, the existence of interfaces or free boundaries, the question as to whether or not the solution can be continued for all time, the asymptotic behavior, both in time and space, and the development of singularities, for instance when the solution ceases to exist after finite time, either through extinction or through blow up.

For the corresponding time-independent equation

$$\Delta v + g(v) = 0,$$

where $g = f \circ \varphi^{-1}$, one of the main questions at present is to determine the relation between the structure of the function g, such as its growth at infinity and at the origin, and the existence and uniqueness of – often positive – solutions either in bounded or in unbounded domains, sometimes under an additional assumption on the rate of decay of solutions as $x \in \mathbf{R}^n$ tends to infinity. In the presence of parameters we are then led to important and subtle global bifurcation problems.

Many qualitative properties of these equations, in both time-dependent and steay state situations, can be described by means of special solutions which possess appropriate symmetry properties. For steady state equations such special solutions illuminate, for example, the character of singular solutions and the behavior of solutions for large x. For time-dependent problems, various *self-similar solutions*, reflecting symmetries in the space-time domain, yield characteristic information about singular behavior as well as asymptotic properties as time becomes large.

The study of special solutions leads to problems in ordinary differential equations going back to the beginning of this century, as in the classical Emden-Fowler theory, motivated by questions in astrophysics. Recently, time neglected. (See the paper by Levine, Payne, Sacks, and Straughan [16] for a recent example where nonlinear dependence on derivatives of u is considered.)

In an effort to understand the effects of first order terms on blow up, Chipot and the second author [5] studied the initial value problem (1.1) on a bounded domain $\Omega \subset \mathbb{R}^n$ (with $\lambda = 1$). Shortly thereafter Kawohl and Peletier [13] and Fila [8] extended and improved some of the results in [5]. The best results so far have been obtained in the case where Ω is a ball in $\mathbb{R}^n$. In this case, for a fixed $\lambda > 0$, there exist nonglobal solutions of (1.1) under each of the following circumstances:

(i) $q < 2p/(p+1)$, and (if $n \geq 3$) $p < (n+2)/(n-2)$;

(ii) $q = 2p/(p+1), n = 1$, and p is sufficiently large;

(iii) $q = 2 < p$, and the radius of Ω is sufficiently large.

On the other hand, (again for a fixed λ) if

(iv) $p \leq q \leq 2$;

then all nonnegative solutions in $L^\infty(\Omega)$ are global; and if

(v) $2p/(p+1) < q < p, \ q \leq 2$,

then a nonnegative solution in $L^\infty(\Omega)$ is gobal if the radius of Ω is sufficiently small, depending on $\|u(0)\|_\infty$. (If we delete the requirement that $q \leq 2$ in (iv) and (v), it is still true that the L^∞ norm of the solution $u(t)$ remains bounded for all $t > 0$, as long as it is defined.) Of course, not all of these results require that Ω be a ball. Also, the papers [8,13] study how the blow up of solutions to (1.1) on a fixed domain Ω depends on the parameter λ. A standard dilation argument renders the parameter constant, but changes the size of the domain Ω, thereby giving conditions (iii) and (v) above.

It is natural, especially in view of how the above results depend on the size of the domain, to study the problem on $\mathbb{R}^n$. In fact, a straightforward maximum principle argument shows that if $\Omega_1 \subset \Omega_2$ and there is a nonnegative blowing up solution on Ω_1, then there is a nonglobal solution on Ω_2. (See Section 4 below for a precise formulation and proof.) Thus on $\mathbb{R}^n$, (1.1) admits nonglobal solutions under conditions (i), (ii), and (iii) above. Nonetheless, it would be interesting to find criteria for blow up on $\mathbb{R}^n$ which are intrinsic to $\mathbb{R}^n$. The purpose of this paper is to prove the existence of blowing up solutions to (1.1) using arguments valid for any sufficiently regular domain, not necessarily bounded.

The proof of blow up in [5] uses energy arguments, and is modeled on Ball's proof of blow up of solutions to (1.2) on a bounded domain [2]. The energy functional for this equation is defined by

$$E(u) = \frac{1}{2}\|\nabla u\|_2^2 - \frac{\lambda}{p+1}\|u\|_{p+1}^{p+1}. \tag{1.3}$$

Unfortunately, Ball's calculation, and even more so the calculation in [5], depend crucially on the domain being bounded. To prove that the solutions to (1.1) on an unbounded domain can be nonglobal, we need to return to Levine's original arguments [14], which do not explicitly depend on the nature of the domain. In the case of equation (1.2), Levine's arguments can be considerably simplified, still without making any restriction on the domain. This simplification is given in Proposition 5.1 of [11]. (As noted in [11], the simplification of Levine's arguments was supplied by Payne. The authors of [11] regret the omission of a reference to [14].) Adapting the proof of this proposition, we obtain the following result.

Theorem 1. (i) *Suppose* $1 < q < 2p/(p+1)$. *Let* $u(t,x)$ *be a solution of* (1.1) *such that*
(a) $u \in C^1([0,T]; L^2(\Omega) \cap L^{p+1}(\Omega))$,
(b) $u \in C([0,T]; H^2(\Omega) \cap H_0^1(\Omega))$,
(c) $u, u_t \geq 0$ *for* $0 \leq t < T$,
(d) $E(\varphi) < 0, (E$ *as defined in* (1.3)$)$,
(e) $-E(\varphi)/\|\varphi\|_2^2$ *is sufficiently large.*
Then $T < \infty$.
(ii) *Suppose* $1 < q = 2p/(p+1)$. *Let* $u(t,x)$ *be a solution of* (1.1) *such that*
(a) $u \in C^1([0,T]; L^2(\Omega) \cap L^{p+1}(\Omega))$,
(b) $u \in C([0,T]; H^2(\Omega) \cap H_0^1(\Omega))$,
(c) $u, u_t \geq 0$ *for* $0 \leq t < T$,
(d) $E(\varphi) < 0, (E$ *as defined in* (1.3)$)$,
(e') $\lambda > \Lambda_1(p) \equiv \frac{2^p(p+1)}{(p-1)^{p+1}}$.
Then $T < \infty$.

Proof. The proof of Theorem 1 will show that the largeness condition (e) can be expressed explicitly as a function of the parameters p, q, and λ, independent of the domain Ω. (See formula (2.7) below, as well as (2.2), (2.5) and the discussion just before formula (2.5).) Note that in Theorem 1 part (ii), if we set $\lambda = 1$, the range of allowable p is the same as in [5]. (See the remark at the end of Section 3 in [5].) Also, the condition (e) in part (i) is a little different from the corresponding condition in [5], where $\|\varphi\|_{p+1}$ was assumed to be sufficiently large. However, if $n = 1$ or 2, or if $n \geq 3$ and $p \leq (n+2)/(n-2)$, then (e), along with the Sobolev embedding $H_0^1(\Omega) \subset L^{p+1}(\Omega)$, implies a largeness condition on $\|\varphi\|_{p+1}$.

Of course it is not all obvious that a solution u of (1.1) satisfying the properties required by Theorem 1 really exists, i.e. that an appropriate initial value φ can be found. As in [5], a natural candidate for φ is a positive solution (with homogeneous Dirichlet boundary conditions on Ω) of the elliptic equation

$$\Delta\varphi - |\Delta\varphi|^q + \mu|\varphi|^{p-1}\varphi = 0, \tag{1.4}$$

with $u > 0$ sufficiently small. Indeed, if $\varphi \in H_0^1(\Omega) \cap L^{p+1}(\Omega)$ is a positive solution of (1.4) with $\mu \leq 2\lambda/(p+1)$, then one sees easily that

$$\varphi > 0, \ \Delta\varphi - |\nabla\varphi|^q + \lambda|\varphi|^{p-1}\varphi > 0, \tag{1.5}$$

$$E(\varphi) < 0. \tag{1.6}$$

Condition (1.5) implies hypothesis (c) of Theorem 1, at least formally. Moreover, if we now set $\Omega = \mathbb{R}^n$, a dilation argument shows that conditions (d) and (e) are easily met. To see this, suppose that $\varphi \in H^1(\mathbb{R}^n) \cap L^{p+1}(\mathbb{R}^n)$ is a positive solution of (1.4), and let $\varphi_\alpha(x) = \alpha^{[(2-q)/(q-1)]}\varphi(\alpha x)$ for all $\alpha > 0$. Then

$$\Delta\varphi_\alpha - |\nabla\varphi_\alpha|^q + \mu\alpha^{[q(p+1)-2p]/(q-1)}|\varphi_\alpha|^{p-1}\varphi_\alpha = 0. \tag{1.7}$$

If $q < 2p/(p+1)$, then the exponent of α in (1.7) is negative; so for $\alpha > 1, \varphi_\alpha$ still satisfies (1.4), but with even smaller μ. Moreover,

$$\|\varphi_\alpha\|_{p+1}^{p+1} = \alpha^{[(p+1)(2-q)/(q-1)-n]}\|\varphi\|_{p+1}^{p+1},$$
$$\|\varphi_\alpha\|_2^2 = \alpha^{[2(2-q)/(q-1)-n]}\|\varphi\|_2^2,$$
$$\|\nabla\varphi_\alpha\|_2^2 = \alpha^{[2/(q-1)-n]}\|\nabla\varphi\|_2^2.$$

Therefore, if $q < 2p/(p+1)$, then for sufficiently large α conditions (d) and (e) are satisfied. In fact, this argument shows that if φ satisfies (1.5) and $q < 2p/(p+1)$, then for sufficiently large α, φ_α satisfies (d) and (e) as well as (1.5).

Thus, finding an example fitting the hypotheses of Theorem 1 reduces to finding a positive solution $\varphi \in H^1(\mathbb{R}^n) \cap L^{p+1}(\mathbb{R}^n)$ of (1.4) which is regular enough so that properties (a), (b), and (c) can be rigorously proved. Recall that if $n = 1$ or 2, or if $n \geq 3$ and $p < (n+2)/(n-2)$, then $H^1(\mathbb{R}^n) \subset L^{p+1}(\mathbb{R}^n)$.

Theorem 2. (i) *Let* $p, q > 1$ *satisfy* $q < 2p/(p+1)$ *and (if* $n \geq 3$*)* $p < (n+2)/(n-2)$*, and let* $\mu > 0$*. Then there exists a positive radially symmetric solution* $\varphi \in H^1(\mathbb{R}^n)$ *of* (1.4) *if and only if one of the following conditions is met:*

(f) $n < 4$ *and* $q < \frac{n+4}{n+2}$;

(g) $n = 4$ *and* $q \leq \frac{n+4}{n+2}$;

(h) $n > 4$.

In all three cases, the positive radially symmetric solution $\varphi \in H^1(\mathbb{R}^n)$ *of* (1.4) *can be chosen such that* $\varphi \in W^{k,s}(\mathbb{R}^n)$ *for* $0 \leq k \leq 3$ *and* $2 \leq s \leq \infty$.

(ii) *Let* $p, q > 1$ *satisfy* $q = 2p/(p+1)$*. Suppose* $n = 1$ *and* $0 < \mu \leq \Lambda_2(p) \equiv \frac{(2p)^p}{(p+1)^{2p+1}}$*. There exists a positive (symmetric) solution* $\varphi \in H^1(\mathbb{R})$ *of* (1.4) *if and only if* $p < 5$*. In this case, the positive (symmetric) solution* $\varphi \in H^1(\mathbb{R})$ *of (1.4) can be chosen such that* $\varphi \in W^{k,s}(\mathbb{R})$ *for* $0 \leq k \leq 3$

and $2 \leq s \leq \infty$. *(On* $\mathbb{R}$, *any positive regular solution of* (1.4) *tending to* 0 *at infinity is symmetric about its maximum.)*

Note that if $q = 2p/(p+1)$, conditions (f) and (g) become $p < 1 + 4/n$ if $n = 1, 2$, or 3, with equality permitted if $n = 4$. This explains the restriction $p < 5$ in part (ii) of Theorem 2.

As mentioned earlier, the works [8,13] treat (1.1) with a factor of λ multiplying the term $|u|^{p-1}u$. If $q \neq 2p/(p+1)$, this factor can be scaled away on $\mathbb{R}^n$ by setting $u_\alpha(t, x) = \alpha^{[(2-q)/(q-1)]}u(\alpha^2 t, \alpha x)$. On the other hand, for the special value $q = 2p/(p+1)$, if u is a solution of (1.1) with the initial value φ, then u_α is still a solution of (1.1), except with the intial value $\varphi_\alpha(x) = \alpha^{[(2-q)/(q-1)]}\varphi(\alpha x)$. In other words, if $q = 2p/(p+1)$, the blow up properties of solutions to (1.1) might depend in an essential way on a factor of λ multiplying the term $|u|^{p-1}u$. This is borne out by the second parts of Theorems 1 and 2.

Finally, we need to put Theorems 1 and 2 together to produce solutions which blow up in finite time. The local theory of the Cauchy problem (1.1) developed in [5] for a bounded domain is in general still valid here, though some aspects need to be modified to include unbounded domains. Basically, the problem is well posed in $W_0^{1,s}(\Omega)$ for large finite s or (if $q < 2$)$L^r(\Omega)$ for large finite r. Moreover, if the initial value φ is in $W^{3,s}(\Omega) \cap W_0^{1,s}(\Omega)$ for all s in some specified interval and satisfies (1.5), where also $\Delta\varphi - |\nabla\varphi|^q + \lambda|\varphi|^{p-1}\varphi$ is in $W_0^{1,s}(\Omega)$, then $u, u_t \geq 0$ throughout the tragectory of the solution. (See Section 4 below for a more precise statement.) However, if the domain is not bounded, it is not clear that these solutions are in $L^2(\Omega)$ so that Theorem 1 can be applied, nor is it clear that nonglobal solutions blow up in $L^\infty x$. These difficulties turn out to be of a minor technical nature, and we prove the following result.

Theorem 3. (i) *Let* $p, q > 1$ *and* $\lambda, \mu > 0$. *Suppose* $q < 2p/(p+1)$ *and (if* $n \geq 3$) $p < (n+2)/(n-2)$. *If* $n \leq 4$, *assume in addition* $q < (n+4)/(n+2)$, *with equality permitted if* $n = 4$. *Let* φ *be a positive radially symmetric solution* $\varphi \in H^1(\mathbb{R}^n)$ *of* (1.4), *as shown to exist by part* (i) *of Theorem 2. If necessary, replace* φ *by* $\varphi_\alpha(x) = \alpha^{[(2-q)/(q-1)]}\varphi(\alpha x)$, *where* α *is large enough so that conditions (d) and (e) of Theorem 1 are met, as well as the second inequality in* (1.5). *(This of course changes the value of* μ *in* (1.4).) *Then the resulting solution* $u(t)$ *of the Cauchy problem* (1.1) *is nonglobal, and* $\|u(t)\|_r \to \infty$ *as* t *approaches the blow up time for all* $r > 1$ *such that* $n(p-1)/2 < r \leq \infty$.

(ii) *Let* $p, q > 1$ *and* $\lambda, \mu > 0$. *Suppose* $q = 2p/(p+1), n = 1, p < 5$, *and* $\lambda > \Lambda_1(p) \equiv \frac{2^p(p+1)}{(p-1)^{p+1}}$. *Let* $\mu > 0$ *satisfy* $\mu \leq \Lambda_2(p) \equiv \frac{(2p)^p}{(p+1)^{2p+1}}$. *(It follows that* $\mu \leq 2\lambda/(p+1)$.) *Let* φ *be a positive radially symmetric solution* $\varphi \in H^1(\mathbb{R})$ *of* (1.4), *as shown to exist by part* (ii) *of Theorem 2. Then the resulting solution* $u(t)$ *of the Cauchy problem* (1.1) *is nonglobal, and* $\|u(t)\|_r \to \infty$ *as* t *approaches the blow up time for all* $r > 1$ *such that*

$(p-1)/2 < r \leq \infty$.

We remark that if $q < 2p/(p+1)$, replacing φ by φ_α for an appropriate choice of α can also produce a stationary, hence global, solution of (1.1). In other words, the same argument we use to show the existence of nonglobal solutions of (1.1) will at the same time construct global solutions. Thus, in contrast to equation (1.2) on $\mathbb{R}^n$ (see[10]), we do not find the phenomenon that all nontrivial positive solutions must be nonglobal for certain values of parameters, at least not in the range of parameters covered by part (i) of Theorem 2. Furthermore, if $n(p-1)/2 > 1$, then the proof of Theorem 3 part (b) in [20] essentially shows that if $\|\varphi\|_{n(p-1)/2}$ is sufficiently small, and if φ belongs to the appropriate space for the Cauchy problem (1.1), (see Section 4 below), then the corresponding solution of (1.1) is global. This suggests that if $q = 2p/(p+1)$ and $n(p-1)/2 < 1$, it might indeed be possible that all positive solutions of (1.1) are nonglobal. (See the survey article [15] for a discussion of this aspect of blow up, and its generalization to other unbounded domains.)

Theorems 1, 2, and 3 are proved respectively in Sections 2, 3, and 4 below; and in the last two sections we give more information than is contained in the statements of the theorems.

We close this section with some remarks on the limitations of our methods. First, the energy calculation in Section 2 requires that $u(t)$ be an L^2 solution. Since the problem is well posed in higher spaces, such as L^r and $W^{1,s}$, this condition seems artificial. Perhaps, it can be relaxed slightly by using the modified energy introduced in [21]. Also, perhaps looking for solutions to the elliptic problem (1.4) is not the best way to meet the hypotheses of Theorem 1. More crucially, it seems that these types of arguments work only if $q \leq 2p/(p+1)$, while the evidence of [8,13] suggests that on $\mathbb{R}^n$ this restriction can be weakened to $q < p$. Hopefully some new approach could prove blow up under these weaker conditions on the parameters.

2. Energy calculations

In this section we prove Theorem 1. Thus, we assume that $1 < q \leq 2p/(p+1)$, and that $u(t,x)$ is a solution of (1.1) satisfying (a), (b), (c), and (d) in the statement of Theorem 1. Integrals in this section are over Ω.

Proposition 2.1. *$E(u(t))$ is a nonincreasing function of $t \in [0,T)$, and*

$$\frac{d}{dt}E(u(t)) = -\int (u_t(t))^2 - \int u_t(t)|\nabla u(t)|^q. \tag{2.1}$$

In particular, $E(u(t)) < 0$ throughout the trajectory of u.

Proof. The calculation in the proof of Lemma 3.1 in [5] is valid in this context. (The terms containing λ cancel.) To justify differentiating $\|\nabla u(t)\|_2^2$, write it as $-\int u(t) \cdot \Delta u(t)$ and take the limit of the difference quotient explicitly.

Proposition 2.2. *Suppose* $q < 2p/(p+1)$. *Let*

$$\alpha = \frac{2p - q(p+1)}{p-1}. \tag{2.2}$$

Then for all $\varepsilon > 0$ *and all* $t \in [0, T)$,

$$\int u(t)|\nabla u(t)|^q \leq \left(\frac{2\lambda}{p+1}\right)^{q/2} \{(\alpha/2)\varepsilon^{-2/\alpha}\|u(t)\|_2^2$$
$$+ [(2-\alpha)/2]\varepsilon^{2/(2-\alpha)}\|u(t)\|_{p+1}^{p+1}\}. \tag{2.3}$$

Proof. Note that the assumption $1 < q < 2p/(p+1)$ implies that $0 < \alpha < 1$. Moreover, α satisfies the relation

$$\frac{2-q}{2} = \frac{\alpha}{2} + \frac{1-\alpha}{p+1}.$$

We apply Hölder's inequality twice, and then use the fact that $E(u(t)) < 0$. More precisely,

$$\int u(t)|\nabla u(t)|^q \leq \|u(t)\|_{2/(2-q)}\| \,|\nabla u(t)|^q\|_{2/q}$$
$$\leq \|u(t)\|_2^\alpha \|u(t)\|_{p+1}^{1-\alpha}\|\nabla u(t)\|_2^q$$
$$\leq \left(\frac{2\lambda}{p+1}\right)^{q/2} \|u(t)\|_2^\alpha \|u(t)\|_{p+1}^{1-\alpha}\|u(t)\|_{p+1}^{q(p+1)/2}$$
$$= \left(\frac{2\lambda}{p+1}\right)^{q/2} \|u(t)\|_2^\alpha \|u(t)\|_{p+1}^{(p+1)[1-(\alpha/2)]}.$$

If $0 < \theta < 1$, and $\varepsilon > 0$, then $xy \leq \theta(x/\varepsilon)^{1/\theta} + (1-\theta)(\varepsilon y)^{1/(1-\theta)}$. Applying this to the last expression above with $\theta = \alpha/2$, we obtain precisely (2.3).

Next, we define the functional $F(u) = \|u\|_2^2$; and we wish to estimate $(d/dt)F(u(t))$ from below. Using first the equation (1.1), then the definition of the energy, and finally (2.3), we find that if $q < 2p/(p+1)$, then for any real β, any $\varepsilon > 0$, and α as defined in (2.2),

instead of (2.4). The hypothesis on λ implies that for some $\beta > 4$, the coefficient of $\|u(t)\|_{p+1}^{p+1}$ in this last inequality is positive. We deduce that

$$\frac{d}{dt}F(u(t)) \geq -\beta E(u(t)). \tag{2.8}$$

Now we set $H(u(t)) = -E(u(t))F(u(t))^{-\beta/4}$; and in analogy with the calculation involving $G(u(t))$ above, one shows easily that $\frac{d}{dt}H(u(t)) \geq 0$. Since $H_0 = H(\varphi) = H(u(0))$ is necessarily positive, we see that $-E(u(t)) \geq H_0 F(u(t))^{\beta/4}$. Putting this into (2.8), we obtain finally that

$$\frac{d}{dt}F(u(t)) \geq -\beta H_0 F(u(t))^{\beta/4},$$

from which we conclude that T must be finite.

3. The elliptic problem

In this section we study (positive) radially symmetric solutions on $\mathbb{R}^n$ of the elliptic equation (1.4), where $q, p > 1$ and $\mu > 0$. Thus, we write $\varphi(x) = w(r)$ $(r = |x|)$, where $w : (0, \infty) \to \mathbb{R}$ satisfies the following ordinary differential equation:

$$w''(r) + \frac{n-1}{r}w'(r) - |-w'(r)|^q + \mu|w(r)|^{p-1}w(r) = 0. \tag{3.1}$$

In this setting n need no longer be an integer, but rather a real number greater than or equal to 1. We suppose that w is a solution of (3.1) in the sense of distributions on $(0, \infty)$, and in particular that the distribution derivative w' belongs to $L_{\text{loc}}^q(0, \infty)$. Since (3.1) is equivalent to

$$(r^{n-1}w')' = r^{n-1}|-w'(r)|^q - \mu r^{n-1}|w(r)|^{p-1}w(r), \tag{3.2}$$

it follows by standard arguments that w is of class C^2 on $(0, \infty)$, with w'' locally absolutely continuous. Differentiating (3.1) then yields that w is of class C^3 on $(0, \infty)$.

It turns out that the necessity of the conditions on q in Theorem 2 can be deduced from lower bounds on the asymptotic behavior of w as $r \to \infty$. In fact, to derive these bounds, it suffices that w be a solution of (3.1) for r sufficiently large. By the same argument as just given, such a solution w is of class C^3 on its interval of definition (r_0, ∞), with $r_0 > 0$.

Throughout this section, we denote $-w'(r)$ by $v(r)$. Also, C denotes a constant whose value may change from formula to formula (or perhaps even within the same formula).

Proposition 3.1. *Let $p, q > 1, \mu > 0$, and $r_0 > 0$. Let w be a positive solution of (3.1) on (r_0, ∞) as described above such that $\liminf_{r \to \infty} w(r) = 0$. It follows that $v(r) = -w'(r) > 0$ for sufficiently large r. Moreover,*

(i) *if $(n-1)(q-1) < 1$, then $q < 2$ and $v(r) \geq Cr^{-1/(q-1)}, w(r) \geq Cr^{-(2-q)/(q-1)}$ as $r \to \infty$;*

(ii) *if $(n-1)(q-1) > 1$, then $n > 2$ and $v(r) \geq Cr^{1-n}$, $w(r) \geq Cr^{2-n}$ as $r \to \infty$;*

(iii) *if $(n-1)(q-1) = 1$, then $n > 2$ and $v(r) \geq Cr^{1-n}(\log r)^{-1/(q-1)}$, $w(r) \geq Cr^{2-n}(\log r)^{-1/(q-1)}$ as $r \to \infty$.*

In particular, if $w \in L^2(r_0, \infty; r^{n-1}dr)$, then one of the conditions (f), (g), or (h) in Theorem 2 must hold.

Proof. Let w be such a positive solution. We first note that if $w'(r) = 0$, then $w''(r) = -\mu w(r)^p < 0$. In other words, any critical point of w must be a local maximum. Since $\lim \inf_{r \to \infty} w(r) = 0$, it follows that $w'(r) < 0$ for r sufficiently large. Since w is now decreasing, we conclude that $\lim_{r \to \infty} w(r) = 0$. By increasing the value of r_0 we may assume that $w'(r) < 0$ for $r \in (r_0, \infty)$.

Since $w(r) \to 0$ as $r \to \infty$, and since $v(r) = -w'(r) > 0$, it follows that

$$w(r) = \int_r^\infty v(s)ds. \tag{3.3}$$

In particular, v is integrable at ∞. Setting $H(r) = r^{n-1}v(r)$, we deduce from (3.2) that

$$\frac{d}{dr}H(r)^{1-q} = (q-1)r^{-(n-1)(q-1)}[1 - \frac{\mu w^p}{v^q}] \leq (q-1)r^{-(n-1)(q-1)}. \tag{3.4}$$

Integrating the inequality in (3.4) results immediately in an upper bound for $H(r)^{1-q}$ as $r \to \infty$. We consider three cases.

Suppose first that $(n-1)(q-1) > 1$. It follows that $H(r)^{1-q}$ is bounded as $r \to \infty$, i.e. $H(r)$ is bounded away from 0 as $r \to \infty$. Thus, there exists a constant $C > 0$ such that $v(r) \geq Cr^{1-n}$ for all $r > 0$. If $n \leq 2$, then v is not integrable at ∞, which is a contradiction. If $n > 2$, then formula (3.3) implies that $w(r) \geq Cr^{2-n}$ as $r \to \infty$.

Suppose next that $(n-1)(q-1) < 1$. We get here that $H(r)^{1-q} \leq Cr^{1-(n-1)(q-1)}$ as $r \to \infty$; and so that $v(r) \geq Cr^{-1/(q-1)}$. If $q \geq 2$, then v is not integrable at ∞, again contradicting (3.3). If $q < 2$, then formula (3.3) implies that $w(r) \geq Cr^{-(2-q)/(q-1)}$ as $r \to \infty$.

Finally, if $(n-1)(q-1) = 1$, then $H(r)^{1-q} \leq C\log r$ as $r \to \infty$. We conclude here that $v(r) \geq Cr^{1-n}(\log r)^{-1/(q-1)}$ if $n < 2$, v is not integrable. If $n = 2$, then $q = 2$; and v is still not integrable. If $n > 2$, putting the estimate for v into (3.3), dividing by $r^{2-n}(\log r)^{-1/(q-1)}$, and applying l'hôpital's rule, we obtain $w(r) \geq Cr^{2-n}(\log r)^{-1/(q-1)}$.

The last statement in the proposition follows in a straightforward way from the lower bounds just obtained. This completes the proof of the proposition.

Looking at formula (3.4), we note that if a solution of (3.1) has the property that $\mu w^p / v^q$ stays less than $1 - \varepsilon$ as $r \to \infty$, for some $\varepsilon > 0$, then, modulo a multiplicative constant, the inequality in (3.4) can be reversed for large r. In this case, all the asymptotic lower bounds given in Proposition 3.1 turn into asymptotic upper bounds; and we get the precise decay rates of $w(r)$ and $w'(r)$ as $r \to \infty$. The equation (3.1) and its derivative then give the decay rates for $w''(r)$ and $w^{(3)}(r)$ as $r \to \infty$.

To prove that such a solution exists, we use a phase plane argument. Since $v(r) = -w'(r)$, the equation (3.1) becomes the system

$$w' = -v,$$
$$v' = -\frac{n-1}{r} v - |v|^q + \mu |w|^{p-1} w. \tag{3.5}$$

Proposition 3.2. *Let $p, q > 1, \mu > 0$; and suppose that $q < 2p/(p+1)$. Let $\alpha > 0$ satisfy*

$$\frac{1}{2-q} < \alpha < \frac{p}{q}. \tag{3.6}$$

Then for sufficiently small $w_0 > 0$, the region $\mathcal{R}$ defined by $0 < w < w_0, 0 < v < w^\alpha$ is (forwardly) invariant under the action of the dynamical system (3.5) for all $r > 0$.

Proof. In the region $v > 0, w > 0$, we see that $w' < 0$. Moreover, if $v = 0$ and $w > 0$, then $w' = 0$ and $v' > 0$. Thus, along the two straight segments forming part of the boundary of $\mathcal{R}$, the vector field determined by (3.5) points into $\mathcal{R}$. It remains to show the same along the part of the boundary coinciding with the graph of $v = w^\alpha$.

Along the curve $v = w^\alpha$, as long as $w > 0$ and $v > 0$, the vector field is given by

$$w' = -w^\alpha < 0,$$
$$v' = -\frac{n-1}{r} w^\alpha - w^{\alpha q} + \mu w^p < 0,$$

this last inequality following for small $w > 0$ since $q\alpha < p$. Thus, along the curve, the vector field points down and to the left. To show that it points into $\mathcal{R}$, we need to show that v'/w' is steeper than the slope of the curve. In other words, we need to check that

$$\frac{\frac{n-1}{r} w^\alpha + w^{\alpha q} - \mu w^p}{w^\alpha} > \alpha w^{\alpha - 1},$$

for all sufficiently small $w > 0$ (independent of $r > 0$). For this we need that $\alpha q < p$ and $\alpha q < 2\alpha - 1$, both of which follow from (3.6). This proves the proposition.

To prove the existence of regular solutions to (1.4), we consider the initial value problem determined by equation (3.1) and the initial conditions

$w(0) = a > 0$ and $w'(0) = 0$. Proposition 4.4 of [5] shows that for every $a > 0$, there is a (unique) solution of class C^2 defined on some maximal interval $[0, R_a)$. Furthermore, if $w(r) > 0$ on $[0, R_a)$, then $R_a = \infty, w'(r) < 0$ on $(0, \infty)$, and $w(r)$ and $w'(r)$ both converge to 0 as $r \to \infty$. Otherwise, we denote by $z(a)$ the smallest (positive) zero of w. By local uniqueness, we see that $w'(z(a)) < 0$, so $w(r)$ is negative just after $z(a)$. Thus, $\{a > 0, z(a) < \infty\}$ is open. If $w(r) > 0$ on $[0, \infty)$, we say that $z(a) = \infty$.

Proposition 3.3. *Let $p, q > 1, \mu > 0$; and suppose that $q < 2p/(p+1)$ and (if $n > 2$) $p < (n+2)/(n-2)$. Then there exists a solution $w \in C^2([0, \infty), \mathbb{R})$ of (3.1) such that $w(r) > 0$ for all $r \geq 0, v(r) \equiv -w'(r) > 0$ for all $r > 0, w'(0) = 0$, and $w(r)^p/v(r)^q$ converges to 0 as r tends to ∞.*

Proof. Let w be the solution to the initial value problem as described above. Since $q < 2p/(p+1)$, Proposition 5.7 in [5] tells us that $z(a) = \infty$ if a is sufficiently small. Since $q < 2p/(p+1)$ and (if $n > 2$) $p < (n+2)/(n-2)$, Proposition 4.6 in [5] tells us that $z(a)$ is finite if a is sufficiently large. (Note that Proposition 3.9 in [11] is valid for noninteger n. Also, here is the only place we use the hypothesis $p < (n+2)/(n-2)$.) Let a_0 be the infimum of the open set $\{a > 0 : z(a) < \infty\}$. Then $0 < a_0 < \infty$, and $z(a_0) = \infty$.

We consider the solution $w(r)$ of the initial value problem with $w(0) = a_0$ and $w'(0) = 0$ as a trajectory $(w(r), v(r))$ in phase plane determined by the dynamical system (3.5). Let α satisfy (3.6) and let $\mathcal{R}$ be the invariant region described in the previous proposition. We claim that the trajectory $(w(r), v(r))$ can never enter the region $\mathcal{R}$. Suppose to the contrary that $(w(r), v(r))$ enters $\mathcal{R}$ at some point. Then by continuous dependence on initial data, since $\mathcal{R}$ is open, the trajectories of other solutions with $w(0) = a$ close to a_0 (and $w'(0) = 0$) will also enter the region $\mathcal{R}$. Since $\mathcal{R}$ is invariant under the dynamical system, it follows that $w(r) > 0$ throughout its interval of existence, i.e that $z(a) = \infty$ for a sufficiently close to a_0. This contradicts the definition of a_0.

We conclude that the trajectory $(w(r), v(r))$ must converge to $(0,0)$ without passing through $\mathcal{R}$. In particular, $v(r) \geq w(r)^\alpha$ as $r \to \infty$. Since $\alpha < p/q$, the desired result follows.

Completion of the proof of Theorem 2 part (i). Let $p, q > 1$ satisfy $q < 2p/(p + 1)$ and (if $n > 2$) $p < (n + 2)/(n - 2)$, and let $\mu > 0$. Let w be the solution of (3.1) described in Proposition 3.3 above. Since $w(r)^p/v(r)^q$ converges to 0 as r tends to ∞, it follows from (3.4) that

$$\frac{d}{dr}H(r)^{1-q} = (q-1)r^{-(n-1)(q-1)}[1 - \frac{\mu w^p}{v^q}] \geq [(q-1)/2]r^{-(n-1)(q-1)}, \quad (3.7)$$

for sufficiently large r. Integrating the inequality in (3.7) yields a lower bound for $H(r)^{1-q}$ as $r \to \infty$. This yields the same estimates as in Proposition 3.1, except with the inequalities reversed. Thus, the conditions (f), (g), and (h) in Theorem 2 each become sufficient to have $w \in L^2(0, \infty; r^{n-1}dr)$;

all $r \geq r_0/q$. Iterating, we get eventually that $u \in C([0, T_\varphi); W^{2,r}(\mathbb{R}^n))$ for all $r \geq 2$. It follows then from (4.4) that $u \in C^1([0, T_\varphi); L^r(\mathbb{R}^n))$ for all $r \geq 2$. By Theorem 1 we may now conclude that $T_\varphi < \infty$, and (as remarked above) $\|u(t)\|_r \to \infty$ as $t \to T_\varphi$ for all $r > 1$ such that $n(p-1)/2 < r < \infty$.

The remaining point is to show that L^∞ norm of the solution blows up. To accomplish this, we show that the integral equation admits local solutions on $C_0(\mathbb{R}^n)$, the space of continuous functions on $\mathbb{R}^n$ tending to 0 at infinity. We use the set up in [19], except that we allow the nonlinear map to have two parts, $J = J_1 + J_2$, as given by (4.2). The space E in [19] is $C_0(\mathbb{R}^n)$, so $J_2 : E \to E$ is Lipschitz on bounded sets. The space E_{J_1} is $C_0^1(\mathbb{R}^n)$, the set of all C^1 functions in $C_0(\mathbb{R}^n)$ all of whose first order derivatives also belong to $C_0(\mathbb{R}^n)$. $J_1 : E_{J_1} \to E$ is clearly Lipschitz on bounded sets. Finally, a simple calculation using the explicit kernel for $e^{t\Delta}$ on $\mathbb{R}^n$ shows that $e^{t\Delta} : E \to E_{J_1}$ with norm bounded by $Ct^{-1/2}$. Formula (2.2) in [19] is verified precisely if $q < 2$. In other words, we have proved the following proposition, which completes the proof of Theorem 3.

Proposition 4.1. *Let $1 < q < 2$, and $p > 1$. For every $\varphi \in C_0(\mathbb{R}^n)$, there is a unique maximal solution $u \in C([0, T_\varphi); C_0(\mathbb{R}^n))$ of (4.1); and if $T_\varphi < \infty$, then $\|u(t)\|_\infty \to \infty$ as $t \to T_\varphi$.*

We conclude this section by making good on the statement, which we made early in the first section of this paper, that blow up on a particular domain implies blow up on any larger domain. To do this, we use the following weak form of the maximum principle for the heat equation. Its proof, which we rapidly sketch, uses some of the ideas from Section 2 of [5], and we are indebted to M. Chipot [4] for the formulation we present.

Lemma 4.2. *Let $2 \leq r < \infty$, and let $\Omega \subset \mathbb{R}^n$ be any open set. Suppose*

(i) $w \in C([0, T_0); L^r(\Omega))$,

(ii) $w \in C^1((0, T_0); L^r(\Omega))$,

(iii) $w(t) \in W_0^{1,r}(\Omega)$, *for all* $t \in (0, T_0)$,

(iv) $w_t - \Delta w \geq -C(|w| + |\nabla w|)$, *where C can be chosen uniformly for* $t \in (0, T)$, $\forall T < T_0$,

(v) $w(0) \geq 0$.

Then $w(t) \geq 0$ for all $t \in [0, T_0)$.

Proof. We define $w_-(t)$ as $\frac{|w(t)| - w(t)}{2} \geq 0$. The assumptions (ii) and (iii) of the lemma are just what we need to prove that

$$\int w_t(t) \cdot (w_-(t))^{r-1} = \frac{-1}{r} \frac{d}{dt} \int (w_-(t))^r,$$

$$\int \Delta w(t) \cdot (w_-(t))^{r-1} = (r-1) \int |\nabla w_-(t)|^2 (w_-(t))^{r-2},$$

where all integrals are over Ω. (See Lemma 7.6 in [7].) Therefore,

$$\frac{1}{r}\frac{d}{dt}\int (w_-(t))^r = \int -[w_t(t) - \Delta w(t)]\cdot (w_-(t))^{r-1}$$

$$-(r-1)\int |\nabla w_-(t)|^2 (w_-(t))^{r-2}$$

$$\leq C\int \{|w_-(t)|^r + |\nabla w_-(t)|\cdot |w_-(t)|^{r-1}\}$$

$$-(r-1)\int |\nabla w_-(t)|^2 (w_-(t))^{r-2} \leq C\int |w_-(t)|^r,$$

where for the last step, we have estimated $|\nabla w_-(t)|\cdot |w_-(t)|$ by $\epsilon|\nabla w_-(t)|^2$ $+ C|w_-(t)|^2$. It follows easily now that if $w_-(0) = 0$, then $w_-(t) = 0$ on each interval $(0,T)$, hence on $(0,T_0)$.

Now let $\Omega_1 \subset \Omega_2 \subset \mathbb{R}^n$ be two domains with the regularity properties described in the beginning of this section. Let u_1 and u_2 be solutions of (4.1), with initial values $\varphi_1 \geq 0$ and $\varphi_2 \geq 0$, satisfying (4.4) and (4.5) on Ω_1 and Ω_2 respectively, for some large value of s. Suppose $\varphi_1 \leq \varphi_2$ on Ω_1. If we extend u_1 to be 0 on $\Omega_2 - \Omega_1$, then $w = u_2 - u_1$ meets the hypotheses of Lemma 4.2 as long as both solutions exist. (Property 4 follows from the embedding theorems.) We conclude that $u_1(t) \leq u_2(t)$ as long as both solutions exist. Thus, if u_2 is global, then so is u_1; and if u_1 has finite existence time, the same is true for u_2.

REFERENCES

1. R.A. Adams, *Sobolev Spaces*, Academic Press, New York, 1975.
2. J.M. Ball, *Remarks on blow-up and nonexistence theorems for nonlinear evolution equations*, Quart. J. Math. Oxford Ser. **28** (1977), 473–486.
3. H. Brezis, *Analyse Fonctionnelle: Théorie et Applications*, Masson, Paris, 1983.
4. M. Chipot, private conversation.
5. M. Chipot and F.B. Weissler, *Some blow-up results for a nonlinear parabolic equation with a gradient term*, SIAM J. Math. Anal. **20** (1989), 886–907.
6. M. Dauge, *Elliptic Boundary Value Problems on Corner Domains*, LNM **1341**, Springer, New York, 1988.
7. D. Gilbarg and N.S. Trudinger, *Elliptic Partial Differential Equations of Second Order*, Springer, New York, 1977.
8. M. Fila, *Remarks on blow up for a nonlinear parabolic equation with a gradient term*, Proc. A.M.S. **11**(1991), 795–802.
9. A. Friedman, *Remarks on nonlinear parabolic equations*, Proc. Symp. in Appl. Math. A.M.S. **13** (1965), 3–23.

10. H. Fujita, *On the blowing up of solutions to the Cauchy problem for* $u_t = \Delta u + u^{1+\alpha}$, J. Fac. Sci. Univ. Tokyo, Sect. I A Math. **13** (1966), 109–124.

11. A. Haraux and F.B. Weissler, *Non-uniqueness for a semilinear initial value problem*, Indiana Univ. Math. J. **31** (1982), 167–189.

12. S. Kaplan, *On the growth of solutions of quasilinear parabolic equations*, Commun. Pure. Appl. Math. **16** (1963), 305–330.

13. B. Kawohl and L.A. Peletier, *Observations on blow up and dead cores for nonlinear parabolic problems*, Math Z. **202** (1989), 207–217.

14. H.A. Levine, *Some nonexistence and instability theorems for solutions of formally parabolic equations of the form* $Pu_t = -Au + \mathcal{F}(u)$, Arch. Rat. Mech. Anal. **51** (1973), 371–386.

15. H.A. Levine, *The role of critical exponents in blow up theorems*, SIAM Review, **32**(1990), 262–288.

16. H.A. Levine, L.E. Payne, P.E. Sacks, and B. Straughan, *Analysis of a convective reaction-diffusion equation (II)*, SIAM J. Math. Anal., **20** (1989), 133–147.

17. B. Straughan, *Instability, Nonexistence and Weighted Energy Methods in Fluid Dynamics and Related Theories*, Research Notes in Mathematics #74, Pitman, Boston, 1982.

18. F.B. Weissler, *Semilinear evolution equations in Banach spaces*, J. Functional Analysis, **32** (1979), 277–296.

19. F.B. Weissler, *Local existence and nonexistence for semilinear parabolic equations in* L^p, Ind. Univ. Math. J. **29** (1980), 79–102.

20. F.B. Weissler, *Existence and non-exsistence of global solutions for a semilinear heat equation*, Israel J. Math. **38** (1981), 29–40.

21. F.B. Weissler, L^p-*energy and blow-up for a semilinear heat equation*, Proceedings of Symposia in Pure Mathematics, **45**, Part 2, Amer. Math Soc., Providence, 1986, 545–551.

22. F.B. Weissler, *Rapidly decaying solutions of an ordinary differential equation with applications to semilinear elliptic and parabolic partial differential equations*, Arch. Rational Mech. Anal. **91** (1986), 247–266.

Liliane Alfonsi
Institut Universitaire
de Technologie de Paris-Sceaux
8, Avenue Cauchy
92330 Sceaux, France

Fred B. Weissler
Centre de Mathématiques
Ecole Normale Supérieure de Cachan
61, Avenue du Président Wilson
94235 Cachan Cedex, France

Shrinking Doughnuts

SIGURD B. ANGENENT

Introduction

Let M^n be a smooth compact oriented manifold, and let $X : M \times [0, T) \to \mathbf{R}^{n+1}$ be a smooth family of immersions of M in $n+1$ dimensional Euclidean space. The orientation of M allows one to define a unique smooth unit normal vector field $\nu_X : M \times [0, T) \to \mathbf{R}^{n+1}$. Given this choice of ν_X, we can define the principal curvatures, $\kappa_1, \ldots, \kappa_n$, of the immersion $X(\cdot, t)$ and the mean curvature $H_X = (\kappa_1 + \ldots + \kappa_n)/n$ in the usual way. By definition, the family of immersions $X(\cdot, t)$ "moves by its mean curvature" if the normal velocity satisfies

$$(1) \qquad \left\langle \frac{\partial X}{\partial t}, \nu_X \right\rangle = nH_X$$

at all $(p, t) \in M \times [0, T)$. Here $\langle x, y \rangle = x_0 y_0 + \cdots + x_n y_n$ denotes the Euclidean inner product on $\mathbf{R}^{n+1}$.

The object of this note is to study some of the similarity solutions of (1), and in particular to show that there exists an embedding of the 2-torus, $X_0 : \mathbf{T}^2 \to \mathbf{R}^3$ for which the corresponding solution to (1) is simply given by $X(p, t) = \sqrt{2(1 - t)} X_0(p)$, i.e., for which the torus will shrink to the origin by dilations, and for which it will become singular at $t = 1$. More precisely, we'll prove the following.

Theorem. *For $n \geq 2$ there exist embeddings $X_n : S^1 \times S^{n-1} \to \mathbf{R}^{n+1}$ for which $X_n(p, t) = \sqrt{2(1 - t)} \cdot X_n(p)$ is a solution of the flow by mean curvature equation.*

We will show how the existence of such solutions can be exploited to study the singularities of general solutions of (1).

Similarity solutions of (1) were studied both by Abresch and Langer [AL], and by Epstein and Weinstein [EW] in the case of curves in the plane. The higher dimensional situation was studied by Huisken in [Hu2], in which paper he also obtained some detailed results on the formation of singularities in the flow by mean curvature problem. One of Huisken's results in [Hu2] states that the only immersed hypersurface with positive mean curvature which gives rise to a similarity solution is the standard n

---

Partially supported by the NSF, under grant nr. DMS - 8801486.

sphere. Quoting numerical evidence for the existence of toroidal similarity solutions, obtained by Matt Grayson, Huisken then points out that his hypothesis concerning the positivity of the mean curvature appears to be necessary. Our existence result confirms this.

Before we discuss the toroidal similarity solutions, we give a short overview of some facts about the flow by mean curvature problem which have been established recently. Our discussion is necessarily incomplete: in particular we won't mention the recent interesting theories of "weak solutions" of Evans and Spruck [ES], and Chen, Giga and Goto [CGG]. One excuse for this omission is that these theories deal with a slightly different problem, e.g. all solutions considered in [ES] or [CGG] are embedded, as long as they are smooth. The section on "the one dimensional case", reflects the contents of my talk at Gregynog.

Acknowledgment. The preprint [Hu2] has inspired some of what follows, and it is a pleasure to thank Bernd Kawohl for sending me a copy of Huisken's work.

The initial value problem

One can regard the equation (1) as a degenerate parabolic partial differential equation for X. The degeneracy is caused by the fact that (1) is invariant under reparametrizations; if $\varphi : M \times [0, T) \to M \times [0, T)$ is a smooth one parameter family of diffeomorphisms of M, then $X(p, t)$ will satisfy (1) if and only if $X^\varphi(p, t) = X(\varphi(p, t), t)$ does so. This shows that the solution to (1) is not unique, and, in fact, that it can at most be unique up to reparametrizations of the form $X^\varphi(p, t) = X(\varphi(p, t), t)$.

The degeneracy can be removed by restricting oneself to a specific class of immersions X. One possible choice is the following. Let $Y : M \to \mathbf{R}^{n+1}$ be a given immersion. Then by the tubular neighborhood theorem there is a small $\epsilon > 0$ such that the map $\sigma : M \times (-\epsilon, \epsilon) \to \mathbf{R}^{n+1}$ given by $\sigma(p, u) = Y(p) + u\nu_Y(p)$ is a local diffeomorphism. Any immersion $X_0 : M \to \mathbf{R}^{n+1}$ which is C^1 close to Y can then be represented as the graph of a C^1 function u_0 on M, i.e. as $X_0(p) = \sigma(p, u_0(p))$. If one has a family of such immersions, then they will be represented by a function $u : M \times [0, T) \to \mathbf{R}$, at least as long as they stay close to the "reference immersion" Y. If one computes the mean curvature and unit normal corresponding to an immersion of the form $X_0(p) = \sigma(p, u_0(p))$, then one finds that, in local coordinates $x_1, \ldots, x_n$ on M, (1) is equivalent to

$$(1') \qquad \frac{\partial u}{\partial t} = g^{ij}(x, u, \partial u)\frac{\partial^2 u}{\partial x_i \partial x_j} + A(x, u, \partial u),$$

where $[g^{ij}]_{1 \le i, j \le n}$ is the inverse of $[g_{ij}]_{1 \le i, j \le n}$, and the g_{ij} are the components of the metric which the immersion $p \mapsto X(p, t) = \sigma(p, u(p, t))$ induces on M. The lower order term $A(x, u, \partial u)$ depends on the reference immersion Y in a fairly complicated way; the term vanishes in the special case when $M = \mathbf{R}^n$ and $Y(x_1, \ldots, x_n) = (x_1, \ldots, x_n, 0)$ is a hyperplane (but this example can only be used to study the evolution of graphs in $\mathbf{R}^{n+1} = \mathbf{R}^n \times \mathbf{R}$ – if M is compact then $Y : M \to \mathbf{R}^{n+1}$ can never be represented as such a graph.) Using the extant theory of quasilinear parabolic partial differential equations, one can now prove a "short time" existence result for (1′), assuming that the initial immersion $X_0 : M \to \mathbf{R}^{n+1}$ has bounded principal curvatures, say. Huisken [Hu1] showed that the solution will exist as long as its principal curvatures remain bounded. On the other hand, one can show that the maximal classical solution which one obtains in this way must become singular within finite time. Naturally, one wants to know *how* this solution can become singular.

Assuming that M is a sphere, that $n \geq 2$, and that the initial immersion is a strictly convex embedding, Huisken [Hu1] showed that the solution remains convex, shrinks to a point, and assumes the shape of a sphere with radius approximately equal to $\sqrt{c(T - t)}$, as $t \uparrow T$. Roughly at the same time Gage and Hamilton [GH] gave a proof of the same fact in the case $n = 1$ (the so called "curve shortening" problem); they showed that a convex curve will shrink to a point and become asymptotically circular as it shrinks. In fact, Gage had shown earlier that the isoperimetric ratio of the curve is decreasing. Gage and Hamilton also observed that, if one starts with a simple closed curve, then the corresponding solution to (1) never develops a self intersection before it becomes singular. This result was then strengthened by Grayson: he showed that, if one starts with any simple closed curve, then the solution actually becomes convex before it can become singular, so that by Gage and Hamilton's result it will eventually shrink to a "round point".

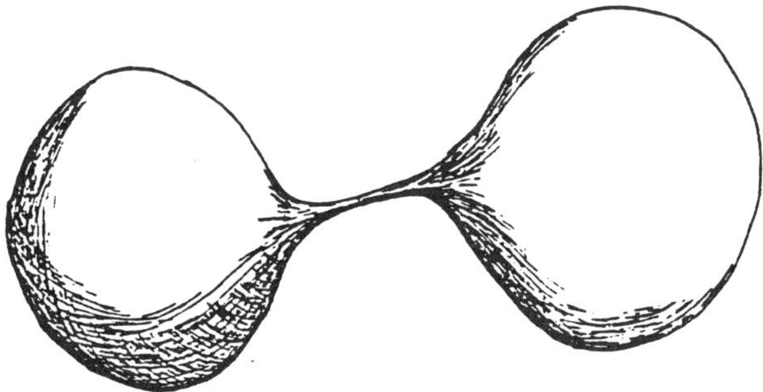

Figure 1.

It was later found that one cannot extend Grayson's theorem to the higher dimensional situation, or, to put it differently, one cannot remove the convexity hypothesis from Huisken's theorem. The following counter example shows why. Assume that M is indeed a sphere, and suppose that the initial immersion X_0 has the shape suggested in Figure 1: two spheres connected by a very thin tube. Intuitively it is clear that the two spherical parts of the surface will evolve slowly (their curvature is not very large), while the thin tube will tend to collapse. Indeed, one of its curvatures (parallel to its axis) is rather small, and the other is very large, and directed "inwards".

It turns out to be nontrivial to prove rigorously that these effects will really cause the "neck" to be pinched off. As far as I know, Matt Grayson [Gr2] was the first to provide such a proof. At this conference Bernd Kawohl has shown us another proof, using the classical maximum principle; in addition he showed us (some of) the output of a numerical simulation done by G. Dziuk, which ought to convince anyone that the thin "neck" in Figure 1 will really break. In [Hu2] Huisken has given the best results in this direction so far, of which I am aware. Assuming that the maximal curvature does not blow up faster than $c(T - t)^{-1/2}$, he shows that the asymptotic shape of the solution near a blow up point is given by a self similar solution. In the case of a two dimensional surface of rotation in $\mathbf{R}^3$, whose shape resembles the surface of Figure 1, Huisken shows that the curvature will actually not blow up faster than $c(T - t)^{-1/2}$; moreover, the shape of the "neck" near a blow up point will converge to a cylinder, after it is magnified by a factor $(T - t)^{-1/2}$.

The case of curves in the plane

It was observed by Gage and Hamilton [GH] that the evolution of plane curves under the (mean) curvature flow can be simplified if one restricts ones attention to convex curves, i.e. to immersed curves without inflection points. On such a curve the angle θ which the tangent makes with a fixed direction, such as the x- axis, is a good coordinate, and the curve is determined up to a translation, if one specifies the curvature k as a function of the angle θ.

Given the curvature $k(\theta)$ one finds the arc length s as a function of θ by integrating $ds = d\theta/k$; a parametrization of the curve is then given by

$$x(\theta) = x_0 + \int_0^\theta \frac{\cos\vartheta}{k(\vartheta)}\, d\vartheta; \qquad y(\theta) = y_0 + \int_0^\theta \frac{\sin\vartheta}{k(\vartheta)}\, d\vartheta,$$

where (x_0, y_0) are the coordinates of the point with $\theta = 0$. Conversely, any positive continuous $2\nu\pi$ periodic function function $k(\theta)$ which satisfies

$$\int_0^{2\nu\pi} \frac{e^{i\theta}}{k(\theta)}\, d\theta = 0$$

defines a closed curve of winding number ν (number of times the tangent runs through the unit circle, as you go around the curve once).

The curve shortening problem turns out to be equivalent to the following parabolic partial differential equation for $k(\theta, t)$

$$(2) \qquad\qquad k_t = k^2 k_{\theta\theta} + k^3,$$

where the variable θ belongs to $\mathbf{R}/2\nu\pi\mathbf{Z}$, if ν is the winding number of the curve; alternatively, $k(\theta, t)$ should satisfy periodic boundary conditions $k(\theta + 2\nu\pi, t) \equiv k(\theta, t)$.

In a different context this equation was studied by A. Friedman and B. McLeod, in [FM], which inspired some of the results obtained in [An2].

The similarity solutions which were found and studied by Abresch and Langer, and by Epstein and Weinstein correspond to the solutions of (2) which one obtains if one tries $k(\theta, t) = f(t)K(\theta)$. One finds that $f(t)$ must be $\{2(1 - t)\}^{-1/2}$, while K should be a solution of

$$K''(\theta) + K(\theta) = \frac{1}{2K(\theta)}.$$

These are not the only special solutions of (2): by looking for time independent solutions of (2) one easily finds that $k(\theta, t) = A\cos(\theta - \theta_0)$ satisfies (2) for any $A > 0, \theta_0 \in \mathbf{R}$. This solution corresponds to a curve which is essentially (i.e. up to translation, rotation and dilation) the graph of $y = -\log\cos x$. Its evolution under the mean curvature flow is given by translating it with constant velocity, parallel to its asymptotes. This solution is now refered to as the "Grim Reaper."

A less obvious solution to (2) is:

$$k(\theta, t) = \sqrt{\cos 2\theta - \coth 2t} \qquad\qquad (-\infty < t < 0, \theta \in \mathbf{R}).$$

The shape of this solution may be described as follows. For $t = -\infty$ the corresponding curve consists of two Grim Reapers, with the same asymptotes, but separated by an infinite distance; as t increases from $-\infty$ to 0, the two Grim Reapers will move toward each other, and for t close to 0, the curve will become asymptotically circular, and shrink to a point. It is not clear to me whether this solution can be obtained as some sort of "similarity solution" of (2).

In the one dimensional case one might hope that Grayson's theorem could be extended to curves with self intersections. Indeed, as we have just mentioned, Abresch and Langer found that there are solutions, analogous to the circle, which shrink in a self similar way to a point. Nevertheless, it can be made quite plausible by intuitive arguments, that curves with self intersections will in general become singular without shrinking to a point: small loops have larger curvature, and hence should contract faster than the remainder of the curve. The sequence in Figure 2 indicates what can

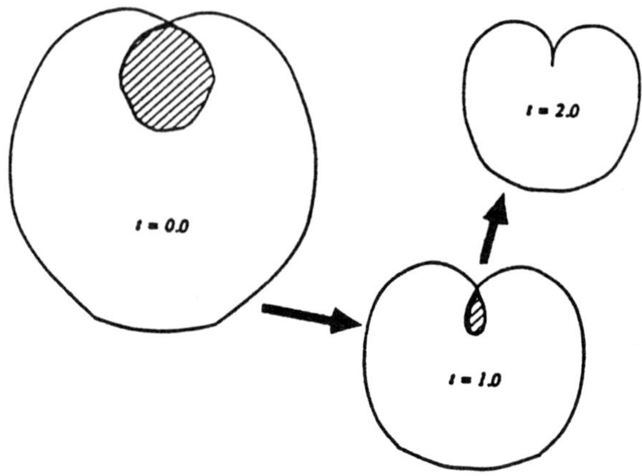

Figure 2. A singularity in the curve shortening problem.

happen. The formation of such singularities was studied by this author in [An1, An2].

In [An2] we considered strictly convex immersions of the circle in $\mathbf{R}^2$, and used equation (2) to study how their corresponding solutions become singular. The results may summarized as follows.

Let $\kappa(t)$ be the maximal curvature of the curve at time t, and assume that the curve becomes singular at time T. *If $\sqrt{T-t}\cdot\kappa(t)$ remains bounded, then the curve will shrink to a point, and its asymptotic shape will be one of the self similar solutions found by Abresch and Langer.* Thus self similar blow up occurs if and only if the curvature blows up like $(T-t)^{-1/2}$.

To describe the situation in which $\kappa(t)\sqrt{T-t}$ is not bounded, we define a "normalized" curve for each $t < T$. Choose a point $P(t)$ on the curve where the curvature attains its maximum, rotate and translate the curve so that this point becomes the origin, so that its tangent becomes horizontal, and so that it is curved upwards at the origin; next magnify the curve so that its curvature at the origin becomes $+1$. We shall call the curve thus obtained $\Gamma(t)$.

It was shown in [An2] that, *if $\sup_{t<T} \sqrt{T-t} \cdot \kappa(t) = \infty$, then there exists a sequence $t_n \uparrow T$ such that the curves $\Gamma(t_n)$ converge to the graph of* $y = -\log\cos x$ $(|x| < \pi/2)$.

Finally, if one defines the *blow up* set of a solution $k(\theta,t)$ of (2) to be the set $\Sigma = \{\theta \in \mathbf{R}/2\nu\pi\mathbf{Z} : \lim_{t\uparrow T} k(\theta,t) = \infty\}$, then one can show that Σ consists of a finite number of intervals, each of which has length at least π (see [GH, An2]). Theorem D of [An2] says that, if $|\Sigma| < 2\pi$, then one actually has $|\Sigma| = \pi$; in this situation the curves $\Gamma(t)$ will converge to the "Grim Reaper," i.e. to the graph of $y = -\log\cos x$. (This is stronger than the previous statement which only said that some subsequence $\Gamma(t_n)$ converges.) Moreover, one obtains the following estimate for the rate with which the curvature blows up:

$$\lim_{t\uparrow T}(T-t)^{1/2+\epsilon}\kappa(t) = \begin{cases} \infty, & \text{if } \epsilon = 0; \\ 0, & \text{for all } \epsilon > 0. \end{cases}$$

The results in [An2] are definitely not the last word on "blow up" for (2). Many questions remain, such as "what is the precise blow up rate?", and "which blow up sets Σ can occur?" (guess: Σ must be the union of disjoint intervals whose lengths are multiples of π). Besides answering such questions one would also like to remove the convexity hypothesis concerning the curves, which allows one to use the convenient equation (2).

Self similar solutions

By definition a self similar solution of (1) is a solution which can be parametrised as $X(p,t) = f(t)X_0(p)$. By inspection one finds that the only $f(t)$'s which can occur here are given by $f(t) = \sqrt{2(T-t)}$; without loss of generality we may assume that $T = 1$. Again, by substituting the *ansatz* $X(p,t) = f(t)X_0(p)$ in (1), one finds that $X(p,t) = \sqrt{2(1-t)}\cdot X_0(p)$ is a solution of (1) if and only if the immersion $X_0 : M \to \mathbf{R}^{n+1}$ satisfies

$$(3) \qquad nH_{X_0}(p) + \frac{1}{2}\langle X_0(p), \nu_{X_0}(p)\rangle = 0$$

for all $p \in M$. Except for the lower order term $\frac{1}{2}\langle X_0(p), \nu_{X_0}(p)\rangle$, this is the equation which determines minimal hypersurfaces. In fact the solutions to (3) are exactly the immersions $X_0 : M \to \mathbf{R}^{n+1}$ at which the functional

$$A(X) = \int_M e^{-|X(p)|^2/4} d\sigma_X^n(p)$$

is stationary (cf. [Hu2, Theorem 3.1]). Here $d\sigma_X^n$ denotes the n dimensional volume element which $X : M \to \mathbf{R}^{n+1}$ induces on M.

Indeed, a straightforward calculation shows that the first variation of $A(X)$ under a normal variation $X(\epsilon, p) = X(p) + \epsilon u(p)\nu_X(p)$ for any given $u \in C^\infty(M)$ is given by

$$(4) \qquad \frac{dA(X+\epsilon u\nu_X)}{d\epsilon}\bigg|_{\epsilon=0} =$$

$$-\int_M e^{-|X(p)|^2/4}[nH_{X_0}(p) + \frac{1}{2}\langle X_0(p), \nu_{X_0}(p)\rangle]u(p)\, d\sigma_X^n(p)$$

In other words, the solutions to (3) are exactly the minimal hypersurfaces in $\mathbf{R}^{n+1}$ with respect to the metric $(ds)^2 = e^{-|x|^2/4n}\{(dx_0)^2 + \ldots + (dx_n)^2\}$.

We shall not consider the most general solution of (3); instead, we shall restrict our attention to hypersurfaces of revolution. This means that we assume that $M = S^{n-1} \times (a, b)$, and that the immersion X_0 has the form $X_0(\omega, s) = x(s)e_0 + r(s)\omega$, i.e. that X_0 is obtained by rotating the plane curve parametrised by $(x(s), r(s))$ around the x_0 axis. We have identified S^{n-1} with the standard unit sphere in $\{0\} \times \mathbf{R}^n \subset \mathbf{R}^{n+1}$, and e_0 denotes the first unit basis vector $(1, 0, \ldots, 0)$ in $\mathbf{R}^{n+1}$.

The functional $A(X)$ will be stationary at the immersion X corresponding to $(x(s), r(s))$ if and only if the curve $\{(x(s), r(s)) : s \in (a, b)\}$ is a geodesic in the upper half plane $\{r > 0\}$ with metric

$$(5) \qquad (ds)^2 = r^{2(n-1)}e^{-(x^2+r^2)/4}\{(dx)^2 + (dr)^2\}.$$

Indeed, the volume element $d\sigma_X^n$ is given by

$$d\sigma_X^n = r(s)^{n-1}\sqrt{x'(s)^2 + r'(s)^2} \cdot ds\, d\omega^{n-1},$$

where $d\omega^{n-1}$ is the volume element on the $n-1$ sphere. Therefore one can write $A(X)$ as

$$A(X) = \mathrm{vol}(S^{n-1})\int_a^b r(s)^{n-1}\sqrt{x'(s)^2 + r'(s)^2}\, ds$$

which up to a multiplicative constant is the length of the curve parametrized by $(x(s), r(s))$ in the metric (5). Thus if $A(X)$ is stationary at some immersion X which comes from a hypersurface of rotation, then the corresponding curve will certainly be a geodesic of (5). Conversely, if one has a geodesic, then the corresponding immersion X will be a stationary point of A under rotationally symmetric variations of X. Using (4) one shows that, if X is rotationally symmetric, then

$$\left.\frac{dA(X + \epsilon u(s, \omega)\nu_X)}{d\epsilon}\right|_{\epsilon=0} = \left.\frac{dA(X + \epsilon\bar{u}(s)\nu_X)}{d\epsilon}\right|_{\epsilon=0}$$

where $\bar{u}(s)$ is the average of $u(s, \omega)$ over S^{n-1}. In other words, if A is stationary at X under symmetric variations, then it is stationary under all variations.

At this point we can formulate the main result of this note.

Theorem. *Let $n \geq 2$. The upper half plane, equipped with the metric (5), has at least one simple closed geodesic which is symmetric with respect to reflection in the r-axis.*

One should observe that the metric (5) is not complete (the r-axis has finite length). One should also note that in the case $n = 1$, which corresponds to the curve shortening problem, the metric extends to a smooth metric on the entire plane. This metric is invariant under rotations, so that its geodesic flow is integrable: all geodesics are either closed, or are densely wound on tori of constant angular momentum. In the case which we shall be considering, $n \geq 2$, the metric admits no obvious symmetry, and a simple minded numerical study of the geodesic flow (done on a Macintosh SE/30 in Turbo Pascal) indicates that the flow does not seem to have a second integral (the energy $r^{n-1}e^{-(x^2+r^2)/8}\sqrt{\dot{x}^2 + \dot{r}^2}$ being the first integral). Figure 3 shows some geodesic curves for the metric (5) with $n = 2$.

To prove our theorem, we shall use a "shooting method". Given any point (x, r) in the upper halfplane, and any angle $\theta \in \mathbf{R}$ there exists a unique geodesic through (x, r) whose tangent at that point is $(\cos\theta, \sin\theta)$. If one parametrizes such geodesics by arc length, then one obtains a flow on the set of unit tangent vectors, the geodesic flow of the metric (5) on the unit tangent bundle. Using Liouville's formula ([DoC, p. 253]) one finds that this flow is given by the following system of ordinary differential equations:

$$(6) \qquad \begin{cases} \dot{x} = \cos\theta \\ \dot{r} = \sin\theta \\ \dot{\theta} = \dfrac{x}{2}\sin\theta + \left(\dfrac{n-1}{r} - \dfrac{r}{2}\right)\cos\theta \end{cases}$$

Let $(x_R, r_R, \theta_R) = \Gamma_R : [0, T(R)) \to \mathbf{R}^3$ be the maximal solution of (6) with initial value $\Gamma_R(0) = (0, R, 0)$, and let $\gamma_R(t) = (x_R(t), r_R(t))$ be the projection of Γ_R on the xr- plane. Then we intend to show that there exist $R > 0, t_R > 0$ such that $\gamma_R([0, t_R])$ is a simple curve in the first quadrant which begins and ends on the r-axis, and whose tangents on the r-axis are horizontal, i.e. perpendicular to the r-axis. Since reflection in the r-axis, $(x, r) \mapsto (-x, r)$, is an isometry for the metric (5), the curve obtained by reflecting $\gamma_R([0, t_R])$ in the r-axis is a closed geodesic whose existence is claimed by the theorem.

Special solutions. The metric (5) has a few simple geodesics whose existence will help us in our proof. They are the following. First, the r-axis is a geodesic: refering to the original flow by mean curvature problem (1), this geodesic corresponds to a plane through the origin. Such a plane is an equilibrium for (1), and hence may be considered as a self similar solution.

It is well known that an n sphere in $\mathbf{R}^{n+1}$ will shrink to its centre by dilations, so there should be a geodesic for (5) corresponding to a sphere; it is given by the circle $x^2 + r^2 = 2n$.

Finally, a cylinder $\mathbf{R} \times S^{n-1}$ centered at the x_0 axis in $\mathbf{R}^{n+1}$ will also shrink by dilations. This is reflected in the fact that the straight line $r = \sqrt{2(n-1)}$ in the upper half plane is a geodesic for (5).

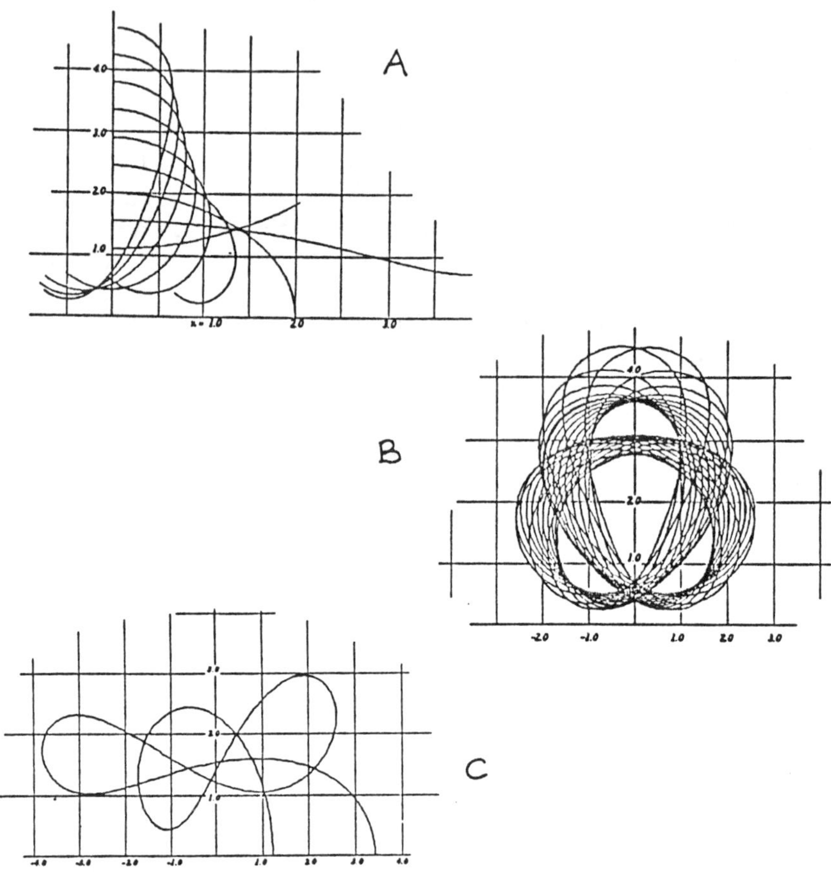

Figure 3a *shows some of the geodesic segments γ_R which are used in the proof of the main theorem.*

Figure 3b *exhibits a geodesic segment which appears to come from a quasi periodic orbit of the system (6).*

Figure 3c *shows a geodesic which when rotated around the x-axis, would lead to a self intersecting immersed 2-sphere in $\mathbf{R}^3$, which generates a similarity solution of the flow by mean curvature equation. By experimenting one easily finds numerical evidence for many more of these solutions.*

General behaviour of Γ_R.   We shall always assume that $R > \sqrt{2(n-1)}$. Under this restriction one has $\theta'(0) < 0$, so that the curve γ_R will initially bend downwards. Then, as long as $x > 0$, $\theta > -\pi/2$ and $r > \sqrt{2(n-1)}$ hold, one will have $\theta' < 0$, so that the curve will be convex (in the common, Euclidean, sense).

Let $t_1 = t_1(R) > 0$ be the first time, if any, at which either $x_R = 0$ or $\theta_R = 0$, or $\theta_R = -\pi$ occurs (if this never happens, then we put $t_1(R) = T(R)$.) On the segment $\gamma_R[(0, t_1(R))]$ one has $0 > \theta > -\pi$, so that this segment is the graph of a function $x = f_R(r)$, defined for $r_R(t_1(R)) < r < R$.

At any point where this function is stationary one has $\theta = -\pi/2$, so that $\theta' = -x/2 < 0$, which implies $f_R'' < 0$. In other words, f_R can only have local maxima, and hence it can have at most one critical point, which must then be a maximum.

Asymptotics as $R \uparrow \infty$.   In this section we show that, for large enough R, the geodesic γ_R will make a sharp downward bend at $(0, R)$, and then follow the r-axis closely for a while, until it intersects the r-axis, somewhere above the line $r = 1/R$.

Since R is large, we put $R = \epsilon^{-1}$, and we introduce the variables $\xi(\tau) = Rx(\epsilon\tau)$, $\rho(\tau) = R(r(\epsilon\tau) - R)$ and $\vartheta(\tau) = \theta(\epsilon\tau)$. They satisfy

$$(7) \qquad \begin{cases} \dot{\xi} = \cos\vartheta \\ \dot{\rho} = \sin\vartheta \\ \dot{\vartheta} = -\dfrac{1}{2}\cos\vartheta + \mathcal{O}(\epsilon^2) \end{cases}$$

while their initial values are given by $\xi(0) = \vartheta(0) = 0, \rho(0) = 1$. For $\epsilon = 0$ the system (7) can be solved explicitly, and one finds that $\vartheta(\tau) = -\arcsin\tanh\frac{\tau}{2}$. Since the solution of (7) depends smoothly on the parameter ϵ, we may conclude the following.

Lemma 1. *There is a $t_2 > 0$ such that for all sufficiently large R one has $T(R) > t_2/R$, while, at $t = t_2/R$, one will have $-\pi/3 \leq \theta_R \leq -\pi/6$, $x_R = \mathcal{O}(1/R)$ and $r_R = R - \mathcal{O}(1/R^2)$.*

As we noted before, for $t \leq t_1(R)$ we have $0 > \theta_R(t) > -\pi$, so that γ_R is a graph. We shall now estimate how far it can get from the r-axis.

Define $\alpha = \frac{\pi}{2} + \theta$, and regard α as a function of r. Then we have

$$(8) \qquad \begin{aligned} \frac{d\alpha}{dr} &= \frac{\theta'}{r'} = \frac{x}{2} + \left(\frac{r}{2} - \frac{n-1}{r}\right)\tan\alpha \\ &\geq \left(\frac{r}{2} - \frac{n-1}{r}\right)\tan\alpha. \end{aligned}$$

Let $R_1 = r_R(t_1)$, and integrate this inequality from any $r < R_1$ with

$\alpha(r') > 0$ for $r < r' < R_1$, to R_1. This shows that

$$\frac{\sin \alpha(R_1)}{\sin \alpha(r)} \geq e^{(R_1^2 - r^2)/2} \left(\frac{r}{R_1}\right)^{n-1},$$

i.e. we find that for all r such that $\theta \in (-\frac{\pi}{2}, 0)$ on (r, R_1) we have:

$$(9) \qquad \sin \alpha(r) \leq \left(\frac{R_1}{r}\right)^{n-1} e^{-(R_1^2 - r^2)/2} \sin \alpha(R_1).$$

In particular we have $\sin \alpha(r) \leq \sin \alpha(R_1) \leq \frac{1}{2}\sqrt{3}$ for all $r \in (\frac{1}{R}, R_1)$ with $\theta \in (-\frac{\pi}{2}, 0)$ on (r, R_1). For such r's one therefore has $\tan \alpha \leq 2 \sin \alpha$. Since $\tan \alpha = -f_R'(r)$, this implies that for all $1/R < r < R_1$ with $\theta \in (-\frac{\pi}{2}, 0)$ on (r, R_1) one has

$$(10) \qquad f_R(r) \leq f_R(R_1) + 2 \int_r^{R_1} \left(\frac{R_1}{r}\right)^{n-1} e^{-(R_1^2 - r^2)/2} dr \leq \frac{C}{R},$$

as $R \to \infty$. This inequality gives us an estimate for the first maximum of $f_R(r)$ which one might encounter if one decreases r, starting at $r = R_1$. We had already observed that f_R can have at most one maximum, so that the estimate (10) implies that we have proved

Lemma 2. *For* $0 < t < t_1$ *one has* $x_R(t) \leq C/R$ *or* $r_R(t) < 1/R$, *if* R *is large enough. The constant* C *does not depend on* R.

The next step in the proof of our theorem is to show that the geodesic γ_R will intersect the r-axis, if R is large enough.

Lemma 3. *For large enough* R *one has* $t_1(R) < T(R)$, *while* $x_R(t_1) = 0$, *and* $r_R(t_1) \geq 1/R$.

Proof. Suppose not. Then there exists a sequence $R_n \uparrow \infty$ for which $x_{R_n}(t) > 0$ at least as long as $r_{R_n}(t) > 1/R_n$. In this situation the functions $f_{R_n}(r)$ are defined for $1/R_n < r < R_n$, on which interval they satisfy

$$(11) \qquad \frac{f''(r)}{1 + f'(r)^2} + \left(\frac{n-1}{r} - \frac{r}{2}\right) f'(r) + \frac{f(r)}{2} = 0.$$

Moreover it follows from Lemma 2 that the $f_{R_n}(r)$ are bounded by $0 < f_{R_n}(r) < C/R_n$. Since the geodesic γ_{R_n} gets arbitrarily close to the r-axis, its tangents also must converge to the r-axis; if for each n there were some point $p_n \in \gamma_{R_n}$ whose tangent made an angle $\theta \in (-\pi + \delta, -\delta)$ for some constant $\delta > 0$, and if these points remained in a compact domain in the upper half plane, then one could extract a convergent subsequence, $p_{n_j} \to p_*$. For large n_j the geodesic $\gamma_{R_{n_j}}$ would have to be close to the geodesic through p_* with the same tangent direction, and hence would have to intersect the r axis somewhere near p_*. We are assuming that this doesn't happen, so that this argument shows us that both $f_{R_n}(r)$ and $f'_{R_n}(r)$ converge uniformly to zero, on compact intervals in $(0, \infty)$.

Since the f_{R_n}'s are solutions of (11), and since we know that f'_{R_n} converges uniformly to zero, it follows from the positivity of the f_{R_n} that there is a constant $C < \infty$ for which one has $|f'_{R_n}(1)| \leq C|f_{R_n}(1)|$. It follows that one can now extract a subsequence $n_1 < n_2 < n_3 < \cdots$ of the integers for which

$$\frac{f_{R_{n_j}}(r)}{f_{R_{n_j}}(1)} \to g(r), \qquad (0 < r < \infty)$$

where the convergence is in C^2 on any compact interval in $(0, \infty)$.

The limit function $g(r)$ is a nontrivial, positive solution of the linearized equation of (11), i.e. of

$$(12) \qquad g''(r) + \left(\frac{n-1}{r} - \frac{r}{2}\right) g'(r) + \frac{g(r)}{2} = 0$$

on $0 < r < \infty$. A short computation will reveal that $h(r) = e^{r^2/8}g(r)$ satisfies

$$(13) \qquad h''(r) + \frac{n-1}{r}h'(r) + \left(\frac{n}{4} - \frac{r^2}{16}\right)h(r) = 0.$$

Up to a constant term the differential operator in this equation is the radial part of the Harmonic Oscillator on $\mathbf{R}^n$, so that it is easily verified that (13) has no positive solutions on $(0, \infty)$. But this contradicts the fact that $h(r) > 0$.

Q. E. D.

So, for large R we have found that $t_1(R) < T(R)$. On the other hand, we have seen that the circle $x^2 + r^2 = 2n$ is a geodesic for the metric (5), so that $t_1(R) = T(R)$ at $R = \sqrt{2n}$. It follows that there is a smallest $R_* \geq \sqrt{2n}$ for which $t_1(R) < \infty$ and $x_R(t_1) = 0$ holds whenever $R > R_*$.

Lemma 4. $\inf_{R_* < R \leq R_*+1} r_R(t_1(R)) > 0$.

Proof. We argue by contradiction again. Assume that $r_{R_n}(t_1(R_n)) \downarrow 0$ for some sequence $R_n \downarrow R_*$. We reparametrize the geodesics so that $r\,d\tau = dt$. In other words, we introduce a new time variable which is related to t via

$$\tau = \int^t \frac{dt}{r(t)}.$$

In this new time variable τ the equations (6) become regular at $r = 0$. Indeed one finds:

$$
(14) \quad
\begin{cases}
\dfrac{dx}{dt} = r\cos\theta, \qquad \dfrac{dr}{dt} = r\sin\theta, \\[2mm]
\dfrac{d\theta}{dt} = \left(n - 1 - \dfrac{r^2}{2}\right)\cos\theta + \dfrac{xr}{2}\sin\theta.
\end{cases}
$$

This system has a line of fixed points, $\ell = \{(x, r, \theta) : r = 0, \theta = -\frac{\pi}{2}\}$. By linearizing (14) one finds that this line is normally hyperbolic.

At $t = t_1(R_n)$ we have $-\frac{\pi}{2} > \theta(t_1(R_n)) > -\pi$, while

$$
\frac{d\theta}{d\tau} = (n - 1)\cos\theta + \mathcal{O}(|x|^2 + |r|^2), \qquad \frac{dx}{d\tau}, \frac{dr}{d\tau} = \mathcal{O}(|x| + |r|).
$$

This means that just before the geodesic γ_{R_n} hits the r-axis, one has $\theta \approx -\frac{\pi}{2}$, with $x, r = o(1)$. Thus the curve Γ_{R_n} converges to the stable manifold of the point $(x, r, \theta) = (0, 0, -\frac{\pi}{2})$ on the line ℓ. But this stable manifold is exactly the "planar solution", i.e. the r-axis. This cannot happen, since γ_{R_n} also converges to γ_{R_*}, a geodesic distinct from the r-axis.

Q. E. D.

This lemma shows us that the γ_R are bounded away from the x axis, as $R \downarrow R_*$. In the following lemma we also obtain a bound in the x direction for the γ_R.

Lemma 5. $\sup(x_R(t) : 0 < t < t_1(R), R_* < R \leq R_* + 1) < \infty.$

Proof. If the lemma is not true, then for some sequence $R_n \downarrow R_*$ the maximal value $\xi_n = \max\{x_R(t) : 0 < t < t_1\}$ becomes unbounded. The geodesic $\gamma_{R_n}([0, t_1(R_n)])$ is a graph $x = f(r)$, with one maximum, and hence it can also be represented as the union of two graphs $r = g_{n,\pm}(x)$ where $0 \leq x \leq \xi_n$, and where $g_{n,-}(x) < g_{n,+}(x)$ for $0 \leq x < \xi_n$. Moreover, we have $\delta \leq g_{n,\pm}(x) \leq R_* + 1$ for all n, and all $x \leq \xi_n$. Finally, the $g_{n,\pm}$'s are strictly monotone, and they satisfy the following differential equation:

$$
(15) \quad \frac{g''(x)}{1 + g'(x)^2} - \frac{x}{2}g'(x) + \frac{n-1}{g(x)} - \frac{g(x)}{2} = 0.
$$

By studying the asymptotic form of the equations (6) for large x and bounded r, one finds that the slopes $g'_{n,\pm}(x)$ must be uniformly bounded on $[0, \xi_n - 1]$. Indeed, if one introduces a new time variable, s, with $ds = x\,dt$, then the equations (6) imply that

$$
(16) \quad
\begin{aligned}
x_s, r_s &= \mathcal{O}\left(\frac{1}{x}\right) \\[2mm]
\theta_s &= \frac{1}{2}\sin\theta + \mathcal{O}\left(\frac{1}{x}\right)
\end{aligned}
$$

hold in the region $\delta \le r \le R_* + 1$. If one neglects the $\mathcal{O}(\frac{1}{s})$ terms, then these equations can be solved explicitly: x and r are constant, and $\theta(s) = -\frac{\pi}{2} - \arcsin \tanh \frac{s}{2}$. If $|g'_{n,\pm}(x)|$ were large at some point $p \in \gamma_{R_n}$, then the corresponding angle θ would be close to $-\frac{\pi}{2}$, and it would follow from the equations (16) that the geodesic γ_{R_n} must have a vertical tangent at a distance $\mathcal{O}(\frac{1}{x})$ from the point p. If we choose p so that its x coordinate does not exceed $\xi_n - 1$, then this is impossible, since the geodesic is a graph $r = g_{n,\pm}(x)$ near p.

Thus we have a bound for the derivatives of the $g_{n,\pm}$'s. It follows from (15) that we then also have a bound for the second, and by induction, all higher order derivatives of $g_{n,\pm}$. After passing to a subsequence, if necessary, one may assume that the $g_{n,\pm}$ converge in $C^{\infty}_{\text{loc}}(\mathbf{R}_+)$ to two functions $g_{\pm}(x)$, both of which again satisfy (15). These two functions must be different, for their values $g_+(0) = R_* \ge \sqrt{2n}$, and $g_-(0)$, at $x = 0$ are different. Indeed, if $g_+(0) = g_-(0)$, then the geodesics γ_{R_n} would be trapped between the horizontal lines $r = g_{n,-}(0)$ and $r = g_{n,+}(0)$, which converge to the line $r = g_+(0)$. This limit line would then also have to be a geodesic, but the only horizontal line which is a geodesic is the line $r = \sqrt{2(n-1)}$. Thus we have $g_-(x) < g_+(x)$ for all $x \ge 0$.

We can write (15) as $g''(x) = F(x, g(x), g'(x))$, where

$$F(x, g, p) = (1 + p^2) \left(\frac{xp}{2} - \frac{n-1}{g} + \frac{g}{2} \right)$$

satisfies

$$\frac{\partial F}{\partial g}(x, g, p) > 0,$$

$$\frac{\partial F}{\partial p}(x, g, p) = \frac{x}{2}(1 + 3p^2) - 2p \left(\frac{n-1}{g} - \frac{g}{2} \right) \ge \frac{x}{4} - C,$$

for all $\delta < g < R_* + 1$, $x \ge 1$, $p \in \mathbf{R}$, and for a sufficiently large constant C. Using the mean value theorem one then finds that the difference $z(x) = g_+(x) - g_-(x)$ satisfies a linear equation of the form $z''(x) - M(x)z'(x) - N(x)z(x) = 0$, in which $M(x) \ge \frac{x}{4} - C$ and $N(x) \ge 0$. Since $z(x) > 0$ this implies the differential inequality $z''(x) - M(x)z'(x) \ge 0$, which one can integrate once, on the interval (x_0, x_1). The result is:

$$z'(x_0) \le \exp \left\{ - \int_{x_0}^{x_1} M(x)dx \right\} z'(x_1) \le C' e^{-x_1^2/8} z'(x_1).$$

Using our bounds on $g'_{\pm}(x)$ and choosing x_1 arbitrarily large, this leads us to the conclusion that $z'(x_0) = 0$, for all $x_0 \ge 0$. Thus (15) does not admit two ordered solutions $g_- < g_+$ on $\mathbf{R}_+$. Since our original assumption led us to two such solutions, it must have been false. $\qquad$ Q. E. D.

So far we have found that the geodesics γ_R stay away from the x axis, and remain uniformly bounded as $R \downarrow R_*$. Therefore the limiting geodesic $\gamma_* = \gamma_{R_*}$ begins and ends on the r-axis, i.e. it goes from $(0, R_*)$ to $(0, r_*)$, where $r_* = r_{R_*}(t_1(R_*))$.

We now claim that $\theta_* = \theta_{R_*}(t_1(R_*)) = -\pi$. We know that $\theta_* \geq -\pi$; if one had strict inequality, then one could vary the height R near R_*, and still obtain a geodesic γ_R, which, in the first quadrant, is a graph $x = f_R(r)$, and which hits the r-axis in finite time. Since one would be able to do this for $R < R_*$, but close to R_*, this would contradict the minimality of R_*. Thus $\theta_* = -\pi$, as claimed. This also completes the proof of the main theorem.

The pinching neck in the flow by mean curvature problem

In this section we'll show how the existence of the shrinking torus leads to the formation of singularities in finite time, for some solutions of the flow by mean curvature problem. We shall restrict our discussion to the most easily visualized case, i.e. to the motion of surfaces in three dimensional space.

We shall consider a solution to the flow by mean curvature problem whose initial value Σ_0 is the surface suggested in Figure 4. The surface is the image of an embedding $X_0 : S^2 \to \mathbf{R}^3$. The relevant features of Σ_0 are:

(i) it encloses two spheres of radius $R > 0$,

(ii) it has a "neck" which is circled by one of the self similar shrinking tori, whose existence we have just established.

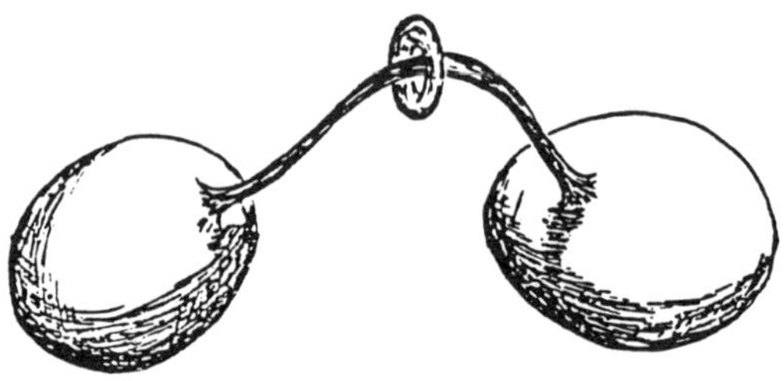

Figure 4.

One can let the two spheres and the torus flow according to their mean curvature; each will shrink in a self similar way to its center. We shall assume that the radius R of the two spheres is so large that the little torus will shrink to a point before the spheres do so. Denote the time at which the torus disappears by T_*.

Let $X : S^2 \times [0, T) \to \mathbf{R}^3$ be the maximal solution of (1) with initial value X_0. Then one must have $T \leq T_*$. Indeed, it follows from the maximum principle for parabolic equations that the four solutions (X, the two spheres and the torus) remain disjoint for as long as they are defined (see [Br, ES]). Thus if the smooth solution X existed for $t > T_*$, then at time $t = T_*$ the surface $X(S^2, T_*)$ still encloses the two spheres, and it must have a neck which is circled by the torus. But this torus just shrank to a point, so the neck must have zero diameter at some point, which shows that the solution X is singular at time T_*.

The final conclusion is that the solution corresponding to the initial value depicted in Figure 4 becomes singular before it can become convex, so that we have one more proof that Grayson's theorem cannot be extended to higher dimensions.

REFERENCES

[AL] U. Abresch and J. Langer, *The normalized curve shortening flow and homothetic solutions*, Journal of Differential Geometry, **23** (1986), 175–196.

[An1] S. B. Angenent, *Parabolic Equations for Curves on Surfaces, I&II*, to appear, Ann. of Math.

[An2] ———, *On the formation of Singularities in the Curve Shortening Problem*, to appear, Journ. Diff. Geom.

[Br] K. A. Brakke, *The Motion of a Surface by its Mean Curvature*, Math. Notes, Princeton, New Jersey, Princeton University Press, 1978.

[CGG] Y.-G. Chen, Y. Giga and S. Goto, *Uniqueness and Existence of Viscosity Solutions of Generalized Mean Curvature Flow Equations*, Hokkaido University preprint, July 1989.

[DoC] Manfredo P. DoCarmo, *Differential Geometry of Curves and Surfaces*, Prentice-Hall, Englewood Cliffs, New Jersey, 1976

[ES] C. Evans and J. Spruck, *Motion of Level Sets by Mean Curvature I&II*, preprint, Spring 1989.

[EW] C. Epstein and M. Weinstein, *A stable manifold theorem for the curve shortening equation*, Communications in pure and applied mathematics, **40** (1987), 119–139.

[FM] A. Friedman and J. B. McLeod, *Blow-up of Solutions of Nonlinear Degenerate Parabolic Equations*, Archive for Rational Mechanics and Analysis, **96** (1986), 55–80.

[GH] M. Gage and R. S. Hamilton, *The heat equation shrinking convex plane curves*, Journal of Differential Geometry, **23** (1986), 69–96.

[Gr1] M. Grayson, *The heat equation shrinks embedded plane curves to round points*, Journal of Differential Geometry, **26** (1987), 285–314.

[Gr2] ———, *Shortening embedded curves*, Annals of Mathematics, **129** (1989), 71–111.

[Gr3] ——— , *A short note on the evolution of a surface by its mean curvature*, preprint, UCSD, February 1988.

[Hu1] G. Huisken, *Flow by mean curvature of convex surfaces into spheres*, Journal of Differential Geometry, **20**(1984), 237–266.

[Hu2] ——— , *Asymptotic behaviour for singulartities of the mean curvature flow*, Research report of the Australian National University, CMA-R49-87, to appear, Journal of Differential Geometry.

Sigurd Angenent
Department of Mathematics
University of Wisconsin at Madison
Madison, Wisconsin 53706

Higher Approximations to Eigenvalues for a Nonlinear Elliptic Problem

F. V. ATKINSON

Abstract

Approximations are obtained for the eigenvalues yielding radial solutions of the Dirichlet problem for $-\Delta u = \lambda u + 3u^5$ in the unit ball in R^3 the exponent "5" being critical for Sobolev embedding. Nodal solutions are considered as well as positive ones.

1. Introduction

The detailed study of the Dirichlet eigenvalue problem

$$- \Delta u = \lambda u + 3u^5 \text{ in } B, \quad u = 0 \text{ on } \delta B, \tag{1.1}$$

where B is the unit ball in R^3, was initiated by Brézis and Nirenberg in a well-known paper [6], along with the investigation of many similar problems. For the lowest eigenvalue λ, that for which $u > 0$ in the interior of the ball, they obtained the striking property that

$$\pi^2/4 < \lambda < \pi^2. \tag{1.2}$$

In this problem $u(0)$ may have any positive value; the limits for λ in (1.2) correspond respectively to the choices $u(0) \to \infty$, and $u(0) \to 0$; we refer again to [6], and to [2,3,4].

In a recent paper of Brézis and Peletier [7], PDE methods were developed for discussing several such problems, and applied in particular to the case of (1.1) in which $\lambda \to (\pi/2)^2 + 0, u(0) \to \infty$; writing $\lambda = \pi^2/4 + \varepsilon$ and $u_\varepsilon(0)$ for the corresponding value of $u(0)$, they showed that

$$\varepsilon u_\varepsilon^2(0) = \pi^3/2 + 0(\varepsilon|\log \varepsilon|) \tag{1.3}$$

as $\varepsilon \to 0$. They obtained also the asymptotic behavior of u_ε at points away from the origin, and dealt with problems in which the critical index 5 ($=$ $(3+2)/(3-2)$) is replaced by $5 - \varepsilon$.

As is well known [13], the problem of positive eigenfunctions in the ball may be simplified by confining attention to radial solutions. Accordingly, one may discuss instead a certain boundary-value problem for an ordinary differential equation, namely,

$$u'' + (2/r)u' + \lambda u + 3u^5 = 0 \tag{1.4}$$

We ask that the solution $u(r)$ satisfy

$$u(0) = \gamma > 0, \quad u'(0) = 0, \tag{1.5}$$

be positive in $(0, 1)$, and also, for the purposes of the Dirichlet problem, satisfy

$$u(1) = 0. \tag{1.6}$$

This leads to the "shooting method" in which γ or λ or both are to be adjusted so that the solution of (1.4,1.5) does in fact satisfy (1.6). The method is, of course, quite distinct from methods based on the variational characterization of eigenvalues and eigenfunctions.

The present paper is devoted, in the first place, to the asymptotic integration problem posed by (1.4),(1.5), without regard to (1.6). The asymptotics are in the sense that $\gamma \to \infty$, while λ is positive and bounded. Having obtained the asymptotic form of $u = u(r) = u(r, \gamma, \lambda)$, one can then turn to the problem of imposing (1.6), together with specification of the number of sign changes of $u(r)$ in $(0, 1)$.

We find it convenient to retain throughout the three variables r, λ, γ, though they are in fact interrelated. We express this relationship in the invariance property

$$u(r, \lambda, \gamma) = \mu u(\mu^2 r, \lambda \mu^{-4}, \gamma \mu^{-1}), \tag{1.7}$$

valid for any positive constant μ.

The principal results of the paper are presented in the next two sections; the main part of the proof occupies Sections 11-12 and is devoted to the problem of asymptotic integration, without regard to any eigenvalue problem. The last sections (Sections 13-14) are concerned with the differentiability of eigenvalues with respect to the "shooting height" γ.

2 . Main Results: (1) Asymptotics of Solutions

We denote by $u(r)$ or $u(r, \gamma)$ or $u(r, \lambda, \gamma)$ the solution of the initial value problem given by (1.4),(1.5). To be more precise, u is to be in $C'[0, \infty) \cap C''(0, \infty)$, to satisfy (1.4) in $(0, \infty)$ and also (1.5). It is easily seen that this solution is well defined. In the first aim of the paper, we are concerned with asymptotic formulae for solutions, valid as $\gamma \to \infty$ with λ, r confined to domains of the form

$$0 < A \leq \lambda \leq B < \infty, \tag{2.1}$$

$$0 \leq r \leq C < \infty. \tag{2.2}$$

To be more specific, we assume that

$$0 < A < \pi^2/4, \tag{2.3}$$

$$C > \sqrt{2/A}, \tag{2.4}$$

$$\gamma > 2. \tag{2.5}$$

The first of these is suggested by the Brézis-Nirenberg result (1.2). The second is designed to ensure that u has at least one zero; we denote this first zero by r_1.

For the purposes of asymptotics we need to consider intervals bounded by the points

$$0, \gamma^{-2}, R, r_1, \infty, \tag{2.6}$$

where R satisfies

$$1/(2\gamma) < R < 2/\gamma. \tag{2.7}$$

The greater part of the work is devoted to matching at the point R a solution over $(0, R]$ with one over $[R, \infty)$. The solution over $(0, R]$ is obtained, approximately, by means of a method of variation of parameters. A similar argument is used by Budd, for a closely related problem; we refer to [8, 9], where other references are given. Over $[R, \infty)$ the equation has an oscillatory character, amenable, as we show in Sections 9 and 10, to the Prüfer transformation.

For approximation over $(0, R]$, or indeed over $(0, r_1)$, the basic fact is that

$$0 < u(r) \leq v(r), \quad 0 \leq r \leq r_1, \tag{2.8}$$

where

$$v(r) = v(r, \gamma) = \gamma/\sqrt{1 + \gamma^4 r^2}, \tag{2.9}$$

which may be identified with $u(r, 0, \gamma)$; a slightly closer comparison function was basic in [3, 4] and also is used in [7]. This motivates the introduction of the functions

$$Z(r) = u(r)/v(r), \tag{2.10}$$

$$Y(r) = 1 - Z(r). \tag{2.11}$$

We formulate an essentially known result as follows.

Proposition 1 *We have*

$$0 \leq Y(r) \leq \lambda r^3/2, \quad 0 \leq r \leq r_1. \tag{2.12}$$

This is very close to Lemma 2.2 (a) of [3] or again to (7.4) of [7], and also proves (2.3). However, this upper bound for $Y(r)$ fails to give the true situation when r is very small. We cite, for the situation that $\gamma \to \infty$, the following proposition.

Proposition 2 *We have*

$$Y(r) = \lambda r^2/6 + 0(r^4\gamma^4), \quad 0 \leq r \leq \gamma^{-2}, \tag{2.13}$$

$$Y(r) = \lambda r^2/2 + 0(r\gamma^{-2}) + 0(r^4), \quad \gamma^{-2} \leq r \leq r_1. \tag{2.14}$$

This last result has the defect that in the transition zone where r is of order γ^{-2}, the error terms in (2.13, 2.14) are of the same order as the main term. For an approximation valid through this zone, the method of variation of parameters yields the following proposition.

Proposition 3 *For $0 \leq r \leq r_1$, we have*

$$Y(r) = \lambda\gamma^{-4}\eta(r\gamma^2) + 0(r^4), \tag{2.15}$$

where $\eta(t)$ is the solution of the initial-value problem

$$\{t^2\eta'(t)(1+t^2)^{-1}\}' + 12t^2\eta(1+t^2)^{-3} = t^2(1+t^2)^{-1}, \tag{2.16}$$

$$\eta(0) = \eta'(0) = 0. \tag{2.17}$$

An explicit expression for η is given in Section 7, along with various estimates.

We pass to our main concern, which is that of approximation to u in the zone $r > R$, which is where the oscillations take place. We use here a method of modified Prüfer transformation. With certain qualifications, we define an angular variable $\theta(r) = \theta(r, \lambda, \gamma)$ by

$$\sqrt{\lambda} \tan \theta(r) = v'/v - u'/u, \tag{2.18}$$

so that zeros of u are characterized by $\theta(r) \equiv \pi/2$ (mod. π). In these terms we give the basic result of this paper as follows.

Theorem 1 *For $R \leq r \leq C$, we have, as $\gamma \to \infty$,*

$$u(r, \lambda, \gamma) = \rho(r, \lambda, \gamma) \cos \theta(r, \lambda, \gamma), \tag{2.19}$$

where $\rho > 0$ satisfies

$$\rho^2(r, \lambda, \gamma) = \gamma^2 \{1 + \gamma^4 r^2)^{-1} \{1 - 16\lambda\gamma^{-4} \log(r\gamma^2) + 0(\gamma^{-4})\}, \tag{2.20}$$

$$
\begin{aligned}
\theta(r, \lambda, \gamma) &= \lambda^{1/2}(r - \pi\gamma^{-2} - 7\pi r^{-2}\gamma^{-6}) \\
&\quad + \lambda\gamma^{-4} F_1(r\sqrt{\lambda}) + \lambda\gamma^{-4} F_2(r\sqrt{\lambda}) + 0(\gamma^{-5}),
\end{aligned} \tag{2.21}
$$

and

$$F_1(s) = \frac{\cos^2 s - \cos^6 s}{s^3} + \frac{3\cos^5 s \sin s}{s^2} + \frac{3\cos^6 s - 15\cos^4 s \sin^2 s}{s} \tag{2.22}$$

$$F_2(s) = (15/8)Si(2s\sqrt{\lambda}) + 6Si(4s\sqrt{\lambda}) + (27/8)Si(6s\sqrt{\lambda}). \tag{2.23}$$

Here $Si(s)$ denotes the sine-integral function

$$Si(s) = \int_0^s t^{-1} \sin t\, dt. \tag{2.24}$$

For details of this function, and numerical tables, we refer to [1].

3. Main Results (2): Asymptotics of Eigenvalues

We now consider radial solutions of (1.1) such that $u(r) = u(r, \gamma, \lambda)$ has precisely $k - 1$ zeros in $(0, 1)$, where $k = 1, 2, ...,$ and vanishes when $r = 1$; such a solution is necessarily radial when $k = 1$ [13], but not necessarily otherwise [9, 12]. We denote by $\lambda_k(\gamma)$ a value such that $u((r, \gamma, \lambda_k)$ has these properties. Thus λ_k is to be a root of the equation

$$\theta(1, \lambda_k, \gamma) = (k - 1/2)\pi. \tag{3.1}$$

We confine attention to λ satisfying (2.1) where $B \geq (k + 1)^2\pi^2$. Our first result relates to the asymptotics of $\lambda_k(\gamma)$.

Theorem 2 *For large γ, $\lambda_k(\gamma)$ exists and satisfies*

$$\sqrt{\lambda_k}\{1 - \pi\gamma^{-2}\} = (k - 1/2)\pi - D_k\gamma^{-4} + 0(\gamma^{-5}), \tag{3.2}$$

where

$$
\begin{aligned}
D_k &= (k - 1/2)^2\pi^2\{(15/8)Si\{(2k - 1)\pi\} \\
&\quad + 6Si\{(4k - 2)\pi\} + (27/8)Si\{(6k - 3)\pi\}\}.
\end{aligned} \tag{3.3}
$$

We remark that $Si(x) > 0$ if $x > 0$ [1] so that $D_k > 0$.

In an alternative formulation, one writes

$$\lambda_k = (k - 1/2)^2 \pi^2 + \epsilon, \tag{3.4}$$

and then finds, by manipulation of (3.2),

$$\epsilon \gamma^2 = (2k - 1)^2 \pi^3 / 2 + \gamma^{-2} \{ 3(k - 1/2)\pi^4 + (2k - 1)D_k \} + 0(\gamma^{-3}). \tag{3.5}$$

The case $k = 1$ was among the problems studied in [7], where (1.3) was proved.

The above are consequences of Theorem 1; we give details of the deduction in Section 12.

In Sections 13-14, we estimate the partial derivatives $\partial\theta/\partial\lambda$, $\partial\theta/\partial\gamma$ with the results (13.29),(14.1). In particular, from the positivity of $\partial\theta/\partial\lambda$ for large γ, it follows that $\lambda_k(\gamma)$ is unique for large γ. Furthermore, we have the following theorem.

Theorem 3 *For large* γ,

$$\partial\lambda_k(\gamma)/\partial\gamma = -(2k - 1)^2 \pi^3 \gamma^{-3} + 0(\gamma^{-4}). \tag{3.6}$$

This follows from (14.2), together with (3.4),(3.5). Thus, in particular, for large increasing γ, $\lambda_k(\gamma)$ decreases monotonely to $(k - 1/2)^2 \pi^2$.

4. A Basic Upper Bound over $(0, r_1)$

For completeness, we sketch the proof of the basic inequality (2.8). To begin with, we note an alternative formulation. For much of our work it is convenient to study the differential equation (1.4) with the change of dependent variable

$$U(r) = ru(r), \tag{4.1}$$

so that U is the solution of

$$U'' + \lambda U + 3r^{-4}U^5 = 0, \quad U(0) = 0, \quad U'(0) = \gamma. \tag{4.2}$$

Thus, if we write

$$V(r) = \gamma r / \sqrt{1 + \gamma^4 r^2}, \tag{4.3}$$

an equivalent result to (2.9) will be that

$$U(r) < V(r), \quad 0 < r < r_1. \tag{4.4}$$

We remark that V is the solution of

$$V'' + 3r^{-4}V^5 = 0, \quad V(0) = 0, \quad V'(0) = \gamma, \tag{4.5}$$

that is to say, of (4.2) with λ replaced by 0.

We have then the following lemma.

Lemma 1 *The bound (4.4) holds.*

Proof. We indicate the main steps briefly and work in terms of the first formulation (2.8). We actually deduce (2.8) from a sharper bound, namely,

$$u(r) < \gamma\{1 + (\gamma^4 + \lambda/3)r^2)^{-1/2} \tag{4.6}$$

obtained in [3]. For this we note the identity

$$(d/dr)\{r^3 u'^2 + r^2 uu' + (\lambda/3)u^2 r^3 + u^6 r^3\} = -(4\lambda/3)r^3 uu',$$

from which we deduce that the expression in the braces on the left is positive for $0 < r < r_1$. Using (1.4), we deduce that

$$u'^2 + uu'/(3r) - uu''/3 > 0,$$

so that $|u'|/\{u^3 r\}$ is increasing. Writing (1.4) in the form

$$(r^2 u')' = -r^2(\lambda u + 3u^5),$$

we deduce that

$$r^2 u' \sim -(r^3/3)(\lambda\gamma + 3\gamma^5),$$

as $r \to 0$, whence

$$u'/u^3 < -r\{\gamma^2 + \lambda/(3\gamma^2)\}.$$

Integrating over $(0, r)$, we get

$$1/2\{\gamma^{-2} - u^{-2}) < -r^2\{\gamma^2 + \lambda/(3\gamma^2)\}/6,$$

which yields (4.6), and so (2.8) or (4.4). This completes the proof. ∎

5. Preliminary Estimates over $(0, r_1)$

We proceed to estimate the discrepancy between U and V, studying for this purpose the differential equation satisfied by their ratio. Equivalently to (2.10), we write

$$Z(r) = U(r)/V(r), \quad r > 0, \tag{5.1}$$

so that, by Lemma 1,

$$0 < Z(r) < 1, \quad 0 < Y(r) < 1, \quad 0 < r < r_1, \tag{5.2}$$

where $Y = 1 - Z$. We prove first the following lemma.

Lemma 2 *We have*

$$0 < Y(r) < \lambda r^2/2, \quad 0 < r \leq r_1, \tag{5.3}$$

and

$$2/\lambda < r_1^2 \tag{5.4}$$

$$r_1^2 < \pi^2/\lambda. \tag{5.5}$$

Proof. We deduce from (4.2), (4.5) that

$$(U'V - UV')' = -\lambda UV + 3r^{-4}(V^5U - U^5V),$$

which is equivalent to

$$(V^2Z')' = -\lambda V^2 Z + 3r^{-4}V^6(Z - Z^5). \tag{5.6}$$

Hence

$$V^2(r)Z'(r) = -\lambda \int_0^r V^2 Z ds + 3 \int_0^r s^{-4}V^6(Z - Z^5)ds, \tag{5.7}$$

and

$$1 - Z(r) = \lambda \int_0^r V^{-2}(s)ds \int_0^s V^2 Z dt - 3 \int_0^r V^{-2}(s)ds \int_0^s t^{-4}V^6(Z - Z^5)dt. \tag{5.8}$$

Here, for $0 < r \leq r_1$, the left is positive, as are also both the iterated integrals on the right. We deduce that

$$1 - Z(r) < \lambda \int_0^r V^{-2}(s)ds \int_0^s V^2 Z dt < \lambda \int_0^r V^{-2}(s)ds \int_0^s V^2 dt < \lambda r^2/2, \tag{5.9}$$

since $V(r)$ is an increasing function.

This proves (5.3). As to (5.4), it is easily seen that the contrary hypothesis to (5.4) would yield a contradiction; the complementary bound (5.5) is a consequence of the Sturm Comparison Theorem applied to (4.2) and is essentially the same as the upper bound in (1.2). A similar result to (5.3) is used in [7] (Section 7); further results of a more general nature are to be found in [4]. ∎

It follows from (5.7), and the fact that $V(r)$ is increasing, that

$$Y'(r) < \lambda r, \quad 0 < r \le r_1. \tag{5.10}$$

It is not obvious that $Y' \ge 0$ in this interval; however, we must have

$$0 \le Y'(r_1) \le \lambda r_1 < \pi\sqrt{\lambda}. \tag{5.11}$$

One may use Lemma 2 combined with iteration in the above integral equations to get further estimates for Y, and for that matter also for Y'. In this way, we can get the results (2.13),(2.14) but not, apparently, a unified result covering the transition zone $r \approx \gamma^{-2}$. We therefore pass to the method of variation of parameters.

6. Improved Approximation to $Z(r)$

With the notation $Y = 1 - Z$ we deduce from (5.6) that $Y(r)$ satisfies the differential equation

$$(V^2 Y')' + \lambda V^2 Y = \lambda V^2 - 12 V^6 r^{-4} Y (1-Y)(1-3Y/2+Y^2-Y^3/4). \tag{6.1}$$

It is convenient to rearrange this as

$$(V^2 Y')' + 12 V^6 r^{-4} Y = \lambda V^2 - \lambda V^2 Y + V^6 r^{-4} g(Y), \tag{6.2}$$

where $g(Y) = 3\{10Y^2 - 10Y^3 + 7Y^4 - Y^5\}$. Clearly, $0 \le g(Y) \le 30Y^2$, whenever $0 \le Y \le 1$, and so

$$g(Y(r)) = 0(Y^2(r)) = 0(r^4), \quad 0 \le r \le r_1, \tag{6.3}$$

by Lemma 2. In addition, Y must satisfy the initial data

$$Y(0) = 0, \quad Y'(0) = 0. \tag{6.4}$$

We solve (6.2) approximately by the method of variation of parameters, treating the right as a known function. We need for this purpose linearly independent solutions of the associated homogeneous equation, that is to say, of the equation

$$(V^2 \phi')' + 12 V^6 r^{-4} \phi = 0. \tag{6.5}$$

This is most conveniently treated with the aid of a scaling of the independent variable. We write

$$t = \gamma^2 r. \tag{6.6}$$

We are thus led to the equation

$$(d/dt)\left\{\frac{t^2\psi'}{1+t^2}\right\} + \frac{12t^2\psi}{(1+t^2)^3} = 0. \tag{6.7}$$

By a change of dependent and independent variables, one may solve this in terms of associated Legendre functions; related calculations are to be found in [8, p. 254]. It will perhaps suffice to present the resulting solutions of (6.7), namely,

$$\psi_1(t) = (1 - t^2)/(1 + t^2), \tag{6.8}$$

$$\psi_2(t) = (1 - 6t^2 + t^4)/(t + t^3). \tag{6.9}$$

We write

$$\phi_j(r) = \psi_j(\gamma^2 r), \quad j = 1, 2, \tag{6.10}$$

for the corresponding solutions of (6.5). It may be verified that

$$V^2(\phi_1'\phi_2 - \phi_1\phi_2') = 1. \tag{6.11}$$

We apply this to the solution of (6.2), (6.4), using the method of variation of parameters. As a first step we get the result of Proposition 3, which we rephrase as follows.

Lemma 3 *For $0 \le r \le r_1$, we have*

$$Y(r) = Y_0(r) + 0\{r^4\}, \tag{6.12}$$

where

$$Y_0(r) = \lambda\gamma^{-4}\eta(r\gamma^2), \tag{6.13}$$

and

$$\eta(x) = \frac{1 - x^2}{1 + x^2}\int_0^x \frac{1 - 6y^2 + y^4}{(1 + y^2)^2}y\,dy - \frac{1 - 6x^2 + x^4}{x + x^3}\int_0^x \frac{y^2(1 - y^2)}{(1 + y^2)^2}dy. \tag{6.14}$$

Proof. Applied to (6.2), the method of variation of parameters yields

$$Y(r) = \lambda\int_0^r J(r,s)ds - \int_0^r J(r,s)f(s)ds, \tag{6.15}$$

where

$$J(r,s) = \{\phi_1(r)\phi_2(s) - \phi_2(r)\phi_1(s)\}\,V^2(s), \tag{6.16}$$

and

$$f(s) = \lambda Y'(s) - \{V(s)/s\}^4 g(Y(s)). \tag{6.17}$$

Here the first term on the right of (6.14) gives that in (6.13). Passing to the last term in (6.14), we note that

$$\phi_1(r)\phi_2(s)V^2(s) = s\frac{(1 - \gamma^4 r^2)(1 - 6\gamma^4 s^2 + \gamma^8 s^4)}{(1 + \gamma^4 r^2)(1 + \gamma^4 s^2)^2} \tag{6.18}$$

so that $|\phi_1(r)\phi_2(s)V^2(s)| \leq s$; similarly,

$$\phi_2(r)\phi_1(s)V^2(s) = \frac{s^2(1 - \gamma^4 s^2)(1 - 6\gamma^4 r^2 + \gamma^8 r^4)}{r(1 + \gamma^4 s^2)^2(1 + \gamma^4 r^2)} \tag{6.19}$$

$$= r\left(\frac{s^2 + \gamma^4 r^2 s^2}{r^2 + \gamma^4 r^2 s^2}\right)\left(\frac{1 - \gamma^4 s^2}{1 + \gamma^4 r^2}\right)\left(\frac{1 - 6\gamma^4 r^2 + \gamma^9 r^4}{(1 + \gamma^4 r^2)}\right), \tag{6.20}$$

from which it follows that $|\phi_2(r)\phi_1(s)V^2(s)| \leq r$, We deduce that

$$|J(r,s)| \leq 2r, \quad 0 \leq s \leq r. \tag{6.21}$$

Also, since $g(Y(s)) = 0(Y^2(s)) = 0(s^4)$,

$$f(s) = 0(s^2) + 0(V^4(s)) = 0(s^2). \tag{6.22}$$

Inserting (6.21,6.22) in (6.14), we get the result of Lemma 3. $\blacksquare$

We put this result in more explicit form by approximating to $Y_0(r)$ for the cases that $r\gamma^2$ is small or large; this entails approximating to $\eta(x)$ for small or large x. Evaluating the integrals in (6.14), we have

$$\eta(x) = \frac{1 - x^2}{1 + x^2}\left\{\frac{x^2}{2} - 4\log(1 + x^2) + \frac{4x^2}{1 + x^2}\right\}$$
$$+ \frac{1 - 6x^2 + x^4}{1 + x^2}\left\{1 - 2\frac{\tan^{-1} x}{x} + \frac{1}{1 + x^2}\right\}. \tag{6.23}$$

For small x we find that

$$\eta(x) = x^2/6 - x^4/15 + 0(x^6), \tag{6.24}$$

and for large x,

$$\eta(x) = x^2/2 - \pi x + 8\log x - 7 + 7\pi x - 16x^{-2}\log x + (1/3)x^{-2} + 0(x^{-4}\log x)). \tag{6.25}$$

Thus, in particular, by Lemma 3,

$$Y(r) = \lambda r^2/6 + 0\{(r\gamma)^4\} \text{ as } r\gamma^2 \to 0, \tag{6.26}$$

and, for $r\gamma^2 > 2$, $r < r_1$,

$$\begin{aligned}
Y_0(r) &= \lambda\{r^2/2 - \pi r\gamma^{-2} + 8\gamma^{-4}\log(r\gamma^2) - 7\gamma^{-4} + 7\pi r^{-1}\gamma^{-6} \\
&\quad - 16\lambda r^{-2}\gamma^{-8}\log(r\gamma^2) + (1/3)r^{-2}\gamma^{-8}\} \\
&\quad + 0\{r^{-4}\gamma^{-12}\log(r\gamma^2)\}.
\end{aligned} \tag{6.27}$$

The expression (6.27) will also represent $Y(r)$ in the stated region subject to an additional error term $0(r^4)$. However, in the application when r is of order $1/\gamma$, the terms on the right starting with $7\gamma^{-4}$ will then be irrelevant.

7.   Further Approximations to $Y(r)$

The error term $0(r^4)$ is (6.12) will now be replaced for the case that $r\gamma^2$ is large. We are concerned with the range $2\gamma^{-2} \leq r \leq r_1$. By (6.15) we have

$$\begin{aligned}
Y(r) &= \lambda \int_0^r J(r,s)ds - \int_0^r J(r,s)\left\{\lambda Y(s) - (V(s)/s)^4 g(Y(s))\right\} ds \\
&= Y_0(r) - \lambda\phi_1(r)\chi_1(r) + \lambda\phi_2(r)\chi_2(r) \\
&\quad + \phi_1(r)\chi_3(r) - \phi_2(r)\chi_4(r),
\end{aligned} \tag{7.1}$$

where

$$\chi_1(r) = \int_0^r \phi_2(s)V^2(s)Y(s)ds, \quad \chi_2(r) = \int_0^r \phi_1(s)V^2(s)Y(s)ds, \tag{7.2}$$

$$\chi_3(r) = \int_0^r \phi_2(s)V^6(s)s^{-4}g(Y(s))ds, \quad \chi_4(r) = \int_0^r \phi_1(s)V^6(s)s^{-4}g(Y(s))ds. \tag{7.3}$$

We proceed to estimate, or approximate to, these four integrals. For $\chi_3(r)$ an order bound will suffice, while for the others approximations will be developed. These approximations all follow a similar pattern. In the integrals we approximate $Y(s)$ by $Y_0(s)$, using Lemma 3, change the variable of integration from s to $\sigma = s\gamma^2$, use a rough bound for $\eta(\sigma)$ over $(0,2)$ and some form of (6.25) over $(2, r\gamma^2)$.

(i) Approximation to χ_1

The procedure just outlined gives

$$\chi_1(r) = \chi_{10}(r) + 0(r^6), \tag{7.4}$$

where

$$\chi_{10}(r) = \int_0^r \phi_2(s)V^2(s)\lambda\gamma^{-4}\eta(s\gamma^2)ds$$

$$= \lambda\gamma^{-8}\int_0^{r\gamma^2} \psi_2(\sigma)\sigma^2(1+\sigma^2)^{-1}\eta(\sigma)ds.$$

Here the integrand is $0(\sigma^3)$ for small σ, and so the interval $(0,2)$ contributes an amount $0(1)$ to the integral. For $\sigma > 2$, the integrand has the form

$$(\sigma + 0(1/\sigma))\{\sigma^2/2 - \pi\sigma + 0(\log\sigma)\} = \sigma^3/2 - \pi\sigma^2 + 0(\sigma\log\sigma).$$

and so we get, for $r > 2\gamma^{-2}$,

$$\chi_{10}(r) = \lambda\gamma^{-8}\left\{(r\gamma^2)^4/8 - \pi(r\gamma^2)^3/3 + 0((r\gamma^2)^2\log(r\gamma^2)) + 0(1)\right\},$$

$$= \lambda\{r^4/8 - \pi r^3\gamma^{-2}/3\} + 0(\{r^2\gamma^{-4}\log(r\gamma^2)\}). \tag{7.5}$$

Since then $\phi_1(r) = -1 + 0(r^{-1}\gamma^{-2})$, we deduce that

$$\phi_1(r)\chi_1(r) = -\lambda r^4/8 + \lambda\pi r^3\gamma^{-2}/3 + 0\{r^2\gamma^{-4}\log(r\gamma^2)\}. \tag{7.6}$$

(ii) Approximation to χ_2

We have in the same way

$$\chi_2(r) = \chi_{20}(r) + 0(\gamma^{-2}r^5), \tag{7.7}$$

where

$$\chi_{20}(r) = \lambda\gamma^{-8}\int_0^{r\gamma^2} \psi_1(\sigma)\sigma^2(1+\sigma^2)^{-1}\eta(\sigma)d\sigma.$$

Again, the integrand is continuous in $[0,2]$ and for $\sigma > 2$ has the form

$$\{-1 + 3\sigma^{-2} + 0(\sigma^{-4})\}\{\sigma^2/2 - \pi\sigma + 8\log\sigma - 7 + 0(\sigma^{-1})\}$$

$$= -\sigma^2/2 + \pi\sigma - 8\log\sigma + 17/2 + 0(\sigma^{-1}).$$

We thus get

$$\chi_{20}(r) = -\lambda r^3\gamma^{-2}/6 + \lambda\pi r^2\gamma^{-4}/2 - \lambda r\gamma^{-6}\{8\log(r\gamma^2) + 33/2\}$$
$$+ 0(\gamma^{-8}\log(r\gamma^2)). \tag{7.8}$$

Since, for $r > 2\gamma^{-2}$, $\phi_2(r) = r\gamma^2 - 7/(r\gamma^2) + 0\{(r\gamma^2)^{-3}\}$, we get that

$$\phi_2(r)\chi_2(r) = -\lambda r^4/6 + \lambda\pi r^3\gamma^{-2}/2$$
$$+ r^2\gamma^{-4}\{-8\log(r\gamma^2)) + 46/3\}$$
$$+ 0(r\gamma^{-6}\log(r\gamma^2)). \tag{7.9}$$

52   F.V. ATKINSON

(iii) An order-bound for χ_3

We use the fact that $|g(Y(s))| \leq (30/4)\lambda^2 s^4$, which leads to

$$\chi_3(s) = 0(\int_0^r \phi_2(s)V^6(s)ds) = 0(\gamma^{-4}r^2). \tag{7.10}$$

Hence

$$\phi_1(r)\chi_3(r) = 0(\gamma^{-4}r^2). \tag{7.11}$$

(iv) Approximation to χ_4

By Section 6 we have $g(Y(s)) = 30Y^2(s) + 0(s^6) = 30Y_0^2(s) + 0(s^6)$, and so $\chi_4(r) = \chi_{40}(r) + 0(\gamma^{-6}r^3)$, where

$$\begin{aligned}
\chi_{40}(r) &= 30\int_0^r \phi_1(s)V^6(s)s^{-4}Y_0^2(s)ds \\
&= 30\lambda^2\gamma^{-8}\int_0^{r\gamma^2} \psi_1(\sigma)\sigma^2(1+\sigma^2)^{-3}\eta^2(\sigma)d\sigma.
\end{aligned}$$

Here the integrand is continuous over $[0,2]$ and for $\sigma > 2$ has the form

$$\{-\sigma^{-4} + 0(\sigma^{-6})\}\{\sigma^2/2 + 0(\sigma)\}^2 = -1/4 + 0(1/\sigma).$$

Hence

$$\chi_{40}(r) = 30\lambda^2\{-r\gamma^{-6}/4 + 0(\gamma^{-8}\log(r\gamma^2))\}, \tag{7.12}$$

and so

$$\phi_2(r)\chi_4(r) = -(15/2)\lambda^2r^2\gamma^{-4} + 0(r\gamma^{-6}\log(r\gamma^2). \tag{7.13}$$

Assembling the results (7.1),(7.50,(7.7),(7.9), and (7.11), we get

$$Y(r) = Y_0(r) - \lambda^2 r^4/24 + \lambda^2\pi r^3/(6\gamma^2) + 0(r^2\gamma^{-4}\log(r\gamma^2)). \tag{7.14}$$

8.   Approximation to Y'

According to the variation of parameters formalism, we may differentiate (7.1) to obtain

$$Y'(r) = Y_0'(r) - \lambda\phi_1'(r)\chi_1(r) + \lambda\phi_2'(r)\chi_2(r) + \phi_1'(r)\chi_3(r) - \phi_2'(r)\chi_4(r). \tag{8.1}$$

We need to approximate to these terms for large $r\lambda^2$. We carry this to the point at which the error is $0(\gamma^{-5})$ when r is of order γ^{-1}. We consider a range

$$\gamma^{-1} \leq r \leq r_1. \tag{8.2}$$

We have first

$$Y_0'(r) = \lambda \gamma^{-2} \eta'(r\gamma^2),$$

and here, for large x,

$$\eta'(x) = x - \pi + 8x^{-1} - 7\pi x^{-2} + 32x^{-3}\log x - (50/3)x^{-3} + 0(x^{-5}\log x), \quad (8.3)$$

while, for small x,

$$\eta'(x) = x/3 - 4x^2/15 + 0(x^4). \quad (8.4)$$

Hence, for $2\gamma^{-2} \le r \le r_1$,

$$\begin{aligned} Y_0'(r) &= \lambda\{r - \pi\gamma^{-2} + 8r^{-1}\gamma^{-4} - 7\pi r^{-2}\gamma^{-6}\} \\ &\quad + 32\lambda r^{-3}\gamma^{-8}\log(r\gamma^2) + 0(r^{-3}\gamma^{-8}). \end{aligned} \quad (8.5)$$

The second, fourth, and fifth terms on the right of (8.1) are treated as error terms, within the chosen degree of approximation to Y'. For the second and fourth terms we note that

$$\phi_1'(r) = 0(r^{-3}\gamma^{-4}),$$

and so, by Section 7,

$$\phi_1'(r)\chi_1(r) = 0(r\gamma^{-4}), \quad \phi_1'(r)\chi_3(r) = 0(r^{-1}\gamma^{-8}),$$

and both of these are $0(r^5)$ subject to (8.2). For the fifth term we note that

$$\phi_2'(r) = \gamma^2 + 7\gamma^{-2}r^{-2} + 0(r^{-4}\gamma^{-6}) = 0(\gamma^2),$$

and so

$$\phi_2'(r)\chi_4(r) = 0\{\gamma^2(r\gamma^{-6} + r^3\gamma^{-6})\} = 0(r^5).$$

Finally, we must approximate to the third term on the right of (8.1). Using (7.7, 7.8), we get

$$\begin{aligned} \phi_2'(r)\chi_2(r) &= \{\gamma^2 + 0(r^{-2}\gamma^{-2})\}\lambda\{-r^3\gamma^{-2}/6 \\ &\quad + \pi r^2\gamma^{-4}/2 + 8r\gamma^{-6}\log(r\gamma^2) + 0(r\gamma^{-6})\} \\ &= -\lambda r^3/6 + \lambda\pi r^2\gamma^{-2}/2 - 8\lambda r\gamma^{-4}\log(r\gamma^2) + 0(r^5). \end{aligned}$$

Assembling these results, we get that

$$Y'(r) = Y_0'(r) - \lambda^2 r^3/6 + \lambda^2\pi r^2\gamma^{-2}/2 - 8\lambda^2 r\gamma^{-4}\log(r\gamma^2) + 0(r^5). \quad (8.6)$$

9. The Prüfer Transformation

We now study the asymptotics of Z, Z' for $r \geq 1/\gamma$ as $\gamma \to \infty$ with the aid of new dependent variables $\sigma(r), \theta(r)$ of polar coordinate type. We set

$$Z(r) = \sigma(r)\cos\theta(r), \quad Z'(r) = -\sqrt{\lambda}\sigma(r)\sin\theta(r), \tag{9.1}$$

and impose the usual conditions to fix $\sigma(r)$, $\theta(r)$ uniquely. Since Z, Z' cannot vanish together, by uniqueness properties of the differential equation (5.6), we may require that $\sigma(r) > 0$, and in view of (8.2) will have

$$\sigma(r) = \sqrt{F}(r), \tag{9.2}$$

where

$$F(r) = Z'^2(r)/\lambda + Z^2(r). \tag{9.3}$$

We require that $\theta(r)$ be continuous, and specify that $\theta(0) = 0$, in view of the results $Z(0) = 1, Z'(0) = 0$. Since $Z(r) > 0$ in $(0, r_1)$, we have $\theta(r) \in (-\pi/2, \pi/2)$ in this interval.

In this section we approximate to $\sigma(r), \theta(r)$ for $1/\gamma \leq r \leq r_1$ by means of the asymptotic expressions for Y, Y' obtained in Sections 6-8. These expressions can then be improved with the aid of differential equations satisfied by $\sigma(r)$ and $\theta(r)$.

In this preliminary approximation we estimate $F(r)$ with an error $0(r^4)$. We have from (8.5),(8.6) that, if

$$1/\gamma \leq r < r_1, \tag{9.4}$$

$$Z'(r) = -Y'(r) = -\lambda(r - \pi\gamma^{-2}) + 0(r^3), \tag{9.5}$$

and from (6.12),(6.27) that

$$Z(r) = 1 - \lambda(r^2/2 - \pi r\gamma^{-2} + 8\gamma^{-4}\log(r\gamma^2)) + 0(r^4). \tag{9.6}$$

Hence

$$F(r) = 1 - 16\lambda\gamma^{-4}\log(r\gamma^2) + 0(r^4). \tag{9.7}$$

In the case of $\theta(r)$ we require an approximation to order $0(r^5)$. We have, for $r < r_1$,

$$\tan\theta = -Z'/(Z\sqrt{\lambda}) = Y'/\{(1 - Y)\sqrt{\lambda}\}. \tag{9.8}$$

Here we must approximate to Y' with error $0(r^5)$, and to Z with error $0(r^4)$, as in (9.6). Also, since $Y' = 0(r)$, $Y = 0(r^2)$, we may replace (9.8) by

$$\tan\theta = Y'(1 + Y)/\sqrt{\lambda} + 0(r^5). \tag{9.9}$$

The requisite improved version of (9.5) is

$$Y'(r) = \lambda\{r - \pi\gamma^{-2} + 8r^{-1}\gamma^{-4} - 7\pi r^{-2}\gamma^{-6} + 32r^{-3}\gamma^{-8}\log(r\gamma^2)\}$$
$$- \lambda^2 r^3/6 + \lambda^2\pi r^2\gamma^{-2}/2 - 8\lambda^2 r\gamma^{-4}\log(r\gamma^2) + 0(r^5). \quad (9.10)$$

Substituting (9.10) and (9.6) in (9.9), we find that

$$\tan\theta(r) = (\lambda^{1/2}\{r - \pi\gamma^{-2} + 8r^{-1}\gamma^{-4} - 7\pi r^{-2}\gamma^{-6} + 32r^{-3}\gamma^{-8}\log(r\gamma^2)\}$$
$$- \lambda^{3/2}\{r^3/6 + \pi r^2\gamma^{-2}/2 - 8r\gamma^{-4}\log(r\gamma^2)\}) \cdot$$
$$\cdot (1 + \lambda(r^2/2 - \pi r\gamma^{-2} + 8\gamma^{-4}\log(r\gamma^2)) + 0(r^5)$$
$$= \lambda^{1/2}\{r - \pi\gamma^{-2} + 8r^{-1}\gamma^{-4} - 7\pi r^{-2}\gamma^{-6} + 32r^{-3}\gamma^{-8}\log(r\gamma^2)\}$$
$$+ \lambda^{3/2}(r^3/3 - \pi r^2\gamma^{-2}) + 0(r^5).$$

Applying the MacLaurin series for the inverse tangent, we get

$$\theta(r) = \lambda^{1/2}\{r - \pi\gamma^{-2} + 8r^{-1}\gamma^{-4} - 7\pi r^{-2}\gamma^{-6} + 32r^{-3}\gamma^{-8}\log(r\gamma^2)\} + 0(r^5),$$
$$(9.11)$$

again subject to (9.4).

In the next section we improve and extend these results.

10. Extension to the range $[\gamma^{-1}, C]$

We take first the case of $\sigma(r)$, or $F(r)$, and prove the following lemma.

Lemma 4 *For large γ and $\gamma^{-1} \leq r \leq C$, we have*

$$F(r) = 1 - 16\lambda\gamma^{-4}\log(r\gamma^2) + 0(\gamma^{-4}). \quad (10.1)$$

Proof.

For $r = 1/\gamma$ this was proved in (9.7). We extend it first to the range $[\gamma^{-1}, r_1)$ by means of the differential equation satisfied by F. Writing (5.6) in the form

$$Z'' + \lambda Z = -(2V'/V)Z' + 3(V/r)^4(Z - Z^5),$$

we may easily verify that

$$\lambda F' = -4(V'/V)Z'^2 + 6(V/r)^4(Z - Z^5)Z'. \quad (10.2)$$

We then integrate (10.2) over $[R, r)$, where $R = \gamma^{-1}$ and r lies in the range $[R, r_1]$. We get

$$\lambda\{F(r) - F(R)\} = -4I_1 - 6I_2, \quad (10.3)$$

where

$$I_1 \;=\; \int_R^r (V'/V)Z'^2 ds, \tag{10.4}$$

$$I_2 \;=\; \int_R^r (V/s)^4 (Z - Z^5) Y' ds. \tag{10.5}$$

We estimate these expressions using weakened forms of (9.5),(9.6), namely,

$$Y'(r) = \lambda r + 0(\gamma^{-2}) + 0(r^3), \tag{10.6}$$

$$Y(r) = \lambda r^2/2 + 0(r\gamma^{-2} + 0(r^4), \tag{10.7}$$

both valid for $R \le r \le r_1$.

We have from (10.6) that $Z'^2(r) = \lambda^2 r^2 + 0(r\gamma^{-2}) + 0(r^4)$. Since $V'(s)/V(s) = 1/(s + \gamma^4 s^3) = \gamma^{-4}s^{-3} + 0(\gamma^{-8}s^{-5})$, we get, for $R \le s \le r_1$,

$$(V'/V)Z'^2 = \lambda^2\gamma^{-4}s^{-1} + 0(s^{-2}\gamma^{-6}) + 0(s\gamma^{-4}),$$

and so

$$I_1 = \lambda^2\gamma^{-4}\{\log r - \log R\} + 0(\gamma^{-4}). \tag{10.8}$$

Turning to (10.5), we have that

$$(V/s)^4 = \gamma^4\{1 + \gamma^4 s^2\}^{-2} = \gamma^{-4}s^{-4} + 0(\gamma^{-8}s^{-6})$$

$$(Z - Z^5) = 4Y + 0(Y^2) = 2\lambda s^2 + 0(s\gamma^{-2}) + 0(s^4),$$

and so, using (10.6), we have

$$(V/s)^4(Z - Z^5)Y' = 2\lambda^2\gamma^{-4}s^{-1} + 0(s^{-2}\gamma^{-6}) + 0(s\gamma^{-4}).$$

We deduce on integration that

$$I_2 = 2\lambda^2\gamma^{-4}\{\log r - \log R\} + 0(R^{-1}\gamma^{-6}) + 0(\gamma^{-4}),$$

since r_1 is bounded.

Combining this with (10.3) and (10.8), we have

$$F(r) - F(R) = -16\lambda\gamma^{-4}\{\log r - \log R\} + 0(\gamma^{-4}).$$

It remains to extend this to the range $r_1 \le r \le C$. Let I_1', I_2' have a similar meaning to (10.4, 10.5), with the integrals now taken over (r_1, r), for some $r \in (r_1, C)$. Applying the argument of (10.3), we need to show that these integrals are of order $0(\gamma^{-4})$, and the integrands will be of this order if Z, Z' are bounded.

The proof of Lemma 4 will thus be completed if we show that

$$Z(r), Z'(r) = 0(1), \quad r_1 \leq r \leq C. \tag{10.9}$$

For this purpose we use an energy-type argument. We define

$$E(r) = U'^2 + \lambda U^2 + r^{-4}U^6, \tag{10.10}$$

the natural energy function for (4.2), from which it follows that

$$E'(r) = -4r^{-5}U^6 \leq 0.$$

We deduce that $E(r) \leq E(r_1)$ for $r \geq r_1$. Here $E(r_1) = U'^2(r_1)$, and since $U = VZ$ and $Z(r_1) = 0$, we have $U'(R_1) = V(r_1)Z'(r_1)$. By (5.11) we thus have

$$U'^2(r) + \lambda U^2(r) \leq \{\lambda r_1 V(r_1)\}^2, \quad r \geq r_1. \tag{10.11}$$

Hence $\lambda V^2(r)Z^2(r) \leq \{\lambda r_1 V(r_1)\}^2$, and since $V(r)$ is increasing, we have $Z^2(r) \leq \lambda r_1^2$, and so, by (5.5),

$$|Z(r)| \leq \pi, \quad r \geq r_1, \tag{10.12}$$

in partial verification of (10.9). In a similar way we have from (10.11) that

$$|U'(r)/V(r)| \leq \pi\sqrt{\lambda}, \quad r \geq r_1,$$

and so

$$|Z'(r)| = |U'(r)/V(r) - Z(r)V'(r)/V(r)| \leq \pi\{\sqrt{\lambda} + V'(r)/V(r)\}.$$

Here

$$V'(r)/V(r) = 1/\{r + \gamma^4 r^3\} \tag{10.13}$$

so that $V'(r)/V(r) < 1/r < \sqrt{\lambda}/2$, for $r \geq r_1$. Hence

$$|Z'(r)| \leq 3\pi\sqrt{\lambda}/2, \quad r \geq r_1.$$

This proves (10.9) and so completes the proof of Lemma 4. ∎

11.   Proof of Theorem 1

We now estimate $\theta(r)$ in the range $r > R$, this argument being the counterpart of that of Section 10, to which it is closely parallel; bounds developed in Section 10 will also play a part in this section.

We need first the differential equation for $\theta(r)$. Using (5.6) and (9.8), we find this to be

$$\theta'(r) = \sqrt{\lambda} - (V'/V)\sin 2\theta(r) - 3r^{-4}V^4\lambda^{-1/2}(1 - Z^4)\cos^2\theta(r). \tag{11.1}$$

We note that this has the property that $\theta'(r) = \sqrt{\lambda} > 0$ when θ is an odd multiple of $\pi/2$, that is to say, when $Z(r)$ or $U(r)$ or $u(r)$ vanishes, so that θ can serve the role of counting the zeros of u.

Integrating (11.1) over (R, r) and using (9.11), we get

$$\theta(r) = \sqrt{\lambda}\{r - \pi/\gamma^2 + 8/(R\gamma^4) - 7\pi/(R^2\gamma^6)\} + 32R^{-3}\gamma^{-8}\log(R\gamma^2)$$

$$- K_1 - K_2 + 0(\gamma^{-5}), \tag{11.2}$$

where

$$K_1 \;\; = \;\; \int_R^r (V'/V)\sin 2\theta(s)ds, \tag{11.3}$$

$$K_2 \;\; = \;\; (3/\sqrt{\lambda}) \int_R^r s^{-4}V^4(1 - Z^4)\cos^2\theta(s)ds. \tag{11.4}$$

We first approximate to these integrals using cruder estimates for $\theta(s)$. In the case of K_1, we note that $|\sin 2\theta| \leq 1$, and so find that

$$|K_1| < 1/(\gamma^4 R^2) = 0(\gamma^{-2}).$$

In the case of K_2 we use the facts that, by Lemma 2, $1 - Z^4(s) = 0(s^2)$ for $0 \leq s \leq r_1$, and that $Z(s) = 0(1)$ for $s > r_1$, by (10.12). Hence we find that $K_2 = 0(\gamma^{-3})$. Using these in (11.2), we have the rough bound

$$\theta(r) = \sqrt{\lambda}r + 0(\gamma^{-2}), \quad R \leq r \leq C.$$

However, if we substitute this in (11.3), we find that $K_1 = 0(\gamma^{-3})$, and so we have

$$\theta(r) = \sqrt{\lambda}(r - \pi/\gamma^2) + 0(\gamma^{-3}), \quad R \leq r \leq C. \tag{11.5}$$

This will yield sufficiently accurate estimates for K_1, K_2.

Starting with the case of K_1, we first use the facts that

$$V^1(s)/V(s) = \gamma^{-4}s^{-3} + 0(\gamma^{-8}s^{-5}), \quad \sin 2\theta(s) = 0(s),$$

which show that

$$K_1 = \gamma^{-4} \int_R^r s^{-3} \sin 2\theta(s)ds + 0(\gamma^{-5}).$$

Here we substitute the result

$$\sin 2\theta(s) = \sin(2s\sqrt{\lambda}) - 2\sqrt{\lambda}\pi\gamma^{-2} \cos(2s\sqrt{\lambda}) + 0(\gamma^{-3})$$

to get

$$K_1 = \gamma^{-4} \int_R^r s^{-3}\{\sin(2s\sqrt{\lambda}) - \sqrt{\lambda}\pi\gamma^{-2} \cos(2s\sqrt{\lambda})\}ds + 0(\gamma^{-5}).$$

This we rewrite in the form

$$\begin{aligned}
K_1 &= \gamma^{-4} \int_R^r s^{-3}\{\sin(2s\sqrt{\lambda}) - 2s\sqrt{\lambda}\}ds \\
&\quad + 2\sqrt{\lambda}\pi\gamma^{-6} \int_R^r s^{-3}\{1 - \cos(2s\sqrt{\lambda})\}ds \\
&\quad + 2\sqrt{\lambda}\gamma^{-4}\{1/R - 1/r) - \pi\sqrt{\lambda}\gamma^{-6}\{1/R^2 - 1/r^2\} \\
&\quad + 0(\gamma^{-5}).
\end{aligned} \tag{11.6}$$

We note that the lower limit R in the first of these integrals can be replaced by 0 without affecting the error term, and that the second integral on the right is of lower order, namely, $0(\gamma^{-6} \log \gamma)$. Hence

$$\begin{aligned}
K_1 &= \gamma^{-4} \int_0^r s^{-3}\{\sin(2\sqrt{\lambda}s) - 2\sqrt{\lambda}s\}ds \\
&\quad + 2\sqrt{\lambda}\gamma^{-4}\{1/R - 1/r) - \pi\sqrt{\lambda}\gamma^{-6}\{1/R^2 - 1/r^2\} \\
&\quad + 0(\gamma^{-5}).
\end{aligned} \tag{11.7}$$

Passing to the case of K_2, we note that

$$(V(s)/s)^4 = \gamma^{-4}s^{-4} + 0(\gamma^{-8}s^{-6}),$$

and that $1 - Z^2(s) = 0(s^2)$, $Z = \sigma \cos \theta$. Since $\sigma = -1 + o(1)$, we rearrange the result as

$$\begin{aligned}
K_2 &= 3/(\sqrt{\lambda}\gamma^4) \int_R^r s^{-4}\{\cos^2 \theta - \cos^6 \theta\}ds \\
&\quad + 3/(\gamma^4\sqrt{\lambda}) \int_R^r s^{-4}(1 - \sigma^4) \cos^6 \theta ds + 0(\gamma^{-5}) \\
&= K_{21} + K_{22} + 0(\gamma^{-5}),
\end{aligned} \tag{11.8}$$

say. Here we use the approximation (11.5). We write $f(x)$ in place of $\cos^2 x - \cos^4 x$, so that

$$f'(x) = -\sin 2x\{1 - 3\cos^4 x\} = 4x + 0(x^3), \quad f''(x) = 0(1).$$

Taylor's theorem then gives

$$f(\theta) = f(s\sqrt{\lambda}) - 4s\pi\gamma^{-2}\sqrt{\lambda} + 0(s\gamma^{-3} + s^3\gamma^{-2} + \gamma^{-4}).$$

This yields

$$K_{21} = 3/(\gamma^4\sqrt{\lambda})\int_R^r s^{-4}f(s\sqrt{\lambda})ds - 6\pi\sqrt{\lambda}\gamma^{-6}\{R^{-2}-r^{-2}\}+0(\gamma^{-5}). \quad (11.9)$$

Since $f(x) = 2x^2 + 0(x^4)$, we have

$$\int_R^r s^{-4}f(s\sqrt{\lambda})ds = \int_0^r s^{-4}\{f(s\sqrt{\lambda}) - 2\lambda s^2\}ds + 2\lambda\{R^{-1} - r^{-1}\} + 0(R)$$

and so

$$K_{21} = 3/(\gamma^4\sqrt{\lambda})\int_0^r s^{-4}\{f(s\sqrt{\lambda}) - 2\lambda s^2\}ds + 6\sqrt{\lambda}\gamma^{-4}(R^{-1} - r^{-1})$$

$$-6\pi\sqrt{\lambda}\gamma^{-6}\{R^{-2} - r^{-2}\} + 0(\gamma^{-5}).$$

In the case of K_{22} we have, by (10.1), (11.5),

$$1 - \sigma^4 = 32\lambda\gamma^{-4}\log(s\gamma^2) + 0(\gamma^{-4}), \quad \cos^6\theta = 1 + 0(\gamma^{-2}),$$

which gives

$$\begin{aligned}
K_{22} &= 96\gamma^{-8}\lambda^{1/2}\int_R^r s^{-4}\log(s\gamma^2)ds + 0(\gamma^{-5}) \\
&= 32\gamma^{-8}\lambda^{1/2}R^{-3}\log(R\gamma^2) + 0(\gamma^{-5}).
\end{aligned}$$

Hence

$$\begin{aligned}
K_2 &= 3/(\sqrt{\lambda}\gamma^4)\int_0^r s^{-4}\{\cos^2(s\sqrt{\lambda}) - \cos^6(s\sqrt{\lambda}) - 2s^2\lambda\}ds \\
&\quad + 6\sqrt{\lambda}\gamma^{-4}\{1/R - 1/r\} - 6\pi\sqrt{\lambda}\gamma^{-6}\{1/R^2 - 1/r^2\} \\
&\quad + 32\gamma^{-8}\lambda^{-1/2}R^{-3}\log(R\gamma^2) + 0(\gamma^{-5}).
\end{aligned}$$

We now combine this result with that of (11.7) and, in the integrals, make the change of variable $t = s\sqrt{\lambda}$. This gives

$$K_1+K_2 = \lambda\gamma^{-4}\int_0^{r\sqrt{\lambda}} h(t)dt + 8\sqrt{\lambda}\gamma^{-4}\{1/R-1/r\} - 7\pi\sqrt{\lambda}\gamma^{-6}\{1/R^2-1/r^2\}$$

$$+32\gamma^{-8}\lambda^{-1/2}R^{-3}\log(R\gamma^2) + 0(\gamma^{-5}),$$

where

$$h(t) = \frac{\sin(2t) - 2t}{t^3} + 3\frac{\cos^2 t - \cos^6 t - 2t^2}{t^4}. \tag{11.10}$$

Substituting in (11.2), we have, for $r \geq R$,

$$\theta(r) = \sqrt{\lambda}\{r - \pi/\gamma^2 + 8\gamma^{-4}r^{-1} - 7\pi r^{-2}\gamma^{-6}\} - \lambda\gamma^{-4}\int_0^{r\sqrt{\lambda}} h(t)dt + 0(\gamma^{-5}). \tag{11.11}$$

It now remains to identify (11.10) with (2.21). By repeated integration by parts, we find that

$$\int_0^s h(t)dt = -F_1(s) + 8/s - 3\int_0^s t^{-1}\{16\sin t\cos^5 t - 20\sin^3 t\cos^3 t\}dt,$$

where $F_1(s)$ is given in (2.22). It may then be verified that

$$16\sin t\cos^5 t - 20\sin^3 t\cos^3 t = (9/8)\sin 6t + 2\sin 4t + (5/8)\sin 2t,$$

so that

$$\int_0^s h(t)dt = -F_1(s)-(15/8)Si(2x)-6Si(4x)-(27/8)Si(6x) = -F_1(s)-F_2(s),$$

in the notation of Section 2. We then get (2.21) from (11.11). This proves Theorem 1.

12. Proof of Theorem 2

It is well known, and elementary, that $\lambda_k(\gamma)$ exists, satisfying

$$0 < \sqrt{\lambda_k} < k\pi, \quad k = 1, 2, ...; \tag{12.1}$$

this upper bound arises from the Sturm Comparison Theorem, applied to (4.2). We note in addition that

$$\sqrt{\lambda_k} > (k - 1/2)\pi, \tag{12.2}$$

at least for large γ. This follows from (11.5), which shows that

$$\theta(1, \lambda, \gamma) - (k - 1/2)\pi$$

changes sign from negative to positive as λ goes from $\{(k - 1/2)\pi\}^2$ to $(k\pi)^2$.

For the second approximation we use (2.21) in the weakened form

$$\theta(1, \lambda, \gamma) = \sqrt{\lambda}(1 - \pi\gamma^{-2}) + 0(\gamma^{-4}). \tag{12.3}$$

Here we have used the boundedness of λ, as ensured by (12.1),(12.2). Hence λ_k satisfies

$$(k - 1/2)\pi = \sqrt{\lambda_k}(1 - \pi\gamma^{-2}) + 0(\gamma^{-4}),$$

and so

$$\sqrt{\lambda_k} = (k - 1/2)\pi(1 + \pi\gamma^{-2}) + 0(\gamma^{-4}). \tag{12.4}$$

Using (2.21) again, we have

$$(k - 1/2)\pi = \sqrt{\lambda_k}(1 - \pi\gamma^{-2}) + \lambda_k\gamma^{-4}\{F_1(\sqrt{\lambda_k}) + \lambda_k F_2(\sqrt{\lambda_k})\} + 0(\gamma^{-5}). \tag{12.5}$$

However, $F_1((k - 1/2)\pi) = 0$, by (2.22), and so $F_1(\sqrt{\lambda_k}) = 0(\gamma^{-2})$, in view of (12.4). Hence this term may be omitted from (12.3). Similarly,

$$F_2(\sqrt{\lambda_k}) = F_2((k - 1/2)\pi + 0(\gamma^{-2}),$$

and here the consequent error in (12.3) is of lower order than the existing error term. We thus derive (3.2), and so prove Theorem 2.

The deduction of Theorem 3, using the results of Sections 13-14, is discussed briefly in Section 3.

13.   Approximation to $\partial\theta/\partial\gamma$

We now commence the study of $\partial\lambda/\partial\gamma$, which involves estimating first $\partial\theta/\partial\gamma$ and then $\partial\theta/\partial\lambda$; it will then be a simple matter to deduce an estimate of $\partial\theta/\partial\lambda$ which will give the result. We first relate $\partial\theta/\partial\gamma$ to the function Z. We write

$$X(r) := \partial Y/\partial\gamma = -\partial Z/\partial\gamma. \tag{13.1}$$

We have from (9.1) that

$$\theta = arg\{Z - iZ'/\sqrt{\lambda}\},$$

from which we deduce that

$$\partial\theta/\partial\gamma = F^{-1}(X'Z - XZ')\sqrt{\lambda}, \tag{13.2}$$

where F is as in (8.2), and $X' = \partial X/\partial r$. Our problem is thus to estimate

$$X'(r)Z(r) - X(r)Z'(r). \tag{13.3}$$

Our procedure will be to estimate the terms in (13.3) when $r = R$, and then to estimate the rate of change of (13.3) for $r \geq R$.

We prove first the following lemma.

Lemma 5 *For large γ we have*

$$X(r) = 2\pi\lambda r\gamma^{-3} + 0(r^3\gamma^{-3}), \quad \gamma^{-2} \le r \le C. \tag{13.4}$$

Proof. We differentiate (6.14) with respect to γ. The result may be written

$$X(r) = \lambda L_1 - L_2 - L_3, \tag{13.5}$$

where L_1, L_2, L_3 denote, respectively, the expressions

$$(\partial/\partial\gamma)\{\gamma^{-4}\eta(\gamma^2 r)\}, \int_0^r [(\partial/\partial\gamma)J(r,s)]\, f(s)ds, \int_0^r J(r,s)\,[(\partial/\partial\gamma)f(s)]\, ds.$$

Here $J(r,s)$ is given by (6.16) and $f(s)$ by (6.17).

For the leading term we have $L_1(r) = 2\gamma^{-3}r\eta'(\gamma^2 r) - 4\gamma^{-5}\eta(\gamma^2 r)$ and hence, using (6.25) and (8.3) we have, for $r \ge \gamma^{-2}$,

$$L_1(r) = 2\pi r\gamma^{-3} + 0\{\gamma^{-5}(\log(r\gamma^2) + 1)\}, \tag{13.6}$$

and, by (6.24) and (8.4), $L_1(r) = 0(r^2\gamma^{-1})$ for $0 \le r \le \gamma^{-2}$. Hence,

$$L_1(r) = 0(r\gamma^{-3}), \quad 0 \le r \le C. \tag{13.7}$$

We show next that

$$L_2 = 0(r^2\gamma^{-5}), \quad 0 \le r \le R. \tag{13.8}$$

For this we need the result that

$$(\partial/\partial\gamma)J(r,s)f(s) = 0\{r^5\gamma^3(1 + \gamma^4 r^2)^{-2}\}. \tag{13.9}$$

In view of (6.22), it will be sufficient to prove that

$$(\partial/\partial\gamma)\phi_1(r)\phi_2(s)V^2(s) = 0\{r^3\gamma^3(1 + \gamma^4 r^2)^{-2}\}, \tag{13.10}$$

$$(\partial/\partial\gamma)\phi_2(r)\phi_1(s)V^2(s) = 0\{r^3\gamma^3(1 + \gamma^4 r^2)^{-2}\}. \tag{13.11}$$

Verification of these is routine; it may be assisted by rewriting the relevant expressions in (6.18),(6.19) in the forms

$$-s\left(1 - \frac{2}{(1+\gamma^4 r^2)}\right)\left(1 - \frac{8\gamma^4 s^2}{(1+\gamma^4 s^2)^2}\right) \tag{13.12}$$

$$-r\left(1 + \frac{s^2 - r^2}{r^2 + \gamma^4 r^2 s^2}\right)\left(1 - \frac{2}{(1+\gamma^4 s^2)}\right)\left(1 - \frac{(8\gamma^4 r^2)}{(1+\gamma^4 r^2)^2}\right), \tag{13.13}$$

64 F.V. ATKINSON

respectively. Integration of (13.9) then gives (13.8).

In the case of L_3, we have, by (6.17),

$$(\partial/\partial\gamma)f(s) = X(s)f_1(s) - g(Y(s))(\partial/\partial\gamma)(V(s)/s)^4, \qquad (13.14)$$

where

$$f_1(s) = \lambda - \{V(s)/s\}^4 g'\{Y(s)\} = 0(1), \qquad (13.15)$$

and

$$g\{Y(s)\}(\partial/\partial\gamma)\{V(s)/s\}^4 = 0(\gamma^{-5}). \qquad (13.16)$$

Since $J(r,s) = 0(r)$, we deduce that

$$L_3 = \int_0^r J(r,s)f_1(s)X(s) + 0(\gamma^{-5}r^2).$$

We thus obtain from (13.2) that

$$X(r) = \lambda L_1 + 0(r^2\gamma^{-5}) + \int_0^r J(r,s)f_1(s)X(s)ds. \qquad (13.17)$$

Here

$$\int_0^r |J(r,s)f_1(s)|ds = 0(r^2), \qquad (13.18)$$

since $J(r,s) = 0(r)$ and $f_1(s) = 0(1)$. Hence, by (13.17) and the Gronwall inequality,

$$X(r) = 0\{r\gamma^{-3} + r^2\gamma^{-5}) = 0(r\gamma^{-3}). \qquad (13.19)$$

Substituting this bound on the right in (13.17), and using (13.6), we get Lemma 5. ∎

We next need a similar result for $X'(r) = \partial^2 Y/\partial r\partial\gamma$. This is Lemma 6.

Lemma 6 *For $r \geq \gamma^{-2}$*

$$X'(r) = 2\pi\lambda\gamma^{-3} + 0(\gamma^{-5}r^{-1} + r^2\gamma^{-3}).$$

Proof. It follows from (13.2) that

$$\begin{aligned}
X'(r) &= \lambda(\partial/\partial r)L_1 - \int_0^r \left[(\partial^2/\partial r\partial\gamma)J(r,s)\right]f(s)ds \\
&\quad - \int_0^r \left[(\partial/\partial r)J(r,s)\right]\left[(\partial/\partial\gamma)f(s)\right]ds \\
&= \lambda L_4 - L_5 - L_6, \qquad (13.20)
\end{aligned}$$

say. Here, we have used the facts that $J(r,s) = 0$, $(\partial/\partial\gamma)J(r,s) = 0$ when $r = s$.

We first estimate the leading term. From (13.3) we have

$$L_4 = 2\gamma^{-1}r\eta''(\gamma^2 r) - 2\gamma^{-3}\eta'(\gamma^2 r). \tag{13.21}$$

We note here that, for $t \geq 1$, $\eta''(t) = 1 + 0(t^{-2})$; this may be deduced from (6.23). Hence, using (8.3), we have

$$\begin{aligned}
L_4 &= 2\gamma^{-1}r\left\{1 + 0(r^{-2}\gamma^{-4})\right\} - 2\gamma^{-3}\left\{\gamma^2 r - \pi + 0(\gamma^{-2}r^{-1})\right\} \\
&= 2\pi\gamma^{-3} + 0(\gamma^{-5}r^{-1}). \tag{13.22}
\end{aligned}$$

We pass to the term L_5 and claim that

$$L_5 = 0\left\{\gamma^3 r^5(1 + \gamma^4 r^2)^{-2}\right\} = 0(r\gamma^{-5}). \tag{13.23}$$

For this we use the fact that

$$f(s)(\partial^2/\partial r\partial\gamma)J(r,s) = 0(\gamma^3 r^5(1 + \gamma^4 r^2)^{-2}). \tag{13.24}$$

The proof follows the same lines as that of (13.9), using (6.22); the details are omitted.

Finally, we estimate L_6. In place of (13.24), we use the result that $(\partial/\partial r)J(r,s) = 0(1)$, which is easily checked by applying the operator $(\partial/\partial r)$ to (13.12),(13.13). For the last factor in the integrand we use the results (13.14)-(13.16), which give

$$(\partial/\partial\gamma)f(s) = 0(X(s)) + 0(\gamma^{-5}) = 0(s\gamma^{-3} + \gamma^{-5})$$

so that $L_6 = 0(r^2\gamma^{-3} + r\gamma^{-5})$. This proves Lemma 6. ∎

In the next stage of the argument we use Lemmas 5 and 6, together with (7.14), (8.6), to approximate to (13.3) when $r = R$. We get

$$\begin{aligned}
X'(R)Z(R) - X(R)Z'(R) &= \left\{2\pi\lambda\gamma^{-3} + 0(\gamma^{-4})\right\}\left\{1 + 0(R^2)\right\} - 0(\gamma^{-5}) \\
&= 2\pi\lambda\gamma^{-3} + 0(\gamma^{-4}). \tag{13.25}
\end{aligned}$$

We claim that in fact

$$X'(r)Z(r) - X(r)Z'(r) = 2\pi\lambda\gamma^{-3} + 0(\gamma^{-4}), \quad R \leq r \leq C, \tag{13.26}$$

or, in view of (13.25), that

$$[X'Z - XZ']_R^r = 0(\gamma^{-4}). \tag{13.27}$$

For this we use the differential equation satisfied by Z, namely,

$$Z'' + 2(V'/V)Z' + \lambda Z = 3(V/r)^4(Z - Z^5),$$

and that satisfied by X, namely,

$$X'' + 2(V'/V)X' + \lambda X = 3(V/r)^4 X(1 - 5Z^4)$$

$$+3\left\{(\partial/\partial\gamma)(V/r)^4\right\}(Z - Z^5) + 2\left\{(\partial/\partial\gamma)(V'/V)\right\}Z'.$$

Combining this with (5.6), we have

$$(X''Z - Z''X) + 2(V'/V)(X'Z - Z'X) = -12(V/r)^4 X Z^4$$

$$-3\left\{(\partial/\partial\gamma)(V/r)^4\right\}Z^2(1 - Z^4) + 2\left\{(\partial/\partial\gamma)(V'/V)\right\}Z'Z,$$

so that

$$\begin{aligned}
\left\{V^2(X'Z - XZ')\right\} &= -12V^6 r^{-4} X Z^4 \\
&\quad - 3V^2\left\{(\partial/\partial\gamma)(V/r)^4\right\}Z^2(1 - Z^4) \\
&\quad - 2V^2\left\{(\partial/\partial\gamma)(V'/V)\right\}X'Z. \qquad (13.28)
\end{aligned}$$

Here we use the estimates

$$V = 0(\gamma^{-1}), \quad (\partial/\partial\gamma)(V/r)^4 = 0(\gamma^{-5}r^{-4}), \quad (\partial/\partial\gamma)(V'/V) = 0(\gamma^{-5}r^{-3}),$$

together with

$$Z = 0(1), \quad 1 - Z^2 = 0(r^2), \quad X = 0(r\gamma^{-3}), \quad X' = 0(\gamma^{-3}),$$

for $r \geq R$. Hence the terms on the right of (13.28) are, respectively, of orders

$$\gamma^{-9}r^{-3}, \quad \gamma^{-7}r^{-2}, \quad \gamma^{-10}r^{-3},$$

of which the second predominates, since $r \geq R$. Thus, integrating over the interval (R, r), we get

$$\left[V^2(X'Z - XZ')\right]_R^r = 0(\gamma^{-7}R^{-1}) = 0(\gamma^{-6}).$$

We thus have, after rearrangement,

$$\begin{aligned}
[X'Z - XZ']_R^r &= \left\{V^2(R)V^{-2}(r) - 1\right\}\left\{X'(R)Z(R)\right. \\
&\quad \left. -X(R)Z'(R)\right\} + 0\left\{V^{-2}(r)\gamma^{-6}\right\}.
\end{aligned}$$

The required result (13.27) now follows in view of the fact that

$$V^2(r) = \gamma^{-2} + 0(\gamma^{-6}r^{-2}) = \gamma^{-2} + 0(\gamma^{-4})$$

for $r \geq R$. This proves (13.26).

Finally, as the main result of this section, we approximate to $\partial\theta/\partial\gamma$.

Lemma 7 *For $R \leq r \leq C$ and large γ,*

$$\partial\theta/\partial\gamma = 2\pi\gamma^{-3}\sqrt{\lambda} + 0(\gamma^{-4}). \tag{13.29}$$

Proof. We have from (8.2) and (9.1) that

$$\partial\theta/\partial\gamma = F^{-1}(X'Z - XZ')\sqrt{\lambda},$$

and here, by the above estimates, for $r \geq \gamma^{-2}$,

$$X'(r)Z(r) - X(r)Z'(r) = 2\pi\gamma^{-3}\lambda + 0(\gamma^{-3}r^2) + 0(\gamma^{-5}r^{-1}).$$

Using Lemma 4, we thus get

$$\partial\theta/\partial\gamma = 2\pi\gamma^{-3}\sqrt{\lambda} + 0(r^2\gamma^{-3}) + 0(\gamma^{-5}r^{-1}).$$

This proves Lemma 7. ∎

14. Estimation of $\partial\theta/\partial\lambda$

Based on results of Section 12, this estimation is very simple. We have the following lemma.

Lemma 8 *For large γ, $R < r \leq C$,*

$$\partial\theta/\partial\lambda = \{1/(2\sqrt{\lambda})\}\{r - \pi\gamma^{-2}\} + 0(r^{-2}\gamma^{-4}) + 0(\gamma^{-3}). \tag{14.1}$$

Proof. We use the invariance property of θ which derives from (1.7), namely, $\theta(r, \lambda, \gamma) = \theta(\mu^2 r, \lambda\mu^{-4}, \gamma\mu^{-1})$. Differentiating with respect to μ and setting $\mu = 1$, we get $2r\partial\theta/\partial r - 4\lambda\partial\theta/\partial\lambda - \gamma\partial\theta/\partial\gamma = 0$, so that

$$\partial\theta/\partial\lambda = \{r/(2\lambda)\}\partial\theta/\partial r - \{\gamma/(4\lambda)\}\partial\theta/\partial\gamma.$$

Here we substitute the results, derived from (11.5), (13.29),

$$\partial\theta/\partial r = \sqrt{\lambda} + 0(r^{-3}\gamma^{-4}), \quad \partial\theta/\partial\gamma = 2\pi\gamma^{-3}\sqrt{\lambda} + 0(\gamma^{-4}).$$

Hence

$$\partial\theta/\partial\lambda = r/(2\sqrt{\lambda}) + 0(r^{-2}\gamma^{-4}) - \pi/\{2\sqrt{\lambda}\gamma^{-2}\} + 0(\gamma^{-3}),$$

which proves the result of the lemma. It follows that if $\theta(r, \lambda, \gamma) > 0$ and $r > 0$ are fixed, this determining λ as a function of γ for large γ, then

$$
\begin{aligned}
\partial\lambda/\partial\gamma &= -(\partial\theta/\partial\gamma)/(\partial\theta/\partial\lambda) \\
&= -\{2\pi\gamma^{-3}\sqrt{\lambda} + 0(\gamma^{-4})\}\{r/(2\sqrt{\lambda}) + 0(\gamma^{-3})\} \\
&= 4\pi\lambda r^{-1}\gamma^{-3} + 0(\gamma^{-4}).
\end{aligned} \tag{14.2}
$$

This is the result needed for Theorem 3. ∎

Acknowledgments

This paper incorporates research done while visiting Argonne National Laboratory (host H. G. Kaper); further work was done while visiting the University of Regensburg (host Prof. R. Mennicken), with support from the Deutsche Forschungsgemeinschaft. The author is grateful for comments from these colleagues, and also for discussions with Dr. C. Budd and Prof. L. A. Peletier, while visiting the University of Leiden. Appreciation is also expressed for the continuing support of the National Science and Engineering Research Council of Canada, under Grant A-3979.

REFERENCES

[1] M. Abramovitz and I. A. Stegun, *Handbook of Mathematical Functions,* Dover Publications, Inc., N.Y., 1965

[2] F. V. Atkinson, H. Brézis, and L. A. Peletier, *Solutions qui changent de signe d'équations elliptiques avec exposant de Sobolev critique,* C. R. Acad. Sci. Paris t. 306, Série II (1988), 711–714

[3] F. V. Atkinson and L. A. Peletier, *Emden-Fowler equations involving critical exponents,* Nonlinear Analysis 10 (1986) 755–776

[4] F. V. Atkinson and L. A. Peletier, *Large solutions of elliptic equations involving critical exponents,* Asymptotic Analysis 1 (1988) 139–160

[5] H. Brézis, *Some variational problems with lack of compactness,* Proc. Sympos. Pure Math. 45 (1986) 165–201

[6] H. Brézis and L. Nirenberg, *Positive solutions of nonlinear elliptic equations involving critical Sobolev exponents,* Comm. Pure Appl. Math. 36 (1983) 437–477

[7] H. Brézis and L. A. Peletier, preprint, Department of Mathematics, University of Leiden (March 1988)

[8] C. Budd, *Semilinear elliptic equations with near critical growth rates,* Proc. Roy. Soc. Edin. 107 (1987) 249–270

[9] C. Budd and J. Norbury, *Symmetry Breaking in Semilinear Elliptic Equations with Critical Exponents,* in Nonlinear Diffusion Equations and Their Equilibrium States, I (W.-M. Ni, L. A. Peletier, and J. Serrin, eds.), Springer-Verlag, New York, 1988

[10] H. Egnell, *Linear and nonlinear elliptic eigenvalue problems*, Uppsala University Department of Mathematics, September 1987

[11] H. Egnell, *Elliptic boundary value problems with singular coefficients and critical nonlinearities*, Uppsala University, Department of Mathematics, U.U.D.M. Report 1987:21 (January 1988)

[12] D. Fortunato and E. Jannelli, *Infinitely many solutions for some nonlinear elliptic problems in symmetrical domains* (to appear)

[13] B. Gidas, W.-M. Ni, and L. Nirenberg, *Symmetry and related properties via the maximum principle*, Comm. Math. Phys. 68 (1979) 209–243

[14] L. A. Peletier and J. Serrin, *Uniqueness of positive solutions of semilinear equations in R^n*, Arch. Rat. Mech. Anal. 81 (1983) 181–197

Department of Mathematics,
University of Toronto,
Toronto, Ontario M5S 1A1,
Canada

Positive Solutions of Emden Equations in Cone-Like Domains

CATHERINE BANDLE

1. Introduction

Let $x \in \mathbb{R}^N$, $N \geq 2$ be a generic point, $S^{N-1} = \{x : |x| = 1\}$ be the $(N-1)$-dimensional unit sphere, $\Omega \subset S^{N-1}$ be an arbitrary domain and denote by $\mathcal{C}$ the cone $\{x : x/|x| \in \Omega\}$.

Consider the nonlinear Dirichlet problem

$$(P) \qquad \Delta u + |x|^\delta u^p = 0 \text{ in } \mathcal{C}, u = 0 \text{ in } \mathcal{C} - \{0\},$$

where $\delta \in \mathbb{R}$ and $p > 1$. We shall be interested in *positive* classical solutions. They might be discontinuous at the origin, and in this case they will be called *singular* solutions and will be denoted by u_S, in contrast to the *regular* solutions $u \in C^2(\mathcal{C}) \cap C^0(\mathcal{C} \cup \partial\mathcal{C}), u \not\equiv 0$.

The solutions of (P) have the following important properties:

—*Scaling invariance*: If $u(x)$ is a solution, then $v(x; \alpha) := \alpha^{(2+\delta)/(p-1)} \cdot u(\alpha x)$, $\alpha > 0$ is again a solution of (P).

—*Inversion Property*: If $u(x)$ is a solution of (P), then we have that $v(y) := |x|^{N-2} u(x), x := y/|y|^2$ solves (P) with δ replaced by $\delta' := -N - 2 - \delta + p(N - 2)$.

—*Representation formula*: If $K(x, y)$ is the Green's function in $\mathcal{C}$, vanishing on $\partial\mathcal{C}$ then

$$u(x) = \int_{\mathcal{C}} K(x, y)|y|^\delta u^p(y)dy.$$

For our purposes it will often be convenient to use the polar coordinates (r, Θ), $r := |x|$ and $\Theta \in S^{N-1}$.

The main question adressed in this note are:

(1) How does the angle at the vertex affect existence and nonexistence of the solutions?

(2) How does the angle influence the local and global behavior of the solutions?

The angle will be measured in terms of the lowest eigenvalue ω of

(1) $\Delta_\Theta \varphi + \omega\varphi = 0$ in Ω, $\varphi = 0$ on $\partial\Omega$,

where Δ_Θ stands for the Laplace–Beltrami operator on S^{N-1}.

2. Special solutions

If we use separation of variables to construct special solutions, it turns out that they must be of the form [BL]

$$(2) \qquad u_S(r, \Theta) = r^{-(2+\delta)/(p-1)} \alpha(\Theta),$$

where α is a positive solution of

$$(3) \quad \Delta_\Theta \alpha + c(N, p, \delta)\alpha + \alpha^p = 0, \ \text{in } \Omega, \alpha = 0 \text{ on } \partial\Omega,$$

$$c(N, p, \delta) = -\frac{\delta + 2}{p - 1}(N - 2 - \frac{\delta + 2}{p - 1}).$$

If we multiply (3) by the first eigenfunction of (1) and integrate, we obtain

$$(4) \qquad \int_\Omega \{-\omega + c(N, p, \delta)\}\alpha\varphi \, d\Theta + \int_\Omega \alpha^p \varphi d\Theta = 0$$

Since φ is of constant sign, a necessary condition for the existence of a separable solution is

$$(5) \qquad c(N, p, \delta) < \omega.$$

This condition can also be expressed as

$$(6) \qquad p < \begin{cases} 1 - \frac{\delta + 2}{\gamma^+}, & \text{if } \delta < -2 \\ 1 - \frac{\delta + 2}{\gamma^-}, & \text{if } \delta > -2 \end{cases}$$

$$\gamma\pm = -\frac{N - 2}{2} \pm \sqrt{\omega + (\frac{N - 2}{2})^2}.$$

It turns out that (5) is also a sufficient condition [BL] for separable solutions to exist.

Lemma 1. *If*

$$p_0 := max\{1 - \frac{\delta + 2}{\gamma^+}, \ 1 - \frac{\delta + 2}{\gamma^-}\} < p < \begin{cases} \infty & \text{if } N = 2, 3 \\ (N + 3)/(N - 1) & \text{if } N > 3, \end{cases}$$

then (P) has separable solutions of the type (2).

3. Nonexistence

It is well known (cf. e.g. [GS]). that in the following cases no solutions to the Emden equation exists in $\mathbb{R}^N$ or $\mathbb{R}^N - \{0\}$

(i) $N = 2, \delta \in R$ and all $p > 1$

(ii) $N \geq 3, \delta > -2$ and $p \leq \frac{N+\delta}{N-2}$,

(iii) $N \geq 3, \delta < -2$ and all $p > 1$.

Similar results hold also in cones.

Theorem 2. [BL, BE] *If* $1 < p \leq p_0$ *(cf. Lemma 1), no regular or singular solutions to* (P) *exist.*

This theorem shows that the presence of angles decreases the ranges of p-values for which no solution exist. This is illustrated in Fig. 1.

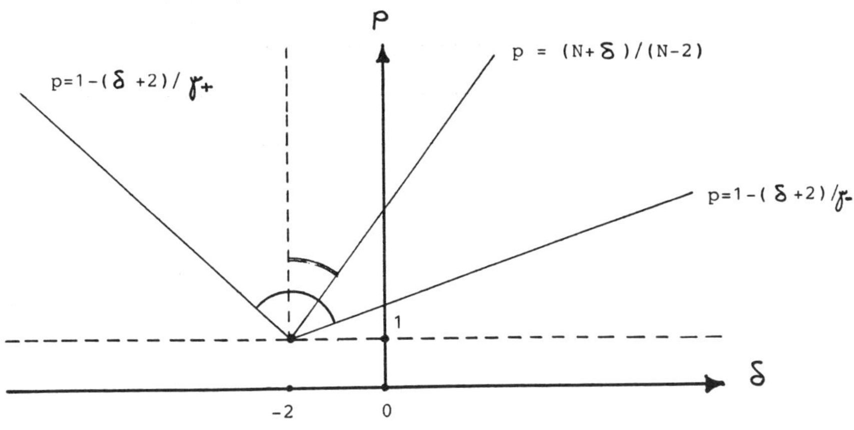

Figure 1.

Remark. Let $\Omega^* \subset S^{N-1}$ be the geodesic sphere of the same measure as Ω. Then by a result of Pólya and Szegö [PS], $\omega(\Omega) \geq \omega(\Omega^*)$. This together with Theorem 2 and Lemma 1 shows that, given vol Ω, the circular cone has largest range of p-values for which no solutions exist.

4. Local behavior

Let $u(x)$ be any regular or singular solution of (P) and put $M(r) := \sup_{|x|=r} u(x)$. The following result is proved in [BE].

Theorem 3. (i) *If* $M(r)r^{(2+\delta)/(p-1)} \to 0$ *as* $r \to 0(\infty)$, *then there exist positive constants* c_1 *and* c_2 *such that* $c_1 r^{\gamma^+} \leq M(r) \leq c_2 r^{\gamma^+}$ $(c_1 r^{\gamma^-} \leq M(r) \leq c_2 r^{\gamma^-}$.

(i') *If* $u = 0(r^{\gamma^+})(u = 0(r^{\gamma^-}))$ *as* $r \to 0(\infty)$, *then* $\lim_{r \to 0(\infty)} r^{-\gamma^{\pm}} u = u_0\varphi(\Theta)$ *uniformly in* Θ, *where* φ *is the first eigenfunction of (1).*

(ii) *If* $M(r)r^{(2+\delta)/(p-1)} \leq c$ *as* $r \to 0(\infty)$, $p < (N+2+2\delta)/(N-2) =:$

$p^*(p > p^*)$ *and* $r_n \to 0(\infty)$ *as* $n \to \infty$, *then there exists a subsequence* $\{r_{n'}\}$, *such that* $r_{n'}^{(2+\delta)/(p-1)} \cdot u(r_{n'}, \Theta) \to \alpha(\Theta)$, *as* $n' \to \infty$, $\alpha(\Theta)$ *being a solution of (3).*

Remarks (1) The case $p = p^*$ is special since the equation is invariant under inversion.

(2) If $N = 2$, (3) has a unique positive solution [NN, Theorem 2.4]. In this case the second statement of Theorem 2 holds for all sequences $\{r_n\}$ tending to zero (infinity).

As already observed in [GS] the asymptotic estimates together with the scaling invariance provides asymptotic gradient estimates.

Corollary 4 [BE], *Let* u *solve* (P).
(i) *If* $u = 0(r^{\gamma^+})$ *as* $r \to 0$, *then* $|\nabla u| = 0(r^{\gamma^+ - 1})$ *as* $r \to 0$.
(ii) *If* $u = 0(r^{\gamma^-})$ *as* $r \to \infty$, *then* $|\nabla u| = 0(r^{\gamma^- - 1})$ *as* $r \to \infty$.

The next results are based on generalized Pohozaev type identity due to Pucci and Serrin [PSe] and on Theorem 4.

Theorem 5 [BE]. *If* $p \neq p^*$, (P) *has no solution such that*

$$r^{(2+\delta)/(p-1)} u(r, \theta) \to 0$$

as $r \to 0$ *and as* $r \to \infty$.

If $p = p^*$ and $\mathcal{C} = \mathbb{R}^N$, such solutions are known to exist. In a recent paper Egnell [E] gave other examples of cones posessing solutions of the above type, such as
(1) $\mathcal{C} = \{x : x_N > 0, dx_N^2 < \sum_{k=1}^{N-1} x_k^2 < Dx_N^2\}, 0 < d < D$ and $\delta > 0$.
(2) $-2 < \delta < 0$ and $\mathcal{C}$ arbitrary.
He was also able to extend the nonexistence result to the critical exponent $p = p^*$ as follows.

Theorem 6. *The statement of Theorem 5 is valid for* $p = p^*$ *if* $\mathcal{C}$ *is contained in a half-space with normal* x_0 *and if either* $\delta = 0$ *or* $\delta \geq 0$ *and* $\mathcal{C}$ *is star-shaped with respect to* $x_0 \in \mathcal{C}$.

Concluding remarks

Up until now little is known concerning the existence of regular solutions, even for sectors $\mathcal{C}_\alpha := \{x \in \mathbb{R}^2 : |x| > 0, 0 < \Theta < \alpha\}$. If no growth conditon at infinity is imposed, Theorems 5 and 6 don't apply, and such solutions might still exist. However in the special case of the sector $\mathcal{C}_\alpha, \pi/2 < \alpha < \pi$, and $\delta = 0$, regular solutions are excluded by the following argument.

According to a result in [GNN, p. 228], any regular solution satisfies $\partial u/\partial e > 0$ for all directions $e = (\cos \Theta_0, \sin \Theta_0)$ with $\alpha - \pi/2 < \theta_0 < \pi/2$ and at any point in C.

But this contradicts the fact, proved in [BE], that

$$\int_0^\alpha u(r, \Theta) \sin(\pi\theta/\alpha)d\Theta \leq cr^{-2/(p-1)}$$

for all $r > 0$.

The question now arises whether there are other solutions besides the separable ones.

It is also not clear if this nonexistence result holds for sectors with $\alpha \leq \pi/2$.

REFERENCES

[BE] C. Bandle and M. Essen, *On the positive solutions of Emden equations in cone-like domains*, Arch. Rat. Mech. Anal., **112** (1990), 319–338.

[BL] C. Bandle and H. A. Levine, *On the existence and nonexistence of global solutions of reaction-diffusion equations in sectorial domains*, Trans. Amer. Math. Soc. **316** (1989), 595–622.

[E] H. Egnell, *Positive solutions of semilinear equations in cones*, (to appear).

[GNN] B. Gidas, W.-M. Ni and L. Nirenberg, *Symmetry and related properties via the maximum principle*, Commun. Math. Phys. **68** (1979), 209–243.

[GS] B. Gidas and J. Spruck, *Global and local behaviour of positive solutions of nonlinear elliptic equations*, Commun. Pure and Appl. Math. **34** (1981), 525–98.

[NN] W. -M. Ni and R. Nussbaum, *Uniqueness and nonuniqueness for positive radial solutions of $\Delta u + f(u, r) = 0$*, Commun. Pure and Appl. Math. **38** (1985), 67–108.

[PSe] P. Pucci and J. Serrin, *A general variational identity*, Indiana Univ. Math. J. **35** (1986), 681–703.

[PS] G. Pólya and G. Szegö, *Isoperimetric inequalities in mathematical physics*, Princeton University Press (1951).

Mathematisches Institut der Universität Basel

Rheinsprung 21,

CH-4051 Basel, Switzerland

Nonlinear Parabolic Equations Arising in Semiconductor and Viscous Droplets Models

FRANCISCO BERNIS

1. Introduction

This paper is devoted to a class of higher order degenerate parabolic equations whose simplest example is the fourth order equation

$$u_t + D(|u|^n\, D^3 u) = 0 \tag{1.1}$$

where $D = \partial/\partial x$, $u = u(x,t)$ is a real-valued function and n is a real parameter, $n > 0$.

This class includes fourth and sixth order equations of the type

$$u_t + D(f(u)\, D^3 u) = 0 \tag{1.2}$$

$$u_t - D(f(u)\, D^5 u) = 0 \tag{1.3}$$

and, more generally, $(2m + 2)^{\text{th}}$-order equations of the form

$$u_t + (-1)^{m-1} D(f(u)\, D^{2m+1} u) = 0. \tag{1.4}$$

With respect to the function f, the relations

$$f(u) = |u|^n f_0(u), \quad f_0 \in C_{\text{loc}}^{1+\alpha}, \ \ 0 < \alpha < 1, \ \ f_0 > 0 \tag{1.5}$$

form an interesting set of hypotheses which include all the models quoted below.

Equation (1.2) arises in modelling the motion of viscous droplets spreading over a solid surface: Ludvikson and Lightfoot [21], Greenspan [12], Greenspan and McCay [13], Hocking [14], Lacey [19] and references therein. These authors take $f(u) = |u|^3$, but also $f(u) = |u|^3 + \beta|u|$ in [13] and $f(u) = |u|^3 + \beta u^2$ in [19]; further, since they assume, on physical grounds, that $u \geq 0$, they replace $|u|$ by u. Hocking [15] introduces in (1.2) an additional second order term of porous media type. Paper [21] dates back to 1968.

Most of this work has been carried out at the Universidad Complutense de Madrid.

The sixth order equation (1.3) with $f(u) = |u|^3$ appears in recent work (from 1986) by King [16, 17] and Tayler and King [24] in a model of oxidation of silicon in semiconductor devices. The last paper includes a simplified and suggestive derivation of both the semiconductor and droplets models in the case $f(u) = |u|^3$.

Smyth and Hill [23] consider nonnegative solutions of (1.4) with $f(u) = u^n$ and perform an asymptotic analysis for small n by the method of singular perturbations. They also give some explicit solutions.

The present paper is a survey on some mathematical research (done and in progress) about the above equations. The reported research is, in part, joint work with Friedman [4] and with Peletier and Williams [5].

For expository reasons our attention is focused on Equation (1.1), although most of the results and methods apply to the more general equation (1.4)–(1.5).

Equation (1.1) can be regarded as a fourth order generalization of the well-known porous media equation (see e.g. [1, 18])

$$u_t - D(|u|^n Du) = 0. \tag{1.6}$$

(This is also the second order case of (1.4) with $f(u) = |u|^n$.) Another fourth order generalization of (1.6) is

$$u_t + D^3(|u|^n Du) = 0$$

which can also be written, up to an unessential constant,

$$u_t + D^4(|u|^{n+1} \operatorname{sgn} u) = 0. \tag{1.7}$$

The properties of (1.1) and (1.7) are extremely different: see Remark 3.1 below. Also the equations

$$u_t + D^2(|u|^n D^2 u) \quad \text{and} \quad u_t + |u|^n D^4 u = 0$$

have very different behaviour: McLeod and Williams [22].

The factor $|u|^n$ in (1.1) introduces a very strong degeneracy. In the last years some "superdegenerate" or strongly degenerate second order parabolic equations of the form

$$u_t = u \Delta u + g(u)$$

$$u_t = u^\alpha \Delta u, \quad \alpha \geq 1$$

$$u_t = u \Delta u - \gamma |\operatorname{grad} u|^2, \quad \gamma \geq 0$$

have been studied: Ughi [25], Dal Passo and Luckhaus [11], Bertsch and Ughi [8], Bertsch, Dal Passo and Ughi [6, 7] and references therein. Some of the results and methods of these papers have a similar flavor to those of equations (1.1), although in many other respects are very different. With respect to the methods, an important difference is that *no comparison principles* are known to us for fourth and higher order parabolic equations (degenerate as well as nondegenerate).

2. Weak Solutions

Let Ω be an open, bounded interval of $\mathbb{R}$ and $T > 0$ a real number. We consider the equation

$$u_t + D(|u|^n D^3 u) = 0 \quad \text{in} \quad Q \equiv \Omega \times (0, T) \tag{2.1}$$

with initial conditions

$$u(\cdot, 0) = u_0, \quad u_0 \in H^1(\Omega) \equiv W^{1,2}(\Omega) \tag{2.2}$$

and boundary conditions

$$Du = D^3 u = 0 \quad \text{on} \quad \partial\Omega \times (0, T). \tag{2.3}$$

We emphasize that in this section u_0 may have changing sign.

In [4] we prove that if $n > 1$ there exists a weak solution u of (2.1)–(2.3), i.e. a function u satisfying the following relations (2.4) to (2.10):

$$u \text{ is Hölder continuous in } \bar{Q} \tag{2.4}$$

$$u_t, Du, D^2 u, D^3 u, D^4 u \text{ belong to } C(P) \tag{2.5}$$

where $P = \bar{Q} - \{u = 0\} - \{t = 0\}$, and

$$|u|^n D^3 u \in L^2(P). \tag{2.6}$$

u satisfies (2.1) in the following sense:

$$\int\int_Q u\,\varphi_t + \int\int_P |u|^n (D^3 u) D\varphi = 0 \tag{2.7}$$

for all $\varphi \in \text{Lip}(\bar{Q})$, $\varphi = 0$ near $t = 0$ and near $t = T$,

$$u(\cdot, 0) = u_0 \quad \text{on} \quad \bar{\Omega} \tag{2.8}$$

$$Du(\cdot,t) \to Du_0 \quad \text{strongly in} \quad L^2(\Omega) \text{ as } t \to 0 \tag{2.9}$$

and

$$u \text{ satisfies (2.3) at the points where } u \neq 0. \tag{2.10}$$

In addition, the weak solution obtained in [4] satisfies the following properties:

$$\int_\Omega u(x,t)dx = \text{const.} = \int_\Omega u_0(x)dx \tag{2.11}$$

$$u \in C([0,T]; H^1(\Omega) \text{ weak}) \tag{2.12}$$

$$\frac{1}{2}\int_\Omega (Du(\cdot,T))^2 + \int\int_P |u|^n(D^3u)^2 \leq \frac{1}{2}\int_\Omega (Du_0)^2. \tag{2.13}$$

Furthermore, (2.12) and (2.13) imply that

$$Du(\cdot,t) \in L^2(\Omega) \quad \text{for all} \quad t \in [0,T] \tag{2.14}$$

$$\sup_{0\leq t\leq T} \int_\Omega (Du(\cdot,t))^2 \leq \int_\Omega (Du_0)^2 \equiv K^2 \tag{2.15}$$

and

$$|u(x,t) - u(y,t)| \leq K|x-y|^{1/2} \quad \text{for all} \quad (x,t),(y,t) \text{ in } \bar{Q}. \tag{2.16}$$

The above concept of solution is very weak; it includes stationary solutions with compact support of the form

$$\text{Constant.} \ (x-b)_+(c-x)_+; \ b,c \in \Omega; \ b < c. \tag{2.17}$$

These solutions are not C^1 at $x = b$ and at $x = c$. Under some additional hypotheses (see [4]), the construction of the solution can be performed so as to obtain $D^2u \in L^2(Q)$, thus excluding (2.17). These additional hypotheses are compatible with a compactly supported initial datum only if $n < 2$.

On the other hand, in the droplets models the solutions of the form (2.17) actually represent stationary droplets (from the physical point of view).

Remark 2.1. In [4] it is also obtained that the weak solution is uniformly Hölder continuous in t with exponent $1/8$.

Remark 2.2. The above results remain true if $0 < n \leq 1$, although the arguments in [4] require some modifications.

Remark 2.3. Since the function $f(u) = |u|^n$ is C^∞ for $u \neq 0$, it follows from (2.5), (2.7) and linear parabolic theory that $u \in C^\infty(\Gamma)$.

Remark 2.4. If we multiply Equation (2.1) by $D^2 u$ and perform formal computations, we obtain the estimate (2.13) with $\leq$ replaced by $=$. In fact, the use of this estimate for appropriate approximating problems is a basic ingredient of the proof of the above results.

Remark 2.5. In [4] it is developed a theory of existence of weak solutions for the general $(2m+2)^{\text{th}}$-order equation (1.4) under the hypotheses (1.5) and $n > 1$. See Remark 2.2 for the case $0 < n \leq 1$.

3. Positivity and Expansion of the Support

Equation (2.1) has a second simple estimate:

$$\int_\Omega |u(\cdot,t)|^{2-n}\big]_{t=0}^{t=T} + (n-1)(n-2)\int\int_\Omega (D^2 u)^2 = 0 \ (\text{``formally''}) \quad (3.1)$$

which is obtained formally multiplying the equation by $|u|^{1-n}\mathrm{sgn}\,u$. Intuitively speaking, the negative power $2-n$ produces a tendency to preserve positivity. Although (3.1) cannot be used directly, it is in the background of the proofs in [4] of the following results for Problem (2.1)–(2.3):

I. *Nonnegativity.* If $u_0 \geq 0$ on Ω there exists (for all $n \geq 1$) a weak solution u such that $u \geq 0$ in Q.

II. *Positivity.* If $n \geq 4$ then $u_0 > 0$ on $\bar{\Omega}$ implies that $u > 0$ in $\bar{Q}$. This solution u is unique and classical for $t > 0$.

III. *Expansion of the support.* Assume that $u_0 \geq 0$ and $n \geq 2$. Then there exists a weak solution u such that the support of $u(\cdot,t)$ is nondecreasing with t (in the sense of set inclusion).

IV. *Point-wise positivity.* Let $u_0 \geq 0$ and $n \geq 4$. Then there exists a weak solution u such that

$$b \in \bar{\Omega} \text{ and } u_0(b) > 0 \Rightarrow u(b,t) > 0 \text{ for all } t \in [0,T].$$

Since the solution of (2.1)–(2.3) may not be unique (except in the case II), the words "there exists a weak solution" may be important. In some cases there are two (or more) solutions with radically different properties.

Remark 3.1. Property I is *not* true for the linear equations $u_t + D^4 u = 0$ and $u_t + (-1)^m D^{2m} u = 0$, $m \geq 2$. Furthermore, it is *also not true* for other higher (≥ 4) order nonlinear *degenerate* parabolic equations; e.g.

$u_t + D^3(|u|^n Du) = 0$. (See also (1.7) in the introduction.) Even in some cases the property of instantaneous change of sign holds. See [2] for precise statements and proofs. On the other hand, the initial boundary value problems for the equations of this remark have a natural concept of solution for which uniqueness holds.

Remark 3.2. It is remarkable that Property I is obtained (for $n \geq 1$) in [4] using approximating problems which do not preserve nonnegativity. Property I can be extended to any $n > 0$ using the approximating problems of [4, Section 6].

Remark 3.3. For the $(2m + 2)^{\text{th}}$-order equation (1.4)–(1.5) the properties I and II are also considered in [4]. The condition $n \geq 4$ is replaced by $n \geq 8/3$ if $m = 2$ and by $n \geq 5/2$ if $m \geq 3$. Properties III and IV are not treated in [4] when the order is ≥ 6.

Remark 3.4. In spite of these positivity results, no comparison principles are known to us for the equations (1.1)–(1.4) if the order is ≥ 4.

4. Solutions with Vertical Interfaces

The proofs of the properties I to IV of Section 3 rely on global hypotheses on u_0 (e.g. $u_0 \geq 0$ in Ω) and on the boundary conditions. Furthermore, the proofs of I, III and IV use approximating problems.

In this section we shall obtain if $n > 10$ a "local" property of preservation of sign which is independent of the boundary conditions and allows u_0 with changing sign. Furthermore, since no approximating problems are used, this property is valid for all the weak solutions (if $n > 10$).

Theorem 4.1. *Let u be a weak solution of (2.1) and (2.2) satisfying (2.4) to (2.16), except that (2.10) and (2.11) are not required. Assume that $n > 10$. Then*

$$b \in \Omega \text{ and } u_0(b) > 0 \Rightarrow u(b,t) > 0 \text{ for all } t \in [0,T].$$

(An analogous statement holds with $u_0(b) < 0$ and $u(b,t) < 0$. The case $T = \infty$ can be included with a straightforward modification).

Proof of Theorem 4.1. Since u is continuous, there exists δ such that $0 < \delta \leq T$,

$$J \equiv [b - \delta, b + \delta] \subset \Omega \tag{4.1}$$

and $u > 0$ in $J \times [0, \delta]$. Assume (for contradiction) that there exists $T_0 \in (0, 1]$ such that

$$u > 0 \text{ in } J \times [0, T_0) \text{ and } u(x_0, T_0) = 0 \text{ for some } x_0 \in J. \qquad (4.2)$$

Let

$$\zeta(x) = (x - q)(r - x) \text{ with } q = b - \delta \text{ and } r = b + \delta \qquad (4.3)$$

$$\xi(x) = (\zeta(x))^s \text{ with } s > 3. \qquad (4.4)$$

(The use of the functions ζ and ξ below is reminiscent of Campanato's method to obtain Caccioppoli-type inequalities.) Let

$$H \equiv J \times [t_1, t_2] \text{ with } 0 < t_1 < t_2 < T_0. \qquad (4.5)$$

Notice that u is smooth in H, since it is positive and H is bounded away from $t = 0$. Multiplying Equation (2.1) by $\xi\, u^{1-n}$ and integrating in H we obtain:

$$A - \int\int_H D^2(\xi u) D^2 u + n \int\int_H D(\xi Du) D^2 u = 0$$

where

$$A = \frac{1}{n - 2} \int_J \xi(x) u(x, t)^{2-n} dx \Big]_{t=t_1}^{t=t_2}. \qquad (4.6)$$

(The boundary terms in the integrations by parts are zero because of the function ξ.) Computing the derivatives of the products it follows that

$$A + (n - 1) \int\int_H \xi(D^2 u)^2 = \int\int_H (\xi'' u - (n - 2)\xi' Du) D^2 u. \qquad (4.7)$$

In what follows C stands for a constant which may be different in different occurrences. Observing that in J

$$\xi' \leq C \zeta^{s-1} \text{ and } \xi'' \leq C \zeta^{s-2}$$

the right-hand side of (4.7) is bounded by

$$C\left(\int\int_H \zeta^{s-2} u |D^2 u| + \int\int_H \zeta^{s-1} |Du| |D^2 u| \right)$$

$$\leq \varepsilon \int\int_H \zeta^s (D^2 u)^2 + C_\varepsilon \int\int_H \zeta^{s-4} u^2 + C_\varepsilon \int\int_H \zeta^{s-2} (Du)^2$$

where Schwarz's and Young's inequalities have been used. The integrals of the last two terms are bounded by a constant because $u \in C(\bar{Q})$, $s > 3$ and

$Du \in L^2(Q)$. Taking a fixed ε small enough (e.g. $\varepsilon = (n-1)/2$) we obtain (recall that $\xi = \zeta^s$)

$$A + \int \int_H \xi (D^2 u)^2 \le C. \tag{4.8}$$

Recalling (4.6) and letting $t_2 \to T_0$ it follows that

$$\int_J \xi(x) u(x, T_0)^{2-n} dx < \infty. \tag{4.9}$$

Taking into account the notations introduced in (4.1) and (4.3), in (4.2) we have that either $x_0 = q$ or $x_0 = r$ or $q < x_0 < r$. Consider the case $x_0 = q$. Then (4.3), (4.4), (4.9) and (2.16) imply that

$$\int_q^r (x-q)^s (r-x)^s (x-q)^{(2-n)/2} dx < \infty.$$

Thus, $s + (2-n)/2 + 1 > 0$, i.e. $n < 2s + 4$. Since this holds for any $s > 3$, we obtain $n \le 10$: contradiction. The case $x_0 = r$ is analogous the case $q < x_0 < r$ implies $n < 4$. Therefore (4.9) and $n > 10$ contradict (4.2) and the theorem is proved.

Remark 4.1. Under the hypotheses of Theorem 4.1, take u_0 such that $u_0 > 0$ on (a_1, a_2) and $u_0 < 0$ on (a_2, a_3). Then $u(a_2, t) = 0$ for all $t \ge 0$, i.e. u has a *"vertical" straight line of zeros* in the (x, t) plane. It is striking that this property holds even if the positive part of u_0 is very large and the negative part very small.

Remark 4.2. Let $[a_1, a_2] \subset \Omega$,

$$u_0 > 0 \text{ on } (a_1, a_2) \text{ and } u_0 = 0 \text{ on } \Omega - (a_1, a_2). \tag{4.10}$$

Then if $n > 10$ there exists a weak solution u (in the sense of Section 2) of Problem (2.1)–(2.3) such that u preserves (4.10) for all time. Thus, u has exactly constant support and exactly vertical interfaces *for all time.* This can be proved combining Theorem 4.1 with an approximating process in which the initial data are negative on $\Omega - [a_1, a_2]$. A similar result for some second order superdegenerate equations (proven by very different methods) appears in [6, 8, 11, 25]. Notice that we are not assuming that u_0 is "flat" at a_1 and a_2 as in the waiting time phenomenon for the porous media equation (1.6); in any case, the solutions of (1.6) do not have vertical interfaces for all time.

Remark 4.3. For the $(2m+2)^{\text{th}}$-order equation (1.4) with $f(u) = |u|^n$ and $m \geq 2$ the condition $n > 10$ is replaced by $n > 2m+4$. Then an analogous of Theorem 4.1 holds.

5. Source-Type Similarity Solutions

We consider the Cauchy problem

$$u_t + D(|u|^n D^3 u) = 0 \quad \text{in} \ \ \mathbb{R} \times (0, \infty) \tag{5.1}$$

$$u(\cdot, t) \to \delta \quad \text{as} \ \ t \to 0 \tag{5.2}$$

where δ stands for the Dirac measure. Since (5.1) is invariant under a group of scaling transformations and δ is homogeneous of degree -1 (in dimension 1), a classical argument (see e.g. [9] or [20]) suggests that the solution of (5.1)–(5.2) is a similarity solution of the form

$$u(x, t) = t^{-b} v(x/t^b), \quad b = 1/(n+4) \tag{5.3}$$

where $v = v(y)$ satisfies the ordinary differential equation

$$|v|^n v''' = b \, y \, v \tag{5.4}$$

and the conditions

$$v \in L^1(\mathbb{R}), \quad \int_{\mathbb{R}} v(y) \, dy = 1. \tag{5.5}$$

Thus, where $v \neq 0$ we have

$$v''' = b \, y \, |v|^{1-n} \operatorname{sgn} v. \tag{5.6}$$

In joint work with Peletier and Williams [5] we prove the following results:

I. *Existence and uniqueness results.* If $0 < n < 3$ there exists a unique even solution v of (5.4)–(5.5) such that v is nonnegative and C^1 in $\mathbb{R}$. This solution is positive on an interval $(-a, a)$ and zero on $\mathbb{R} - (-a, a)$. Furthermore, at $y = a$ v behaves

$$\text{as} \ (a - y)^2 \ \text{if} \ 0 < n < 3/2, \quad \text{and}$$

$$\text{as} \ (a - y)^{3/n} \ \text{if} \ 3/2 < n < 3.$$

(If $n = 3/2$ v'' behaves as a logarithm at $y = a$.) In all cases v is *not* C^2 at $y = a$. Notice that v'' is bounded if and only if $n < 3/2$, while $v'' \in L^2$ if and only if $n < 2$.

Coming back to Equation (5.1), the corresponding solutions have strictly increasing support and the regularity at the interfaces depends on n in a delicate way.

Remark 5.1. For $n = 1$ the solution v has the explicit formula (Smyth and Hill [23])

$$v(y) = \frac{1}{120}(a^2 - y^2)^2 \ \ \text{if} \ \ |y| \le a, \ a = (225/2)^{1/5}.$$

The solution v of Result I has no explicit formula for other values of n, as far as we know.

II. *Nonexistence results.* If $n \ge 3$ there is no solution of (5.4)–(5.5). Thus, there is no solution of (5.1)–(5.2) of the form (5.3). It seems likely that the words "of the form (5.3)" can be dropped, although this has not been proved (to our knowledge). Brezis and Friedman [10] prove nonexistence theorems for some second order parabolic equations with δ as initial condition. It would be interesting to investigate if there is some relation between these two kinds of nonexistence results.

Remark 5.2. Consider the $(2m+2)^{\text{th}}$-order equation (1.4) with $f(u) = |u|^n$. This equation has a source-type, nonnegative similarity solution of class C^m if and only if $n < (2m + 1)/m$.

Remark 5.3. If $0 < n < 1$ it is proved in [3] that (5.6)–(5.5) has a C^3 solution. This solution also has bounded support, but it changes sign infinitely many times. Via (5.3) we obtain a corresponding solution of (5.1)–(5.2). For this and other reasons we have the conjecture that (at least if $0 < n < 1$) the initial-boundary value problems for Equation (1.1) have changing sign solutions which are much more regular than the nonnegative ones.

Remark 5.4. The source-type or fundamental solution of the linear equation $u_t + D^4 u = 0$ changes sign infinitely many times. This can be checked by a variety of classical methods, e.g. by Fourier transform methods or by using the equation (5.6) with $n = 0$, i.e. the equation $v''' = (1/4)y\,v$.

Remark 5.5. It is a widely recognized heuristic principle that the source-type solution gives the asymptotic behavior as $t \rightarrow \infty$ (and other properties) of the reasonable solutions of the Cauchy problems. It seems very difficult to *prove* a theorem of this kind for Equation (1.1). Up to now we do not know neither comparison principles nor a way to obtain useful estimates for *differences* of two solutions. Furthermore, it seems that there is a deep problem of nonuniqueness.

Acknowledgments: The author thanks M. Bertsch, J.R. King, A.A. Lacey, J.B. McLeod, J.R. Ockendon, A.B. Tayler, H.F. Weinberger and S.M. Williams for useful discussions.

This work is partially supported by the Spanish CICYT project PB86/ 0485.

REFERENCES

[1] D.G. Aronson, *The porous media equation,* in *Nonlinear Diffusion Problems,* Montecatini, 1985. Ed. by A. Fasano and M. Primicerio, Lecture Notes in Mathematics, **1224**, Springer-Verlag, Berlin, 1986, pp. 1–46.

[2] F. Bernis, *Change of sign of the solutions to some parabolic problems,* in *Nonlinear Analysis and Applications,* Arlington, Texas, 1986. Ed. by V. Lakshmikantham, Marcel Dekker, New York, 1987, pp. 75–82.

[3] F. Bernis, *Source-type solutions of fourth order degenerate parabolic equations,* in *Nonlinear Diffusion Equations and Their Equilibrium States,* Berkeley, California, 1986. Ed. by W.-M. Ni, L.A. Peletier and J. Serrin, Springer-Verlag, New York, 1988, **I**, pp. 123–146.

[4] F. Bernis and A. Freidman, *Higher order nonlinear degenerate parabolic equations,* J. Differential Equations **83**(1990), 179–206.

[5] F. Bernis, L.A. Peletier and S.M. Williams, *Source type solutions of a fourth order nonlinear degenerate parabolic equation,* Preprint, University of Leiden, 1990.

[6] M. Bertsch, R. Dal Passo and M. Ughi, *Nonuniqueness and irregularity results for a nonlinear degenerate parabolic equation,* in *Nonlinear Diffusion Equations and Their Equilibrium States,* Berkeley, California, 1986. Ed. by W.-M. Ni, L.A. Peletier and J. Serrin, Springer-Verlag, New York, 1988, **I**, pp. 147–159.

[7] M. Bertsch, R. Dal Passo and M. Ughi, *Nonuniqueness of solutions of a degenerate parabolic equation,* Ann. Mat. Pura Appl., to appear.

[8] M. Bertsch and M. Ughi, *Positivity properties of viscosity solutions of a degenerate parabolic equation,* Nonlinear Anal. **14**(1990), 571–592.

[9] G.W. Bluman and J.D. Cole, *Similarity Methods for Differential Equations,* Springer-Verlag, New York, 1974.

[10] H. Brezis and A. Friedman, *Nonlinear parabolic equations involving measures as initial conditions*, J. Math. Pures Appl. **62**(1983), 73–97.

[11] R. Dal Passo and S. Luckhaus, *A degenerate diffusion problem not in divergence form*, J. Differential Equations **69**(1987), 1–14.

[12] H.P. Greenspan, *On the motion of a small viscous droplet that wets a surface*, J. Fluid Mech. **84**(1978), 125–143.

[13] H.P. Greenspan and B.M. McCay, *On the wetting of a surface by a very viscous fluid*, Stud. Appl. Math. **64**(1981), 95–112.

[14] L.M. Hocking, *Sliding and spreading of thin two-dimensional drops*, Quart. J. Mech. Appl. Math. **34**(1981), 37–55.

[15] L.M. Hocking, *The spreading of a thin drop by gravity and capillarity*, Quart. J. Mech. Appl. Math. **36**(1983), 55–69.

[16] J.R. King, *Mathematical aspects of semiconductor process modelling*, Thesis, University of Oxford, 1986.

[17] J.R. King, *The isolation oxidation of silicon: the reaction-controlled case*, SIAM J. Appl. Math. **49**(1989), 1064–1080.

[18] B.F. Knerr, *The porous medium equation in one dimension*, Trans. Amer. Math. Soc. **234**(1977), 381–415.

[19] A.A. Lacey, *The motion with slip of a thin viscous droplet over a solid surface*, Stud. Appl. Math. **67**(1982), 217–230.

[20] J.D. Logan, *Applied Mathematics: A Contemporary Approach*, John Wiley and Sons, New York, 1987.

[21] V. Ludvikson and E.N. Lightfoot, *Deformation of advancing menisci*, Amer. Inst. Chem. Engrg. J. **14**(1968), 674–677.

[22] J.B. McLeod and S.M. Williams, private communication.

[23] N.F. Smyth and J.M. Hill, *High-order nonlinear diffusion*, IMA J. Appl. Math. **40**(1988), 73–86.

[24] A.B. Tayler and J.R. King, *Free boundaries in semi-conductor fabrication*, in *Free Boundary Problems: Theory and Applications*, Irsee, Bavaria, Germany, 1987. Ed. by K.H. Hoffman and J. Sprekels, **II**, Pitman Research Notes in Mathematics 186, Longman, Harlow, 1990.

[25] M. Ughi, *A degenerate parabolic equation modelling the spread of an epidemic*, Ann. Mat. Pura Appl. **143**(1986), 385–400.

Departamento de Matematicas
Universidad Autonoma de Madrid
28049 Madrid, Spain

A Parabolic Equation with a Mean-Curvature Type Operator

MICHIEL BERTSCH and ROBERTA DAL PASSO

The classical mean-curvature operator arises in geometry, and is given by

$$Lu = \operatorname{div}\left(\frac{Du}{\sqrt{1 + |Du|^2}}\right).$$

Mathematically this operator is of particular interest because of its degeneracy for large gradients. The same type of operators arises also in thermodynamical context in the theory of convex free energy functionals, which are asymptotically linear as the gradient tends to infinity [12]; in particular this leads to the parabolic equation $u_t = Lu$ [6]. Observe that here u is supposed to be a *function*, i.e. u is single-valued, in contrast to the geometric origin of the operator L, where one considers *surfaces* which may correspond very well to multi-valued functions u.

In this paper we consider the parabolic equation simply as a nonlinear diffusion equation, i.e. an equation for a function $u(x, t)$, in which the diffusion coefficient vanishes as the gradient tends to ∞. Since no PDE theory seems to be available in the literature for this sort of parabolic operators, we begin with the case of one space dimension only.

It turns out to be more instructive to consider a slightly more general equation, in which there is an additional nonlinearity in u itself:

$$u_t = (\phi(u)\psi(u_x))_x \qquad \text{in } \mathbf{R} \times \mathbf{R}^+. \tag{1}$$

Here ϕ is a smooth, strictly positive function on $\mathbf{R}$, and $\psi : \mathbf{R} \to \mathbf{R}$ is a smooth, odd function which satisfies

$$\psi' > 0 \text{ in } \mathbf{R}, \quad \text{and} \quad \lim_{s \to \infty} \psi(s) = 1. \tag{2}$$

Observe that in case of the mean-curvature equation

$$\psi(s) = \frac{s}{\sqrt{1 + s^2}}. \tag{3}$$

We also note that for example the p-Laplace operator falls outside this class of equations; indeed it would correspond to the function $\psi(s) = |s|^{p-2}s$, which satisfies only the condition $\psi'(s) \to 0$ as $s \to \infty$ (if $1 < p < 2$) but not the stronger condition $\lim_{s \to \infty} \psi(s) < \infty$.

We shall study the Cauchy problem related to equation (1):

$$u(x, 0) = u_0(x) \qquad \text{for } x \in \mathbf{R}, \tag{4}$$

where u_0 is a bounded function on $\mathbf{R}$. In most of the paper we shall assume that u_0 is increasing.

The only results in the literature which we are aware of, concern the nonlinear semigroup approach. Blanc [4,5] proves for a slightly more general equation the existence of a semigroup solution, which may be discontinuous. In fact he derives several necessary or sufficient conditions for the continuity of these solutions.

A PDE theory for this equation does not exist at all, and it is the purpose of this contribution to present some results in this direction. Apart from an existence result, these results concern mainly the hyperbolic behaviour of the solutions, i.e. the degeneracy in the gradient is so strong that discontinuous solutions may exist, and solutions behave near a discontinuity to some extent like solutions of the first order equations

$$u_t = \psi(\pm\infty)(\phi(u))_x = \pm(\phi(u))_x. \tag{5}$$

More in particular we shall discuss shock waves, Oleinik's entropy condition, and jump conditions at the shocks.

Definition of a solution

The first nontrivial problem is how to define a weak solution. We have mentioned the existence of discontinuous solutions, i.e. the derivative u_x is a measure, and so we have to give a meaning to the expression $\psi(u_x)$.

One of the basic questions is whether one allows solutions like

$$u(x, t) = u_0(x) = \begin{cases} 1 & \text{if } x > 0 \\ 0 & \text{if } x < 0. \end{cases}$$

To understand this question a little better one could try to find other solutions with the same initial function. Indeed, if $\phi = 1$ in $\mathbf{R}$, and, for example, ψ is given by (3), self-similar solutions of the form

$$u(x, t) = \begin{cases} \sqrt{t}g(x/\sqrt{t}) & \text{for } x < 0, 0 < t \le t_0 \\ 1 - \sqrt{t}g(x/\sqrt{t}) & \text{for } x > 0, 0 < t \le t_0, \end{cases}$$

are constructed in [6], where

$$0 < g(0) < \infty, \quad t_0 = \frac{1}{4g^2(0)}, \quad g'(0) = \infty, \quad g(\infty) = g'(\infty) = 0.$$

Hence $u(x, t)$ is discontinuous at $x = 0$ if $t < t_0$, and becomes continuous at $t = t_0$. Observe that this solution is qualitatively different from the piecewise constant solution, in the sense that, since $g'(0) = \infty$, it satisfies the condition

$$\lim_{x \to 0} u_x(x,t) = \infty \quad \text{for } 0 < t < t_0.$$

We could interpret this property as the continuity of the function $\psi(u_x(x,t))$, when we use the convention that $\psi(u_x) = \psi(\infty) = 1$ at points where $u(x+,t) > u(x-,t)$. This motivates the following definition of a solution, which is satisfied by the self-similar solution, but not by the piecewise constant one.

Definition 1. A solution of the problem (1)-(4) is a function

$$u \in BV_{loc}(\mathbf{R} \times [0,\infty)) \cap L^\infty(\mathbf{R} \times \mathbf{R}^+)$$

such that:

1. for any $t > 0$ $u(x,t) \in BV_{loc}(\mathbf{R})$ and there exists a continuous function $\Psi : \mathbf{R} \times [0,\infty) \to \mathbf{R}$ such that for any $x \in \mathbf{R}$ and $t \geq 0$

$$\begin{aligned}
\Psi(x,t) &= \lim_{h \to 0} \psi\left(\frac{u(x+h,t) - u(x-,t)}{h}\right) \\
&= \lim_{h \to 0} \psi\left(\frac{u(x+h,t) - u(x+,t)}{h}\right),
\end{aligned} \tag{6}$$

where $u(x\pm,t)$ denote one-sided limits;

2. for any $\chi \in C^1(\mathbf{R} \times [0,\infty))$ with compact support

$$\iint_{\mathbf{R} \times \mathbf{R}^+} (u\chi_t - \phi(u)\Psi\chi_x)\, dx\, dt = -\int_{\mathbf{R}} u_0(x)\chi(x,0)\, dx.$$

Existence of a solution

To prove the existence of a solution, we regularize the equation, replacing the function ψ by

$$\psi_\epsilon(s) = \psi(s) + \epsilon s.$$

The corrisponding unique classical solution we denote by $u_\epsilon(x,t)$.

Lemma 1. u_ϵ *is uniformly bounded in* $BV_{loc}(\mathbf{R} \times [0,\infty))$.

The result follows from local uniform bounds of the partial derivatives of u_ϵ in L^1. The proof of these integral estimates follows from straightforward manipulations with the equation [2].

By Lemma 1 there exist a function $u \in L^1_{loc}(\mathbf{R} \times \mathbf{R}^+)$ and a sequence $\{u_{\epsilon_n}\}$ which converges to u in L^1_{loc} as $n \to \infty$. It turns out that, if u_0 is increasing, u is a solution of problem (1)-(4), and u does not depend on the choice of the sequence:

Theorem 1 ([2]). *Let u_0 be bounded and increasing in $\mathbf{R}$, such that $\psi(u_0') \in C(\mathbf{R})$ (where we use the convention that $\psi(u_0'(x_0)) = 1$ if u_0 is discontinuous at x_0), and $\psi(u_0'(x)) \to 0$ as $x \to \pm\infty$. Then there exists a function u such that*

$$u_\epsilon \to u \quad in \quad L^1_{loc}(\mathbf{R} \times [0,\infty)),$$

and u is a solution of problem (1)-(4).

We guess that the condition that u_0 is increasing is not essential (we hope to come back to this in a future paper), but let us explain where it comes from.

By the maximum principle, $u_{\epsilon x} > 0$ in $\mathbf{R} \times \mathbf{R}^+$. Hence we may introduce a new independent variable

$$y = u_\epsilon(x, t). \tag{7}$$

We define the function

$$v_\epsilon(y, t) = \psi_\epsilon(u_{\epsilon x}(x, t)) \quad \text{for } a < y < b, \quad t > 0,$$

where a and b are defined by

$$a = \lim_{x \to -\infty} u_0(x), \qquad b = \lim_{x \to \infty} u_0(x).$$

Since $u_{\epsilon x}$ vanishes as $x \to \pm\infty$, we may extend v_ϵ by

$$v_\epsilon(a, t) = v_\epsilon(b, t) = 0.$$

It follows from a staightforward calculation that v_ϵ is a solution of the problem

$$(I_\epsilon) \quad \begin{cases} c_\epsilon(v)_t = (\phi(y)v)_{yy} & \text{in } (a, b) \times \mathbf{R}^+ \\ v(a, t) = v(b, t) = 0 & \text{for } t > 0 \\ c_\epsilon(v(y, 0)) = w_{0\epsilon}(y) & \text{for } a < y < b, \end{cases}$$

where the function c_ϵ is defined by

$$c_\epsilon(v) = -\frac{1}{\psi_\epsilon^{-1}(v)} \quad \text{for } v > 0,$$

and where the choice of the initial function $w_{0\epsilon}$ is obvious.

Problem I_ϵ is a rather standard parabolic initial-boundary value problem (the only minor singularity in the problem is the fact that $c_\epsilon(s) \to -\infty$ as $s \to 0+$), and classical methods, like integral estimates and the maximum principle, can be used to study its solution. Of course we are particularly interested in what happens in the limit $\epsilon \to 0$. Indeed it follows at once from the definition of c_ϵ that

$$c(s) = \lim_{\epsilon \to 0+} c_\epsilon(s) = \begin{cases} -\dfrac{1}{\psi^{-1}(s)} & \text{for } 0 < s < 1 \\ 0 & \text{for } s \geq 1. \end{cases}$$

Hence we arrive formally at the limit problem

$$(I) \quad \begin{cases} c(v)_t = (\phi(y)v)_{yy} & \text{in } (a,b) \times \mathbf{R}^+ \\ v(a,t) = v(b,t) = 0 & \text{for } t > 0 \\ c(v(y,0)) = w_0(y) & \text{for } a < y < b. \end{cases}$$

Observe that, by the definition of c, the equation for v is parabolic at points where $0 < v < 1$ but elliptic at points where $v > 1$, and we say that Problem I is of *elliptic-parabolic* type. Elliptic-parabolic equations of slightly different type have been studied in [1,7,8,10].

Of course one may wonder, in view of the original problem in terms of the x-variable, if the set where $v > 1$ will have any importance at all for the solutions which we are studying. In the following section we shall see that this set is crucial for the study of the qualitative properties of the solutions.

The existence proof is based on the study of the functions v_ϵ. In particular the functions $c_\epsilon(v_\epsilon)$ turn out to be locally uniformly continuous in $(a,b) \times \mathbf{R}^+$, which enables us to conclude the continuity of the function $\psi(u_x(x,t))'$ (or rather of the function $\Psi(x,t)$ in Definition 1). The fact that any sequence converges to the same limit u relies on the uniqueness of the solution of Problem I. For the details we refer to [2].

The shocks

In this section we shall give a formal explanation of our results concerning the possible discontinuities of the solutions, and the importance of the entropy condition. In order to make the arguments more transparent, the discussion will be based on formal considerations about the elliptic-parabolic Problem I. For rigorous proofs of our results we refer to [2].

Let y, $v(y,t)$, and $c(v)$ be defined as in the former section. We begin with the question which we asked before, namely if the set of points where $v > 1$ may be non-empty. The answer is yes. To understand this, we write the equation for v in the form

$$c'(v)v_t = \phi v_{yy} + 2\phi' v_y + \phi'' v.$$

Suppose for the moment that $v(\cdot, t_0)$ has a maximum at the point $y_0 \in (a,b)$, and that $v(y_0, t_0) = 1$. If $\phi''(y_0) \leq 0$ it is unlikely that $v(y_0, \cdot)$ will become larger than 1 for $t > t_0$. However, if $\phi''(y_0) > 0$, this does not seem to be impossible at all. This leads to the following result.

Theorem 2. *Let $\phi''(y_0) > 0$ for some y_0. Then there exist smooth initial functions $u_0(x)$ such that the solution $u(x,t)$, constructed in Theorem 1, becomes discontinuous in some point (x_0, t_0). If, on the other hand, $\phi'' \leq 0$ in $\mathbf{R}$, then for any smooth initial function the corresponding solution is smooth.*

Remark. The similarity solution which we have mentioned before in the case of constant ϕ is not in contradiction with Theorem 2, since in that case the initial function is not smooth. Actually one can prove that in the case that ϕ is constant, the solution becomes smooth after a finite time $t_0(u_0)$. Furthermore it turns out that

$$t_0(u_0) = 0 \text{ for any } u_0 \iff \int_0^\infty s\psi'(s)\, ds = \infty.$$

This integral condition is related to earlier results on mean-curvature type equations of both elliptic and parabolic type [9,11,13], but is of no value when ϕ is not constant.

Now we consider more closely what happens near a discontinuity of a solution. Suppose for example that at some time t_0 the solution $u(x, t_0)$ is discontinous at a point x_0:

$$u^+ = u(x_0+, t_0) > u^- = u(x_0-, t_0). \tag{8}$$

Hence

$$v(y, t_0) \geq 1 \text{ and } c(v(y, t_0)) = 0 \text{ for } u^- < y < u^+, \tag{9}$$

and the equation for v is elliptic for these values of y:

$$(\phi v)_{yy} = 0 \quad \text{for } u^- < y < u^+ \text{ and } t = t_0. \tag{10}$$

Assuming that the function v is continuous with respect to y, we know from (8) that

$$v(u^-, t_0) = v(u^+, t_0) = 1. \tag{11}$$

Hence we can solve $v(\cdot, t_0)$ explicitly from (10-11) in the interval $[u^-, u^+]$:

$$v(y, t_0) = \frac{1}{\phi(y)} \left(\frac{\phi(u^+) - \phi(u^-)}{u^+ - u^-}(y - u^-) + \phi(u^-) \right). \tag{12}$$

Combining this with (9), we find that necessarily

$$\phi(y) \leq \frac{\phi(u^+) - \phi(u^-)}{u^+ - u^-}(y - u^-) + \phi(u^-), \tag{13}$$

i.e., if the solution u, constructed in Theorem 1, has a discontinuity as in (8), then the function ϕ satisfies condition (13), which is nothing else than Oleinik's *entropy condition* for the first order equations (5).

As for the first order equations, the entropy condition turns out to be necessary for uniqueness, i.e., there are solutions of (1)-(4) which do not satisfy the entropy condition, and since all solutions in Theorem 1 do satisfy this condition, this implies nonuniqueness. Again the construction of the solutions which do not satisfy the entropy construction, is based on the use of the y variable [2].

In the case of first order conservation laws, much is known about the behaviour of the characteristics near a shock wave. So let us study the level curves of u in a neighbourhood of a discontinuity.

The speed of the level line is given by

$$-\frac{u_t}{u_x} = -(\phi v)_y. \tag{13}$$

As in the first order case, the speed of the shock wave is

$$-\frac{\phi(u^+) - \phi(u^-)}{u^+ - u^-},$$

which, in view of (12), is equal to

$$-(\phi v)_y(u^-, t) = -(\phi v)_y(u^+, t). \tag{14}$$

Hence, assuming that $(\phi v)_y$ is continuous in y, we conclude from (13) and (14) that the level lines of u arrive (or leave) the shock waves *tangentially*. In some sense this means that our parabolic problem is slightly more regular than the first order conservation laws.

Again the proof is given in [2], under some restrictions however on the function ψ, which come from the proof of the continuity of the function $(\phi v)_y$ with respect to y.

Open problems and conjectures

The first open problem is to prove the existence of a solution without the monotonicity condition on u_0. Observe that it would be sufficient to prove the continuity of the limit function $\Psi(x, t)$, which we guess to be true. As soon as we know that, the singular sets where $u_x = +\infty$ and the ones where $u_x = -\infty$ are decoupled, which allows us to use the transformation to the y-variable, and all the results of this paper carry over to the more general case. We hope to come back to this problem in the near future.

Completely open seems to be the case of severable space variables. Also in that case the major problem is the continuity of the function Ψ. However we have some doubts if Ψ is always continuous. If not, we have to modify the definition of a solution, i.e. we have to give a different meaning to the term $\psi(u_x)$ (which is certainly possible).

An intriguing question is whether the entropy condition is not only necessary, but also sufficient for uniqueness. Some preliminary calculations indicate that this is indeed the case.

Finally we mention that our techniques can be applied to the more general equation

$$u_t = (f(u, u_x))_x \, ,$$

where $f(u, p)$ is odd with respect to p and

$$f(u, s) \to \phi(s) \quad \text{as } s \to \infty.$$

The details will be worked out in a forthcoming paper [3].

Added in proof. Recently one of the authors (Dal Passo) has proved uniqueness in the class of (not necessarily monotone) solutions which satisfy the entropy condition [14].

REFERENCES

[1] H.W. Alt and S. Luckhaus, *Quasilinear elliptic-parabolic differential equations*, Math. Zeitschrift, **183** (1983), 311-341.

[2] M. Bertsch and R. Dal Passo, *Hyperbolic phenomena in a strongly degenerate parabolic equation*, to appear.

[3] M. Bertsch, R. Dal Passo and J. Hulshof, in preparation.

[4] Ph. Blanc, *Sur une classe d'équations paraboliques dégénerées a une dimension d'espace possédant des solutions discontinues*, Thesis University of Lausanne, 1989

[5] Ph. Blanc, *Existence de solutions pour des équations paraboliques*, Comptes Rendues Acad. Sciences Paris, **310** (1990), 53-56.

[6] D.S. Cohen, T.S. Hagan, R.L. Northcut and Ph. Rosenau, *Delayed diffusion due to flux limitation*, Physics Letters, (in press).

[7] E. Di Benedetto and R. Gariepy, *Local behaviour of solutions of an elliptic-parabolic equation*, Archive Rat. Mech. Anal., **97** (1987), 1-18.

[8] C.J. van Duyn and L.A. Peletier, *Nonstationary filtration in partially saturated porous media*, Archive Rat. Mech. Anal., **78** (1982), 173-198.

[9] B. Franchi, E. Lanconelli and J. Serrin, "Existence and uniqueness of ground state solutions of quasilinear elliptic equations", in: *Nonlinear Diffusion Equations and their Equilibrium States - I* (eds. W.M. Ni, L.A. Peletier and J. Serrin), Math. Sciences Research Inst. Publications, **12** (1988), Springer, 187-252.

[10] J. Hulshof, *An elliptic-parabolic problem: the interface*, Thesis University of Leiden, 1986.

[11] A.V. Ivanov, *Quasilinear degenerate and nonuniformly elliptic and parabolic equations of second order*, Proc. Steklov Inst. Math., (1984) Issue 1.

[12] Ph. Rosenau, *Free energy functionals at the high gradient limit*, Phys. Review A, to appear.

[13] J. Serrin, *The problem of Dirichlet for quasilinear elliptic differential equations with many independent variables*, Phil. Trans. Royal Soc. London, Sect. A, **262** (1969), 413-496.

[14] R. Dal Passo, *Uniqueness of the entropy solution of a strongly degenerate parabolic equation*, to appear.

Michiel Bertsch
Dipartimento di Matematica
Università di Roma "Tor Vergata"
Via Fontanile di Carcaricola
00133 Roma
Italy

Roberta Dal Passo
Istituto per le Applicazioni del Calcolo (IAC)
Viale del Policlinico 137
00161 Roma
Italy

Heat Flows and Relaxed Energies
for Harmonic Maps

FABRICE BETHUEL, JEAN-MICHEL CORON
JEAN-MICHEL GHIDAGLIA, ALAIN SOYEUR

Abstract

In this paper we construct weak solutions for the heat flow associated with relaxed energies for harmonic maps between B^3 and S^2. Nonuniqueness results for such solutions are also given.

0. Introduction

Recently very much attention has been drawn to the study of the heat flow for harmonic maps between Riemannian manifolds (see e.g. M. Struwe [19] and the references therein). The starting point was a result of Eells and Sampson [10] which was obtained by considering strong solutions for the heat flow of harmonic maps. Without restrictive hypotheses on the target manifold, these solutions may not exist (J.M. Coron and J.M. Ghidaglia [8], W.Y. Ding [9] and Y. Chen and D.W. Ding [5]). This has led to consider weak solutions as done first by M. Struwe [17], [18] and by Y. Chen and M. Struwe in [6]. However these solutions may not be unique as shown by J.M. Coron in [7].

Recall that the energy of a map u from $B^3 = \{x \in \mathbb{R}^3; \mid x \mid < 1\}$ into $S^2 = \partial B^3$ is

$$E_0(u) = \int \mid \nabla u \mid^2 = \int_{B^3} u^i_j u^i_j \, dx. \tag{0.1}$$

Harmonic maps are critical points of $E_0(u)$; they satisfy the Euler equation

$$-\Delta u = u \mid \nabla u \mid^2 \text{ in } B^3. \tag{0.2}$$

Conversely a map $u \in H^1(B^3; S^2)$ satisfying (0.2) in the distributional sense is termed *weakly harmonic*. Among the weakly harmonic maps in $H^1(B^3; S^2)$ there are the so-called E_0-*stationary maps* which correspond to variations on the domain. More precisely, u is E_0−stationary if u is weakly harmonic and for every smooth and compactly supported vector field $X : B^3 \to \mathbb{R}^3$,

$$\frac{d}{d\epsilon} E_0(u(\cdot + \epsilon X(\cdot)))\bigg|_{\epsilon=0} = 0. \tag{0.3}$$

One sees easily that (0.3) is equivalent to

$$2(u_i u_j)_i = (u_i u_i)_j \text{ for any } j \text{ in } \{1, 2, 3\}. \tag{0.4}$$

In [1], F. Bethuel, H. Brezis and J.M. Coron have introduced modified energies :

$$E_\lambda(u) = E_0(u) + 8\pi\lambda L(u) \tag{0.5}$$

where $\lambda \in [0, 1]$, and

$$L(u) = \operatorname*{Sup}_{\xi \in C_0^\infty(B^3, \mathbb{R}); |\nabla\xi|_{L^\infty} \leq 1} \int D(u) \cdot \nabla\xi \tag{0.6}$$

with $D(u) = (u \cdot (u_2 \times u_3), u \cdot (u_3 \times u_1), u \cdot (u_1 \times u_2))$. The scalar $L(u)$ has been defined by H. Brezis, J.M. Coron and E.H. Lieb in [3] where a more operational formula for computing this quantity is also given. As a consequence of this last formula $L(u)$ depends only on the singularities of the map u. It follows in particular that a critical point of E_λ satisfies (0.2) (see [1]). Hence the heat flow associated with E_λ is independent of λ and reads

$$\frac{\partial u}{\partial t} - \Delta u = u \mid \nabla u \mid^2 . \tag{0.7}$$

Equation (0.7) is complemented with boundary and initial conditions, namely

$$u(x, t) = \gamma(x), \ t \geq 0, \ x \in \partial B^3, \tag{0.8}$$
$$u(x, 0) = u^0(x), \ x \in B^3, \tag{0.9}$$

where u^0 and $\gamma \in H^1(B^3; \mathcal{S}^2)$ and $\gamma = u^0$ on ∂B^3.

Definition 0.1. Let u^0 and γ be given as above. A weak solution for the heat flow of E_λ is a map

$$u \in H^1(B^3 \times (0, T); \mathcal{S}^2), \ \forall T < \infty \tag{0.10}$$

satisfying (0.7) is the weak sense and (0.8)-(0.9) in the trace sense together with the inequalities

$$E_\lambda(u(t)) + \int_0^t \int \left|\frac{\partial u}{\partial t}\right|^2 \leq E_\lambda(u^0), \ \forall t \geq 0. \tag{0.11}$$

In this paper we prove the following results.

Theorem 1. *Let u^0, $\gamma \in H^1(B^3; \mathcal{S}^2)$ with $u^0 = \gamma$ on ∂B^3 and let $\lambda \in [0, 1]$. There exists a weak solution to (0.7)-(0.9).*

Theorem 2. ($\lambda = 0$). *Let u^0 and γ^0 as in Theorem 1. Assume that u^0 is weakly harmonic but not E_0-stationnary then (0.7)-(0.9) has infinitely many weak solutions.*

Finally Theorem 3 is the analogue of Theorem 2 in the cases $\lambda \neq 0$, see Section 3.

Remarks 0.1. a. Theorem 1 for $\lambda = 0$ was proved by Y. Chen in [4], by J. Keller, J. Rubinstein and P. Sternberg in [14] and by Y. Chen and M. Struwe in [6]. *b.* Theorem 2 was proved in [7] but the proof given here is more elementary.

1. Proof of Theorem 1

In order to construct a weak solution to (0.7) satisfying (0.8), (0.9) and (0.11) we proceed as K. Horihata and N. Kikuchi in [13]. For $h \in (0, 1)$, we define the sequence $(u^n)_{n \in \mathbb{N}}$ as follows. We assume that u^{n-1}, $n \geq 1$, is known and denote by $\mathcal{E}_n$ the functional :

$$\mathcal{E}_n(v) = E_\lambda(v) + \int \frac{|v - u^{n-1}|^2}{h}.$$

Then
$$u^n \text{ minimizes } \mathcal{E}_n(v) \text{ under the constraint } v \in H^1_\gamma = $$
$$= \{w : B^3 \to \mathcal{S}^2; \ w \in H^1(B^3, \mathcal{S}^2), \ w = \gamma \text{ on } \partial B^3\}. \tag{1.1}$$

Problem (1.1) possesses at least one solution (see [1]) which satisfies

$$\frac{u^n - u^{n-1}}{h} - \Delta u^n = \lambda^n u^n \tag{1.2}$$

where

$$\lambda^n = |\nabla u^n|^2 + \frac{1 - u^{n-1}.u^n}{h} = |\nabla u^n|^2 + \frac{|u^n - u^{n-1}|^2}{2h}.$$

Let us now construct two mappings $\bar{u}^h : B^3 \times [0, \infty) \to \mathcal{S}^2$ and $u^h : B^3 \times [0, \infty) \to B^3$ by setting for $(n-1)h < t \leq nh$:

$$\bar{u}^h(x, t) = u^n(x),$$
$$u^h(x, t) = \frac{t - (n-1)h}{h} u^n(x) + \frac{nh - t}{h} u^{n-1}(x).$$

With these notations, (1.2) reads

$$\frac{\partial u^h}{\partial t} - \Delta \bar{u}^h = (|\nabla \bar{u}^h|^2 + \frac{h}{2}\left|\frac{\partial u^h}{\partial t}\right|^2)\bar{u}^h. \tag{1.3}$$

Now since u^h solves (1.1), $\mathcal{E}_n(u^n) \le \mathcal{E}_n(u^{n-1})$:

$$E_\lambda(u^n) + \int \frac{|u^n - u^{n-1}|^2}{h} \le E_\lambda(u^{n-1}); \qquad (1.4)$$

adding these inequalities we obtain (as in [13])

$$E_\lambda(u^N) + \sum_{n=1}^{N} \int \frac{|u^n - u^{n-1}|^2}{h} \le E_\lambda(u^0). \qquad (1.5)$$

In terms of $\bar{u}^h$ and u^h, (1.5) reads

$$E_\lambda(\bar{u}^h(Nh)) + \int_0^{Nh} \int \left|\frac{\partial u^h}{\partial t}\right|^2 \le E_\lambda(u^0) \qquad (1.6)$$

which implies that

$$\begin{cases} \displaystyle\int_0^{+\infty} \int \left|\frac{\partial u^h}{\partial t}\right|^2 \le E_\lambda(u^0), \\[2mm] E_\lambda(\bar{u}^h(t)) \le E_\lambda(u^0), \ \forall t \ge 0. \end{cases} \qquad (1.7)$$

Since $L(v) \ge 0$ for every v, it follows from (1.7) that the family $(u^h)_{h>0}$ (resp. $(\bar{u}^h_i)_{h>0}$) is bounded in $H^1(B^3 \times (0,T))^3$ (resp. $L^2(B^3 \times (0,T))^3$ $\forall i \in \{1,2,3\}$), $\forall T < \infty$. Therefore, up to the extraction of subfamilies, we can assume that for every $T < \infty$,

$$\begin{cases} u^h \rightharpoonup u \text{ in } H^1(B^3 \times (0,T))^3, \\ u^h \longrightarrow u \text{ in } L^2(B^3 \times (0,T))^3 \text{ and a.e.,} \end{cases} \qquad (1.8)$$

and

$$\bar{u}^h_i \rightharpoonup \varphi_i \text{ in } L^2(B^3 \times (0,T))^3, \ \forall i \in \{1,2,3\}, \qquad (1.9)$$

where $\rightharpoonup$ stands for weak convergence, while $\longrightarrow$ stands for strong convergence.

Now our goal is to pass to the limit when $h \longrightarrow 0$ in (1.3). Proceeding as in [4], [14] and [15], we take the cross product of (1.3) with $\bar{u}^h$; this leads to

$$\frac{\partial u^h}{\partial t} \times \bar{u}^h - (\bar{u}^h_i \times \bar{u}^h)_i = 0. \qquad (1.10)$$

On the other hand, according to the definition of u^h and $\bar{u}^h$, we have $|u^h - \bar{u}^h| \le h \left|\frac{\partial u^h}{\partial t}\right|$, so that (see (1.7)):

$$\int_0^T \int |u^h - \bar{u}^h|^2 \le h^2 E_\lambda(u^0). \qquad (1.11)$$

Hence using (1.8) we find that for every $T < \infty$,

$$\bar{u}^h \longrightarrow u \text{ in } L^2(B^3 \times]0,T[)^3, \tag{1.12}$$

and therefore

$$\varphi_i = u_i. \tag{1.13}$$

It follows then from (1.8), (1.9), (1.10) and (1.13) that

$$\frac{\partial u}{\partial t} \times u - (u_i \times u)_i = 0, \tag{1.14}$$

and

$$| u | = 1 \text{ a.e.} \tag{1.15}$$

Now from (1.14) and (1.15) (see [4], [14], [15]) we deduce that

$$\frac{\partial u}{\partial t} - \Delta u = | \nabla u |^2 u,$$

and finally using (1.6)

$$E_\lambda(u(t)) + \int_0^t \int \left| \frac{\partial u}{\partial t} \right|^2 \leq E_\lambda(u^0).$$

This completes the proof of Theorem 1.

2. Proof of Theorem 2

We assume here that $\lambda = 0$. Let $u^0 : B^3 \to S^2$ be weakly harmonic $(-\Delta u^0 = | \nabla u^0 |^2 u^0)$ which is not E_0-stationnary. That means

$$\exists j_0 \in \{1,2,3\}, \; 2(u_i^0 u_{j_0}^0)_i \neq (u_i u_i)_{j_0}. \tag{2.1}$$

In order to prove Theorem 2, it is sufficient to show that the weak solution to (0.7)-(0.9) constructed in the previous section is not constant with respect to time (see [7]).

Let X be a smooth compactly supported vector field on B^3 and let u_ϵ^n be defined by $u_\epsilon^n(x) = u^n(x + \epsilon X(x))$. Since u^n minimizes $\mathcal{E}_n$ we have $\frac{d}{d\epsilon}\mathcal{E}_n(u_\epsilon^n)_{|\epsilon=0} = 0$. Hence for every $i \in \{1,2,3\}$ we have

$$2u_i^n \frac{u^n - u^{n-1}}{h} + (u_j^n u_j^n)_i - 2(u_j^n u_i^n)_j = 0.$$

These relations read also

$$2u_i^h \frac{\partial u_i^h}{\partial t} + (\bar{u}_i^h \bar{u}_j^h)_j - 2(\bar{u}_j^h \bar{u}_i^h)_j = 0. \tag{2.2}$$

Let us assume by contradiction

$$u(x,t) = u^0(x), \ \forall x \in B^3, \ \forall t \geq 0. \tag{2.3}$$

We claim that

$$\begin{cases} \nabla \bar{u}^h \longrightarrow \nabla u \ \text{in} \ L^2(B^3 \times (0,T))^9, \\ \dfrac{\partial u^h}{\partial t} \longrightarrow 0 \left(= \dfrac{\partial u}{\partial t} \right) \ \text{in} \ L^2(B^3 \times (0,T))^3. \end{cases} \tag{2.4}$$

We postpone the proof of (2.4) and we complete the proof of Theorem 2. Passing to the limit in (2.2) we obtain

$$(u_j^0 u_j^0)_i = 2(u_j^0 u_i^0)_j,$$

which contradicts (2.1) and proves that (2.3) cannot hold true.

It remains to show (2.4). According to (1.6) we have, for $t = nh$,

$$E_0(\bar{u}^h(t)) + \int_0^t \int \left| \frac{\partial u^h}{\partial t} \right|^2 \leq E_0(u^0).$$

Since the left-hand side of this inequality is lower semicontinuous, we deduce, for every t,

$$E_0(u(t)) + \int_0^t \int \left| \frac{\partial u}{\partial t} \right|^2 \leq E_0(u^0). \tag{2.5}$$

But according to (2.3), (2.5) is an equality and this proves (2.4).

Remark 1.1. Let $\omega : S^2 \simeq \mathbf{C} \cup \{\infty\} \to S^2 \simeq \mathbf{C} \cup \{\infty\}$ be a rational fraction of $z \in \mathbf{C}$. The function $u^0 : u^0(x) = \omega(x/\mid x \mid)$ is weakly harmonic and (2.1) is equivalent to

$$\mathrm{I\!R}^3 \ni \int_{S^2} \mid \nabla \omega \mid^2 \vec{s} d\sigma(\vec{s}) \neq \vec{O}.$$

3. The case $\lambda \in \,]0,1[$

In order to define E_λ-stationary mappings, we define, for u in $H^1(B^3; S^2)$, the set $C(u)$ (see [11], [12]),

$$C(u) = \{\text{one dimensional rectifiable current } c = \tau(\mathcal{C}, \theta, \zeta)$$
$$\text{such that the mass } M(c) \text{ of } c \text{ is equal to}$$
$$L(u) \text{ and } -4\pi \partial c = \mathit{div} \ D(u) \text{ in } \mathcal{D}'(B^3)\}.$$

Theorem 1 in [12] asserts that $C(u) \neq \emptyset$. We now may define E_λ-stationary mappings.

Definition 3.1. We say that u is E_λ-stationary if there exists $c \in C(u)$ such that for every $X \in \mathcal{C}_0^\infty(B^3, \mathbb{R}^3)$

$$\int (\frac{1}{2} \mid \nabla u \mid^2 X_i^i - u_i u_j X_j^i) dx +$$
$$+ 4\pi\lambda \int \zeta^i \zeta^j X_j^i \theta d\mathcal{H}^1 = 0. \tag{3.1}$$

The following result is a generalization of Theorem 2.

Theorem 3. *Let $u^0 \in H^1(B^3, \mathcal{S}^2)$ be weakly harmonic. If u^0 is not E_λ-stationary ($\lambda \in]0,1[$) then there exist infinitely many weak solutions for the flow associated to E_λ.*

Proof of Theorem 3. As for Theorem 2, it is sufficient to show that (2.3) does not hold true. With the notations of the previous sections, there exists a sequence of currents (c^n) with $c^n = \tau(\mathcal{C}^n, \theta^n, \zeta^n) \in C(u^n)$ such that for any $X \in \mathcal{C}_0^\infty(B^3; \mathbb{R}^3)$:

$$2 \int \frac{u^n - u^{n-1}}{h} u_i^n X^i + \int \mid \nabla u^n \mid^2 X_i^i - 2 \int u_i^n u_j^n X_j^i dx$$
$$+ 8\pi\lambda \int \zeta_n^i \zeta_n^i X_j^i \theta^n d\mathcal{H}^1 = 0. \tag{3.2}$$

The proof proceeds again by contradiction : we assume that $u(t) = u^0$ i.e. we assume (2.3). We claim that (2.4) holds true again for $\lambda \in (0,1)$. The proof follows the outlines of the corresponding proof in Section 2. We need the following

Lemma 3.1. *Let λ be fixed in $[0,1)$. If v_n weakly converges to v in H^1 and $E_\lambda(v_n)$ converge to $E_\lambda(v)$ then v_n converges strongly in H^1.*

Proof of Lemma 3.1. It is sufficient to show that $E_0(v_n)$ converges to $E_0(v)$. This follows from the fact that E_1 is weakly lower continuous on H^1 (see [1]) and the identity :

$$E_\lambda(v_n) = (1 - \lambda)E_0(v_n) + \lambda E_1(v_n).$$

We now complete the proof of Theorem 3. Let us pass to the limit in (3.2). Let t_0 be in $(0,\infty)$ and h be a sequence converging to zero such that

$$\nabla \bar{u}^h(t_0) \longrightarrow \nabla u^0 \text{ in } L^2(B^3)^9, \tag{3.3}$$
$$\bar{u}^h(t_0) \longrightarrow u^0 \text{ a.e. in } B^3, \tag{3.4}$$
$$\frac{\partial u^h}{\partial t}(t_0) \longrightarrow 0 \text{ in } L^2(B^3)^3. \tag{3.5}$$

For sake of simplicity, we write u^h instead of $\bar{u}^h(t_0)$ and c^h instead of c^n (where $(n-1)h \leq t_0 < nh$). We also denote by $\mathcal{D}_1(B^3)$ the set of 1-dimensional currents on B^3 equipped with the weak topology. Let us prove that, for a subsequence of the c^h, there exists a 1-dimensional rectifiable current c on B^3 such that

$$c^h \rightharpoonup c \text{ in } \mathcal{D}_1(B^3). \tag{3.6}$$

We recall that for v and w in $H^1(B^3; \mathcal{S}^2)$ the scalar $L(v, w)$ is defined as follows (see [1])

$$L(v, w) = \sup_{\substack{\xi \in C_0^\infty(B^3, \mathbb{R}) \\ |\nabla \xi|_{L^\infty} \leq 1}} \int (D(v) - D(w)) \cdot \nabla \xi.$$

According to [1], there exists a constant A such that

$$L(v, w) \leq A \parallel \nabla v - \nabla w \parallel_{L^2} (\parallel \nabla v \parallel_{L^2} + \parallel \nabla w \parallel_{L^2}),$$
$$\forall v, w \in H^1(B^3, \mathcal{S}^2). \tag{3.7}$$

Let $C(v, w)$ be the set of 1-dimensional rectifiable currents c such that

$$4\pi \partial c = div(D(v) - D(w)) \text{ and } M(c) = L(v, w).$$

It is shown in [12] that $C(v, w) \neq \emptyset$ (in [12] it is assumed that either v or w has a finite number of singular points, however this assumption can be removed using theorem 4bis in [2]). Let $\tilde{c}^h$ be given in $C(\bar{u}^h, u^0)$ and let $c_0 \in C(u)$. We set $\bar{c}^h = c^h + \tilde{c}^h - c_0$ and compute :

$$-4\pi \partial \bar{c}^h = \operatorname{div} D(\bar{u}^h) - \operatorname{div} D(\bar{u}^h)$$
$$+ \operatorname{div} D(u^0) - \operatorname{div} D(u^0) = 0, \tag{3.8}$$

$$M(\bar{c}^h) \leq L(\bar{u}^h) + L(\bar{u}^h, u^0) + L(u^0)$$
$$\leq 2L(u^0) + 2L(\bar{u}^h, u^0). \tag{3.9}$$

Using (3.3), (3.7) and (3.9) we infer that the sequence $M(\bar{c}^h)$ is bounded which, combined with (3.8) and a classical compactness theorem due to Federer and Fleming (see [16, Theorem 27.3]), implies that (for a subsequence) there exists a 1-dimensional integral current $\bar{c}$ such that

$$\bar{c}^h \rightharpoonup \bar{c} \text{ in } \mathcal{D}_1(B^3). \tag{3.10}$$

On the other hand, using (3.3) and (3.7), we have

$$M(\bar{c}^h) = L(\bar{u}^h, u^0) \to 0. \tag{3.11}$$

We set $c = \bar{c} + c_0$. Clearly, c is a 1-dimensional rectifiable current which, thanks to (3.10) and (3.11), satisfies (3.6). By weak lower semicontinuity of M, (3.3), (3.6) and (3.7)

$$M(c) \leq \lim \inf M(c^h) = L(u^0) + \lim \inf (L(\bar{u}^h) - L(u^0))$$
$$\leq L(u^0) + \lim \inf L(u^0, \bar{u}^h) = L(u^0). \tag{3.12}$$

Since $\bar{u}^h \to u^0$ in $H^1(B^3, \mathcal{S}^2)$ we have

$$\mathrm{div} D(\bar{u}^h) \rightharpoonup \mathrm{div} D(u^0) \text{ in } \mathcal{D}'(B^3). \tag{3.13}$$

From (3.13) and (3.6), we deduce

$$-4\pi \partial c = \mathrm{div}\ D(u^0). \tag{3.14}$$

Next, using (3.14) and (3.12) we see that

$$c \in C(u^0). \tag{3.15}$$

We write $c = \tau(\mathcal{C}, \theta, \zeta)$. In order to complete the proof of Theorem 3, it suffices to check that u^0 is E_λ-stationary, which follows from (3.2), (3.3), (3.4), (3.5) and

$$\zeta_i^h \zeta_j^h \theta^h d\mathcal{H}^1 \rightharpoonup \zeta_i \zeta_j \theta d\mathcal{H}^1 \text{ in } \mathcal{D}'(B^3). \tag{3.16}$$

We now prove (3.16). Since $M(c^h)$ is bounded we have (see (3.6))

$$\zeta_i^h \theta^h d\mathcal{H}^1 \rightharpoonup \zeta_i \theta d\mathcal{H}^1 \text{ in } \mathcal{M}(B^3) \tag{3.17}$$

where $\mathcal{M}(B^3)$ denotes the set of Radon measures on B^3 endowed with its weak topology.

Using the fact that $M(c^h)$ is bounded, we may assume that

$$\theta^h d\mathcal{H}^1 \rightharpoonup \theta^* d\mathcal{H}^1 \text{ in } \mathcal{M}(B^3). \tag{3.18}$$

From (3.17) and (3.18) we obtain

$$\theta d\mathcal{H}^1 \leq \theta^* d\mathcal{H}^1. \tag{3.19}$$

On the other hand (3.12) and (3.15) imply

$$M(\theta^* d\mathcal{H}^1) \leq \lim \inf M(\theta^h d\mathcal{H}^1) = M(\theta d\mathcal{H}^1)$$

which, combined with (3.19) gives

$$\theta d\mathcal{H}^1 = \theta^* d\mathcal{H}^1. \tag{3.20}$$

Since $\mid \zeta_i^h \mid \leq 1$, there exists a borelian function ρ_i such that

$$(\zeta_i^h)^2 \theta^h d\mathcal{H}^1 \rightharpoonup \rho_i \theta d\mathcal{H}^1 \text{ in } \mathcal{M}(B^3). \tag{3.21}$$

From (3.21) and (3.17) we obtain easily

$$\zeta_i^2 \theta d\mathcal{H}^1 \leq \rho_i \theta d\mathcal{H}^1. \tag{3.22}$$

Since $\zeta_i^h \zeta_i^h \theta^h d\mathcal{H}^1 = \theta^h d\mathcal{H}^1$, we have using (3.21) and (3.22) (for $i = 1, 2, 3$)

$$(\zeta_i^h)^2 \theta^h d\mathcal{H}^1 \rightharpoonup \rho_i \theta d\mathcal{H}^1 = \zeta_i^2 \theta d\mathcal{H}^1 \text{ in } \mathcal{M}(B^3). \tag{3.23}$$

Proceeding as in the proof of (3.23), we have for $i = 1, 2, 3$,

$$(R_{ij}\zeta_j^h)^2 \theta^h d\mathcal{H}^1 \rightharpoonup (R_{ij}\zeta_j)^2 \theta d\mathcal{H}^1 \text{ in } \mathcal{M}(B^3)$$
$$\text{for any } R = (R_{ij}) \text{ in } SO(3). \tag{3.24}$$

Finally (3.16) readily follows from (3.24).

Acknowledgements. We thank M. Struwe for having drawn our attention to [13].

REFERENCES

[1] F. Bethuel, H. Brezis, J.M. Coron, *Relaxed energies for harmonic maps*, Variational Problems, Paris June 1988, H. Berestycki, J.M. Coron, I. Ekeland Eds, Birkhäuser.

[2] F. Bethuel, X. Zheng, *Density of smooth functions between two manifolds in Sobolev spaces*, J. Func. Anal. **80**(1988), 60–75.

[3] H. Brezis, J.M. Coron, E.H. Lieb, *Harmonic maps with defects*, Comm. Math. Phys. **107** (1986), 649–705.

[4] Y. Chen, *Weak solutions to the evolution problem of harmonic maps*, Math. Z. **201** (1989), 69–74.

[5] Y. Chen and W.Y. Ding, *Blow-up and global existence for heat flows for harmonic maps*, Inventiones Math. **99** (1990), 567–578.

[6] Y. Chen and M. Struwe, *Existence and partial regularity result for the heat flow for harmonic maps*, Math. Z. **201** (1989), 83–103.

[7] J.M. Coron, *Nonuniqueness for the heat flow of harmonic maps*, Annales IHP, Analyse Non Linéaire, to appear.

[8] J.M. Coron and J.M. Ghidaglia, *Explosion en temps fini pour le flot des applications harmoniques*, C.R. Acad. Sci. Paris **308** (1989), 339–344.

[9] W.Y. Ding, *Blow-up of solutions of heat flows for harmonic maps*, preprint.

[10] J. Eells and J.H. Sampson, *Harmonic mappings of Riemannian manifolds*, Amer. J. Math. **86** (1964), 109–160.

[11] M. Giaquinta, G. Modica, J. Soucek, *Cartesian currents and variational problems for mappings into spheres*, preprints, 1989.

[12] M. Giaquinta, G. Modica, J. Soucek, *The Dirichlet energy of mappings with values into the sphere*, preprint, 1989.

[13] K. Horihata and N. Kikuchi, *A construction of solutions satisfying a Caccioppoli inequality for nonlinear parabolic equations associated to a variational functional of harmonic type*, preprint.

[14] J. Keller, J. Rubinstein, P. Sternberg, *Reaction-diffusion processes and evolution to harmonic maps*, preprint.

[15] J. Shatah, *Weak solutions and development of singularities of the $SU(2)$ σ-model*, Comm. Pure Appl. Math. **41** (1988), 459–469.

[16] L. Simon, *Lectures on geometric measure theory*, Proc. of the Centre for Mathematical Analysis, Australian National University **3** (1983).

[17] M. Struwe, *On the evolution of harmonic maps of Riemannian surfaces*, Comment. Math. Helv. **60** (1985), 558–581.

[18] M. Struwe, *On the evolution of harmonic maps in higher dimensions*, J. Diff. Geom. **28** (1988), 485–502.

[19] M. Struwe, *The evolution of harmonic maps : existence, partial regularity and singularities*, this volume.

Fabrice Bethuel
Lab. de Mathématiques et Modélisation
CERMA-ENPC, La Courtine
93167 Noisy le Grand Cedex
France

Jean-Michel Ghidaglia
Lab. de Mathématiques et Modélisation
CMLA-ENS, 61 Av. du Président Wilson
94235 Cachan Cedex
France

Jean-Michel Coron
Lab. d'Analyse Numérique
CNRS et Univ. Paris-Sud,
Bât. 425
91405 Orsay Cedex, France

Alain Soyeur
Lab. d'Analyse Numérique
CNRS et Univ. Paris-Sud
Bât. 425
91405 Orsay Cedex, France

Local Existence and Uniqueness
of Positive Solutions of the Equation
$$\Delta u + (1 + \varepsilon\varphi(r))u^{\frac{n+2}{n-2}} = 0,\ \text{in}\ \mathbb{R}^n$$
and a Related Equation

GABRIELLE BIANCHI[1] and HENRIK EGNELL[2]

0. Introduction

We will consider radial solutions of the following problem:

$$(0.1) \qquad \begin{cases} \Delta u + (1 + \varepsilon\varphi)u^p = 0, & \text{in } \mathbf{R}^n \\ u > 0,\ \text{in } \mathbf{R}^n,\quad \text{and}\ u(x) = O(|x|^{2-n}),\ \text{as}\ x \to \infty \end{cases} ,$$

where $p = \frac{n+2}{n-2}$, φ is a continuous radial function and ε is a small real parameter. This equation and its generalizations have been studied in a large numbers of papers (see for example [BE], [DN], [LN], [NI1&2]).

Transforming (0.1) to a problem on the sphere $\mathbf{S}^n$, using the stereographic projection one can show that (0.1) becomes

$$(0.1') \qquad \Delta_{\mathbf{S}^n} v - \frac{(n-2)n}{4}v + (1 + \varepsilon\tilde{\varphi})v^p = 0,\ v > 0 \quad \text{on } \mathbf{S}^n ,$$

where $\Delta_{\mathbf{S}^n}$ denotes the Laplace Beltrami operator on $\mathbf{S}^n$. The regularity of $\tilde{\varphi}$ at the north pole is the same as that of $\varphi(\frac{x}{|x|^2})$ at the origin. In problem (0.1') it is often assumed that $\tilde{\varphi}$ is continuous or is in $C^\infty(\mathbf{S}^n)$, while in problem (0.1), we do not impose any regularity at infinity (although below we will impose some growth conditions at infinity). Also a solution of (0.1) with slower decay than the prescribed r^{2-n}, will after the transformation to (0.1') be singular at the northpole.

On the sphere radial functions correspond to those invariant under rotation around the axis through the north-south poles. Thus they only depend on one variable, the angle θ. The correspondence between r and θ is given by $r = \frac{\sin(\theta)}{1-\cos(\theta)}$, $0 < \theta \le \pi$. Restricted to this class of functions we have $\Delta_{\mathbf{S}^n} = \partial_\theta^2 + \frac{n-1}{\tan(\theta)}\partial_\theta$.

Problem (0.1'), with $K = 1 + \varepsilon\tilde{\varphi}$ a general function on the sphere $\mathbf{S}^n$, is called the Kazdan-Warner problem [KW] and is of interest in Differential

---

[1]Supported by a CNR fellowship
[2]Supported in part by NSF Grant DMS-8914778

Geometry. It has been studied in for example [BC], [CY] and [ES]. In Section 3 we will reformulate our result for problem (0.1), to problem (0.1').

If $\varepsilon = 0$, then $U_\lambda(r) = \lambda\left(1 + \frac{\lambda^{p-1}}{n(n-2)}r^2\right)^{\frac{2-n}{2}}$ with $\lambda > 0$, are the only radial solutions of (0.1). This follows from the uniqueness of the corresponding ODE initial value problem starting at 0 (e.g. (1.1) with $\varepsilon = 0$).

If $\varepsilon \neq 0$, then by the Pohozaev identity, a necessary condition for u to be a solution of (0.1) is that (e.g. [DN] Lemma 3.7)

$$(0.2) \qquad \mathcal{A}(u) = \int_0^\infty r\varphi'(r)u^{p+1}\, r^{n-1}\, dr = 0 \, .$$

Thus, if φ is not a constant and either $\varphi' \geq 0$ or $\varphi' \leq 0$, then there are no solutions if ε is nonzero. On the other hand if for example $\varphi(0) = \varphi(\infty)$, then there is always a solution of (0.1) for small ε (cf [BE]).

Before we proceed let us introduce some notation. First we define a function a on the positive real line by $a(\lambda) = \mathcal{A}(U_\lambda)$. The following space of functions will be important $L^\infty_{U_\lambda^{-1}} = \{\, u \,:\, \|u\| = \sup |uU_\lambda^{-1}| < \infty \,\}$. Throughout this paper $\|\cdot\|$ will denote the norm of $L^\infty_{U_\lambda^{-1}}$. Note that for any positive λ_1 and λ_2 we have $L^\infty_{U_{\lambda_1}^{-1}} = L^\infty_{U_{\lambda_2}^{-1}}$. Let $B_R(v) = \{\, u \,:\, \|u-v\| < R \,\}$ be a ball centered at v with radius R.

Assume that $\{u_\varepsilon\}$ is a family of solutions of (0.1) for small ε, that $r\varphi'$ is bounded and that $u_\varepsilon \to U \neq 0$, in $\mathcal{R}_0^{1,2}$, as $\varepsilon \to 0$. Then U solves (0.1) with $\varepsilon = 0$, and thus by uniqueness $U = U_{\lambda_0}$ for some positive λ_0. Furthermore, we must have $a(\lambda_0) = \mathcal{A}(U_{\lambda_0}) = \lim_{\varepsilon\to 0} \mathcal{A}(u_\varepsilon) = 0$.

Here $\mathcal{R}_0^{1,2}$ is the completion of the set of smooth radial functions with compact support in the norm $\|\nabla \cdot\|_2$. The Sobolev imbedding Theorem states that the imbedding $\mathcal{R}_0^{1,2} \hookrightarrow L^{p+1}$ is continuous (remember $p = \frac{n+2}{n-2}$).

In Theorem I in the present paper we will prove a converse of this observation. However we need an additional "transversality condition".

Theorem I. *Assume that $\varphi \in C^1$ and that $\sup_{1\leq r<\infty} |r^{\delta-1}\varphi'(r)| < \infty$ for some $\delta > 0$.*
If $a(\lambda_0) = 0$ and $a'(\lambda_0) \neq 0$, then there are $R > 0$ and $\varepsilon_0 > 0$ such that (0.1) has a unique branch $\{u_\varepsilon\}$, $0 < |\varepsilon| < \varepsilon_0$ of solutions in $B_R(U_{\lambda_0})$, such that $u_\varepsilon \to u_0 = U_{\lambda_0}$, as $\varepsilon \to 0$.
Furthermore, this branch is continuous in $L^\infty_{U_{\lambda_0}^{-1}}$, and thus also in $\mathcal{R}_0^{1,2} \cap C^2$.

Note that the conditions in the theorem are such that we can allow $\varphi \sim \pm r^\alpha$ with $\alpha \in [0,2)$, as $r \to \infty$. Thus we can get results in cases where direct variational methods do not work.

Using the same technique as in the proof of Theorem I we obtain a similar type of result for the problem of finding radial solutions of

$$(0.3) \qquad \begin{cases} \Delta u + u^p + \varepsilon\varphi = 0\,, & \text{in } \mathbf{R}^n \\[2mm] u > 0\,, \text{ in } \mathbf{R}^n \quad \text{and } u(x) = O(|x|^{2-n})\,, \text{ as } x \to \infty \, . \end{cases}$$

Here p is the same exponent as above, i.e. $p = \frac{n+2}{n-2}$.

In [EK], it was proved that if $\varepsilon\varphi \geq 0$, and φ satisfies some regularity conditions, then (0.3) has at least two solutions for small ε. One of these solutions comes from a branch of solutions containing the trivial solution $u \equiv 0$ and $\varepsilon = 0$. While the other is a "bifurcation" (in some weak sense) from a solution U_{λ_0}.

Again using a Pohozaev type identity, it follows that under some regularity conditions on φ, a necessary condition for u to solve (0.3) is that

$$(0.4) \qquad \mathcal{B}(u) = \int_0^\infty u r^{\frac{n-2}{2}} (r^{\frac{n+2}{2}} \varphi)' \, dr = 0 \ .$$

Repeating the argument above one finds that a necessary condition to have a solution set $\{u_\varepsilon\}$ such that $u_\varepsilon \to U \neq 0$ in $\mathcal{R}_0^{1,2}$, as $\varepsilon \to 0$, is that the function $b(\lambda) = \mathcal{B}(U_\lambda)$ has a zero.

Theorem II. *Assume that $\varphi \in C^1$ and $|\varphi(r)| + |r\varphi'(r)| = O(r^{-n-\delta})$, as $r \to \infty$, for some $\delta > 0$.*
If $b(\lambda_0) = 0$ and $b'(\lambda_0) \neq 0$, then there are $R > 0$ and $\varepsilon_0 > 0$, such that (0.3) has a unique branch of solutions u_ε, $0 < |\varepsilon| < \varepsilon_0$ in $B_R(U_{\lambda_0})$, such that $u_\varepsilon \to u_0 = U_{\lambda_0}$, as $\varepsilon \to 0$.
Furthermore, this branch is continuous in $L^\infty_{U^{-1}_{\lambda_0}}$, and thus also in $\mathcal{R}_0^{1,2} \cap C^2$.

The proof of this theorem is very similar to that of Theorem I. We will give some details in Section 2.

When writing up the manuscript we learned that Prof. W.-Y. Ding has similar results which also covers nonradial situations [DI]. He proved the result for the problem on the sphere (0.3). His approach is different and we believe better. His idea is to use a Liapunov-Schmidt reduction. Then we can solve the bifurcation equation provided φ satisfies $n + 1$ necessary conditions. In the radial case n of these conditions are automatically satisfied. Since Prof. Ding's results are not published we have included them here in Section 3. We give the proof of the result in $\mathbf{R}^n$.

However, since our method is different and perhaps gives more insight into the behaviour of the solution, we believe it is still of some interest. It is especially useful when constructing concrete examples.

One of the main features of the perturbation results in the present paper, is that it gives some insight into the difficult question of understanding the relation of the function $K = 1 + \varepsilon\varphi$ and the number of solutions of (0.1).

For example, in Section 3, we will construct a smooth nontrivial function φ , tending to zero at infinity, so that the lower bound of the number of solutions of (0.1) tends to infinity as $\varepsilon \to 0$. However it is not clear that we can find a fixed $\varepsilon \neq 0$ for which (0.1) has infinitely many solutions. Transformed to the sphere this example gives a function $\tilde{\varphi}$ which is zero at the north pole and can be choosen to be flat of any order less than $n - 1$, at the north pole and smooth elsewhere.

Also we will study the relation between the growth of φ at infinity and the existence of a solution of (0.1).

Acknowledgement. The authors wish to thank Prof. W.-Y. Ding for communicating his results, presented in Section 3 and for a stimulating discussion. We also like to acknowledge Grant mg27201 from Minnesota Supercomputer Institute. During our research we have used numerical experiments to guide our analytical efforts.

1. Proof of Theorem I. The result will be proved using the contraction mapping principle. Throughout this section we will assume that φ satisfies the hypotesis in Theorem I.

Before we proceed let us introduce the set of functions

$$\mathbf{X}_\lambda(d) = \{u \; : \; \frac{1}{d}U_\lambda \leq u \leq dU_\lambda \,, \text{ and } \mathcal{A}(u) = 0 \,\} \,,$$

which is a closed subset of $L^\infty_{U_\lambda^{-1}}$. Here $\mathcal{A}$ is defined as in (0.2), we also recall that $a(\lambda) = \mathcal{A}(U_\lambda)$.

Consider the initial value problem corresponding to (0.1)

$$(1.1) \qquad \begin{cases} u'' + \dfrac{n-1}{r}u' + (1 + \varepsilon\varphi)u^p = 0 \,, \\ u(0) = \lambda \,, \qquad u'(0) = 0 \,. \end{cases}$$

If we assume that φ has compact support, then a solution of this initial value problem can behave in three different ways, depending on $\varepsilon\varphi$ and λ, as r increase. It can have a finite zero, it can decay as $r^{\frac{2-n}{2}}$ or as r^{2-n}. We are interested in finding the λ's for which (0.1) has a global positive solution with fast decay (i.e. $\sim r^{2-n}$).

A solution of (1.1) has a useful representation formula.

Lemma 1. *If u solves (1.1) and $u > 0$ in $[0,r]$, then it satisfies the identity*

$$u(r) = \cfrac{\lambda}{\left(1 + (\frac{\lambda^{p-1}}{n(n-2)} + \frac{\varepsilon\varphi(0)\lambda^{p-1}}{n(n-2)} + \lambda^{\frac{p-1}{2}}G(r,u))r^2\right)^{\frac{n-2}{2}}} \,, \qquad \text{where}$$

$$G(r,u) = \frac{2\varepsilon}{n-2}\int_0^r s^{-3}\int_0^s u^{-\frac{p+3}{2}}t^{1-n}\int_0^t x\varphi'u^{p+1}x^{n-1}\,dx\,dt\,ds \,.$$

Furthermore, if u is continuous and satisfies the integral equation in some interval $[0,r_0]$, then u solves (1.1) in the same interval.

Proof. We will not prove the first part of the Lemma here. However, the proof is almost identical to that of Lemma 1' given in Section 2. The idea of the proof of Lemma 1 comes from a paper by Atkinson and Peletier [AP], their idea was also used in [EK].

Next we will prove that a solution of the integral equation is a solution of (1.1). Since a solution u of (1.1) is unique and satisfies the integral equation we only need to prove that u is the only solution of the integral equation.

Assume that $v \not\equiv u$ is another solution of the integral equation. Let $[0, r_0]$ be the maximal interval on which u and v coincide. Note that $u(0) = v(0) = \lambda$ and both functions are continuous so the interval must be closed.

Take δ positive and small. Then using the fact that u and v solves the integral equation we get

$$\sup_{r\in[0, r_0+\delta]} |u(r) - v(r)| \le \delta C \sup_{r\in[0, r_0+\delta]} |u(r) - v(r)| \,.$$

Here C only depends on the supremum norm (on $[0, r_0 + \delta]$) of u, v, u^{-1} and v^{-1}, λ and the dimension. Thus choosing δ small enough we find that u and v coincide on $[0, r_0+\delta]$. Therefore, we conclude that u and v coincide on the domain where they exist.

Note that the integral equation holds only as long as u stays positive. When u tends to zero $G(r, u)$ tends to infinity and the formula breaks down. It can happen that u has a finite zero. $\qquad\square$

Lemma 2. *If $u \in \mathbf{X}_\lambda(d)$, then $|G(r, u(r))| < C|\epsilon|$, uniformly in $r > 0$, where C only depends on n, d, λ and φ. Here δ is the number given in Theorem I.*

Proof of Lemma 2. In the proof we use the fact that $\frac{1}{d}U_\lambda \le u \le dU_\lambda$ repeatedly.

If $t \le 1$ we obtain

$$\left| \int_0^t x\varphi' u^{p+1} x^{n-1} \, dx \right| \le C \int_0^t x^n \, dx \le C t^{n+1} \,.$$

Since $\mathcal{A}(u) = 0$, we get for $t \ge 1$

$$\left| \int_0^t \cdots \right| = \left| \int_t^\infty x^{\delta-1}\varphi' u^{p+1} x^{n+1-\delta} \, dx \right| \le C \int_t^\infty x^{1-\delta-n} \le C t^{2-n-\delta} \,.$$

Combining the last two estimates yields

$$\left| \int_0^t x\varphi' u^{p+1} x^{n-1} \, dx \right| \le C \min(t^{2-n-\delta}, t^{n+1}) \,.$$

From this we get the estimate

$$\left| \int_0^s u^{-\frac{p+3}{2}} t^{1-n} \int_0^t x\varphi' u^{p+1} x^{n-1} \, dx \, dt \right| \le C \min(s^{2-\delta}, s^3) \,.$$

Thus, finally we obtain

$$|G(r,u)| \leq |\varepsilon| C \int_0^r \min(s^{-\delta-1}, 1)\, ds \leq C|\varepsilon| . \qquad \square$$

For λ positive and a function z with small supremum norm we define the function

$$\mathcal{F}(z,\lambda) = \mathcal{A}\left(\frac{\lambda}{\left(1 + (\frac{\lambda^{p-1}}{n(n-2)} + z(r))r^2\right)^{\frac{n-2}{2}}} \right) .$$

Lemma 3. *If $a(\lambda_0) = 0$ and $a'(\lambda_0) \neq 0$, then there is a positive δ such that we can solve $\mathcal{F}(z,\lambda) = 0$ for a unique C^1 function $\lambda : B_\delta \to \mathbf{R}$, in $B_\delta \times (\lambda_0 - \delta, \lambda_0 + \delta)$, where $B_\delta = \{ z \in L^\infty : \|z\|_\infty < \delta \}$.*

Proof. This is a direct consequence of the implicit function theorem applied to the C^1 function

$$\mathcal{F} : B_{\delta'} \times [\rho, \infty) \to \mathbf{R} ,$$

for $\rho = \rho(\delta') > 0$ small. By the assumptions $\mathcal{F}(0, \lambda_0) = a(\lambda_0) = 0$ and $\mathcal{F}'_\lambda(0, \lambda_0) = a'(\lambda_0) \neq 0$, and the claim follows. $\qquad \square$

We will find $\varepsilon_0 > 0$ and $d_0 > 1$, so that we can construct a family of mappings $\mathcal{T}_\varepsilon : \mathbf{X}_{\lambda_0}(d_0) \to \mathbf{X}_{\lambda_0}(d_0)$, for all $|\varepsilon| < \varepsilon_0$.

The mapping $\mathcal{T}_\varepsilon$ is constructed as follows. Take $u \in \mathbf{X}_{\lambda_0}(d_0)$ and define

$$u_1(r) = \frac{\lambda}{\left(1 + (\frac{\lambda^{p-1}}{n(n-2)} + \frac{\varepsilon\varphi(0)\lambda^{p-1}}{n(n-2)} + \lambda^{\frac{p-1}{2}} G(r,u))r^2\right)^{\frac{n-2}{2}}} ,$$

where $\lambda = u(0) \in [\frac{1}{d_0}\lambda_0, d_0\lambda_0]$. If $|\varepsilon|$ is small enough, this is possible by Lemma 2.

Furthermore by Lemma 2 and Lemma 3, if $|\varepsilon| > 0$ and $d_0 > 1$ are small enough we can find a unique λ_1 near λ_0, such that

$$\mathcal{A}\left(\frac{\lambda_1}{\left(1 + (\frac{\lambda_1^{p-1}}{n(n-2)} + \frac{\varepsilon\varphi(0)\lambda^{p-1}}{n(n-2)} + \lambda^{\frac{p-1}{2}} G(r,u))r^2\right)^{\frac{n-2}{2}}} \right) = 0 .$$

Now we define

$$\mathcal{T}_\varepsilon(u) = \frac{\lambda_1}{\left(1 + (\frac{\lambda_1^{p-1}}{n(n-2)} + \frac{\varepsilon\varphi(0)\lambda^{p-1}}{n(n-2)} + \lambda^{\frac{p-1}{2}} G(r,u))r^2\right)^{\frac{n-2}{2}}} .$$

If ε is small enough we get $\mathcal{T}_\varepsilon(\mathbf{X}_{\lambda_0}(d_0)) \subset \mathbf{X}_{\lambda_0}(d_0)$. To see this we just note that $\lambda_1 \to \lambda_0$ and $\frac{\varepsilon\varphi(0)u(0)^{p-1}}{n(n-2)} + u(0)^{\frac{p-1}{2}}G(r,u) \to 0$, uniformly in $r \in \mathbf{R}_+$ and $u \in \mathbf{X}_{\lambda_0}(d_0)$, as $\varepsilon \to 0$.

Lemma 4. *The mapping* $\mathcal{T}_\varepsilon : \mathbf{X}_{\lambda_0}(d_0) \to \mathbf{X}_{\lambda_0}(d_0)$, *is a contraction provided* $|\varepsilon|$ *is small enough. The norm is that of* $L^\infty_{U_{\lambda_0}^{-1}}$.

Lemma 4 will be proved using

Lemma 5. *If* $u, v \in \mathbf{X}_{\lambda_0}(d_0)$, *then*

$$|G(r,v) - G(r,u)| < C|\varepsilon|\,\|u - v\| \, ,$$

uniformly in $r \geq 0$, *where* C *only depends on* n, d_0, λ *and* φ.

Proof of Lemma 5. As in the proof of Lemma 2, we can assume that $\lambda_0 = 1$. Put $v = u + \eta$, then we have

$$\frac{n-2}{2}G(r,u+\eta) =$$

$$\varepsilon\int_0^r s^{-3}\int_0^s (u+\eta)^{-\frac{p+3}{2}}t^{1-n}\int_0^t x\varphi'(u+\eta)^{p+1}x^{n-1}\,dx\,dt\,ds =$$

$$\frac{n-2}{2}G(r,u)+$$

$$\varepsilon(p+1)\int_0^r s^{-3}\int_0^s u^{-\frac{p+3}{2}}t^{1-n}\int_0^t x\varphi'u^p\eta x^{n-1}\,dx\,dt\,ds-$$

$$\varepsilon\frac{p+3}{2}\int_0^r s^{-3}\int_0^s u^{-\frac{p+5}{2}}\eta t^{1-n}\int_0^t x\varphi'u^{p+1}x^{n-1}\,dx\,dt\,ds+$$

$$\varepsilon O(\|\eta\|^2) \, .$$

Since $\|u - v\| \leq C$ for all $u, v \in \mathbf{X}_1(d)$, we can control the last term. The other terms can be estimated as in the proof of Lemma 2. Note that $\frac{1}{d_0}U_1 \leq u \leq d_0 U_1$ and that $|\eta| \leq U_1\|\eta\|$. $\qquad\square$

Proof of Lemma 4. Take $u, v \in \mathbf{X}_{\lambda_0}(d)$, then we want to estimate $\|\mathcal{T}_\varepsilon(u) - \mathcal{T}_\varepsilon(v)\|$ in terms of $\|u - v\|$.

First we note that $|u(0) - v(0)| \leq \lambda_0\|u - v\|$, and thus by Lemma 2 and Lemma 5, we get

$$|u(0)^{\frac{p-1}{2}}G(r,u) - v(0)^{\frac{p-1}{2}}G(r,v)| \leq$$
$$C\|u - v\|\,|G(r,u)| + C|G(r,u) - G(r,v)| \leq C|\varepsilon|\,\|u - v\| \, .$$

By Lemma 3 and Lemma 5, we get for ε small

$$|\lambda(u(0)^{\frac{p-1}{2}}G(r,u)) - \lambda(v(0)^{\frac{p-1}{2}}G(r,v))| \leq$$

$$\leq \sup_{\|g\|_\infty \leq C|\varepsilon|}|\nabla\lambda(g)|\,|u(0)^{\frac{p-1}{2}}G(r,u) - v(0)^{\frac{p-1}{2}}G(r,v)| \leq C|\varepsilon|\|u - v\| \, .$$

Thus combining the estimates above we get

$$\|\mathcal{T}_\varepsilon(u) - \mathcal{T}_\varepsilon(v)\| \leq C|\varepsilon|\|u - v\|, \quad \text{for all } u, v \in \mathbf{X}_{\lambda_0}(d_0).$$

If we choose ε small enough we conclude that $\mathcal{T}_\varepsilon : \mathbf{X}_{\lambda_0}(d_0) \to \mathbf{X}_{\lambda_0}(d_0)$ is a contraction. $\qquad\square$

Proof of Theorem I. To finish the proof of Theorem I we note that $U_{\lambda_0} \in \mathbf{X}_{\lambda_0}(d_0)$ and thus the set is non empty. Furthermore, by the above $\mathcal{T}_\varepsilon : \mathbf{X}_{\lambda_0}(d_0) \to \mathbf{X}_{\lambda_0}(d_0)$ is a contraction provided $d_0 > 1$ and ε are small enough. By the contraction mapping theorem it follows that $\mathcal{T}_\varepsilon$ has unique fixed point u_ε in $\mathbf{X}_{\lambda_0}(d_0)$. By Lemma 1, this is a solution of problem (0.1).

Next we will show that $(-\varepsilon_0, \varepsilon_0) \ni \varepsilon \to u_\varepsilon \in L^\infty_{U^{-1}_{\lambda_0}}$ is continuous. Take $\varepsilon' \in (-\varepsilon_0, \varepsilon_0)$. Since u_ε all solve (0.1), $|\varphi| \leq C\min(1, r^\alpha)$, some $\alpha < 2$ it follows that Δu_ε are uniformly bounded in L^q for all $q \geq 1$. Combining this fact, Sobolev embedding theorems and the uniqueness of the solution in $\mathbf{X}_{\lambda_0}(d)$. It follows that $u_\varepsilon \to u_{\varepsilon'}$ in $C^{1,\alpha}(\mathbf{R}^n)$ for any $\alpha < 1$, as $\varepsilon \to \varepsilon'$.

This only proves that u_ε is continuous in $L^\infty_{U^{-1}_{\lambda_0}}$ locally. To see that we have the right behaviour near infinity we can apply the Kelvin Transform.

The Kelvin Transform $v_\varepsilon(r) = r^{2-n}u_\varepsilon(\frac{1}{r})$ solves (0.1) with φ replaced by $\varphi(\frac{1}{r})$. Furthermore the Kelvin transform is an isometry on $\mathcal{R}_0^{1,2}$. Thus the continuity of the path follows from the results above.

Finally we need to replace $\mathbf{X}_{\lambda_0}(d_0)$, by $B_R(U_{\lambda_0})$. This is easy since the constraint $\mathcal{A}(u) = 0$ is a necessary condition for the existence of a solution of (0.1) if $\varepsilon \neq 0$. $\qquad\square$

2. Proof of Theorem II

The proof is very similar to that of Theorem I. Therefore we will omit many details. Throughout this section we will assume that φ satisfies the hypothesis given in Theorem II.

In order to prove the result for problem (0.3), we need to introduce some notation. The set of functions $\mathbf{X}_\lambda(d)$ is replaced by $\mathbf{Y}_\lambda(d) = \{u : \frac{1}{d}U_\lambda \leq u \leq dU_\lambda, \mathcal{B}(u) = 0\}$. Where $\mathcal{B}$ is as defined in (0.4) and as before we define $b(\lambda) = \mathcal{B}(U_\lambda)$.

The initial value problem corresponding to (0.3) is

$$(2.1) \qquad \begin{cases} u'' + \dfrac{n-1}{r}u' + u^p + \varepsilon\varphi = 0, \\ u(0) = \lambda, \qquad u'(0) = 0. \end{cases}$$

Lemma 1'. *If u solves (2.1) and u is positive in $[0, r]$, then it satisfies* *the identity*

$$u(r) = \frac{\lambda}{\left(1 + \left(\frac{\lambda^{p-1}}{n(n-2)} + \frac{\varepsilon\varphi(0)\lambda^{-1}}{n(n-2)} + \lambda^{\frac{p-1}{2}}H(r,u)\right)r^2\right)^{\frac{n-2}{2}}}, \qquad \text{where}$$

$$H(r,u)\frac{2\varepsilon}{n-2} =$$

$$\int_0^r s^{-3} \int_0^s u^{-\frac{p+3}{2}} t^{1-n} \left((p+1)\int_0^t (x^{\frac{n+2}{2}}\varphi)' u x^{\frac{n-2}{2}}\, dx - p\varphi u t^n\right) dt\, ds\ .$$

Furthermore, if u is continuous and satisfies the integral equation in some interval $[0, r_0]$, then u solves (1.1) in the same interval.

Proof. The proof of the second part follows by uniqueness as in the proof of Lemma 1 in Section 1. However, the estimate of the innermost integral in the definition of H has to be done with care, as in the proof of Lemma 2' below.

The proof of the first part is very similar to that of Lemma 1 in this paper (which is proved in [BE]) or Lemma 4.6 in [EK]. As mentioned earlier the ideas originates from the paper [AP].

Define the function

$$(2.2) \qquad F(r) = -\left(r^3(r^{n-2}u)'(r^{n-2}u)^{\frac{-n}{n-2}}\right)'\ .$$

Carrying out the differentiation on the right side in (2.2) and using equation (2.1) to substitute for u'', yield

$$(2.3) \qquad \begin{aligned} F(r) &= u^{-\frac{p+3}{2}}r^{1-n}(E(r) + \varepsilon\varphi r^n u)\ , \qquad \text{where} \\ E(r) &= \frac{n}{n-2}r^n(u')^2 + nr^{n-1}uu' + r^n u^{p+1}\ . \end{aligned}$$

Next we will use the equation to modify the function E. First we note that $E(0) = 0$, again using the equation to substitute for u'' we get

$$E'(r) = -\varepsilon(p+1)\varphi r^{\frac{n+2}{2}}(r^{\frac{n-2}{2}}u)'\ .$$

Integrating by part yields

$$E(r) = \varepsilon(p+1)\int_0^r ur^{\frac{n-2}{2}}\, d(r^{\frac{n+2}{2}}\varphi(r)) - \varepsilon(p+1)r^n u\varphi\ .$$

Thus we find that

$$F(r) = u^{-\frac{p+3}{2}}r^{1-n}\left(\varepsilon(p+1)\int_0^r ur^{\frac{n-2}{2}}\, d(r^{\frac{n+2}{2}}\varphi(r)) - \varepsilon p r^n u\varphi\right)\ .$$

Put $v = r^{n-2}u$, then (2.2) becomes

$$(2.4) \qquad\qquad\qquad (r^3 v' v^{\frac{-n}{n-2}})' = -F(r) \, .$$

To integrate (2.4) we need the estimate $|F(r)| < Cr^\alpha$, for some $\alpha > 1$, as $r \to 0$. To prove this we can argue exactly as in the proof of Lemma 2' below. In fact the proof shows that the estimate holds with $\alpha = 2$. Furthermore from equation (2.1) we get that $v(r) = r^{n-2}(\lambda - \frac{\lambda^p + \varepsilon\varphi(0)}{2n} r^2 + o(r^2))$, as $r \to 0$. Thus for small r we have the following asymptotic estimates.

$$\frac{r^3 v'}{v^{\frac{n}{n-2}}} = \frac{n-2}{\lambda^{\frac{2}{n-2}}} + o(1)$$

$$\frac{1}{v^{\frac{2}{n-2}}} - \frac{1}{r^2 \lambda^{\frac{2}{n-2}}} = \frac{1}{n(n-2)}(\lambda^p + \varepsilon\varphi(0))\lambda^{-\frac{n}{n-2}} + o(1) \, .$$

Hence integrating (2.4) yields

$$\frac{v'}{v^{\frac{n}{n-2}}} - \frac{n-2}{\lambda^{\frac{2}{n-2}} r^3} = -\frac{1}{r^3} \int_0^r F(s)\, ds \, ,$$

$$\frac{1}{v^{\frac{2}{n-2}}} - \frac{1}{r^2 \lambda^{\frac{2}{n-2}}} = H(r, u) + \frac{(\lambda^p + \varepsilon\varphi(0))\lambda^{-\frac{n}{n-2}}}{n(n-2)} \, .$$

Solving for v and going back to u gives the formula. $\qquad\qquad\square$

Lemma 2'. *If $u \in \mathbf{Y}_{\lambda_0}(d)$, then $|H(r, u(r))| < C|\varepsilon|$ uniformly in r, where C only depends on n, d, λ_0 and φ.*

Proof. First let us recall that the "absolute value of a measure" is defined by $|d\mu| = (d\mu)^+ + (d\mu)^-$, where $(d\mu)^+$ and $(d\mu)^-$ is the positive and negative part in the Jordan decomposition of the measure.

For $t \leq 1$ we have

$$\int_0^t u(x)x^{\frac{n-2}{2}}\, d(x^{\frac{n+2}{2}}\varphi(x)) =$$

$$\varphi(0)u(0)\int_0^t x^{\frac{n-2}{2}}\, d(x^{\frac{n+2}{2}}) + \varphi(0)\int_0^t (u(x) - u(0))x^{\frac{n-2}{2}}\, d(x^{\frac{n+2}{2}}) +$$

$$u(0)\int_0^t x^{\frac{n-2}{2}}\, d((\varphi(x) - \varphi(0))x^{\frac{n+2}{2}}) +$$

$$\int_0^t x^{\frac{n-2}{2}}(u(x) - u(0))\, d((\varphi(x) - \varphi(0))x^{\frac{n+2}{2}}) \, .$$

Thus we obtain

$$\left| (p+1) \int_0^t u(x)x^{\frac{n-2}{2}} \, d(x^{\frac{n+2}{2}}\varphi(x)) - p\varphi u t^n \right| \le$$

$$C \left(\int_0^t |u(x) - u(0)| x^{n-1} \, dx + \int_0^t x^{\frac{n-2}{2}} |d((\varphi(x) - \varphi(0))x^{\frac{n+2}{2}})| + \right.$$

$$\left. + \int_0^t x^{\frac{n-2}{2}} |u(x) - u(0)| \, |d((\varphi(x) - \varphi(0))x^{\frac{n+2}{2}})| \right) +$$

$$p|\varphi(0)u(0) - \varphi(t)u(t)| \, t^n \le C t^{n+1} \ .$$

For $t \ge 1$ we use the fact that $\mathcal{B}(u) = 0$ and the assumptions on φ to get

$$\left| \int_0^t \cdots \right| = \left| \int_t^\infty u x^{\frac{n-2}{2}} \, d(\varphi x^{\frac{n+2}{2}}) \right| \le C t^{2-n-\delta} \ , \qquad \text{and}$$

$$|p\varphi u t^n| \le C t^{2-n-\delta} \ .$$

Collecting the estimates above we get

$$\left| (p+1) \int_0^t u x^{\frac{n-2}{2}} \, d(x^{\frac{n+2}{2}}\varphi) - p\varphi u t^n \right| \le C \min(t^{n+1}, \, t^{2-n-\delta}) \ .$$

The rest of the proof is exactly the same as that of Lemma 2. $\qquad \square$

The rest of the proof of Theorem II is a straight forward repetition of the proof of Theorem I. We omit these details.

3. The result of W.-Y. Ding, some examples and remarks

The results on the sphere.

If the radial function $u = u(r)$ solves (0.1), then the rotational invariant function $v(\theta) = (1 - \cos(\theta))^{\frac{2-n}{2}} u(\frac{\sin(\theta)}{1-\cos(\theta)})$ solves (0.1'). If we write out the equation in θ we get the Neumann boundary value problem

$$\begin{cases} v'' + \dfrac{n-1}{\tan(\theta)} v' - \dfrac{(n-2)n}{4} v + (1 + \epsilon\tilde{\varphi})v^p = 0, \, v > 0 \quad \text{in } (0,\pi) \ , \\ v'(0) = v'(\pi) = 0 \quad . \end{cases}$$

If $\epsilon = 0$, then we have the one parameter family of solutions that corresponds to U_λ

$$V_\mu(\theta) = \frac{(n(n-2)\mu)^{\frac{n-2}{4}}}{((1+\mu) + (1-\mu)\cos(\theta))^{\frac{n-2}{2}}} \ ,$$

for any μ positive. Note that this function concentrates at the south pole if $\mu \to 0$ and at the nort pole if $\mu \to \infty$.

Transforming the functions $\mathcal{A}$ and a gives

$$\tilde{\mathcal{A}}(v) = \int_0^\pi \sin(\theta) \left(\frac{\partial \tilde{\varphi}}{\partial \theta}\right) v(\theta)^{p+1}(\sin \theta)^{n-1} \, d\theta$$

$$\tilde{a}(\mu) = \tilde{\mathcal{A}}(V_\mu) \qquad \text{for } \mu > 0 \,.$$

Thus a necessary condition for $v = v(\theta)$ to solve (0.1') with $\varepsilon \neq 0$ is that $\tilde{\mathcal{A}}(v) = 0$. This is a special case of the necessary conditions obtained by Kazdan and Warner.

Now we obtain the following consequence of Theorem I.

Theorem I'. *Assume that $\tilde{\varphi} \in C^1(\mathbf{S}^n)$. If $\tilde{a}(\mu_0) = 0$ and $\tilde{a}'(\mu_0) \neq 0$, then there exist $R > 0$ and $\varepsilon_0 > 0$ such that (0.1') has a unique branch $\{v_\varepsilon\}$, $0 < |\varepsilon| < \varepsilon_0$ of solutions in $\{ v \in C^2(\mathbf{S}^n) : \|v - V_{\mu_0}\|_\infty \leq R \}$, such that $v_\varepsilon \to u_0 = V_{\mu_0}$, as $\varepsilon \to 0$. Furthermore, this branch is continuous in $C^2(\mathbf{S}^n)$.*

As before the condition that $\tilde{a}(\mu_0) = 0$ is necessary, for the existence of a branch of solutions bifurcating from V_{μ_0}. For the converse we need the "transversality condition" $\tilde{a}'(\mu_0) \neq 0$.

The general approach of W.-Y. Ding.

The following more general approach is due to W.-Y. Ding [DI]. His proof was given for the transformed problem on the sphere. Although the proof is more involved and less elegant when carried out in $\mathbf{R}^n$, we decided to do so to see the connection with our previous approach.

The idea is to use a Liapunov-Schmidt reduction in the same way as in bifurcation theory. This method also works in nonradial situations. For this reason we will introduce some additional necessary conditions for the solution of (0.1).

Proposition 3.1. *If $u \in \mathcal{D}_0^{1,2}$ satisfies $\Delta u + f(x, u) = 0$ in $\mathbf{R}^n$ and $F(x, u)$ is integrable in $\mathbf{R}^n$, then u must satisfy the following n conditions: $\int \tilde{\nabla} F(x, u) \, dx = 0$. Here $\tilde{\partial}_i$ regards u as a constant and $F(x, u) = \int_0^u f(x, s) \, ds$.*

These conditions where introduced by W.-M. Ni in [NI2]. In the special case when $f(x, u) = K(x)u^p$ we get $\int u^{p+1} \nabla K \, dx = 0$. Here $\mathcal{D}_0^{1,2}$ is the completion of C_0^∞ in the norm $\|\nabla \cdot \|_2$.

Proof. Multiplying the equation by $\partial_i u$ and integrating by part in a finite domain Ω yields

$$\int_\Omega \tilde{\partial}_i F(x, u) \, dx + \int_{\partial\Omega} \left(\left(\frac{1}{2}|\nabla u|^2 - F(x, u)\right) d\nu_i - \partial_i u \partial_\nu u \, d\nu \right) = 0 \,.$$

Take Ω to be a ball B_R and let R tend to infinity along an appropriately choosen sequence. Then the boundary terms vanish and the result follows. $\square$

Thus we now have $n+1$ necessary conditions for the existence of a branch of solutions. When transformed back to the sphere these conditions become the $n + 1$ Kazdan-Warner conditions (one for each coordinate direction). Define the function $\mathbf{R}^{n+1} \ni (y, \lambda) \to A(y, \lambda) \in \mathbf{R}^{n+1}$ by

$$A(y, \lambda) = (\partial_\lambda, \nabla_y) \int \frac{U_{\lambda,y}^{p+1}}{p+1} \varphi \, dx = \int (x \cdot \nabla\varphi, \nabla\varphi) U_{\lambda,y}^{p+1} \, dx \ .$$

Here $U_{\lambda,y} = U_\lambda(x - y)$. Now a necessary condition to have a branch of solutions starting at U_{λ_0, y_0} is that $A(\lambda_0, y_0) = 0$. As before this can also be proved to be sufficient if the zero of A is nondegenerate.

Theorem 3.2. *Assume that φ is C^1 and bounded in $\mathbf{R}^n$. If $A(y_0, \lambda_0) = 0$ and $\nabla_{y,\lambda} A$ is nonsingular at λ_0, y_0, then there is a unique continuous branch of solutions starting at U_{λ_0, y_0}.*

Remark. The nondegeneracy condition on A at the zero can be weakened, but then we loose uniqueness in general.

Sketch of the Proof. We want to solve $\Delta u + (1 + \varepsilon\varphi)u^p = 0$, where $u = U_{\lambda,y} + \eta$ and η is small compared to $U_{\lambda,y}$. This gives us the equation

$$\mathcal{L}_{\lambda,y}\eta = \varepsilon\varphi U_{\lambda,y} + g(x, \eta) \,,$$

where $\mathcal{L}_{\lambda,y} = -U_{\lambda,y}^{1-p}(\Delta + pU_{\lambda,y}^{p-1})$ and $g(x, \eta) = (1+\varepsilon\varphi)U_{\lambda,y}(1+\eta U_{\lambda,y}^{-1})^p - (1 + \varepsilon\varphi)U_{\lambda,y} - p\eta$.

We will consider $\mathcal{L}_{\lambda,y}$ as an operator on a subset of $L^2(U_{\lambda,y}^{p-1} \, dx)$. Then the spectrum is discrete since $\mathcal{D}_0^{1,2} \hookrightarrow L^2(U_{\lambda,y}^{p-1} \, dx)$ is compact.

Since $\Delta U_{\lambda,y} + U_{\lambda,y}^p = 0$, differentiating the equation with respect to λ and y yields $(\partial_\lambda, \nabla_y)U_{\lambda,y} \subset \operatorname{Ker} \mathcal{L}_{\lambda,y}$. Some more work using separation of variables show that $\dim(\operatorname{Ker} \mathcal{L}_{\lambda,y}) = n + 1$. Let P be the projection of $L^2(U_{\lambda,y}^{p-1} \, dx)$ onto $\operatorname{Ker}(\mathcal{L}_{\lambda,y})$ and put $Q = I - P$. Then we can rewrite the equation as follows

$$\mathcal{L}_{\lambda,y}(Q\eta) = Q(\varepsilon\varphi U_{\lambda,y} + g(x, \eta)) \,, \quad P(\varepsilon\varphi U_{\lambda,y} + g(x, \eta)) = 0 \ .$$

We will look for solutions in $L^\infty_{U_{\lambda_0, y_0}^{-1}} \subset L^2(U_{\lambda_0, y_0}^{p-1} \, dx)$, with η small (in the same space).

The first equation can always be solved for $Q\eta$. Substituting into the second equation, we find that we only need to find $P\eta$ satisfying the $n + 1$ dimensional system $P(\varepsilon\varphi U_{\lambda,y} + g(x, Q\eta + P\eta)) = 0$. Since $A(\lambda_0, y_0) = 0$ we have $\varphi U_{\lambda_0, y_0} \perp \operatorname{Ker}(\mathcal{L}_{\lambda_0, y_0})$. Thus by the assumption that the zero is nondegenerate, we can always find a unique solution (λ, y) of this system near λ_0, y_0, if ε is small enough. $\square$

Asymptotic behaviour.

It follows from the proof that the solutions we obtained in Theorem I&II have the asymptotic behaviour $r^{n-2}u_\varepsilon(r) \to C \in (0,\infty)$, as $r \to \infty$. In fact our proof only works if the resulting solution has this behaviour. This means that we must have certain restrictions on φ near infinity.

Below we write $f \sim g$ if there are positive constants C_1, C_2 such that $C_1 f < g < C_2 f$.

In Theorem I the asymptotic condition on φ at infinity is that $|\varphi'| < Cr^{1-\delta}$ for some positive δ. Thus we can allow φ to have the asymptotic behaviour $\varphi \sim \pm r^\alpha$ for some $\alpha < 2$, as $r \to \infty$. We can always define φ in a suitable way for r small, so that $a(\lambda) = \mathcal{A}(U_\lambda)$ has a nontrivial zero. Hence for each $\alpha < 2$, Theorem I yields a function φ so that $\varphi \sim r^\alpha$ or $\varphi \sim -r^\alpha$ for large r and so that (0.1) has a solution for small ε.

By Theorem 3.41 in [NI1], we can not have a solution of (0.1) if for some positive constant C, $1 + \varepsilon\varphi \geq Cr^2$ for large r. In fact we can not have a solution that tends to zero at all. Thus the condition is best possible if $1 + \varepsilon\varphi$ is nonnegative for large r.

Next let us see what happens if φ becomes large negative. Integrating (1.1) yields

$$u(r) = \int_r^\infty t^{1-n} \int_0^t s^{n-1}(1 + \varepsilon\varphi(s))u^p(s)\, ds\, dt =$$
$$= \frac{1}{n-2} \int_0^\infty \max(r,s)^{2-n}(1 + \varepsilon\varphi(s))u^p(s)s^{n-1}\, ds\ .$$

Let us assume that $u(r) \sim r^{2-n}$, and that at least one of $\int_0^\infty \varphi_\pm(s)s^{-3}\, ds$ is finite ($\varphi_\pm$ denotes the positive and negative part of φ). Then monotone convergence yields: $r^{n-2}u(r) \to \frac{1}{n-2}\int_0^\infty (1+\varepsilon\varphi(s))u^p(s)s^{n-1}\, ds$, as $r \to \infty$. Thus if one of $\varphi_\pm(s)s^{-3}$ is integrable and $u \sim r^{2-n}$, then we must have $|\varphi|s^{-3}$ integrable.

From this we immediately find that if $\varepsilon\varphi(s) \leq -Cs^2$ (or $\varepsilon\varphi(s) \geq Cs^2$) for large s, then we can not have $u(r) \sim r^{2-n}$.

Let us remark that it is possible to find a function φ such that $\varphi \to -\infty$ very fast, as $r \to \infty$ and such that (0.1) has a solution. However by the above, such a solution must decay faster than r^{2-n}, if $\varepsilon\varphi \leq -Cr^2$ for large r.

Connections with the variational approach.

Many approaches to (0.1) or (0.1') are variational (eg [BC], [BE], [CY] and [ES]).

Let $\mathcal{E}_{\varepsilon\varphi}(u) = \frac{\|\nabla u\|_2^2}{\|u\|_{p+1,1+\varepsilon\varphi}^2}$, on $\mathcal{R}_0^{1,2}$. Then it is well known that for any $\lambda > 0$ we have $\inf_{u \in \mathcal{R}_0^{1,2}} \mathcal{E}_0(u) = \mathcal{E}_0(U_\lambda)$.

If u_ε is a branch of solutions obtained in Theorem I, then it follows that $\mathcal{E}_{\varepsilon\varphi}(u_\varepsilon) \to \mathcal{E}_0(U_\lambda)$, as $\varepsilon \to 0$. Thus although we might choose φ bounded and smooth so that the lower bound of the number of solutions tends to infinity as $\varepsilon \to 0$, they all have small energy. The existence of such a φ is proved below.

Note that we need not have $\mathcal{E}_{\varepsilon\varphi}(u_\varepsilon) = \inf_u \mathcal{E}_{\varepsilon\varphi}(u)$ for any of the branches. If for example φ is continuous and not just the constant function, $\varepsilon > 0$ and $\varphi(0) = \varphi(\infty) \geq \varphi(r)$ for all positive r, then the infimum is not attained.

Multiplicity results.

The function $a(\lambda)$ and its derivative can be rewritten as follows:

$$a(\lambda) = -\int_0^\infty \varphi\left(\frac{r}{\lambda^\alpha}\right)g'(r)\,dr \ ,$$

$$(3.1) \qquad a'(\lambda) = -\alpha\lambda^{-1}\int_0^\infty \varphi\left(\frac{r}{\lambda^\alpha}\right)(rg'(r))'\,dr \ , \quad \text{where}$$

$$g(r) = r^n\left(1 + \frac{r^2}{n(n-2)}\right)^{-n} \ , \quad \alpha = \frac{2}{n-2} \ .$$

It is easy to verify that g' has only one positive zero at $r_0 = \sqrt{n(n-2)}$, and $(rg'(r))' < 0$ at $r = r_0$.

Choose a nonnegative C^∞ function ψ with support in a small neigborhood of r_0 such that the mapping $\lambda \to \int_0^\infty \psi(\frac{r}{\lambda^\alpha})g'(r)\,dr$ has a simple zero λ_1 near $\lambda = 1$, and the derivative with respect to λ is negative near λ_1. The existence of such a function follows from the properties of g and (3.1).

Put

$$(3.2) \qquad \varphi(r) = \sum_{k=0}^\infty a_k\psi(b_k r) , \quad \text{with } a_k = M^{-k}, \ b_k = M^{-\beta k}.$$

We will choose $\beta > \frac{1}{n-1}$ and M will be a large real number to be determined below. Thus, with this choice of φ we get

$$a(\lambda) = -\sum_{k=0}^\infty a_k \int_0^\infty \psi\left(\frac{b_k}{\lambda^\alpha}r\right)g'(r)\,dr \ .$$

We will prove that if M is taken large then a has an infinite number of simple zeros $\{\lambda_k\}$ such that $\lambda_k^\alpha \sim b_k = M^{-k\beta}$.

There is a number c only depending on g and ψ such that

$$\lambda \to a_N \int_0^\infty \psi\left(\frac{b_N}{\lambda^\alpha}r\right)g'(r)\,dr$$

takes values above ca_N and below $-ca_N$ in a neighborhood of $\lambda_N = b_N^{\frac{1}{\alpha}}$.

Next we will estimate the other terms in the series to see that they are much smaller than the leading term for λ near λ_N, provided M is large.

In the estimates below we use the fact that $g'(r) \sim r^{n-1}$, as $r \to 0$ and $g'(r) \sim r^{-1-n}$, as $r \to \infty$.

$$\left| \sum_{k=0}^{N-1} a_k \int_0^\infty \psi(\frac{b_k}{\lambda^\alpha} r) g'(r)\, dr \right| \leq C \sum_{k=0}^{N-1} M^{-k} M^{\beta(k-N)(n-1)}$$

$$\leq C a_N M^{1-\beta(n-1)} \ll c a_N,$$

provided we choose M large enough. Here we used the fact that $\beta > \frac{1}{n-1}$. Also note that the size of M does not depend on N.

The other terms are estimated the same way

$$\left| \sum_{k=N+1}^\infty a_k \int_0^\infty \psi(\frac{b_k}{\lambda^\alpha} r) g'(r)\, dr \right| \leq C \sum_{k=N+1}^\infty M^{-k} M^{\beta(N-k)(n+1)}$$

$$\leq C a_N M^{-1-\beta(n+1)} \ll c a_N,$$

provided M is large enough.

This proves that φ has an infinite number of zeros. The next question is: are they all simple? The derivative of a is given by

$$a'(\lambda) = -\alpha \lambda^{-1} \sum_{k=0}^\infty a_k \int_0^\infty \psi(\frac{b_k}{\lambda^\alpha} r)(rg'(r))'\, dr.$$

The function $(rg'(r))'$ has the same asymptotics as g' at the origin and at infinity. Thus exactly the same argument as above shows that we can choose M large so that the N'th summand dominates near the zero λ_N. Thus for λ near λ_N we have

$$(3.3) \qquad a'(\lambda) \approx -\frac{\alpha}{\lambda} \int_0^\infty a_N \psi(\frac{b_N}{\lambda^\alpha} r)(rg'(r))'\, dr \approx c M^{N(\beta \frac{n-2}{2} - 1)}$$

where c is positive constant independent of N. Thus we find that all zeros are simple. Note that a' is positive at the zeros we have constructed so there must be other consecutive zeros.

Thus with this choice of φ in equation (0.1) we find that the lower bound of the number of solutions of (0.1) tends to infinity as $\varepsilon \to 0$. Note that we can not conclude that there is an infinite number of solutions if ε is small. The reason is that we do not have a uniform estimate of the intervals $(-\varepsilon_k, \varepsilon_k)$, where the k'th branch of solutions exist.

Note that if we choose $\beta \geq \frac{2}{n-2} > \frac{1}{n-1}$, then $a'(\lambda_k) > 0$ is uniformly bounded away from zero. Thus in this case we might get a uniform estimate of the ε_k's. However, it is not clear if this is indeed possible.

Let us see what this means for the problem on the sphere. We only need to check the regularity of $\tilde{\varphi}$ at the nortpole. This regularity is the same as that of $\varphi(\frac{1}{r})$ at the origin. Hence $\tilde{\varphi} \in C^\infty(\mathbf{S}^n \setminus \{N.P.\})$ and $|\tilde{\varphi}(\theta)| \leq \theta^\gamma$, where $\gamma - \frac{1}{\beta} < n - 1$.

Let us remark that these solutions all have small energy. This example shows that the number of solutions depends very heavily on the shape of φ.

For problem (0.3) we can do a similar construction. However we can only get the following weaker result. Given N there is a smooth function φ with compact support such that (0.3) has at least N solutions for small ε. As above we have

$$b(\lambda) = -\int_0^\infty \varphi(\frac{r}{\lambda^\alpha}) h'(r)\, dr\,, \quad b'(\lambda) = -\frac{\alpha}{\lambda} \int_0^\infty \varphi(\frac{r}{\lambda^\alpha})(rh'(r))'\, dr\,,$$

where $h(r) = r^{\frac{n-2}{2}} \left(1 + \frac{r^2}{n(n-2)}\right)^{\frac{2-n}{2}}\,, \quad \alpha = \frac{2}{n-2}\,.$

Solutions with finite zero or slow decay.

Let $u_{\lambda,\varepsilon}$ be a solution of the initial value problem (1.1). Furthermore, assume that φ' has compact support. Then it is well known that $u_{\lambda,\varepsilon}$ falls into one of the following categories.

SD: The solution has slow decay: $u_{\lambda,\varepsilon}(r) \sim r^{\frac{2-n}{2}}$.

FD: The solution has fast decay: $u_{\lambda,\varepsilon}(r) \sim r^{2-n}$.

FZ: The solution has a finite zero r_0. In this case we define the solution to be zero in $[r_0, \infty)$.

If ε is small enough it follows that $u_{\lambda,\varepsilon}$ is FZ if $a(\lambda) > 0$ and SD if $a(\lambda) < 0$. This can be seen from the integral representation formula given in Lemma 1 and the fact that $u_{\lambda,\varepsilon} \to U_\lambda$ uniformly on compact sets, as $\varepsilon \to 0$. Thus in some sence the shooting parameters λ that give a FD solution is the boundary between the parameters giving *FZ* and *SD* solutions.

REFERENCES

[AP] F.V. ATKINSON, L.A. PELETIER, *Emden-Fowler Equations Involving Critical Exponents.* Nonlinear Analysis T.M.A. **10** (1986), 755 – 766.

[BC] A. BAHRI, J.-M. CORON, *The Scalar-Curvature Problem on the Standard three-dimensional Sphere.* To appear in J. Funct. Anal..

[BE] G. BIANCHI, H. EGNELL, work in progress.

[CY] A. CHANG, P. YANG, work in progress.

[DI] W.-Y. DING, Unpublished work.

[DN] W.-Y. DING, W.-M. NI, *On the Elliptic Equation* $\Delta u + K u^{\frac{n+2}{n-2}} = 0$ *and Related Topics.* Duke Math. J. **52** (1985), 485 – 486.

[EK]   H. EGNELL, I. KAJ, *Positive Global Solutions of a Nonhomogeneous Semilinear Elliptic Equation.* To appear in J. Math. Pure Appl..

[ES]   J. ESCOBAR, R. SCHOEN, *Conformal Metrics with Prescribed Scalar Curvature.* Invent. Math. **86** (1986), 243 – 254.

[KW]   J. KAZDAN, F. WARNER, *Existence and Conformal Deformations of Metrics with Prescribed Gaussian and Scalar Curvature.* Ann. of Math. **101** (1975), 317 – 331.

[LN]   Y. LI, W.-M. NI, *On the Conformal Scalar Curvature in* $\mathbf{R}^n$. Duke Math. J. **57** (1988), 895 – 924.

[NI1] W.-M. NI, *On the Elliptic Equation* $\Delta u + K(x)u^{\frac{n+2}{n-2}} = 0$, *its Generalizations, and Applications in Geometry.* Indiana Univ. Math. J. **31** (1982), 493 – 529.

[NI2] W.-M. NI, *Some Aspects of Semilinear Elliptic Equations on* $\mathbf{R}^n$. *Nonlinear Diffusion Equations and Their Equilibrium States II*, 171 – 205. Ed. W.-M. Ni. Springer Verlag (1988).

Gabriele Bianchi   Henrik Egnell
I.A.G.A.-CNR   School of Mathematics
Florence, Italy   University of Minnesota
                          Minneapolis, MN 55455
                          present address:
                          Dept. of Mathematics
                          Uppsala University
                          Sweden

Singularities of Solutions
of a Class of Quasilinear
Equations in Divergence Form

MARIE-FRANÇOISE BIDAUT-VERON

0. Introduction

In Section 1 of this paper we study the behavior near the origin of the positive nonradial solutions of the doubly nonlinear N-dimensional partial differential equation

$$(1) \qquad Qu + u^q = 0,$$

where

$$(2) \qquad Qu = \operatorname{div} A(x, u, \nabla u),$$

in $B'_R = B_R/\{0\}$, with $B_R = \{x \in \mathbb{R}^N \mid |\alpha| < R\}$.

We assume that A is continuous in $\Omega \times \mathbb{R}^+ \times \mathbb{R}^N$ and satisfies

$$(3) \qquad |A(x, z, s)| \leqq a|s|^{p-1} + bz^{p-1} + c,$$

$$(4) \qquad sA(x, z, s) \geqq |s|^p - dz^p - e,$$

where $1 < p \leqq N$, and a, b, c, d, e are given positive constants.

We denote by μ the fundamental p-harmonic function in $\mathbb{R}^N$:

$$(5)(8) \qquad \mu(x) = \begin{cases} \frac{p-1}{N-p}(N\omega_N)^{-1/(p-1)}|x|^{(p-N)/(p-1)} & \text{for } p < N, \\ (N\omega_N)^{-1/(N-1)}\operatorname{Log}(1/|x|) & \text{for } p = N, \end{cases}$$

where $\omega_N = |B_1|$.

For equation (1) two critical values appear when $p < N$:

$$(6) \qquad q_1 = N(p-1)/(N-p), \quad q_2 = (Np - N + p)/(N - p)$$

and we set $q_1 = q_2 = +\infty$ when $p = N$. Our result concerns the subcritical case $q < q_1$; it extends Lions results [8] relative to the Laplace operator $(p = 2, q = \Delta)$. First we prove:

Theorem 1. *Suppose* $1 < p \leq N$, $p - 1 < q < q_1$. *Let* $u \in C^0(B'_R)$ *with* $\nabla u \in L^p_{\mathrm{loc}}(B'_R)$ *be a nonnegative solution of equation* (1) *in* $\mathbf{D}'(B'_R)$. *Then either* $\lim_{x \to 0} u(x)$ *exists and the extended* u *satisfies* (1) *in* $\mathbf{D}'(B_R)$; *or there are* $\alpha_1, \alpha_2 > 0$ *such that* $\alpha_1 \leq u(x)/\mu(x) \leq \alpha_2$ *near* 0, *and there is a* $\beta > 0$ *such that*

$$(7) \qquad\qquad Qu + u^q + \beta \delta_0 = 0 \quad in \ \mathbf{D}'(B_R),$$

where δ_0 *is the Dirac mass at* 0.

Typical examples of such Q are the operators

$$(8) \quad u \to Q_{\varepsilon,p} u = \mathrm{div}\, A_{\varepsilon,p}(\nabla u) = \mathrm{div}\big((\varepsilon + |\nabla u|^2)^{(p-2)/2} \nabla u\big), \ \varepsilon = 0, 1;$$

note that $Q_{0,p} = \Delta_p$ is the p-Laplace operator; and $Q_{1,p}$ arises in the cracking of plates and the modelling of blast furnaces with $1 < p \leq 2$. In that case we can extend easily the scaling method of Guedda & Veron [6]:

Theorem 2. *We make the assumptions of Theorem 1, with* $Q = Q_{\varepsilon,p}$ *given by* (8). *If* u *is not regular at* 0, *then there is an* $\alpha > 0$ *such that*

$$(9) \qquad\qquad \lim_{x \to 0} u(x)/\mu(x) = \alpha$$

and $\beta = \alpha^{p-1}$.

In Section 2 we deal more generally with the behavior near 0 of positive supersolutions of the equation $Qu = 0$, where Q is given by (2), (3), (4). Our main result is the following:

Theorem 3. *Suppose* $1 < p < N$. *Assume that* $u \in C^0(B'_R)$, $\nabla u \in L^p_{\mathrm{loc}}(B'_R)$, $Qu \in L^1_{\mathrm{loc}}(B'_R)$ *in the sense of* $\mathbf{D}'(B'_R)$ *and*

$$(10) \qquad\qquad u \geq 0, \quad Qu \leq 0 \ \text{a.e. in} \ B'_R.$$

Let $g(x) = -Qu(x)$, *a.e. in* B'_R. *Then* $g \in L^1_{\mathrm{loc}}(B_R)$ *and there is a* $\beta \geq 0$ *such that*

$$(11) \qquad\qquad Qu + g + \beta \delta_0 = 0 \quad in \ \mathbf{D}'(B_R),$$

and $u^{p-1} \in M^{N/(N-p)}_{\mathrm{loc}}(B_R)$, $|\nabla u|^{p-1} \in M^{N/(N-1)}_{\mathrm{loc}}(B_R)$.

This theorem extends the result of Brezis & Lions [4] relative to the Laplace operator. When $p = N$ we can give also some results, see [3].

When $p = 1$, which is the case of the mean curvature operator $u \to Q_{1,2}u = \mathrm{div}(1 + |\nabla u|^2)^{-1/2}\nabla u$, a partial result remains, that is $y \in L^1_{\mathrm{loc}}(B_R)$.

In Section 3 we study the behavior near infinity for equation (1), when $Q = Q_{\varepsilon,p}$. Obviously it depends on ε, since now the important thing is the behavior of $A_{\varepsilon,p}(s)$ near $s = 0$. Our results are new, even in the case of the Laplace operator:

Theorem 4. *Let* $\Omega_R = R^N/\bar{B}_R$ *and* $u \in C^1(\Omega_R)$, $u \geqq 0$ *in* Ω_R.
1) *Suppose that* $1 < p \leqq N$, $p - 1 < q < q_1$ *and*

$$(12) \qquad \Delta_p u + u^q = 0 \quad in \quad \mathbf{D}'(\Omega_R),$$

then $u \equiv 0$.

2) *Suppose that* $1 < p$, $1 < q < N/(N-2)$ $(1 < q < +\infty$ *if* $N = 2)$ *and*

$$(13) \qquad \mathrm{div}\big((1 + |\nabla u|^2)^{(p-2)/2}\nabla u\big) + u^q = 0 \quad in \quad \mathbf{D}'(\Omega_R),$$

then $u \equiv 0$.

1. Isolated Singularity for Equation 1

In order to prove Theorem 1, we first give an estimate of u near the singularity.

Proposition 1. *Suppose* $1 < p \leqq N$. *Assume that* $u \in C^0(B'_R)$, $\nabla u \in L^p_{\mathrm{loc}}(B'_R)$, $Qu \in L^1_{\mathrm{loc}}(B'_R)$ *and*

$$(14) \qquad u \geqq 0, \quad -Qu = g \geqq 0, \quad a.e. \ in \ B'_R.$$

Then $u^\gamma \in L^1_{\mathrm{loc}}(B_R)$ *for any* $\gamma \in (0, q_1)$, *and there is a* C *such that, for any small* σ,

$$(15) \qquad \int_{B_\sigma} u^\gamma \, dx \leqq \begin{cases} C\,\sigma^{N - \frac{N-p}{p-1}\gamma}, & if \ p < N, \\ C\,\sigma^N |\mathrm{Log}\,\sigma|^\gamma, & if \ p = N. \end{cases}$$

Proof. Here we use the techniques of Serrin [11, 12] to estimate the minimum of u on the sphere ∂B_σ, and then the weak Harnack inequality. From (3), (14), for any $\phi \in W^{1,\infty}(B'_R)$ with compact support, we have

$$(16) \qquad \int_{B_R} g\,\phi\,dx = \int_{B_R} A(x, u, \nabla u)\,\nabla\phi\,dx.$$

Let $\rho \in (0, R/2)$ and $\bar{u} = u - 2\max_{|x|=\rho} u(x)$. For any $\sigma \in (0, \rho)$ we set $m(\sigma) = \min_{|x|=\sigma} \bar{u}(x)$. Suppose first that $m(\sigma) > 0$ and take $\phi = v(\sigma) - m(\sigma)\eta$ in (16), where $\eta \in \mathbf{D}(B_{2\rho})$, $0 \leqq \eta \leqq 1$, $\eta(x) = 1$ on B_ρ, and

$$(17) \quad v(\sigma)(x) = \begin{cases} 0 & \text{if } \sigma < |x| < \rho \text{ and } \bar{u}(x) \leq 0, \text{ or if } |x| \geq \rho, \\ u(x) & \text{if } 0 \leq \bar{u}(x) \leq m(\sigma) \text{ and } \sigma < |x| < \rho, \\ m(\sigma) & \text{if } \bar{u}(x) > m(\sigma) \text{ and } \sigma < |x| < \rho, \text{ or if } |x| \leq \sigma. \end{cases}$$

As $g(m(\sigma) - v(\sigma)) \geqq 0$, we get

$$\int_{B_\rho} A(x, u, \nabla u)\nabla(v(\sigma))dx \leq m(\sigma) \int_{B_R} ((A(x, u, \nabla u).\nabla \eta + g(1 - \eta))dx$$
$$= m(\sigma)K$$

where K does not depend on σ. Hence from (17), (4) we get as in [12]

$$\int_{B_\rho} |\nabla(v(\sigma))|^p dx \leqq d \int_{B_\rho} v(\sigma)^p dx + e|B_\rho| + m(\sigma)K,$$

and from Sobolev inequality

$$\int_{B_\rho} v(\sigma)^p dx \leqq |B_\rho|^{p/N} \|v(\sigma)\|^p_{L^{\frac{pN}{N-p}}(B_\rho)} \leqq c_{N,p}|B_\rho|^{p/N} \int_{B_\rho} |\nabla v(\sigma)|^p dx;$$

hence choosing ρ small enough, there is a C independent of σ such that

$$C(1 + m(\sigma)) \geqq \int_{B_\rho} |\nabla v(\sigma)|^p dx \geqq m(\sigma)^p \, \mathrm{cap}_p \, B_\sigma,$$

and

$$\mathrm{cap}_p \, B_\sigma = \begin{cases} \omega_N((N-p)/(p-1))^{p-1}\sigma^{N-p}, & \text{if } p < N, \\ \omega_N(\mathrm{Log}(\rho/\sigma))^{1-N}, & \text{if } p = N, \end{cases}$$

hence by return to u, there is another C such that for any $\sigma \in B_\rho$,

$$(18) \qquad \min_{|x|=\sigma} u(x) \leqq \begin{cases} C(1 + \sigma^{(p-N)/(p-1)}), & \text{if } p < N, \\ C(1 + |\mathrm{Log}\,\sigma|), & \text{if } p = N, \end{cases}$$

which is also trivial when $m(\sigma) = 0$.

Now from (14) and [15] u satisfies a weak Harnack inequality in B'_ρ: for any $\gamma \in (0, q_1)$ there is a $\bar{c} = \bar{c}(\gamma, N, p, A)$ such that, for any ball $B_{3r}(x_0) \subset B'_\rho$,

$$(19) \qquad r^{-N/\gamma}\left(\int_{B_{2r}(x_0)} u^\gamma \, dx\right)^{1/\gamma} \leqq \bar{c} \min_{B_r(x_0)} u(x),$$

hence we get (15) from (18), (19) by a covering argument, see [3]. $\qquad\square$

Proof of Theorem 1. Equation (1) may be written in the form

$$(20) \qquad Qu + w\,u^{p-1} = 0 \quad \text{in} \ \ B'_R,$$

where $w = u^{q+1-p}$. Now $u \in C^0(B'_R)$, hence $Qu \in L^1_{\mathrm{loc}}(B'_R)$ and Proposition 1 applies. Since $q < q_1$ we can find a $\delta > 0$ such that $w \in L^{N/(p-\delta)}_{\mathrm{loc}}(B_R)$. Hence the conclusion follows directly from Serrin's results [12]. $\qquad\square$

Proof of Theorem 2. The proof has been given in [6] for $Q_{0,p} = \Delta_p$. We recall it for convenience and extend it to $Q_{1,p}$: let $\alpha = \limsup_{x\to 0} u(x)/\mu(x)$, then $\alpha \geq \alpha_2 > 0$. There are sequences $r_n \searrow 0$ and x_n such that $|x_n| = r_n$, $u(x_n) = \max_{|x|=r_n} u(x)$ and $\lim u(x_n)/\mu(x_n) = \alpha$. Define now

$$(21) \qquad u_n(\xi) = u(r_n\,\xi)/\mu(r_n), \quad \text{for} \ \ 0 < |\xi| < R/r_n,$$

then u_n satisfies the equation

$$(22) \qquad \mathrm{div}\, a_n^{1-p} A_{\varepsilon,p}(a_n\,\nabla u_n) + c_n\,u_n^q = 0 \quad \text{in} \ \ B'_{R/r_n},$$

where $a_n = r_n^{-1}\mu(r_n) \xrightarrow{n\to+\infty} +\infty$, and $c_n = r_n^p \mu(r_n)^{q+1-p} \xrightarrow{n\to+\infty} 0$.

Now with [5], [14] we deduce from Theorem 1 some estimates for u: there are $c \geq 0$ and $\alpha \in (0,1)$ such that for $0 < |x| < |x'| < R/2$,

$$(23) \qquad \begin{cases} |\nabla u(x)| \leqq c|x|^{-1}(|\mu(x)| + 1), \\[2mm] |\nabla u(x) - \nabla u(x')| \leqq c|x - x'|^\alpha\,|x|^{-1-\alpha}(|\mu(x)| + 1), \end{cases}$$

hence $(u_n)_{n\in\mathbb{N}}$ is relatively compact in the C^1_{loc}-topology of $\mathbb{R}^N/\{0\}$; then there is a subsequence $(u_\nu)_{\nu\in\mathbb{N}}$ which converges in this topology to a non-negative function $w \in C^1(\mathbb{R}^N/\{0\})$. Now w is a p-harmonic function in $\mathbb{R}^N/\{0\}$; indeed $a_n^{1-p} A_{\varepsilon,p}(\alpha_n\,\nabla u_n)$ is bounded on any compact of $\mathbb{R}^N/\{0\}$ and converges to $|\nabla u|^{p-2}\nabla u$ everywhere (consider separately the cases $\nabla u(x) \neq 0$ and $\nabla u(x) = 0$, when $\varepsilon = 1$). When $p = N$, w is constant from [7], hence $w = \alpha$; when $p < N$, we get $0 < \alpha_1 \leqq w(x)/\mu(x) \leqq \alpha_2$ near 0 from Theorem 1, hence from [7], w/μ is constant, and $w = \frac{\alpha}{\mu(1)}\mu$. In any case $\alpha = \lim u_n(\xi)$ uniformly for $|\xi| = 1$. Then for any $\varepsilon > 0$ there is a $n(\varepsilon)$ such that $\min_{|x|=r_n} u(x)/\mu(x) \geqq \alpha - \varepsilon$ for any $n \geq n(\varepsilon)$; comparing u and $(\alpha - \varepsilon)\mu$ over any annulus $\{x \in \mathbb{R}^N \mid r_{n+k} < |x| < r_n\}$ for $n \geq n(\varepsilon)$, $k > 0$, we get $u \geqq (\alpha - \varepsilon)\mu$ in $B'_{r_{n(\varepsilon)}}$; then $\liminf_{x\to 0} u(x)/\mu(x) = \alpha$, hence

$\lim_{x\to 0} u(x)/\mu(x) = \alpha$. Now multiplying (20) by any $\phi \in \mathbf{D}(B_R)$ we integrate over $\{x \in \mathbb{R}^N \mid \lambda < |x| < R\}$ and get $\beta = \alpha^{p-1}$ classically as $\lambda \to 0$. $\square$

Remark. We can give some estimates of $u - \alpha\mu$ near 0 as in [5], see [3].

Remark. Theorem 2 can be extended to operators Q of the form $Qu = \operatorname{div} A(\nabla u)$, where A satisfies Tolksdorff's conditions [14] and such that

$$(24) \qquad \lim_{\substack{t\to+\infty \\ \sigma\to s}} t^{1-p} A(t\sigma) = |s|^{p-2}s, \quad \forall s \in \mathbb{R}^N/\{0\}.$$

2. Supersolutions of Equation $Qu = 0$

Here prove Theorem 3 in several steps.

Proposition 2. *Suppose* $1 \leqq p < N$. *Assume that* $u \in C^0(B_R')$, $\nabla u \in L^p_{\mathrm{loc}}(B_R')$, $Qu \in L^1_{\mathrm{loc}}(B_R')$, *and* $u \geq 0$, $-Qu = g \geqq 0$, *a.e. in* B_R'. *Then* $g \in L^1_{\mathrm{loc}}(B_R)$; *moreover for any* $\rho > 0$ *small enough, we have for any* $\eta \in \mathbf{D}(B_{2\rho})$, $0 \leqq \eta \leqq 1$, $\eta = 1$ *in* B_ρ,

$$(25) \qquad \int_{B_R} g\,\eta^p\,dx \leqq \int_{B_R} A(x, u, \nabla u)\nabla(\eta^p)dx;$$

and there is a $c_\rho > 0$ *and a* $k_\rho \geqq 0$ *such that for any* $k \geqq k_\rho$ *and* $\alpha \in (0, 1]$,

$$(26) \qquad \int_{B_\rho \cap \{k<u<k+\alpha\}} |\nabla u|^p dx \leqq \alpha\, c_\rho + (d+e) \int_{B_\rho \cap \{k<u<k+\alpha\}} u^p\,dx.$$

Proof. For any $\phi \in W^{1,\infty}(B_R')$ with compact support in B_R', we have as in Proposition 1

$$(27) \qquad \int_{B_R} g\,\phi\,dx = \int_{B_R} A(x, u, \nabla u)\nabla\phi\,dx.$$

Set $p_{k,\alpha}(t) = \min(1, (t-k)^+/\alpha)$ for any $t \geq 0$ and $k \geq 0$, $\alpha > 0$. Let $0 < \rho < R/2$ and $\varepsilon < \rho/2$. Let $\eta \in \mathbf{D}(B_{2\rho})$ as above, and $\zeta_\varepsilon = \xi_\varepsilon \eta$, with $\xi_\varepsilon \in C^\infty(B_R)$, $0 \leqq \xi_\varepsilon \leqq 1$, $\xi_\varepsilon(x) = 0$ if $|x| < \varepsilon$, $\xi_\varepsilon(x) = 1$ if $|x| > 2\varepsilon$,

$|\nabla \xi_\varepsilon| \leq M/\varepsilon$. Then the function $\phi = (1 - p_{k,\alpha}(u))\zeta_\varepsilon^p$ is admissible in (27) and we get from (4)

$$(28) \qquad \int_{B_R} g(1 - p_{k,\alpha}(u))\zeta_\varepsilon^p \, dx + \frac{1}{\alpha} \int_{\{k<u<k+\alpha\}} |\zeta_\varepsilon \nabla u|^p dx$$

$$\leq \int_{B_R} (1 - p_{k,\alpha}(u))A(x,u,\nabla u)\nabla(\zeta_\varepsilon^p) dx$$

$$+ \frac{1}{\alpha} \int_{\{k<u<k+\alpha\}} \zeta_\varepsilon(du^p + e) dx.$$

Taking first $\alpha = 1$ and adding the equalities for integer $k = 0, 1, \ldots, n$, we have $\sum_{k=0}^n (1 - p_{k,1}(t)) = (n + 1 - t)^+$, hence

$$\int_{\{u<n+1\}} g(n + 1 - u)\zeta_\varepsilon^p + \int_{\{u<n+1\}} |\zeta_\varepsilon \nabla u|^p dx$$

$$\leq \int_{\{u<n+1\}} (n + 1 - u)A(x,u,\nabla u)\nabla(\eta^p) dx$$

$$+ p \int_{\{u<n+1\}\cap B_{2\varepsilon}} (n + 1 - u)|A(x,u,\nabla u)| \, |\nabla \xi_\varepsilon| dx$$

$$+ d \int_{\{u<n+1\}} (\zeta_\varepsilon u)^p dx + e|B_{2\rho}|.$$

Now let $v_n = \min(u, n + 1)$, then from Sobolev injection

$$\int_{\{u<n+1\}} (\zeta_\varepsilon u)^p dx \leq \int_{B_{2\rho}} (\zeta_\varepsilon v_n)^p dx \leq c_{N,p}|B_{2\rho}|^{p/N} \int_{B_{2\rho}} |\nabla(\zeta_\varepsilon v_n)|^p dx$$

$$\leq 2^{p-1}c_{N,p}|B_{2\rho}|^{p/N}\left(\int_{\{u<n+1\}} |(\zeta_\varepsilon \nabla u)|^p dx + p \int_{B_{2\varepsilon}} v_n^p |\nabla \xi_\varepsilon|^p dx \right.$$

$$\left. + \int_{B_{2\rho}} v_n^p \nabla(\eta^p) dx \right).$$

Let us fix p such that $2^{p-1}c_{N,p}|B_{2\rho}|^{p/N}d \leq \frac{1}{2}$ and take $n \geq k_\rho$ where $k_\rho \in \mathbb{N}^*$ satisfies $\max_{\rho \leq |x| \leq 2\rho} u(x) < k_\rho$; then $v_n = u$ when $\nabla(\eta^p) \neq 0$, hence we get

$$\int_{\{u<n+1\}} g(n + 1 - u)\zeta_\varepsilon^p dx + \frac{1}{2} \int_{\{u<n+1\}} |\zeta_\varepsilon \nabla u|^p dx$$

$$\leq \int_{\{u<n+1\}} (n + 1 - u)A(x,u,\nabla u)\nabla(\eta^p) dx$$

$$+ p(n + 1) \int_{\{u<n+1\}\cap B_{2\varepsilon}} |A(x,u,\nabla u)| \, |\nabla \xi_\varepsilon| dx$$

$$+ p(n + 1)^p \omega_N M^p \varepsilon^{N-p} + \int_{B_\rho} u^p \nabla(\eta^p) dx + e|B_{2\rho}|.$$

Now for any real $h > 0$ we have $n + 1 - u(x) > (n+1)h/(h+1)$ a.e. in $\{u < (n+1)/(h+1)\}$; hence dividing by $n+1$ we have from (3)

$$\frac{h}{h+1}\int_{\{u<\frac{n+1}{h+1}\}} g\,\zeta_\varepsilon^p\,dx + \frac{1}{2(n+1)}\int_{\{u<n+1\}} |\zeta_\varepsilon \nabla u|^p dx$$
$$\leqq \int_{\{u<n+1\}}\left(1 - \frac{u}{n+1}\right)A(x,u,\nabla u)\nabla(\eta^p)dx$$
$$+ pa\int_{\{u<n+1\}\cap B_{2\varepsilon}} |\zeta_\varepsilon \nabla u|^{p-1}|\nabla\xi_\varepsilon|dx$$
$$+ p\,2^N\omega_N\,M(b(n+1)^{p-1}+c)\varepsilon^{N-1}$$
$$+ p(n+1)^{p-1}2^N\,\omega_N\,M^p\,\varepsilon^{N-p}$$
$$+ \frac{1}{n+1}\left(\int_{B_\rho} u^p\,\nabla(\eta^p)dx + e|B_{2\rho}|\right);$$

hence from Hölder's inequality if $p > 1$, or directly if $p = 1$, we get an inequality of the form

$$(29)\qquad \frac{h}{h+1}\int_{\{u<\frac{n+1}{h+1}\}} g\,\zeta_\varepsilon^p\,dx + \frac{1}{4(n+1)}\int_{\{u<n+1\}} |\zeta_\varepsilon \nabla u|^p dx$$
$$\leqq \int_{\{u<n+1\}}\left(1 - \frac{u}{n+1}\right)A(x,u,\nabla u)\nabla(\eta^p)dx$$
$$+ c_1(N,p,A,M)\varepsilon^{N-p} + \frac{1}{n+1}c_2(N,p,u,\rho,\eta).$$

Now we let successively $\varepsilon \to 0$, $n \to +\infty$, $h \to 1$. With Fatou's Lemma we get $g\eta^p \in L^1(B_{2\rho})$ and it satisfies (25); hence $g \in L^1_{\text{loc}}(B_R)$. Moreover from (29), for any $n \geqq k_\rho$, we have an estimate of the form

$$(30)\qquad \int_{\{u<n+1\}\cap B_\rho} |\nabla u|^p dx \leqq (n+1)c_3(N,p,u,\rho,\eta,A).$$

Next take any reals $k \geqq k_\rho$ and $0 < \alpha \leqq 1$ in (28), then

$$\frac{1}{\alpha}\int_{\{k<u<k+\alpha\}} |\zeta_\varepsilon \nabla u|^p dx \leqq \int_{B_R} |A(x,u,\nabla u)|\,|\nabla(\eta)^p|dx$$
$$+ p\int_{\{u<k+1\}\cap B_{2\varepsilon}} (a|\nabla u|^{p-1}+bu^{p-1}+c)|\nabla\xi_\varepsilon|dx$$
$$+ \frac{d+e}{\alpha}\int_{\{k<u<k+\alpha\}} \zeta_\varepsilon u^p\,dx,$$

and from (30)

$$\int_{\{u<k+1\}\cap B_{2\epsilon}} |\nabla u|^{p-1}|\nabla \xi_\epsilon|dx \leqq (k+2)c_3^{(p-1)/p}\left(\int_{B_{2\epsilon}} |\nabla \xi_\epsilon|^p\right)^{1/p},$$

hence we get (26) when $\epsilon \to 0$, since $\{k < u < k+\alpha\} \cap B_{2\rho} \subset B_\rho$. $\qquad\square$

Now we will use the inequality (26) when $p > 1$ to prove Theorem 3.

Proposition 3. *Under the assumptions of Theorem 3, u satisfies equation* (11).

Proof. Let us estimate $|\nabla u|^{p-1}$ near the origin. Let $\rho > 0$ be small enough. For any $\sigma < \rho$ and any $0 < \delta < p - 1$ we have

$$(31) \quad \int_{B_\sigma} |\nabla u|^{p-1}dx = \int_{B_\sigma} \left|\frac{\nabla u}{(1+u)^{(\delta+1)}}\right|^{p-1} (1+u)^{(\delta+1)(p-1)/p}dx$$

$$\leqq \left(\int_{B_\sigma} \frac{|\nabla u|^p}{(1+u)^{\delta+1}}dx\right)^{(p-1)/p} \left(\int_{B_\sigma} (1+u)^{(\delta+1)/(p-1)}dx\right)^{1/p}.$$

Now from (26) and (20),

$$(32) \quad \int_{B_\rho} \frac{|\nabla u|^p}{(1+u)^{\delta+1}}dx = \int_{\{u<k_\rho\}\cap B_\rho} \frac{|\nabla u|^p}{(1+u)^{\delta+1}}dx$$

$$+ \sum_{k=k_\rho}^{+\infty} \int_{\{k<u<k+1\}\cap B_\rho} \frac{|\nabla u|^p}{(1+u)^{\delta+1}}dx$$

$$\leqq (k_\rho + 1)c_3 + \sum_{k=k_\rho}^{+\infty} \frac{1}{(1+k)^{\delta+1}}\left(c_\rho + (d+e)\int_{k<u<k+1} u^p dx\right)$$

$$\leqq (k_\rho + 1)c_3 + c_\rho \sum_{k=k_\rho}^{+\infty} \frac{1}{(1+k)^{\delta+1}}$$

$$+ (d+e)\int_{B_\rho} (1+u)^{p-1-\delta}dx < +\infty$$

since $u^{p-1} \in L^1(B_\rho)$ from Proposition 1. Assume also that $\delta < p/(N-p)$; then $(\delta + 1)(p - 1) < q_1$, hence from (31) and Proposition 1, $|\nabla u|^{p-1} \in L^1_{\text{loc}}(B_R)$ and there is a C_δ such that, for small σ,

$$(33) \quad \int_{B_\sigma} |\nabla u|^{p-1}dx \leqq C_\delta\, \sigma^{1-\delta(N-p)/p}.$$

Now from (3) we have $A(x, u, \nabla u) \in L^1_{\text{loc}}(B_R)$, hence we can define the distribution

$$T = -\text{div}(A(x, u, \nabla u)) - g \quad \text{in } \mathbf{D}'(B_R).$$

As in [4], we have $T = \sum_{|r| \leq m} \beta_r \, D^r \delta_0$. Let $\psi \in \mathbf{D}(B_r)$ be such that $(-1)^r D^r \psi(0) = \beta_r$ for every $|r| \leq m$, and $\psi_\sigma(x) = \psi(x/\sigma)$. Then

$$\langle T, \psi_\sigma \rangle = \sum_{|r| \leq m} \beta_r^2 \sigma^{-r} = \int_{B_R} A(x, u, \nabla u) \nabla \psi_\sigma \, dx - \int_{B_R} g \, \psi_\sigma \, dx.$$

As $g \in L^1_{\text{loc}}(B_R)$ we get from (3), (15), (33), with another C_δ

$$\sum_{|r| \leq m} \beta_r^2 \varepsilon^{-r} \leq C_\delta \, \sigma^{-\delta(N-p)/p};$$

now $(\delta(N - p)/p) < 1$, hence $\beta_r = 0$ when $|r| \geq 1$ and u satisfies (11). Finally, for any $\eta \in \mathbf{D}(B_{2\rho})$, $0 \leq \eta \leq 1$, $\eta = 1$ in B_ρ, we have

$$\langle T, \eta^p \rangle = \beta_0 = \int_{B_R} A(x, u, \nabla u) \nabla(\eta^p) dx - \int_{B_R} g \, \eta^p \, dx,$$

and hence $\beta_0 \geq 0$ from (25). $\qquad\qquad\qquad\qquad\qquad\qquad\qquad\qquad\Box$

At last we show the estimates for u and ∇u in Marcinkiewicz spaces. Benilan [1] has proved that if $u \in W^{1,p}_0(B_R)$ satisfies the inequality

$$(34) \qquad \int_{\{k < u < k + \alpha\}} |\nabla u|^p dx \leq c\alpha, \quad \forall k \geq 0, \, \forall \alpha > 0,$$

then $u^{p-1} \in M^{N/(N-p)}(B_R)$ and $|\nabla u|^{p-1} \in M^{N/(N-1)}(B_R)$.

His idea is to obtain a differential inequality for the function $k \to |\{u > k\}|$. Here we follow his method, with two technical difficulties: the function u may not be in $L^p(B_R)$ and even in $L^1(B_R)$, moreover (34) is replaced by (26), which makes the situation more complicated. [1] is still unpublished, henceforth for the sake of understanding, we give here his proof in our extended form.

Proposition 4. *Under the assumptions of Theorem 3, $u^{p-1} \in M^{N/(N-p)}_{\text{loc}}(B_R)$.*

Proof. First we show that a suitable power of $(1 + u)$ is in $W^{11}_{\text{loc}}(B_R)$. Let $\gamma \in (0, (p-1)/p)$; as $u \in W^{1,p}_{\text{loc}}(B'_R)$, we have by the chain rule $(1+u)^\gamma \in W^{1,p}_{\text{loc}}(B'_R)$, and

$$(35) \qquad \nabla((1 + u)^\gamma) = \gamma(1 + u)^{\gamma-1} \nabla u \quad \text{in } L^p_{\text{loc}}(B'_R).$$

Taking $\delta = p - 1 - p\gamma$ in (32) we get $(1+u)^\gamma \in L^p_{\mathrm{loc}}(B_R)$; now $\gamma < (N-1)(p-1)/(N-p) < q_1$, hence $(1+u)^\gamma \in L^1_{\mathrm{loc}}(B_R)$. Let θ be the gradient of $(1+u)^\gamma$ in $\mathbf{D}'(B_R)$. Then

$$\theta = \gamma(1+u)^{\gamma-1}\nabla u + \sum_{|r|\leqq m} \alpha_r \, D^r \, \delta_0.$$

Defining ψ_σ as in Proposition 3, we have

$$\langle \theta, \psi_\sigma \rangle = \sum_{|r|\leqq m} \alpha_r^2 \sigma^{-r} + \gamma \int_{B_R} (1+u)^{\gamma-1}\nabla u \, \psi_\sigma \, dx$$

$$= -\int_{B_R} (1+u)^\gamma \nabla \psi_\sigma \, dx,$$

hence from (15) there is a C such that, for small σ,

$$\sum_{|r|\leqq m} \alpha_r^2 \sigma^{-r} \leqq C\left(\sigma^{N-1-\frac{N-p}{p-1}\gamma} + \|(1+u)^{\gamma-1}\nabla u\|_{L^p(B_{R/2})}\sigma^{N(p-1)/p} \right).$$

From the choice of γ, we get $\alpha_r = 0$, $\forall |r| \leqq m$. Hence $(1+u)^\gamma \in W^{11}_{\mathrm{loc}}(B_R)$. Let $\rho > 0$ small enough as in Proposition 2; from Sobolev injection we have for any $w \in W^{11}_0(B_\rho)$, any $k \geqq 0$, $\alpha > 0$,

$$\|p_{k,\alpha}(w)\|_{L^{N/(N-1)}(B_\rho)} \leqq \frac{c_N}{\alpha} \int_{k<w<k+\alpha} |\nabla w| \, dx$$

and we verify easily, since $\gamma < 1$, that

$$p_{k,\alpha}(1+u) \leqq p_{k^\gamma,(k+\alpha)^\gamma - k^\gamma}((1+u)^\gamma) \quad \text{a.e. in } B_R,$$

hence for large k $(k \geqq k_\rho + 1)$ and $\alpha \in (0,1]$,

$$\|p_{k,\alpha}(1+u)\|_{L^{N/(N-1)}(B_\rho)}$$
$$\leqq \frac{c_N}{(k+\alpha)^\gamma - k^\gamma} \int_{B_\rho \cap \{k<1+u<k+\alpha\}} \gamma(1+u)^{\gamma-1}|\nabla u| \, dx$$
$$\leqq \frac{2\gamma(k+1)^{\gamma-1}c_N}{(k+\alpha)^\gamma - k^\gamma} \int_{B_\rho \cap \{k<1+u<k+\alpha\}} |\nabla u| \, dx$$

hence

$$(36) \qquad \|p_{k,\alpha}(1+u)\|_{L^{N/(N-1)}(B_\rho)} \leqq \frac{2c_N}{\alpha} \int_{B_\rho \cap \{k<1+u<k+\alpha\}} |\nabla u| \, dx.$$

Then as in [1] we define $\Phi(k) = |B_\rho \cap \{k < 1 + u\}|$ and get from (26) and (36), from Hölder's inequality,

$$\|p_{k,\alpha}(1+u)\|^p_{L^{N/(N-1)}(B_\rho)}$$
$$\leqq \frac{2^p c_N^p}{\alpha} \left(\frac{\Phi(k) - \Phi(k+\alpha)}{\alpha}\right)^{p-1} (c_\rho \alpha + (d+e)k^p(\Phi(k) - \Phi(k+\alpha))),$$

hence going to the limit when $\alpha \to 0$, we get for large k a differential inequality of the form

$$(37) \qquad \Phi(k)^{p(N-1)/N(p-1)} \leqq K(-\Phi'(k) + (-k\Phi'(k))^{p/(p-1)}),$$

where $K = K(N, p, \rho, A)$. This implies that Φ satisfies the estimate

$$(38) \qquad \Phi(k) \leqq C\,k^{-q_1} = C\,k^{-N(p-1)/(N-p)},$$

with $C = C(n, p, \rho, A)$. Indeed for any $x \in \mathbb{R}$ the equation $|x|^{p(N-1)/N(p-1)} + Kt - K|kt|^{p/(p-1)} = 0$ has only one solution t, set $t = H(x)$; suppose that Φ has not a compact support and compare Φ to the solution y of the Cauchy problem $y(k_\rho) = \Phi(k_\rho) > 0$ and $y' = H(y)$, that is

$$(39) \qquad \begin{cases} |y(k)|^{p(N-1)/N(p-1)} = K(-y'(k) + |ky'(k)|^{p/(p-1)}), \\[2mm] y'(k) < 0. \end{cases}$$

It is easy to see that H is a nonincreasing function on $\mathbb{R}^+$, hence $0 < \Phi \leqq y$ on $[k_\rho, +\infty)$; and $\lim_{k \to +\infty} y(k) = 0$, by contradiction. Suppose first that $g(k) = k^{q_1} y(k)$ is monotone for large k, then either it has a finite limit and (38) is proved, or $g(k)$ increases to $+\infty$, hence $g'(k) \geqq 0$, hence $|y'(k)| \leqq q_1 y(k)/k$,

$$|ky'(k)|^{p/(p-1)} = (q_1 y(k))^{p/(p-1)} = o(y(k)^{p(N-1)/N(p-1)}),$$

and $y(k)^{p(N-1)/N(p-1)} \leqq -2K\,y'(k)$ for large k; hence by integration we get (38). If $g(k)$ is not monotone for large k it has an infinity of maximum points s, where $|y'(s)| = q_1 y(s)/s$, $y(s)^{p(N-1)/N(p-1)} \leqq K(q_1 y(s)/s + (q_1 y(s))^{p/(p-1)})$, hence $g(s) \leqq (2Kq_1)^{q_1}$ for large s and (38) is proved. And (38) implies $u^{p-1} \in M^{N/(N-p)}(B_\rho)$, hence $u^{p-1} \in M^{N/(N-p)}_{\mathrm{loc}}(B_R)$.

Proposition 5. *Under the assumptions of Theorem 3, $|\nabla u|^{p-1} \in M^{N/(N-1)}_{\mathrm{loc}}(B_R)$.*

Proof. Here we define $K_\lambda = B_\rho \cap \{\lambda < |\nabla u|\}$ for any $\lambda > 0$. Following Proposition 3 and [1] we first estimate $|\nabla u|^{p-1}$ on K_λ. Let $0 < \delta < \min(p-1, p(N-p))$, and take any $h \geq k_\rho$; then

$$(40) \quad \lambda^{p-1}|K_\lambda| \leq \int_{K_\lambda} |\nabla u|^{p-1}dx$$

$$\leq \int_{K_\lambda \cap \{u<h\}} |\nabla u|^{p-1}dx + \int_{B_\rho \cap \{u>h\}} |\nabla u|^{p-1}dx;$$

from (30) we get

$$(41) \quad \int_{K_\lambda \cap \{u<h\}} |\nabla u|^{p-1}dx \leq |K_\lambda|^{1/p}((h+1)c_3)^{(p-1)/p};$$

moreover as in (31),

$$(42) \quad \int_{B_\rho \cap \{u>h\}} |\nabla u|^{p-1}dx \leq \left(\int_{B_\rho \cap \{u>h\}} \frac{|\nabla u|^p}{(1+u)^{\delta+1}}dx \right)^{(p-1)/p}$$

$$\cdot \left(\int_{B_\rho \cap \{u>h\}} (1+u)^{(\delta+1)(p-1)}dx \right)^{1/p};$$

now from (38), for any $\alpha < q_1$,

$$(43) \quad \int_{B_\rho \cap \{u>h\}} (1+u)^\alpha dx = -\int_{h+1}^{+\infty} t^\alpha \Phi'(t)dt$$

$$= (h+1)^\alpha \Phi(h+1) + \alpha \int_{h+1}^{+\infty} t^{\alpha-1}\Phi(t)\, dt$$

$$\leq C\frac{q_1}{q_1-\alpha}(1+h)^{\alpha-q_1};$$

let

$$\psi(h) = \int_{B_\rho \cap \{u>h\}} \frac{|\nabla u|^p}{(1+u)^{\delta+1}}dx$$

and

$$H(h) = \int_{B_\rho \cap \{u>h\}} (1+u)^{p-1-\delta}dx;$$

from (26) we deduce, when $\alpha \to 0$, that

$$-\psi'(h) \leq c_\rho(1+h)^{-(\delta+1)} - (d+e)H'(h),$$

hence by integration between h and $+\infty$, from (32) and (43) with $\alpha = p - 1 - \delta$,

$$\psi(h) \leqq c_\rho \delta^{-1}(1+h)^{-\delta} + (d+e)H(h) \leqq M(1+h)^{-\delta},$$

where $M = M(N, p, \rho, \delta, A)$; hence from (42) and (43) with $\alpha = (\delta+1)(p-1)$,

$$(44) \qquad \int_{K_\lambda \cap \{u > h\}} |\nabla u|^{p-1} dx \leqq M(1+h)^{(1-p)/(N-p)},$$

with another M, and from (40), (41) there is a C such that for any $\lambda > 0$ and $h \geqq k_\rho$,

$$\lambda^{p-1}|K_\lambda| \leqq C\big(|K_\lambda|^{1/p}(1+h)^{(p-1)/p} + (1+h)^{(1-p)/(N-p)}\big);$$

now $\lim_{\lambda \to 0} |K_\lambda| = 0$ since $|\nabla u|^{p-1} \in L^1(B_\rho)$, hence by minimization in h we get

$$\lambda^{p-1}|K_\lambda| \leqq C \frac{N}{N-p}\left(\frac{N-p}{N}\right)^{p/N} |K_\lambda|^{1/N},$$

hence with another C

$$(45) \qquad |K_\lambda| \leqq C \lambda^{(1-p)N/(N-1)} \quad \text{for any} \quad \lambda > 0,$$

hence $|\nabla u|^{p-1}$ (in the sense a.e. in B_R) is in $M^{N/(N-1)}(B_\rho)$, hence in $M^{N/(N-1)}_{\text{loc}}(B_R)$. $\qquad \Box$

Remark. From Proposition 5 we have in particular the estimate

$$(46) \qquad \int_{B_\sigma} |\nabla u|^{p-1} dx \leqq C_\rho \sigma \quad \text{for any} \quad \sigma < \rho < R,$$

which improves (33).

3. Behavior Near Infinity

Here we prove Theorem 4. The idea is to prove that if u is positive, we can construct a radial solution v of the same equation, such that $0 < v < u$, and then prove that the radial problem does not admit any positive solution.

Proof of Theorem 4. Let $u \in C^1(\Omega_R)$ be any solution of the equation

$$(47) \qquad Q_{\varepsilon,p}u + u^q = 0 \quad \text{in} \quad \mathbf{D}'(\Omega_R),$$

such that u is nonnegative and does not vanish identically on Ω; then u is positive everywhere in Ω_R, by the strong maximum principle [14], [16]. Let $\rho \in \mathbb{R}$ and $n \in \mathbb{N}$ be fixed, with $R < \rho < n$. By minimization we construct a sequence $(u_{n,k})_{k\in\mathbb{N}}$ of radial functions satisfying $u_{n,0} \equiv 0$ and, for any $k \geq 1$,

$$(48) \qquad \begin{cases} -Q_{\varepsilon,p}u_{n,k} = |u_{n,k-1}|^{q-1}u_{n,k-1} & \text{for } \rho < |x| < n, \\ u_{n,k}(x) = m & \text{for } |x| = \rho, \\ u_{n,k}(x) = 0 & \text{for } |x| = n, \end{cases}$$

where $m = \min_{|x|=\rho} u > 0$. Then $u_{n,k} > 0$ for $\rho < x < n$, and from the classical maximum principle we get

$$(49) \qquad u_{n,k} \leq u_{n,k+1} \leq u \quad \text{for} \quad \rho < |x| < n.$$

Now from (48), (49) and [13], $(u_{n,k})_{k\in\mathbb{N}}$ is bounded in $C^{1,\alpha}(\{\rho \leq |x| \leq n\})$ for some $\alpha \in (0,1)$, hence it converges strongly in $C^1(\{\rho \leq |x| \leq n\})$ to a radial function u_n such that $u_n \leq u$ for $\rho < |x| < n$, and

$$(50) \qquad \begin{cases} -Q_{\varepsilon,p}u_n = u_n^q & \text{for } \rho < |x| < n, \\ u_n(x) = m & \text{for } |x| = \rho, \\ u_n(x) = 0 & \text{for } |x| = n. \end{cases}$$

Consider now the sequence $(u_n)_{n\in\mathbb{N}}$; it is bounded in $C^{1,\alpha}(K)$ for any compact $K \subset \bar{\Omega}_\rho$, and $u_n \leq u_{n+1}$ for $\rho < |x| < n$, hence $(u_n)_{n\in\mathbb{N}}$ converges strongly in $C^1_{\mathrm{loc}}(\bar{\Omega}_\rho)$ to a nonnegative radial function v such that

$$(51) \qquad \begin{cases} -Q_{\varepsilon,p}v = v^q & \text{for } |x| > \rho, \\ v(x) = m & \text{for } x = |\rho|. \end{cases}$$

Under any of the two assumptions of Theorem 4, such a solution v cannot exist: in the case $\varepsilon = 0$, $q < q_1$, see [3]; in the case $\varepsilon = 1$, $q < N/(N-2)$, we can apply [9], [10] since $(1+s^2)^{(p-2)/2}$ is bounded near 0. Hence we get a contradiction. $\qquad\qquad\square$

REFERENCES

[1] Ph. Benilan, *personnal communication* (1988).

[2] Ph. Benilan, H. Brezis and M.G. Crandall, *A semilinear elliptic equation in $L^1(\mathbb{R}^N)$*, Ann. Scuola Norm Sup. Pisa, **2**(1975), 523–555.

[3] M.F. Bidaut-Veron, *Local and global behavior of solutions of quasilinear equations of Emden–Fowler type*, Arch. for Rat. Mech. Anal. **107** (4) (1989), 293–324.

[4] H. Brezis and P.L. Lions, *A note on isolated singularities for linear elliptic equations*, Jl. Math. Anal. and Appl. **7A**(1981), 263–266.

[5] A. Friedman and L. Veron, *Singular solutions of some quasilinear elliptic equations*, Arch. for Rat. Mech. Anal. **96**(1986), 259–287.

[6] M. Guedda and L. Veron, *Local and global properties of solutions of quasilinear equations*, Jl. Diff. Eq. **76**(1988), 159–189.

[7] S. Kichenassamy and L. Veron, *Singular solutions of the p-Laplace equation*, Math. Ann. **275**(1986), 599–615.

[8] P.L. Lions, *Isolated singularities in semilinear problems*, Jl. Diff. Eq. **38**(1980), 441–450.

[9] W.M. Ni and J. Serrin, *Existence and nonexistence theorems for ground states of quasilinear partial differential equations: the anomalous case*, Acad. Naz. dei. Lincei **77**(1986), 231–257.

[10] P. Pucci and J. Serrin, *Continuation and limit properties for solutions of strongly nonlinear second order differential equations*, to appear.

[11] J. Serrin, *Local behavior of solutions of quasilinear equations*, Acta Math., **111**(1964), 247–302.

[12] J. Serrin, *Isolated singularities of solutions of quasilinear equations*, Acta Math. **113**(1965), 219–240.

[13] P. Tolksdorff, *On the Dirichlet problem for quasilinear equations in domains with conical boundary points*, Comm. in Part. Diff. Eq. **8** (7) (1983), 773–817.

[14] P. Tolksdorff, *Regularity for a more general class of quasilinear elliptic equations*, Jl. Diff. Eq. **51**(1984), 126–150.

[15] N.S. Trudinger, *On Harnack type inequalities and their applications to quasi-linear elliptic equations*, Comm. in Pure and Appl. Math. **20**(1967), 721–747.

[16] J.L. Vazquez, *A strong maximum principle for some quasilinear elliptic equations*, Appl. Math. Opt. **12**(1984), 191–202.

Département de Mathématiques
Faculté des Sciences
Parc de Grandmont
37200 Tours, FRANCE

An Existence Result Via L^s Regularity for Some Nonlinear Elliptic Equations

LUCIO BOCCARDO and FRANÇOIS MURAT

Abstract

This note is concerned with the existence and L^s-regularity of the solutions of the equation

$$\begin{cases} A(u) = -\operatorname{div}(c(x) + \varphi(u)) & \text{in } \Omega \\ u = 0 & \text{on } \partial\Omega \end{cases}$$

where A is a Leray–Lions operator defined from $W_0^{1,p}(\Omega)$ into its dual, $\varphi: \mathbb{R} \to \mathbb{R}^N$ is a continuous function satisfying $|\varphi(s)| \le c_0(1 + |s|^\gamma)$ and $c(x)$ belongs to $(L^q(\Omega))^N$. When $p' \le q < \frac{N}{p-1}$ and $0 \le \gamma \le [q(p-1)]^*$, we prove the existence of a solution u which belongs to $W_0^{1,p}(\Omega) \cap L^{[q(p-1)]^*}(\Omega)$. This L^s-regularity result (which implies that $\varphi(u)$ belongs to $(L^1(\Omega))^N$) is an important step in the proof of the existence of a solution.

1. The Result: Statement and Comments

This note deals with the existence and L^s-regularity of solutions of the following nonlinear elliptic equation

$$(1.1) \qquad \begin{cases} A(u) = -\operatorname{div}(c(x) + \varphi(u)) & \text{in } \mathcal{D}'(\Omega) \\ u = 0 & \text{on } \partial\Omega. \end{cases}$$

Here Ω is a bounded open subset of $\mathbb{R}^N$ (no smoothness is assumed on its boundary $\partial\Omega$), p and p' are given real numbers with

$$(1.2) \qquad 1 < p,\ p' < +\infty, \quad \frac{1}{p} + \frac{1}{p'} = 1$$

and A is a Leray–Lions operator from $W_0^{1,p}(\Omega)$ into its dual, defined by

$$(1.3) \qquad A(u) = -\operatorname{div} a(x, u, Du),$$

where $a: \Omega \times \mathbb{R} \times \mathbb{R}^N \to \mathbb{R}^N$ is a Caratheodory function which satisfies, for some $\alpha > 0$, $\beta > 0$:

$$(1.4) \quad \begin{cases} |a(x,s,\xi)| \leq \beta(1 + |s|^{p-1} + |\xi|^{p-1}) \\ a(x,s,\xi) \cdot \xi \geq \alpha|\xi|^p \\ [a(x,s,\xi) - a(x,s,\eta)] \cdot [\xi - \eta] > 0, \quad \forall \xi \neq \eta. \end{cases}$$

Concerning the right-hand side of (1.1) we assume that

$$(1.5) \qquad\qquad c(x) \in (L^q(\Omega))^N$$

$$(1.6) \qquad\qquad \varphi \in (C^0(\mathbb{R}))^N$$

$$(1.7) \qquad\qquad |\varphi(s)| \leq c_0(1 + |s|^\gamma), \quad c_0 \in \mathbb{R}^+$$

where the exponents q and γ will be specified below.

Denoting by r^* the real number

$$r^* = \frac{rN}{N - r}, \quad \text{where} \quad 1 \leq r < N,$$

the result of this note is the following:

Theorem 1. *Assume that (1.2)–(1.7) hold true. Whenever the exponents q and γ satisfy*

$$(1.8) \qquad\qquad p' \leq q < \frac{N}{p-1}$$

$$(1.9) \qquad\qquad 0 \leq \gamma \leq [q(p-1)]^*$$

there exists at least one solution u of (1.1) which satisfies

$$(1.10) \qquad\qquad u \in W_0^{1,p}(\Omega) \cap L^{[q(p-1)]^*}(\Omega).$$

Remark 1. Note that $A(u)$ belongs to $W^{-1,p'}(\Omega)$ when u belongs to $W_0^{1,p}(\Omega)$, and that $\varphi(u)$ belongs to $(L^1(\Omega))^N$ when $u \in L^{[q(p-1)]^*}(\Omega)$ while φ satisfies (1.6), (1.7) with an exponent γ bounded by (1.9). Any term in equation (1.1) has thus a meaning in distributional sense in the setting of Theorem 1.

We refer the reader to [BDGM1], [BDGM2] for the proof of the existence of a "renormalized solution" in the case where $q = p'$ and where no growth condition is assumed on φ. In this case the renormalized solution u only belongs to $W_0^{1,p}(\Omega)$ and in general $\varphi(u)$ does not belong to $(L^1(\Omega))^N$.

Remark 2. In (1.8) q is assumed to satisfy $q \geq p'$ in order for $\operatorname{div} c(x)$ to belong to $W^{-1,p'}(\Omega)$. The case where $q > \frac{N}{p-1}$, φ satisfying only (1.6) but no growth condition, and the case where $q = \frac{N}{p-1}$, φ satisfying (1.6) and the growth condition (1.7) for a given γ, $0 \leq \gamma < +\infty$, are easier since in these cases u belongs to $L^\infty(\Omega)$ or to $L^s(\Omega)$ for any $1 \leq s < +\infty$. They are treated in [BG2]. Note finally that (1.8) implies $p < N$.

Remark 3. Theorem 1 unifies and improves two results obtained in [BG2]:

(i) Theorem 2 of [BG2] gave the result of the present Theorem 1 under the further assumption that $0 \leq \gamma < [q(p-1)]^*/p'$. Note that $1/p' < 1$, so that (1.9) allows φ to grow faster.

(ii) Theorem 5 of [BG2] gave the result of the present Theorem 1 in the quasilinear case (i.e. when $p = 2$ and $a(x,s,\xi) = B(x,s)\xi$ for some matrix B) assuming $0 \leq \gamma < q^*$. Note that for $p = 2$ equality $\gamma = q^*$ is allowed to hold in (1.9) even when the operator is nonlinear with respect to Du.

Remark 4. Theorem 1 also improves the existence results of a standard weak solution obtained via renormalization in [BDGM1], [BDGM2]. In these papers the existence of a weak solution is proved when φ satisfies (1.7) and (1.9) as well as a further condition, which restricts the class of allowed φ's (see Theorem 1.2 in [BDGM1] and Section 1.6 in [BDGM2]).

The proof of Theorem 1 which we give in Section 2 below can be described as follows. In the first step we consider an approximation φ^ε of φ which belongs to $(C^0(\mathbb{R}) \cap (L^\infty(\mathbb{R}))^N$. Classical results ([LL]) then imply the existence of a sequence of solutions $u^\varepsilon \in W_0^{1,p}(\Omega)$, which is bounded in $W_0^{1,p}(\Omega)$. In a second step we prove that u^ε is actually relatively compact in the strong topology of $W_0^{1,p}(\Omega)$. This result, which is the first key point of the proof of Theorem 1, is achieved through the use of nonlinear (with respect to u^ε) test functions, following exactly the proof of [BDGM1], [BDGM2]. In the third step we prove that under assumptions (1.8) and (1.9), u^ε is relatively compact in the strong topology of $L^{[q(p-1)]^*}(\Omega)$. This result is the second key point of the proof of Theorem 1. It is obtained by an improvement of the proof of [BG1], [BG2]: in these papers u^ε was only proved to be bounded in $L^{[q(p-1)]^*}(\Omega)$, while we prove here also the equiintegrability of u^ε in $L^{[q(p-1)]^*}(\Omega)$ by the use of refined test functions. It is then easy to pass to the limit in ε and to obtain Theorem 1.

2. Proof of Theorem 1

Step 1. Let $T_m(t)$ be the truncation to level $m > 0$ defined by:

$$T_m(t) = \begin{cases} t & \text{if } |t| \leq m \\ mt/|t| & \text{if } |t| \geq m. \end{cases}$$

Define an approximation φ^ε of φ by:

$$\varphi^\varepsilon(t) = \varphi(T_{1/\varepsilon}(t)) \quad \text{for } \varepsilon > 0$$

and an approximation $c^\varepsilon(x)$ of $c(x)$ such that

$$\begin{cases} c^\varepsilon \in (L^\infty(\Omega))^N \\ c^\varepsilon \to c \quad \text{strongly in } (L^q(\Omega))^N. \end{cases}$$

Consider the nonlinear elliptic equation

$$(2.1) \quad \begin{cases} u^\varepsilon \in W_0^{1,p}(\Omega) \\ A(u^\varepsilon) = -\operatorname{div}(a(x, u^\varepsilon, Du^\varepsilon)) = -\operatorname{div}(c^\varepsilon(x) + \varphi^\varepsilon(u^\varepsilon)) \quad \text{in } \mathcal{D}'(\Omega). \end{cases}$$

Since φ^ε lies in $(C^0(\mathbb{R}) \cap L^\infty(\mathbb{R}))^N$, a classical result due to Leray and Lions [LL] implies that (2.1) has at least one solution. This solution u^ε belongs to $L^\infty(\Omega)$ for any $\varepsilon > 0$ fixed ([S], [BG1]).

Moreover multiplying (2.1) by u^ε and integrating by parts we obtain

$$(2.2) \qquad \int_\Omega a(x, u^\varepsilon, Du^\varepsilon) Du^\varepsilon = \int_\Omega c^\varepsilon(x) Du^\varepsilon + \int_\Omega \varphi^\varepsilon(u^\varepsilon) Du^\varepsilon.$$

Denoting by $\psi^\varepsilon : \mathbb{R} \to \mathbb{R}^N$ the function $\psi^\varepsilon(t) = \int_0^t \varphi^\varepsilon(s) ds$ we have by Stokes' Theorem

$$\int_\Omega \varphi^\varepsilon(u^\varepsilon) Du^\varepsilon = \int_\Omega \operatorname{div}(\psi^\varepsilon(u^\varepsilon)) = \int_{\partial\Omega} \psi^\varepsilon(u^\varepsilon) n = 0.$$

(Note that all the above computations are licit since φ^ε is bounded.) It is then easily deduced from (2.2) and from the coerciveness (1.4) that u^ε is bounded in $W_0^{1,p}(\Omega)$, since $c(x)$ belongs to $(L^{p'}(\Omega))^N$ (see (1.8)). Thus for some subsequence (again denoted by u^ε) there exists u in $W_0^{1,p}(\Omega)$ such that

$$(2.3) \qquad u^\varepsilon \rightharpoonup u \quad \text{weakly in } W_0^{1,p}(\Omega) \text{ and a.e. in } \Omega.$$

Step 2. We are here in position to perform the second step of the proof of [BDGM2] (see also Theorem 2.1 of [BDGM1] in the case where the operator A is linear). This proof (which will not be reproduced here) uses successively the test functions $u^\varepsilon - T_k(u^\varepsilon)$ and $T_i(u^\varepsilon - T_j(u))$ and gives the following compactness result:

$$(2.4) \qquad\qquad u^\varepsilon \to u \quad \text{strongly in} \quad W_0^{1,p}(\Omega).$$

Step 3. This is the most original part of the present proof: we will prove that

$$(2.5) \qquad\qquad u^\varepsilon \to u \quad \text{strongly in} \quad L^{[q(p-1)]^*}(\Omega).$$

Defining G_m by

$$G_m(t) = t - T_m(t)$$

we consider the test function

$$(2.6) \qquad\qquad w = |G_m(u^\varepsilon)|^r G_m(u^\varepsilon)$$

where

$$(2.7) \qquad\qquad r = \frac{N[q(p-1)-p]}{N - q(p-1)}.$$

Note that in view of (1.8) we have $0 \le r < +\infty$; the case $r = 0$ corresponds to $q = p'$.

The use of such test functions traces back to the work of G. Stampacchia [S] who used the test function $w = G_m(u^\varepsilon)$ to prove the L^∞ regularity of the solutions of the linear second order equation. On the other hand, using the test function $w = |u|^r u = |G_0(u)|^r G_0(u)$ with the same r as in (2.7) allowed the authors of [BG1] to prove the $L^{[q(p-1)]^*}$-regularity of solutions of equation (1.1) when $\varphi = 0$.

We mix here the two ideas, choosing $w = |G_m(u^\varepsilon)|^r G_m(u^\varepsilon)$ in (2.1). This allows us to prove (2.5), which is stronger than to prove an $L^{[q(p-1)]^*}(\Omega)$ estimate on Ω.

Multiplying (2.1) by w defined by (2.6) yields, using the coerciveness:

$$\alpha \int_\Omega |G_m(u^\varepsilon)|^r |DG_m(u^\varepsilon)|^p \le \int_\Omega |G_m(u^\varepsilon)|^r a^\varepsilon(x, u^\varepsilon, Du^\varepsilon) DG_m(u^\varepsilon)$$

$$(2.8) \qquad\qquad = \int_\Omega |G_m(u^\varepsilon)|^r c^\varepsilon(x) DG_m(u^\varepsilon)$$

$$+ \int_\Omega |G_m(u^\varepsilon)|^r \varphi^\varepsilon(u^\varepsilon) DG_m(u^\varepsilon).$$

The last term of (2.8) is equal to zero since defining $\psi_m^\varepsilon : \mathbb{R} \to \mathbb{R}^N$ by $\psi_m^\varepsilon(t) = \int_0^t |G_m(s)|^r \varphi^\varepsilon(s)ds$ we have (note that the computations are licit since $u^\varepsilon \in L^\infty(\Omega)$ for $\varepsilon > 0$ fixed)

$$(2.9) \qquad \int_\Omega |G_m(u^\varepsilon)|^r \varphi^\varepsilon(u^\varepsilon)DG_m(u^\varepsilon) = \int_\Omega |G_m(u^\varepsilon)|^r \varphi^\varepsilon(u^\varepsilon)Du^\varepsilon$$

$$= \int_\Omega \operatorname{div}(\psi_m^\varepsilon(u^\varepsilon)) = \int_{\partial\Omega} \psi_m^\varepsilon(u^\varepsilon)n = 0.$$

Define now

$$A_m^\varepsilon = \{x \in \Omega : |u^\varepsilon(x)| > m\};$$

using Young's inequality in the first term of the right-hand side of (2.8) yields:

$$(2.10) \int_\Omega |G_m(u^\varepsilon)|^r c^\varepsilon(x)DG_m(u^\varepsilon)$$

$$= \int_{A_m^\varepsilon} |G_m(u^\varepsilon)|^{r/p}DG_m(u^\varepsilon)|G_m(u^\varepsilon)|^{r(1-\frac{1}{p})}c^\varepsilon(x)$$

$$\leq \frac{\alpha}{2} \int_\Omega |G_m(u^\varepsilon)|^r |DG_m(u^\varepsilon)|^p + K_\alpha \int_{A_m^\varepsilon} |G_m(u^\varepsilon)|^r |c^\varepsilon(x)|^{p'}.$$

The first term of the last line is absorbed by the left-hand side of (2.8) while the second term is estimated by Cauchy–Schwartz inequality with exponents $s = q/p'$, $s' = q/(q - p')$ (note that $s \geq 1$ in view of (1.8); the case $s = 1$ $(s' = \infty)$ corresponds to $r = 0$ (see above) and does not give trouble in the estimate):

$$\int_{A_m^\varepsilon} |c^\varepsilon(x)|^{p'} |G_m(u^\varepsilon)|^r \leq \left(\int_{A_m^\varepsilon} |c^\varepsilon(x)|^q \right)^{\frac{p'}{q}} \left(\int_\Omega |G_m(u^\varepsilon)|^{\frac{rq}{q-p'}} \right)^{\frac{q-p'}{q}}.$$

Note now that $p < N$ because of (1.8) and that r defined by (2.7) satisfies:

$$(2.11) \qquad \frac{rq}{q - p'} = [q(p - 1)]^* = \left(\frac{r}{p} + 1 \right)p^*.$$

Using Sobolev's inequality in the left-hand side of (2.8):

$$(2.12) \qquad \int_\Omega |G_m(u^\varepsilon)|^r |DG_m(u^\varepsilon)|^p$$

$$= \left(\frac{r}{p} + 1 \right)^{-p} \int_\Omega (D[|G_m(u^\varepsilon)|^{r/p}G_m(u^\varepsilon)])^p$$

$$\geq K_{\Omega,p} \left(\int_\Omega |G_m(u^\varepsilon)|^{(\frac{r}{p}+1)p^*} \right)^{\frac{p}{p^*}},$$

we finally deduce from (2.8) that

$$(2.13) \qquad \int_\Omega |G_m(u^\varepsilon)|^{[q(p-1)]^*} \leq K_0 \left(\int_{A_m^\varepsilon} |c^\varepsilon(x)|^q \right)^{\frac{N}{N-q(p-1)}}.$$

For any measurable subset E of Ω we have, since $|t| \leq |G_m(t)| + m$:

$$(2.14) \quad \int_E |u^\varepsilon|^{[q(p-1)]^*} \leq K_1 \int_E |G_m(u^\varepsilon)|^{[q(p-1)]^*} + K_1 \int_E m^{[q(p-1)]^*}$$

$$\leq K_1 K_0 \left(\int_{A_m^\varepsilon} |c^\varepsilon(x)|^q \right)^{\frac{N}{N-q(p-1)}} + K_1 m^{[q(p-1)]^*} |E|.$$

Since c^ε is compact in $(L^q(\Omega))^N$ and since the boundedness of u^ε in $L^p(\Omega)$ implies that the measure of A_m^ε is small (uniformly in ε) when m is large, convenient choices first of m, then of the measure $|E|$ of E, show that $|u^\varepsilon|^{[q(p-1)]^*}$ is equi-integrable (uniformly in ε). Since u^ε is already known to converge almost everywhere, Vitali's Theorem implies (2.5).

Step 4. It is now easy to pass to the limit in each term of (2.1). Indeed

$$a(x, u^\varepsilon, Du^\varepsilon) \to a(x, u, Du) \quad \text{strongly in} \quad (L^{p'}(\Omega))^N$$

in view of (2.4). On the other hand, the growth condition (1.7), (1.9) on φ, the definition of φ^ε and the strong convergence (2.5) imply that

$$\varphi^\varepsilon(u^\varepsilon) \to \varphi(u) \quad \text{strongly in} \quad (L^1(\Omega))^N.$$

This completes the proof of Theorem 1.

Acknowledgment. This work was done as Lucio Boccardo was visiting INRIA. Both authors thank this institution for supporting his visit.

REFERENCES

[BDGM1] L. Boccardo, I. Diaz, D. Giachetti, and F. Murat, *Existence of a solution for a weaker form of a nonlinear elliptic equation,* in *Recent advances in nonlinear elliptic and parabolic problems (Proceedings, Nancy 1988),* P. Benilan, M. Chipot, L.C. Evans, and M. Pierre eds., Pitman Research Notes in Mathematics, **208**(1989), 229–246.

[BDGM2] L. Boccardo, I. Diaz, D. Giachetti, and F. Murat, *Existence and regularity of renormalized solutions for some elliptic problems involving derivatives of nonlinear terms,* to appear.

[BG1] L. Boccardo and D. Giachetti, *Alcune osservazioni sulla regolarità delle soluzioni di problemi fortemente non lineari e applicazioni,* Ric. Mat. **34**(1985), 309–323.

[BG2] L. Boccardo and D. Giachetti, *Existence results via regularity for some nonlinear elliptic problems,* Comm. P.D.E. **14**(1989), 663–680.

[LL] J. Leray and J.L. Lions, *Quelques résultats de Visik sur les problèmes elliptiques non linéaires par la méthode de Minty–Browder,* Bull. Soc. Math. France **93**(1965), 97–107.

[S] G. Stampacchia, *Le problème de Dirichlet pour les équations elliptiques du second ordre à coefficients discontinus,* Ann. Inst. Fourier Grenoble **15**(1965), 189–258.

Dipartimento di Matematica
Università di Roma I
Piazzale A. Moro 2
00185 Roma

Laboratoire d'Analyse Numérique
Université Paris VI
4 Place Jussieu
75252 Paris Cedex 05

Identifying a Time Dependent Unknown Coefficient in a Nonlinear Heat Equation

J.R. CANNON[1], PAUL DUCHATEAU[2] and KEN STEUBE

Abstract

In this paper we show how an unknown coefficient inverse problem may be reformulated as a problem involving a trace type functional partial differential equation. The trace type functional problem can be solved in a relatively straightforward manner and the solution shown to lead to a solution for the inverse problem. This approach allows us to investigate the inverse problem with various alternative overspecified conditions.

0. Introduction

For T a generic positive constant let $Q_T \equiv \{x > 0,\ 0 < t < T\}$. Let B_T denote the Banach space of functions $u = u(x,t)$ such that u, u_x belong to $C[Q_T]$ and for which $\|u\|_T < \infty$ where,

$$\|u\|_T \equiv \sup_{Q_T} |u| + \sup_{Q_T} |u_x| .$$

We are interested in solving and unknown coefficient inverse problem in which we are to find unknown functions $u = u(x,t)$ and $a = a(t)$ satisfying,

$$
\begin{aligned}
u_t(x,t) - u_{xx}(x,t) &= a(t)F\big[u(x,t)\big] \text{ in } Q_T \\
u(x,0) &= 0, \qquad x > 0, \\
u_x(0,t) &= g(t), \qquad 0 < t < T, \\
u(0,t) &= f(t), \qquad 0 < t < T, \\
u(x,t) & \text{ bounded on } Q_T .
\end{aligned}
\tag{0.1}
$$

Problem (0.1) is an example of an unknown ingredient inverse problem. Such problems have received considerable attention in the literature over the past 25 years; c.f. [1] to [9]. These references show a diverse selection of ad hoc approaches for treating such problems. It is our aim here to

---

[1]Supported in part by NSF grant DMS-8901301.

[2]Supported in part by ONR contract number N00014-85-K-0224.

show that a uniform approach may be feasible, at least in the cases where the unknown ingredient(s) are functions of only one variable. We illustrate this approach by applying it to the example (0.1). References [14] to [17] provide a survey of the technique.

Note that if the coefficient $a = a(t)$ is known then a well posed initial boundary value problem (IBVP) for the unknown function $u(x,t)$ remains when one or the other of the boundary conditions at $x = 0$ is removed from (0.1). It seems plausible that by including the extra boundary condition which appears in (0.1) it then becomes possible to simultaneously determine the two unknown functions $u(x,t)$ and $a(t)$. However, it is far from evident that such an overspecification uniquely determines the function pair $\{u, a\}$ nor is it clear that there are not alternative overspecifications which are preferable. We shall show that the inverse problem (0.1) is well posed in the sense that there is exactly one pair of functions $a = a(t)$ and $u = u(x,t)$ that satisfies all of the conditions of (0.1). In addition we shall consider inverse problems based on alternative overspecifications.

The problem (0.1) includes two boundary conditions imposed at the single point $x = 0$. In any physical realization of the situation modeled by (0.1) it would be impossible to simultaneously satisfy both conditions for arbitrary functions $f(t)$ and $g(t)$. We interpret this to mean that one of the conditions is *controlled* while the other is simply *observed*. For example, we could force $u_x(0,t)$ to equal the arbitrary prescribed function $g(t)$ while observing the values of $u(0,t)$ and recording them as the function $f(t)$. Conversely we could control the values of $u(0,t)$ and observe the values of $u_x(0,t)$. On the face of it there is nothing to identify which of the conditions should be controlled and which should be observed.

Let us use the term "direct" problem to refer to the well posed IBVP that remains when the observed boundary condition is deleted from (0.1) (but of course the controlled boundary condition is retained). For each admissible choice of the coefficient $a(t)$, the direct problem has a unique solution $u = u(x,t;a)$. Here the notation is intended to indicate the dependence of $u(x,t)$ on $a(t)$. Then to solve the inverse problem we must find a coefficient $a(t)$ which produces a solution $u(x,t;a)$ for the direct problem that satisfies the observed boundary condition.

Of course it may happen that if the controlled boundary condition in the direct problem is not *properly* controlled then no such $a(t)$ can be found. Alternatively, it may be the case that more than one coefficient $a(t)$ produces a solution $u(x,t;a)$ for the direct problem which also satisfies the observed boundary condition. This situation could arise if the unknown coefficient $a(t)$ is not sufficiently sensitive to the quantity measured in the observed boundary condition.

In the sections to follow we propose to show that in the inverse problem (0.1), it is the Dirichlet condition $u(0,t) = f(t)$ that must be the observed condition and it is the Neumann condition that must be controlled. We shall also show that other overspecifications are allowable and that it is possible to assess the relative sensitivity of $a(t)$ to these various alternative overspecifications. We shall also show that it is possible to ascertain how to properly control the controlled boundary condition.

1. The Direct IBVP

In section 2 we are going to show that the inverse problem (0.1) may be solved by reformulating it as a so-called trace type functional problem. One of the boundary conditions (in this case it is the Dirichlet boundary condition) is used to eliminate the unknown coefficient $a(t)$ from the partial differential equation and to replace it with a coefficient that is a (trace type) functional of the solution $u(x,t)$. The resulting problem is referred to as a trace type functional problem. We shall solve this problem and show that its solution leads immediately to a solution of the original inverse problem. In order to do all of this we shall need to obtain certain information about the solution of the following IBVP.

$$u_t(x,t) - u_{xx}(x,t) = a(t)F\big[u(x,t)\big] \quad \text{in } Q_T$$
$$u(x,0) = 0 \qquad \text{for } x > 0,$$
$$-u_x(0,t) = g(t) \qquad \text{for } 0 < t < T, \tag{1.1}$$
$$u(x,t) \text{ bounded on } Q_T.$$

Here we suppose $a(t), F(u)$ and $g(t)$ are all given and that there exist positive constants a_1, F_0, F_1 and δ such that,

$$i) \ g \in C[0,\infty) \text{ with } g(0) = 0$$
$$ii) \ F \in C(-\infty,\infty) \text{ and} \tag{1.2}$$
$$0 < F_0 \leq F(z) \leq F_1 \text{ for } |z| \leq \delta$$
$$iii) \ a \in C[0,T] \text{ with } \big|a(t)\big| \leq a_1 \text{ for } 0 \leq t \leq T.$$

Now we shall investigate the problem (1.1) and some of the properties of its solution.

Theorem 1.1 *Under the assumptions (1.2), the problem (1.1) has a unique classical solution $u = u(x,t)$.*

Proof. Let $U(x,t)$ satisfy,

$$U_t(x,t) - U_{xx}(x,t) = a_1 F_1 \quad \text{in } Q_T$$
$$U(x,0) = 0, \qquad x > 0,$$
$$-U_x(0,t) = g(t), \quad 0 < t < T,$$
$$U(x,t) \text{ bounded on } Q_T.$$

Then,

$$[\partial_{xx} - \partial_t]U(x,t) + a(t)F[U] = a(t)F[U] - a_1 F_1 < 0$$

and it follows that $U(x,t)$ is an upper solution for (1.1). Similarly, if $V(x,t)$ solves,

$$V_t(x,t) - V_{xx}(x,t) = -a_1 F_1 \quad \text{in } Q_T$$
$$V(x,0) = 0 \qquad x > 0,$$
$$-V_x(0,t) = g(t), \qquad 0 < t < T,$$
$$V(x,t) \text{ bounded in } Q_T,$$

then $V(x,t)$ is a lower solution for (1.1). The maximum principle implies that $U > V$ in Q_T and then it follows by monotonicity that there exists a unique classical solution for (1.1). Moreover,

$$V(x,t) \leq u(x,t) \leq U(x,t) \quad \text{in } Q_T. \tag{1.3}$$

We easily find that

$$U(x,t) = 2 \int_0^t K(x,t-\tau)g(\tau)d\tau + a_1 F_1 t,$$

where,

$$K(x,t) \equiv (4\pi t)^{-1/2} \exp[-x^2/4t], \ t > 0.$$

If we add to the assumptions (1.2), the condition that,

$$g \in C^1[0,T], \text{ with } g(0) = 0, \text{ and } g'(t) > 0 \text{ for } t > 0 \tag{1.4}$$

then,

$$0 \leq U(x,t) \leq \{g(T) + a_1 F_1\}T \equiv C_1(T).$$

In addition, (1.3) implies that,

$$|u(x,t)| \leq C_1(T) \quad \text{in } Q_T \tag{1.5}$$

Since $C_1(0) = 0$, we can choose the parameter $T > 0$ such that, $C_1(T) = \delta$. This ensure that under the conditions (1.2) and (1.4) we have $F[u(x,t)] \geq F_0 > 0$, for (x,t) in Q_T and $u(x,t)$ solving (1.1).

Now add the following assumptions to (1.2) and (1.4),

$$F(z),\ F'(z),\ F''(z) \text{ all continuous for } |z| \leq C_1(T) = \delta,$$
$$\text{and } |F'(z)| \leq F_2 \qquad \text{for } |z| \leq C_1(T) \tag{1.6}$$
$$|F''(z)| \leq F_3 \qquad \text{for } |z| \leq C_1(T)$$

Then, under the prevailing assumptions,

$$w(x,t) \equiv u_x(x,t)$$

satisfies,

$$w_t(x,t) - w_{xx}(x,t) = a(t)F'\big[(x,t)\big]w(x,t) \quad \text{in } Q_T$$
$$w(x,0) = 0, \qquad x > 0,$$
$$w(0,t) = -g(t), 0 < t < T,$$
$$w(x,t) \text{ bounded in } Q_T.$$

Then, using the maximum principle, we can show

$$0 \geq w(x,t) \geq 2\exp[a_1 F_2 t] \int_0^t K_x(x,t-\tau)g(\tau)\,d\tau \text{ in } Q_T.$$

That is,

$$\sup_{Q_T} \big|u_x(x,t)\big| \leq C_2(T) \tag{1.7}$$

where,

$$C_2(T) \equiv 2\,g(T)\exp[a_1 F_2 T](T/\pi)^{1/2}$$

Note that if $u(x,t)$ solves (1.1), then we can write,

$$u(x,t) = 2\int_0^t K(x,t-\tau)g(\tau)d\tau + \iint_{Q_t} N(x,y,t-\tau)a(\tau)F\big[u(y,\tau)\big]\,dyd\tau$$

where,

$$N(x,y,t) = K(x-y,t) + K(x+y,t).$$

Then,

$$u_{xx}(x,t) = -2\int_0^t K_\tau(x,t-\tau)g(\tau)d\tau$$
$$+ \iint_{Q_t} N_{yy}(x,y,t-\tau)a(\tau)F\big[u(y,\tau)\big]\,dyd\tau$$

where we have used the facts,

$$K_{xx}(x,t) = K_t(x,t), \quad N_{xx}(x,y,t) = N_{yy}(x,y,t).$$

Integration by parts leads to,

$$U_{xx}(x,t) = 2\int_0^t K(x,t-\tau)g'(\tau)\,d\tau$$
$$- \iint_{Q_T} N_y(x,y,t-\tau)a(\tau)F'[u]u_y(y,\tau)\,dy d\tau$$

It follows then, that

$$\sup_{Q_T} |u_{xx}| \le T \sup_{[0,T]} |g'| + a_1 F_2 C_2(T)(4T/\pi)^{1/2} \equiv C_3(T) \tag{1.8}$$

2. Existence of a Solution to the Trace Type Functional Problem

Suppose that $u(x,t)$, $a(t)$ together solve the inverse problem (0.1). Then, formally,

$$a(t)F\big[u(0,t)\big] = u_t(0,t) - u_{xx}(0,t)$$
$$= f'(t) - u_{xx}(0,t),$$

where we have used the Dirichlet condition at $x = 0$ to replace $u_t(0,t)$ with $f'(t)$. We can solve for $a(t)$ to obtain

$$a(t) = \frac{f'(t) - u_{xx}(0,t)}{F\big[u(0,t)\big]}.$$

The right side of this expression involves the terms $u_{xx}(0,t)$ and $u(0,t)$ which may be viewed as "traces on the boundary $x = 0$" of the solution $u(x,t)$. Thus we refer to the expression as a TRACE TYPE FUNCTIONAL of $u = u(x,t)$. We are motivated then to consider the following IBVP for the unknown function $u = u(x,t)$,

$$u_t(x,t) - u_{xx}(x,t) = a[u;t]F\big[u(x,t)\big] \text{ in } Q_T$$
$$u(x,0) = 0, \qquad x > 0, \tag{2.1}$$
$$u_x(0,t) = g(t), \quad 0 < t < T,$$
$$u(x,t) \text{ bounded on } Q_T,$$

where,

$$a[u;t] \equiv \frac{f'(t) - u_{xx}(0,t)}{F[u(0,t)]} \quad \text{for } 0 \le t \le T. \tag{2.2}$$

We refer to the IBVP (2.1) as a trace type functional problem (TTF problem). Our interest in this trace type functional problem lies in the following,

Theorem 2.1 *Suppose $u(x,t)$ solves the TTF problem (2.1) for $a = a[u;t]$ given by (2.2) for $f(t)$ such that,*

$$f(t) \in C^1[0,T] \text{ with } f(0) = 0. \tag{2.3}$$

Then the function pair $\{u, a\}$ solves the inverse problem (0.1).

Proof. Suppose $u = u(x,t)$ solves (2.1). Then in particular at $x = 0$, the equation reduces to,

$$u_t(0,t) = f'(t)$$

that is,

$$u(0,t) = f(t) + C.$$

But $u(0,0) = 0$ and $f(0) = 0$ so that $C = 0$ and all of the conditions of the inverse problem (0.1) are satisfied.

Note that since we do not know a-priori that the solution of the TTF problem satisfies the condition $u(0,t) = f(t)$, it is essential that the expression $F[u(0,t)]$ appears in the denominator in the definition (2.2) instead of the expression $F[f(t)]$. If $F[f(t)]$ were in the denominator in (2.2), then at $x = 0$ the equation

$$u_t(x,t) - u_{xx}(x,t) = a(t)F[u(x,t)]$$

reduces to,

$$u_t(0,t) - u_{xx}(0,t) = \frac{f'(t) - u_{xx}(0,t)}{F[f(t)]} F[u(0,t)].$$

Obviously, this does not lead to $u_t(0,t) = f'(t)$.

Now we prove a stability result that says that if we solve the direct problem (1.1) with coefficient $a(t)$ satisfying the condition (1.2)iii, and then use this solution $u(x,t;a)$ to generate a new coefficient $a(t)$ according to the recipe (2.2), then this new coefficient also satisfies the condition (1.2)iii.

Theorem 2.2 *here exists a positive value for the parameter T such that if $u(x,t)$ solves the IBVP (1.1) under the hypotheses (1.2), and if $f(t)$*

satisfies (2.3) then the function $a[u;t]$ computed from u via the recipe (2.2) satisfies (1.2)iii.

It will be convenient at this point to introduce the notation,

$$f_1 \equiv \sup_{[0,T]} |f(t)|, \ \ f_2 \equiv \sup_{[0,T]} |f'(t)|, \ .$$

In addition, we can suppose that the constant a_1 appearing in (1.2)iii satisfies,

$$a_1 F_0 > f_2. \tag{2.4}$$

Proof. Since $u = u(x,t)$ is a classical solution to (1.1), $a[u;t]$ given by (2.2) is continuous as a function of t on $[0,T]$. It follows from (1.2), (2.2), and (1.8) that,

$$\sup_{[0,T]} |a[u;t]| \le \frac{f_2 + C_3(T)}{F_0}$$

Since $C_3(0) = 0$, and a_1 satisfies (2.4) it follows at once that for T sufficiently small,

$$\sup_{[0,T]} |a[u;t]| \le a_1 \tag{2.5}$$

i.e., $a[u;t]$ satisfies (1.2)iii. We have previously chosen $T > 0$ sufficiently small that $C_1(T) = \delta$ and now we simply decrease T still further if necessary so that (2.5) holds as well.

We proceed now to show that the trace type functional problem (2.1) has a unique solution. We suppose that the hypotheses (1.2), (1.4), and (1.6) are in effect.

Let,

$$a_1(t) \equiv 0 \quad \text{for } 0 \le t \le T$$

and for $k = 1, 2, \ldots$

$$u_k(x,t) = \text{unique solution of (1.1) for } a(t) = a_k(t) , \tag{2.6}$$

where

$$a_{k+1}(t) \equiv a[u_k;t].$$

We are going to need several estimates in order to show that the sequence of functions defined in (2.6) is convergent.

Note first that,

$$u_{k+1}(x,t) - u_k(x,t) = \iint_{Q_t} N(x,y,t-\tau)\big[a_{k+1}(\tau)F(u_{k+1}) - a_k(\tau)F(u_k)\big]\,dyd\tau$$

$$= \iint_{Q_t} N(x,y,t-\tau)\big\{a_{k+1}(\tau) - a_k(\tau)\big\}F(u_{k+1})\,dyd\tau$$

$$+ \iint_{Q_t} N(x,y,t-\tau)a_k(\tau)\big[F(u_{k+1}) - F(u_k)\big]\,dyd\tau \ .$$

Then,

$$\sup_{Q_T} |u_{k+1} - u_k| \leq F_1 T \sup_{[0,T]} |a_{k+1} - a_k| + a_1 F_2 T \sup_{Q_T} |u_{k+1} - u_k|$$

and it follows immediately that,

$$\sup_{Q_T} |u_{k+1} - u_k| \leq M_1(T) \sup_{[0,T]} |a_{k+1} - a_k|$$

where,

$$M_1(T) \equiv F_1 T/(1 - a_1 F_2 T).$$

Note that if T is not already sufficiently small, then we can decrease T still further so as to have,

$$T \leq (2a_1 F_2)^{-1}.$$

Then $M_1(T) \leq 2F_1 T$ and we have

$$\sup_{Q_T} |u_{k+1} - u_k| \leq 2F_1 T \sup_{[0,T]} |a_{k+1} - a_k| \ . \tag{2.7}$$

Similarly,

$$\partial_x\big\{u_{k+1}(x,t) - u_k(x,t)\big\} = \iint_{Q_t} N_x(x,y,t-\tau)\big[a_{k+1}(\tau)F(u_{k+1}) - a_k(\tau)F(u_k)\big]\,dyd\tau$$

and hence,

$$\sup_{Q_T} \big|\partial_x\{u_{k+1} - u_k\}\big| \leq F_1(4T/\pi)^{1/2} \sup_{[0,T]} |a_{k+1} - a_k|$$

$$+ a_1 F_2(4T/\pi)^{1/2} T \sup_{Q_T} |u_{k+1} - u_k| \ .$$

Then using (2.7) on the right side of this last inequality leads to,

$$\sup_{Q_T} \big|\partial_x\{u_{k+1} - u_k\}\big| \leq M_2(T) \sup_{[0,T]} |a_{k+1} - a_k| \tag{2.8}$$

where,

$$M_2(T) \equiv (4T/\pi)^{1/2} 2F_1 \ .$$

Finally,

$$\partial_{xx}\{u_{k+1}(x,t) - u_k(x,t)\} = \iint_{Q_t} N_{yy}(x,y,t-\tau)[a_{k+1}(\tau)F(u_{k+1}) - a_k(\tau)F(u_k)]\, dy d\tau$$

$$= -\iint_{Q_t} N_y(x,y,t-\tau)\{a_{k+1}(\tau) - a_k(\tau)F'(u_{k+1})\}\partial_y u_{k+1}(y,\tau)\, dy d\tau$$

$$-\iint_{Q_t} N_y(x,y,t-\tau)\{a_k(\tau)[F'(u_{k+1}) - F'(u_k)]\}\partial_y u_{k+1}(y,\tau)\, dy d\tau$$

$$-\iint_{Q_t} N_y(x,y,t-\tau)a_k(\tau)F'(u_k)\{\partial_y u_{k+1}(y,\tau) - \partial_y u_k(y,\tau)\}\, dy d\tau \ .$$

This leads to the result,

$$\sup_{Q_T}\left|\partial_{xx}\{u_{k+1} - u_k\}\right| \le (4T/\pi)^{1/2}\Big\{ F_1\|\partial_x u_{k+1}\| \sup_{[0,T]} |a_{k+1} - a_k|$$

$$+ a_1 F_3\|\partial_x u_{k+1}\| \sup_{Q_T} |u_{k+1} - u_k|$$

$$+ a_1 F_2 \sup_{Q_T} |\partial_x\{u_{k+1} - u_k\}|\Big\} \ .$$

Now using the estimate (1.7), we obtain,

$$\sup_{Q_T}\left|\partial_{xx}\{u_{k+1} - u_k\}\right| \le (4T/\pi)^{1/2}\Big\{ F_1 C_2(T) \sup_{[0,T]} |a_{k+1} - a_k|$$

$$+ a_1 F_3 C_2(T) \sup_{Q_T} |u_{k+1} - u_k|$$

$$+ a_1 F_2 \sup_{Q_T} |\partial_x\{u_{k+1} - u_k\}|\Big\} \ .$$

Combined with (2.7) and (2.8) this implies,

$$\sup_{Q_T}\left|\partial_{xx}\{u_{k+1} - u_k\}\right| \le M_3(T) \sup_{[0,T]} |a_{k+1} - a_k| \ , \tag{2.9}$$

where

$$M_3(T) \equiv (4T/\pi)^{1/2}\big[F_1 C_2(T) + a_1 F_3 C_2(T) 2F_1 T + a_1 F_2 M_2(T)\big]$$

$$= (4T/\pi)\big[2g(T) \exp[a_1 F_2 T](F_1 + 2a_1 F_3 F_1 T) + 2a_1 F_2 F_1\big]$$

$$\equiv T\, C_4(T).$$

Now using (2.6) and (2.2), we find that,

$$\sup_{[0,T]} |a_{k+1} - a_k|F_0 \le \sup_{Q_T}\left|\partial_{xx}\{u_k - u_{k-1}\}\right| + a_1 F_1 \sup_{Q_T} |u_k - u_{k-1}|$$

That is,

$$\sup_{[0,T]} |a_{k+1} - a_k| \le T\, C_5(T) \sup_{[0,T]} |a_k - a_{k-1}| \tag{2.10}$$

where

$$C_5(T) \equiv \left[C_4(T) + 2a_1 F_1^2 \right] / F_0.$$

Now it is clearly possible (decreasing T still further, if necessary) to choose $T > 0$ such that for any θ, $0 < \theta < 1$,

$$T < \frac{F_0 \theta}{C_4(T) + 2a_1 F_1^2}.$$

Then $T C_5(T) \equiv \theta < 1$ and hence

$$\sup_{[0,T]} |a_{k+1} - a_k| \le \theta \sup_{[0,T]} |a_k - a_{k-1}|$$

$$\le \theta^{k-1} \sup_{[0,T]} |a_1 - a_0| \equiv M_5 \theta^{k-1}.$$

Using this fact, together with (2.7) and (2.8), we can easily show that $\{u_k(x,t)\}$ is a Cauchy sequence in the Banach space B_T and that $\{a_k(t)\}$ is a Cauchy sequence in the Banach space $C[0,T]$ (equipped with the sup norm). Let the unique limit points of these two sequences be denoted respectively by $u(x,t)$ and $a(t)$. Note that for each $k = 1, 2, \ldots$, we can write

$$u_k(x,t) = 2 \int_0^t K(x, t-\tau) g(\tau) d\tau + \iint_{Q_t} N(x, y, t-\tau) a_k(\tau) F[u_k] \, dy d\tau \ .$$

That is, denoting the first integral on the right by $v(x,t)$,

$$u_k(x,t) = v(x,t) + \iint_{Q_t} N(x, y, t-\tau) a(\tau) F\big[u(y,\tau)\big] \, dy d\tau$$

$$+ \iint_{Q_t} N(x, y, t-\tau) a_k(\tau) \big\{ F[u_k] - F[u] \big\} \, dy d\tau$$

$$+ \iint_{Q_t} N(x, y, t-\tau) \big\{ a_k(\tau) - a(\tau) \big\} F[u] \, dy d\tau$$

Then, letting k tend to infinity, it follows that $u(x,t)$ satisfies

$$u(x,t) = v(x,t) + \iint_{Q_t} N(x, y, t-\tau) a(\tau) F\big[u(y,\tau)\big] \, dy d\tau;$$

i.e., $u(x,t)$ solves (1.1). In addition,

$$
\begin{aligned}
a[u;t]F\big[u(0,t)\big] &= f'(t) - \partial_{xx}u(0,t) \\
&= f'(t) - \partial_{xx}u_k(0,t) + \partial_{xx}u_k(0,t) - \partial_{xx}u(0,t) \\
&= F\big[u_k(0,t)\big]a_k(t) + \partial_{xx}u_k(0,t) - \partial_{xx}u(0,t).
\end{aligned}
$$

Letting k tend to infinity on the right side of this last expression, we obtain $a[u;t] = a(t)$. Thus, $u(x,t)$ solves the trace type functional problem (2.1).

We have proved:

Theorem 2.3. *Let the data functions $f(t), g(t), F(u)$ satisfy the assumptions (1.2), (1.4), (1.6), (2.3), and let (2.4) hold. Then for $T > 0$ sufficiently small, there exist functions $u = u(x,t)$ in B_T and $a(t) = a[u;t]$ given by (2.2) such that the trace type functional problem (2.1) is solved by u and a.*

According to Theorem 2.1 this same pair of functions $u(x,t)$ and $a(t)$ solve the inverse problem (0.1).

3. Remarks

Note that the inverse problem (0.1) has been reformulated as a TTF problem by using the Dirichlet boundary condition to eliminate the unknown coefficient $a(t)$ from the partial differential equation in (0.1) and replace it with the TTF term $a[u;t]$. Then the inverse problem has been replaced by the problem (2.1), an IBVP of standard form containing the coefficient $a[u;t]$ which is of nonstandard type. Using TTF problems as a means of solving associated inverse problems was begun in [10,11,12]. More recent results can be found in references [14] to [17].

Using the Dirichlet boundary condition in (0.1) in order to eliminate the unknown coefficient $a(t)$ from the partial differential equation identifies the Dirichlet condition as the overspecified condition in the inverse problem. The Neumann condition is then considered to be part of the direct problem; i.e., it is the controlled condition. Note that in order for the reformulation of the inverse problem as a TTF problem to succeed in this example it is necessary that the (known) source function $F(u)$ be bounded away from zero on some interval. There is no loss in generality in supposing that this interval is the set $|u| < \delta$, since we can accommodate any interval of the form $|u - u_0| < \delta$ by replacing the homogeneous initial condition in (0.1) by the inhomogeneous condition $u(x,0) = u_0$. Then we can ensure that $F[u(x,t)]$ is bounded away from zero for (x,t) in Q_T and $u(x,t)$ solving (1.1) by proper control of the controlled boundary condition (i.e., by controlling the data function $g(t)$ in the Neumann condition at $x = 0$).

There does not appear to be any natural way in which the Neumann condition in this example could be used to eliminate the unknown coefficient from the partial differential equation; i.e., the Dirichlet condition appears to be the preferred condition for the role of the overspecified condition. It is typical of the unknown ingredient inverse problems involving overspecified data measured on the boundary that one of the boundary conditions in the inverse problem seems to be better suited than the other for use in this elimination procedure where the inverse problem is transformed to a TTF problem. See [16] and [17] for additional examples.

This raises the question of whether there should be a preferred boundary condition to use as the overspecified condition when solving inverse problems computationally by output least squares methods [13]. Here the overspecified condition is used to define an objective functional whose argument is the solution $u(x,t)$ of the direct problem. One then seeks to optimize the objective functional over some set of admissible coefficients $a(t)$.

In this regard, it is interesting to consider the use of alternative overspecified conditions in place of one of the boundary conditions at $x = 0$ in an inverse problem like (0.1). It will be convenient to consider the problem on a bounded set $Q_T \equiv \{(x,t) : 0 < x < 1, 0 < t < T\}$. We add a boundary condition at $x = 1$, say the Neumann condition $u_x(1,t) = 0$, $0 < t < T$, and replace the Neumann condition at $x = 0$ with the condition,

$$\int_0^1 u(x,t)\, dx = E(t), \qquad t > 0 \, . \tag{3.1}$$

If we use this condition in place of the Neumann boundary condition at $x = 0$, then the inverse problem consists in finding unknown functions $u(x,t)$ and $a(t)$ satisfying (3.1) and,

$$\begin{aligned}
u_t(x,t) - u_{xx}(x,t) &= a(t)F\big[u(x,t)\big] \text{ in } Q_T \\
u(x,0) &= 0, \qquad 0 < x < 1, \\
u(0,t) &= f(t), \qquad 0 < t < T, \\
u_x(1,t) &= 0, \qquad 0 < t < T.
\end{aligned} \tag{3.2}$$

Here $E(t), f(t)$ and $F(u)$ are assumed to be given data.

It follows from (3.1) that,

$$\int_0^1 u_t(x,t)\, dx = E'(t), \qquad t > 0 \, ,$$

and then integrating the partial differential equation in (3.2) with respect to x from 0 to 1 leads to the result,

$$E'(t) + u_x(0,t) = a(t) \int_0^1 F\big[u(x,t)\big]\, dx \, .$$

Now we can eliminate $a(t)$ from the inverse problem, leaving the following trace type functional problem for the unknown function $u = u(x,t)$,

$$u_t(x,t) - u_{xx}(x,t) = a[u;t]F\big[u(x,t)\big] \text{ in } Q_T$$
$$u(x,0) = 0, \qquad 0 < x < 1,$$
$$u(0,t) = f(t), \qquad 0 < t < T, \qquad\qquad (3.3)$$
$$u_x(1,t) = 0, \qquad 0 < t < T,$$

where

$$a[u;t] \equiv \frac{E'(t) + u_x(0,t)}{\int_0^1 F\big[u(x,t)\big]\,dx}. \qquad\qquad (3.4)$$

Then (3.3) is a trace type functional problem for $u(x,t)$ in which the trace type functional (3.4) is "milder" than the trace type functional (2.2) which arises when the overspecified data is measured on the boundary. That is, (3.4) involves a lower order trace operator than does (2.2). This suggests that the inverse problem based on the overspecification (3.1) is more stable than the problem based on the boundary measurement of $u(0,t)$. Of course it may not be experimentally feasible to measure the data in (3.1).

Instead of replacing the Neumann condition at $x = 0$, if we use (3.1) to replace the Dirichlet boundary condition in the inverse problem then we are led to the following TTF problem

$$u_t(x,t) - u_{xx}(x,t) = a[u;t]F\big[u(x,t)\big] \text{ in } Q_T$$
$$u(x,0) = 0, \qquad 0 < x < 1,$$
$$u(0,t) = g(t), \qquad 0 < t < T, \qquad\qquad (3.5)$$
$$u_x(1,t) = 0, \qquad 0 < t < T,$$

where in this case,

$$a[u;t] \equiv \frac{E'(t) + g(t)}{\int_0^1 F\big[u(x,t)\big]\,dx}. \qquad\qquad (3.6)$$

The problem (3.5) is now no longer a trace type functional problem. The partial differential equation involves the nonstandard coefficient $a[u;t]$ which is functionally dependent on the solution $u(x,t)$, but this is not a *trace* type functional. Evidently $a[u;t]$ in (3.6) is milder still than the coefficient given by (3.4) and the corresponding inverse problem may be expected to be more stable than the previous one; the stablest inverse problem for the identification of the unknown coefficient $a(t)$ is obtained when the flux on the boundary is controlled and the data in (3.1) is observed.

In both of these last two problems we can proceed as in section 2 to prove the analogues of the equivalence result Theorem 2.1, the stability result Theorem 2.2, and the existence result Theorem 2.3. Alternatively, these problems are special cases of a more general result found in [14].

Finally, the analysis of this problem provides a framework for dealing with the more difficult inverse problem of identifying an unknown source term in a heat equation. That is, the problem of finding unknown functions $u = u(x,t)$ and $F = F(z)$ that satisfy,

$$\begin{aligned} u_t(x,t) - u_{xx}(x,t) &= F\big[u(x,t)\big] \text{ in } Q_T \\ u(x,0) &= 0, \qquad x > 0, \\ u_x(0,t) &= g(t), \quad 0 < t < T, \\ u(0,t) &= f(t), \qquad 0 < t < T. \end{aligned} \qquad (3.7)$$

Here, if $\{u, F\}$ is a solution pair of the inverse problem (3.7), then

$$\begin{aligned} F\big[u(0,t)\big] &= u_t(0,t) - u_{xx}(0,t) \\ &= f'(t) - u_{xx}(0,t), \end{aligned}$$

and

$$F(z) = f'\big(p(z)\big) - u_{xx}\big(0, p(z)\big)$$

where

$$p(z) = \big[u(0, \cdot)\big]^{-1}(z).$$

Then the trace type functional formulation of the inverse problem (3.7) is,

$$\begin{aligned} u_t(x,t) - u_{xx}(x,t) &= S[f; u]\big(u(x,t)\big) \text{ in } Q_T \\ u(x,0) &= 0, \qquad x > 0, \\ u_x(0,t) &= g(t), \quad 0 < t < T, \end{aligned} \qquad (3.8)$$

where

$$S[f; u](z) \equiv f'\big(p(z)\big) - u_{xx}\big(0, p(z)\big) \qquad (3.9)$$

and

$$p(z) = \big[u(0, \cdot)\big]^{-1}(z). \qquad (3.10)$$

Evidently, the Dirichlet condition is the observed condition and the Neumann condition $u_x(0,t)$, is the controlled condition. In this example we must control the function $g(t)$ in such a way as to ensure that $u(0, \cdot)$ is invertible in order for the reformulation to succeed. Proper control of the controlled boundary condition is more critical in this inverse problem than it was in the identification of the coefficient $a(t)$ in (0.1).

Since we do not know, a-priori, that the solution of the TTF problem (3.8) satisfies the condition $u(0,t) = f(t)$, we are not entitled to suppose that $p(z) = f^{-1}(z)$. We prove a-posteriori that the solution of (3.8) satisfies $u(0,t) = f(t)$ and hence $u(x,t)$ together with $S[f;u](z)$ solves the inverse problem (3.7).

The expression $u_{xx}(0,\cdot)$ which appears in the source term $S[f;u](z)$ is a trace type functional of the unknown solution. The term $p(z)$ is also a trace type functional of the unknown solution, it is the inverse of the restriction, $u(0,t)$. Then $S[f;u](z)$ is an example of a "recursive" trace type functional. The trace type functionals defined in (2.2), (3.4), and (3.6) are examples of "simple" trace type functionals of u since there is no trace type functional which appears as the argument of another trace type functional of the same unknown function.

A TTF problem is harder to analyze than a standard type of nonlinear problem in partial differential equations. Not only do the properties of the ingredients of the TTF equation (e.g. coefficients, source terms) affect the properties of the solution, in addition the properties of the solution influence the behavior of the TTF ingredients of the equation. In a recursive trace type functional problem the affect of the solution properties on the behavior of the solution ingredients is much harder to analyze than it is for a simple type problem. Nevertheless, the simple type problems treated here may provide insight into the analysis of the more difficult problem (3.8).

REFERENCES

[1] Jones, B. Frank, *Determination of a coefficient in a parabolic differential equation, Part I: existence and uniqueness*, J. of Math. and Mech., **11** 6, (1962).

[2] ___, *Various methods for finding unknown coefficients in parabolic differential equations*, CPAM,**XVI** (1963).

[3] Cannon, J. R., *Determination of an unknown coefficient in a parabolic differential equation*, Duke Math. J.,302 (1963).

[4] ___, and DuChateau, P., *Determining unknown coefficients in a nonlinear heat conduction problem*, SIAM JAP, **24** 3 (1973).

[5] ___, and ___, *Determination of the conductivity of an isotropic medium*, JMAA, **48** 3 (1974).

[6] ___, and ___, *An inverse problem for a nonlinear diffusion equation*, SIAM JAP, **39** 2(1980).

[7] Pilant, M. and Rundell, W., *An inverse problem for a nonlinear parabolic equation*, Comm. PDE,114 (1986).

[8] Prilepko, A. I., and Orlovskii, D. G., *Determination of the evolution parameter of an equation and inverse problems of mathematical physics*, Diff. Urav., **21** 1 (1985).

[9] Budak, B. M., and Iskenderov, A. D., *On a class of converse boundary value problems with unknown coefficients*, Dokl Akad Nauk SSSR, **170** 1 (1967).

[10] Cannon, J. R., and DuChateau, P., *An inverse problem for an unknown source term in a heat equation*, JMAA, **752** (1980).

[11] ___, and ___, *Weak solutions $u(x,t)$ to a partial differential equation with coefficients that depend on $u[y, \phi_j(t, u(x,t))]$ $j = 1, \ldots, k$*, JDE, **423** (1981).

[12] ___, and ___, *An inverse problem for an unknown source term in a wave equation*, SIAM J Appl. Math., **453** (1983).

[13] Chavent, G., and Lemonnier, P., *Identification de la nonlinéarité d'une équation parabolique quailinéaire*, Appl. Math. and Opt., **1** 2(1974).

[14] Cannon, J. R., and Yin, Hong-Ming, *A class of nonlinear, nonclassical parabolic equations*, JDE, **78** 1989).

[15] ___, and ___, *A class of multidimensional nonclassical parabolic problems*, Differential Equations and Applications, 1989.

[16] Cannon, J. R., DuChateau, P., and Steube, K., *Unknown ingredient inverse problems and trace type functional differential equations*, (submitted for publication).

[17] ___, ___, and ___, *Trace type functional differential equations and the identification of hydraulic properties of porous media*, (submitted for publication).

J. R. Cannon
Department of Mathematics
Lamar University
Beaumont, Texas 77710

Paul DuChateau
Ken Steube
Department of Mathematics
Colorado State University
Fort Collins, Colorado 80523

On the Structure of Solutions
for Some Semilinear Elliptic Equations

KUO-SHUNG CHENG

1. Introduction

In this paper we shall review some recent results about the structure of the solution sets of equations

$$(1.1) \qquad \Delta u + K e^{2u} = 0$$

and

$$(1.2) \qquad \Delta u + K u^{\sigma} = 0, \quad u > 0$$

in $\mathbb{R}^n$, $n \geq 2$, where $\Delta = \sum_{i=1}^{n} \partial^2/\partial x_i^2$, $K \neq 0$ is a given locally Hölder continuous function on $\mathbb{R}^n$ and $\sigma > 1$ is a constant. We shall consider mainly two types of solutions: $u_c(x)$ denotes a solution of (1.1) or (1.2) satisfying

$$(1.3) \qquad u_c(x) \to c \quad \text{as} \quad |x| \to \infty$$

or

$$(1.4) \qquad u_c(x) = c \log |x| + O(1) \quad \text{at} \quad \infty$$

and $U(x)$ denotes the "maximal solution"

$$(1.5) \qquad U(x) = \sup\{u(x) \mid u \text{ is an entire solution of } (1.1)\}$$

and

$$(1.6) \qquad U(x) = \sup\{u(x) \mid u \text{ is a positive solution of } (1.2)\}.$$

We shall consider mainly three questions:

(a) When do solutions u_c and U exist?

Supported in part by the National Science Council of the Republic of China

(b) What are the relations between K and U?

(c) When do we have $u = u_c$ or $u = U$ for every solution u of (1.1) or (1.2)?

Equations (1.1) and (1.2) arise from physics, geometry and other branches of applied mathematics. Let (M, g) be a given Riemannian manifold of dimension n and K be a given function on M. The following question has been raised: can we find a new metric g_1 on M such that K is the scalar curvature (Gaussian curvature if $n = 2$) of g_1 and g_1 is conformal to g (that is, $g_1 = hg$ for some positive smooth function h in M)? For $n = 2$, we write $h = e^{2u}$, then this is equivalent to the problem of solving the elliptic equation

$$(1.7) \qquad \Delta_g u - k + K e^{2u} = 0$$

on M, where Δ_g and k are the Laplace-Beltrami operator and the Gaussian curvature on M in the g-metric respectively. For $n \geq 3$, we write $h = u^{4/(n-2)}$, $u > 0$. Then the question is equivalent to the problem of solving the elliptic equation

$$(1.8) \qquad \frac{4(n-1)}{n-2} \Delta_g u - ku + K u^{(n+2)/(n-2)} = 0$$

on M, where Δ_g is as before and k is the scalar curvature on M in the g-metric. In case M is compact, equations (1.7) and (1.8) have been considered by many authors and we refer the reader to the monograph by Kazdan [K] for details and references. In case M is complete and non-compact, the first natural case seems to be $M = \mathbb{R}^n$ and g is the usual Euclidean metric. In this case, equations (1.7) and (1.8) reduce to equations (1.1) and (1.2) respectively with proper scaling for (1.8). When $n = 3$ and $K(x) = 1/(1 + |x|^2)$, (1.1) and (1.2) were proposed by Eddington and Matukuma [M] respectively to model globular clusters of stars. For recent references, see Batt, Faltenbacher and Horst [BFH] and Ni and Yotsutani [NY].

The nonexistence problem concerning equations (1.1) and (1.2) has been studied by many authors, Ahl'fors [A], Wittich [W], Keller [Ke], Osserman [Os], Sattinger [S], Oleinik [O], Ni [N1, N2], McOwen [Mc], Lin [L] and Cheng and Lin [CLj], to name a few. These results basically contain the following.

Theorem 1.1 *If $K \leq 0$ in $\mathbb{R}^n$ and $K \leq -C|x|^{-2}$ near ∞, then equations* (1.1) *and* (1.2) *do not possess any solution on $\mathbb{R}^n$.*

The first existence result concerning equation (1.1) and (1.2) seem due to Ni [N1, N2]. He established the following.

Theorem 1.2 *If $K \neq 0$ and $K \leq 0$ in $\mathbb{R}^2$ with $K(x) \geq -C|x|^{-p}$ near ∞ for some constant $p > 2$, then equation* (1.1) *($n = 2$) possesses infinitely* *many solutions in $\mathbb{R}^2$. More precisely, for every sufficiently small $c > 0$, there exists a solution u_c of (1.1) in $\mathbb{R}^2$ satisfying (1.4).*

Theorem 1.3 *Suppose that $|K(x)| \leq C|x_1|^{-p}$ for $|x_1|$ large and uniformly in x_2 for some $p > 2$, where $x = (x_1, x_2)$ and $x_1 \in \mathbb{R}^m$, $x_2 \in \mathbb{R}^{n-m}$, $m \geq 3$. Then (1.1) and (1.2) possess infinitely many bounded solutions.*

Later on, Theorem 1.2 was improved by McOwen [Mc] and Theorem 1.3 was improved by Naito [Na], Kusano and Oharu [KO]. We shall omit the details of these improvements.

In the following section we shall present some recent results about equations (1.1) and (1.2).

2. Main Results

Concerning the maximal solution U, Cheng and Ni [CN1] essentially established

Theorem 2.1 *Suppose that*
(i) $K \leq 0$ *in $\mathbb{R}^n$ and there exists a sequence of bounded smooth domains $\{\Omega_i\}$ such that $\mathbb{R}^n = U_1^\infty \Omega_i$, $\bar{\Omega}_i \subseteq \Omega_{i+1}$, and $K < 0$ on $\partial\Omega_i$, $i = 1, 2, \ldots$, and*
(ii) *equations (1.1) and (1.2) possess a solution v on $\mathbb{R}^n$.*
Then the functions U in (1.5) and (1.6) are well-defined everywhere in $\mathbb{R}^n$ and are a solution of (1.1) and (1.2) on $\mathbb{R}^n$ respectively.

Concerning the type of solutions u_c, we have

Theorem 2.2 (Ni [N2], McOwen [Mc], Cheng and Ni [CN1]) *Suppose that $K \leq 0$ in $\mathbb{R}^2$, and that $C_1|x|^{-p} \leq |K(x)| \leq C_2|x|^{-p}$ for $|x|$ large for some constants $C_2 \geq C_1 > 0$ and $p > 2$. Then for each $c \in (0, (p-2)/2)$, (1.1) possesses a unique solution u_c on $\mathbb{R}^2$ satisfying (1.4).*

Theorem 2.3 (Cheng and Wang [CW]) *Suppose that $K \leq 0$ in $\mathbb{R}^2$ and satisfies one of the conditions:*

(i) $|K(x)| \leq G(|x|)$ *in $\mathbb{R}^2$ for some G and $\int_1^\infty r(\log r)^{\sigma+1} G(r)\, dr < \infty$,*

(ii) $\int_{\mathbb{R}^2} |K(x)|[\log(1+|x|^2)]^{\sigma+1}\, dx < \infty$ *and $|K(x)|[\log(1+|x|^2)]^\sigma \leq C$ for all $x \in \mathbb{R}^2$ for some constant $C > 0$.*

Then for every $c \in (0, \infty)$, (1.2) possesses a unique positive solution u_c in $\mathbb{R}^2$ satisfying (1.4).

Theorem 2.4 (Cheng and Ni [CN2]) *Suppose that K satisfies one of the following conditions:*

(i) $|K(x) \leq G(|x|)$ *in $\mathbb{R}^n$, $n \geq 3$, for some G and $\int_0^\infty tG(t)\, dt < \infty$,*

(ii) $\int_{\mathbb{R}^n} |K(y)|/|y|^{n-2}dy < \infty$ *and $M(x)|x|^2(\mathcal{G}(x))^{2/(n-2)} \to 0$ as $|x| \to$

∞, *where*

$$M(x) = \sup_{|y-x|<|x|/2} |K(y)|$$

and

$$\mathcal{G}(x) = \int_{|y-x|<|x|/2} |K(y)|/|y|^{n-2} dy.$$

Then we have

    (α) *There exists a constant $c_0 > 0$ such that for every $c \in (0, c_0]$, (1.2) possesses a solution u_c in $\mathbb{R}^n$ satisfying (1.3).*

    (β) *If, in addition, $K \leq 0$ on $\mathbb{R}^n$, then for every $c \in (0, \infty)$, (1.2) possess a unique solution u_c on $\mathbb{R}^n$ satisfying (1.3).*

    (γ) *If, in addition, $K \geq 0$ on $\mathbb{R}^n$, then there exists a constant $c_0 > 0$ such that, for every $c \in (0, c_0)$, (1.2) possesses a solution u_c satisfying (1.3) and possesses no solution u_c satisfying (1.3) for $c > c_0$.*

    (δ) *Let u be a bounded solution of (1.2). Then there exists a constant $c > 0$ such that $u = u_c$, that is, $u(x) \rightarrow c$ as $|x| \rightarrow \infty$.*

Theorem 2.5 (Cheng and Lin [CLt]) *Under the same assumptions as in* Theorem 2.4, *then the statements in Theorem 2.4 still hold provided that equation (1.2) is replaced by equation (1.1) and corresponding intervals $(0, c_0]$, $(0, \infty)$ and $(0, c_0)$ are replaced by $(-\infty, c_0)$, $(-\infty, \infty)$ and $(-\infty, c_0)$ respectively, where c_0 is a real number now.*

Now we come to the structure of the solution sets of equations (1.1) and (1.2). We have

Theorem 2.6 (Cheng and Ni [CN1]) *Suppose that $K \leq 0$ in $\mathbb{R}^2$, and that $K \sim -|x|^{-p}$ near ∞ for some $p > 2$. Then the following conclusions hold.*

    (i) *For each $c \in (0, (p-2)/2)$, (1.1) possesses a unique solution u_c satisfying (1.4).*

    (ii) *Let u be a solution of (1.1) on $\mathbb{R}^2$. Then either $u = U$ where U is given by (1.5) or $u = u_c$ for some $c \in (0, (p-2)/2)$ where u_c is given by (i) above.*

    (iii) *If $(p-2)/2 > c_1 > c_2 > 0$ then $u_{c_1} > u_{c_2}$ in $\mathbb{R}^2$. Furthermore, the asymptotic behavior of U near ∞ is given by*

$$U(x) = \frac{p-2}{2} \log |x| - \log \log |x| + O(1) \quad near \quad \infty.$$

Theorem 2.7 (Cheng and Lin [CLt]) *Suppose that $K \leq 0$ in $\mathbb{R}^n$, $n \geq 3$, and that $K(x) \sim -|x|^{-p}$ near ∞ for some $p > 2$. Then the following conclusions hold.*

    (i) *For every $c \in (-\infty, \infty)$, (1.1) possesses a unique solution u_c satisfying (1.3).*

    (ii) *Let u be a solution of (1.1) on $\mathbb{R}^n$, $n \geq 3$. Then either $u = U$ where U is given by (1.5) or $u = u_c$ for some $c \in (-\infty, \infty)$ where u_c is given by (i) above.*

    (iii) *If $+\infty > c_1 > c_2 > -\infty$ then $u_{c_1} > u_{c_2}$ in $\mathbb{R}^2$. Furthermore, the asymptotic behavior of U near ∞ is given by*

$$(2.1) \qquad U(x) \sim |x|^{(p-2)/(\sigma-1)} \quad near \quad \infty.$$

Theorem 2.8 (Cheng and Ni [CN2]) *Under the same assumptions as in* Theorem 2.7, *then the following conclusions hold.*

    (i) *For every $c \in (0, \infty)$, (1.2) possesses a unique solution u_c satisfying (1.3).*

    (ii) *Let u be a solution of (1.2). Then either $u = U$ where U is given by (1.6) or $u = u_c$ for some $c \in (0, \infty)$ where u_c is given by (i) above.*

    (iii) *If $\infty > c_1 > c_2 > 0$ then $u_{c_1} > u_{c_2}$ in $\mathbb{R}^n$. Furthermore the asymptotic behavior of U is given by*

$$U(x) = \frac{p-2}{2} \log |x| + O(1) \quad at \quad \infty.$$

REFERENCES

[A] Ahl'fors, L. V., *An extension of Schwarz's lemma* Trans. Amer. Math. Soc. **43** (1938), 359-364 .

[BFH] Batt, J., Faltenbacher, W. and Horst, E.,*Stationary spherically symmetric models in stellar dynamics*, Arch. Rational Mech. Anal. **93** (1986), 159-183 .

[CLj] Cheng, K.-S. and Lin, J.-T., *On the elliptic equations $\Delta u = K u^\sigma$ and $\Delta u = K e^{2u}$*, Trans. Amer. Math. Soc. **304** (1987) 639-668 .

[CLt] Cheng, K.-S. and Lin, T.-C., *The structure of solutions of a semilinear elliptic equation*, to appear in Trans. Amer. Math. Soc.

[CN1] Cheng, K.-S. and Ni, W.-M., *On the structure of the conformal Gaussian curvature equation on $\mathbb{R}^2$* , to appear in Duke Math. J.

[CN2] Cheng, K.-S. and Ni, W.-M., (In preparation) .

[CW] Cheng, K.-S. and Wang, J.-N., *On the classification of solutions of a semilinear elliptic equation*, to appear in J. Nonlinear Analy. Theory, Method and Application .

[K] Kazdan, J., *Prescribing the curvature of a Riemannian manifold*, NSF-CBMS Regional Conference Lecture Notes **57** (1985) .

[Ke] Keller, J. B., *On the solutions of $\Delta u = f(u)$*, Comm. Pure Appl. Math. **10** (1957), 503-510. .

[KO] Kusan, T. and Oharu, S., *Bounded entire solutions of second order semilinear elliptic equation with application to a parabolic initial value problem*, Indiana Univ. Math. J. **34** (1985), 85-89 .

[L] Lin, F. -H., *On the elliptic equation $D_i[a_{ij}(x)D_jU] - k(x)U + K(x)U^p = 0$*, Proc. Amer. Math. Soc. **95** (1985), 219-226 .

[M] Matukuma, T., *Sur la dynamigue des amas globulaires*, Proc. Imp. Acad. **6** (1930), 133-136 .

[Mc] McOwen, R., *On the equation $\Delta u + Ke^{2u} = f$ and prescribed negative curvature in $\mathbb{R}^2$* J. Math. Anal. Appl. **103** (1984), 365-370 .

[Na] Naito M., *A note on bounded positive entire solutions of semilinear elliptic equations* Hiroshima Math. J. **14** (1984), 211-214 .

[N1] Ni, W.-M. *On the elliptic equation $\Delta u + Ku^{(n+2)/(n-2)} = 0$, its generalizations and applications in geometry*, Indiana Univ. Math. J. **31** (1982), 493-529.

[N2] Ni, W.-M., *On the elliptic equation $\Delta u + Ke^{2u} = 0$ and conformal metrics with prescribed Gaussian curvatures*, Invent. Math. **66** (1982), 343-352.

[NY] Ni, W. -M. and Yotsutani, S., *Semilinear elliptic equations of Matukuma-type and related topics*, Japan J. Appl. Math. **5** (1988), 1-32 .

[O] Oleinik, O. A., *On the equation $\Delta u + k(x)e^u = 0$* Russian Math. Surveys **33** (1978), 243-244. .

[Os] Osserman, R., *On the inequality $\Delta u \geq f(u)$* Pacific J. Math. **7** (1957), 1641-1647 .

[S] Sattinger, D. H., *Conformal metrics in $\mathbb{R}^2$ with prescribed curvature*, Indiana Univ. Math. J. **22**, (1972), 1-4 .

[W] Wittich, H., *Ganze Losungen der Differential gleichung $\Delta u = e^u$*, Math. Z. **49** (1944), 579-582.

Institute of Applied Mathematics
National Chung Cheng University
Chiayi, Taiwan 62117
Republic of China

A Note on Boundary Regularity
for Certain Degenerate Parabolic Equations

E. DIBENEDETTO, J. MANFREDI and V. VESPRI

1. Introduction

Let Ω be an open set in $\mathbf{R}^N$, $N \geq 1$ of boundary $\partial\Omega$, and for $0 < T < \infty$, set $\Omega_T \equiv \Omega \times (0, T]$. Let

$$(1.1) \qquad u \in L^\infty\left((\varepsilon, T]; L^r(\Omega)\right) \cap L^p\left((\varepsilon, T] : W^{1,p}(\Omega)\right) ; \qquad \forall \varepsilon \in (0, T),$$

be a weak solution of the problem

$$(1.2) \qquad \begin{cases} u_t = \operatorname{div}(|\nabla u|^{p-2}\nabla u) & \text{in} \quad \Omega_T \\ u = g \text{ on } \partial\Omega \times (\varepsilon, T]. \end{cases}$$

Here $r \geq 1$ and $p > 1$ are subject to the condition

$$(1.3) \qquad p > \frac{2N}{N + r}$$

and ∇u denotes the gradient of u with respect to the space variables only.

In this note we assume g and $\partial\Omega$ are smooth and prove that

$$(1.3) \qquad \nabla u \in L^\infty(\Omega \times (\varepsilon, T]) \qquad \forall \varepsilon \in (0, T],$$

That is, weak solutions of (1) are Lipschitz continuous up to the boundary.

For background and references on this type of equations see [DB-F1], [DB-F2] and [C-DB], where interior $C^{1,\alpha}$–regularity and boundary C^α–regularity are established. See also [Ch] for the significance of the restriction on p.

To obtain a local bound for $|\nabla u|$ near the lateral boundary of Ω_T, we proceed in two steps. The first and key one is to establish a bound for the normal derivative of u at the boundary. This will follow from the explicit construction of a barrier. Armed with this bound, in the second step we adapt some arguments in [DB-F1] to obtain (1.4). It is at this stage that the condition (1.3) enters. The arguments reduce essentially to *interior* estimates. In this context it was shown in [C-DB] that if $r = 2$ then one can find a sup-bound for the spatial gradient of u. Following the remarks of [Di-H], Choe [Ch] has observed that indeed (1.3) suffices for all $r \geq 1$.

---

[1] Partially supported by NSF grants DMS-8802883 and DMS-8901524.

The arguments in the cited references hold for systems. Because of the application of various comparison principles the results of this note are valid only for equations.

Another consequence of (1.3) is that if the boundary datum g is bounded in $\partial\Omega \times (\varepsilon, T]$ then the solution u is locally bounded in $\Omega \times (\varepsilon, T]$. This follows from the results of [**Di-H**]. Fix $\varepsilon \in (0, T]$ and set

$$(1.5) \qquad |||u|||_\varepsilon \equiv \operatorname*{ess\,sup}_{\varepsilon \leq \tau \leq T} \|u(\cdot, \tau)\|_{r,\Omega} + \|\nabla u\|_{p,\Omega\times(\varepsilon,T]} \ .$$

Then there exists a constant γ depending upon ε, N, p, $|||u|||_\varepsilon$ such that

$$(1.6) \qquad \|u\|_{\Omega\times(\varepsilon,T]} \leq \gamma \ .$$

For this result we refer to [**Di-H**] where also it is computed a precise dependence of γ upon the indicated quantities.

We will assume troughout that $g : \partial\Omega \times (\varepsilon, T) \to R$ is of class C^2 in the space variables and of class C^1 in the t–variable; moreover for every $\varepsilon \in (0, T]$ fixed, there exists a constant γ
depending upon $\partial\Omega$, ε, p, N, such that

$$(1.7) \qquad \| \, g \, , \, g_{x_i} \, , \, g_{x_i x_j} \, , \, g_t \, \|_{\infty, \partial\Omega\times(\varepsilon,T]} \leq \gamma \qquad \forall i, j = 1, 2, \ldots, N-1 \ .$$

A boundary gradient bound will be derived in terms of the quantities in (1.5)–(1.7). Accordingly we say that a constant γ depends upon the data if it can be determined apriori only in terms of these quantities as well as p and the dimension N.

2. The barrier

Fix $\varepsilon \in (0, T)$ and a boundary point $(x_o, t_o) \in \partial\Omega \times (\varepsilon, T]$, which after a translation we may assume to coincide with $(0, 0)$. Following the technique of [**Ch-D**], to prove (1.4) it will suffice to assume that

1. the boundary $\partial\Omega$ is flat near $x_o = 0$; that is, for some positive R

$$\partial\Omega \cap B_R(0) = \big\{ x \colon |x| < R \text{ and } x_N = 0 \big\},$$

2. The solution u is smooth in a neighborhood of $(0, 0)$, up to $x_N = 0$

and to prove that the derivative u_{x_N} is bounded at $x_N = 0$ by a constant γ that depends only upon the data.

Since we will be working in a neiborhood of $(0,0)$, it would be sufficient to have the quantities in (1.5)–(1.7) be redefined and finite in a neighborhood of such a point.

For example we may consider the cylinder

$$Q_R^+ \equiv \{|x| < R\} \times \{-R, 0\} \cap \{x_N > 0\},$$

where $R > 0$ is sufficienlty small that $Q_R^+ \subset \Omega \times (\epsilon, T]$. Consider also the portion of lateral boundary given by

$$S \equiv Q_R \cap \{x_N = 0\}.$$

We will now describe an explicit barrier for the function

$$(x,t) \to v(x,t) \equiv u(x,t) - g(0,0)$$

nearby $(0,0)$. Let A, θ, k be positive constants to be chosen, and let $y = (0, \ldots, 0, -1)$. Introduce the function

$$\eta_k(x,t) = \exp\big(-k(|x - y| - 1)\big) \exp(A\theta k^p t),$$

and the set

$$\mathcal{N}_k = \Big\{(x,t) : x_N > 0, \, 1 < |x - y| < 1 + 1/k, \, \frac{1}{A\theta k^p} < t < 0\Big\}.$$

We assume k is so large that $\mathcal{N}_k \subset Q_R^+$. Our barrier is given by

$$\Phi_k(x,t) = A\big(1 - \eta_k(x,t)\big) + \sum_{i=1}^{N-1} b_i x_i + C \sum_{i=1}^{N-1} x_i^2,$$

where $A > 0$, b_i and $C > 0$ are constants to be determined later only in terms of the quantities in (1.5)–(1.7).

Lemma. *For each choice of A, b_i and C there exists k and θ such that*

$$\frac{\partial \Phi_k}{\partial t} - \operatorname{div}(|\nabla \Phi_k|^{p-2} \nabla \Phi_k) > 0$$

in the set $\mathcal{N}_k$.

Proof. By direct calculation in $\mathcal{N}_k$

$$(2.1) \qquad \nabla \Phi_k = k A \eta_k \frac{x - y}{|x - y|} + \mathbf{b} + 2C\bar{x}$$

where $\mathbf{b} \equiv (b_1, b_2, \ldots, b_{N-1})$ and $\bar{x} \equiv (x_1, x_2, \ldots, x_{N-1})$, $x \equiv (\bar{x}, x_n)$. Since

$$e^{-1} \leq \eta_k \leq e,$$

we have

$$(2.2) \qquad \frac{1}{2} k A \eta_k \le |\nabla \Phi_k| \le 2 k A \eta_k \,.$$

Also

$$\Phi_{k,x_i x_j} = -k^2 A \eta_k \frac{(x-y)_i (x-y)_j}{|x-y|^2} + 2C\bar{\delta}_{ij} + k A \eta_k \left(\frac{\delta_{ij}}{|x-y|} - \frac{(x-y)_i(x-y)_j}{|x-y|^3} \right)$$

where δ_{ij} is the Kronecker delta for $i,j = 1, 2, \ldots, N$ and $\bar{\delta}_{ij}$ is the Kronecker delta for $i,j = 1, 2, \ldots, N-1$. Using (2.1) and (2.2)

$$(2.3) \qquad \frac{\Phi_{k,x_i} \Phi_{k,x_i x_j} \Phi_{k,x_j}}{|\nabla \Phi_k|^2} = -k^2 A \eta_k + Dk \,,$$

where $k > 1$ and D is a constant depending upon C and $|\mathbf{b}|$. From this

$$\operatorname{div}\left(|\nabla \Phi_k|^{p-2} \nabla \Phi_k\right) = |\nabla \Phi_k|^{p-2} \left(\Delta \Phi_k + (p-2)\frac{\Phi_{k,x_i} \Phi_{k,x_i x_j} \Phi_{k,x_j}}{|\nabla \Phi_k|^2} \right)$$

$$\le -(A\eta_k)^{p-1} k^p + D\left(A, C, |\mathbf{b}|\right) k^{p-1} \,,$$

and

$$\frac{\partial}{\partial t} \Phi_k - \operatorname{div}\left(|\nabla \Phi_k|^{p-2} \nabla \Phi_k\right) \ge A\eta_k k^p \left\{ -\theta + (A\eta_k)^{p-2} - \frac{\gamma}{k} \right\} \,.$$

To proceede set

$$A = 2e^2 \left(\|u\|_{\infty, Q_R^+} + \left\| \frac{\partial g}{\partial t} \right\|_{\infty, S} \right),$$

$$b_i = \frac{\partial g}{\partial x_i}(0,0) \qquad i = 1, \ldots, N-1 \quad \text{and}$$

$$C = \|g\|_{\infty, S} + \|\nabla g\|_{\infty, S} + \|\nabla^2 g\|_{\infty, S}.$$

For this choice of constants select k and θ according to the Lemma. Then we have

$$v \le \Phi_k$$

on the parabolic boundary of $\mathcal{N}_k$. Since Φ_k is a supersolution, the comparison principle gives

$$v \le \Phi_k$$

on the whole $\mathcal{N}_k$. In particular

$$\frac{v(0, \ldots, 0, h, 0) - v(0, \ldots, 0)}{h} \le \frac{\Phi_k(0, \ldots, 0, h, 0) - \Phi_k(0, \ldots, 0)}{h} \le Ak.$$

Analogously, by considering

$$\Psi_k = -A(1 - \eta_k) + \sum_{i=1}^{N-1} b_i x_i - C \sum_{i-1}^{N-1} x_i^2$$

we deduce

$$\frac{v(0, \ldots, 0, h, 0) - v(0, \ldots, 0)}{h} \geq -Ak.$$

Therefore we have

$$(2.4) \qquad \left| \frac{u(0, \ldots, 0, h, 0) - u(0, \ldots, 0)}{h} \right| \leq Ak$$

for $h \in (0, \frac{1}{k})$ and

$$\frac{\partial u}{\partial x_N} \in L^\infty(\partial\Omega \times (\epsilon, T]).$$

Since the tangential derivatives are smooth on $\partial\Omega$ we conclude

$$(2.5) \qquad \nabla u \in L^\infty(\partial\Omega \times (\epsilon, T]).$$

3. The gradient bound

We shall now indicate how to obtain (1.4) from (2.5). The argument follows closely the one given in [DB-F1] to prove the interior gradient bound. Thus, we just indicate the few modifications needed.

First of all, without lost of generality, we may assume that

$$u \in L^p\big([s, T]; W^{2, N+1}(\Omega)\big) \cap W^{1, p}\big([s, T]; L^p(\Omega)\big)$$

by the difference quotient argument indicated in p. 95 of [DB-F1]. Thus ∇u is continuous up to the boundary of Ω. Take a number λ such that

$$(3.1) \qquad |\nabla u(x, t)| < \lambda \text{ for } (x, t) \in \partial\Omega \times (\epsilon, T)$$

and set

$$w_\alpha(x, t) = \big[(|\nabla u(x, t)|^2 - \lambda^2)_+\big]^\alpha$$

if $\alpha > 0$. For $\alpha = 0$ we let

$$w_o(x, t) = \begin{cases} 1 & \text{if } |\nabla u(x, t)|^2 \geq \lambda^2 + 1 \\ 0 & \text{if } |\nabla u(x, t)|^2 < \lambda^2 \\ |\nabla u(x, t)|^2 - \lambda^2 & \text{otherwise.} \end{cases}$$

The point is that $\phi = u_{x_j} w_\alpha \zeta^2$, where ζ is a standard cut-off function, is now an admissible test function, since w_α vanishes on $\partial\Omega \times (\epsilon, T)$. Similarly

$$\phi = u_{x_j} |\nabla u|^{(p/2)-2} \left(|\nabla u|^{p/2} - \lambda^{p/2} \right)_+ \zeta^2$$

is a good test function. By repeating the proof of Theorem 2.1 in [DB-F1] with these new test functions we obtain (1.4).

REFERENCES

[C-DB] Y.Z. Chen, E. Dibenedetto, Boundary regularity for non-linear degenerate parabolic systems, Jour. fur die Reine und Agnew. Math. 395(1989), pp. 102–131.

[Ch] H. Choe, Hölder regularity for the gradient of solutions of certain singular parabolic equations. Preprint 1989.

[DB-F1] E.Dibenedetto, A. Friedman, Regularity of solutions of nonlinear degenerate parabolic systems, Jour. fur die Reine und Angew. Math. 349 (1984), pp. 83–128.

[DB-F2] E.DiBenedetto, A. Friedman, Hölder estimates for nonlinear degenerate parabolic systems, Jour. fur die Reine und Angewandte Math. 357 (1985), pp. 1–22.

[Di-H] E.DiBenedetto, M.A. Herrero Non–negative solutions of the evolution p-laplacian equation. Cauchy problem and initial traces when $1 < p < 2$. Archive for Rat. Mech. (1990)

E. DiBenedetto   J. Manfredi   V. Vespri
Dept. Math.   Dept. Math. & Statistics   Dept. Math.
Northwestern Univ.   Northwestern Univ.   Univ. of Pittsburgh
Evanston, IL 60208   Pittsburgh, PA 15260   Evanston, IL 60208

The Quenching Problem
on the N–dimensional Ball

MAREK FILA, JOSEPHUS HULSHOF, PAVOL QUITTNER

1. Introduction

Consider the problem

$$(P) \quad \begin{cases} u_t = \Delta u - u^{-\beta} & x \in \Omega, t > 0, \\ u(x,t) = 1 & x \in \partial\Omega, t > 0, \\ u(x,0) = u_o(x) \equiv 1 & x \in \overline{\Omega} \end{cases}$$

where $\beta > 0$ and $\Omega = B_R(0) := \{x \in \mathbb{R}^N ; |x| < R\}$. It is known ([AW]) that there is a positive number $R_o = R_o(N, \beta)$ such that u exists globally if $R < R_o$ while for $R > R_o$ the solution u reaches zero in a finite time T (it quenches). The only point x_o for which $u(x_o, t) \to 0$ as $t \to T$ is $x_o = 0$ (see [AK]).

The quenching rate (as $t \to T$) of the solution u was studied in [G1–2] and [FH]. In [G1] Guo showed that

$$\lim_{t \to T} u(x,t) \cdot (T-t)^{-\frac{1}{\beta+1}} = (\beta+1)^{\frac{1}{\beta+1}} \tag{1.1}$$

uniformly for $|x| \le C(T-t)^{\frac{1}{2}}$ for any positive constant C, provided $N = 1$, $\beta \ge 3$. This result was improved in [FH] where (1.1) is shown for $N = 1$, $\beta \ge 1$. In [G1], [FH] also more general initial data were considered. In [G2] Guo derived (1.1) for $N > 1$, $\beta > 1$ under the additional assumption that $R > \max(R_o, L)$ where

$$L = \left[\frac{2}{\beta+1}\left(\frac{2}{\beta+1} + N - 2\right)\right]^{\frac{1}{2}}. \tag{1.2}$$

In this paper we observe that in fact $R_o \ge L$, and we prove (1.1) in all remaining cases ($N = 1$, $\beta < 1$ and $N > 1$, $\beta \le 1$).

In [L1] the question was posed whether the solution of (P) can quench in infinite time when $R = R_o$ in more than one dimension. In one space dimension this is impossible because of the existence of a positive equilibrium for $R = R_o$. A partial answer to this question was given in [FK2] where it is shown that quenching in infinite time cannot occur if $N = 2$, $\beta > 1$ or $N = 3$, $\beta > 3$ (for more general initial data and arbitrary convex domains). We remark that the result in [FK2] was derived without any a priori knowledge about the stationary solutions.

Here we give the complete answer to the question concerning quenching in infinite time for $\Omega = B_R(0)$, $R = R_o$, $u_o \equiv 1$. We show first that u exists globally. Then we prove that u quenches in infinite time if

$$3 \le N \le 9, \quad 0 < \beta \le \frac{\nu}{2-\nu}, \quad \nu := \frac{N}{2} - \sqrt{N-1} \quad \text{or} \quad N > 9, \quad \beta > 0.$$
$$(1.3)$$

This follows from the absence of a strictly positive stationary solution in this case, while a singular stationary solution φ with $\varphi(0) = 0$ exists and $u(\cdot, t) \to \varphi$ as $t \to \infty$. Moreover, we have an explicit formula for R_o if (1.3) holds, namely $R_o = L$. On the other hand, if (1.3) is not satisfied there exist infinitely many positive equilibria less than 1 in Ω for $R = L$, $N > 1$. We now have $R_o > L$, and a unique positive equilibrium exists for $R = R_o$. Thus quenching in infinite time cannot occur if

$$N = 2, \quad \beta > 0 \quad \text{or} \quad 3 \le N \le 9, \quad \beta > \frac{\nu}{2-\nu}. \qquad (1.4)$$

We also examine the problem (P) with $0 < u_o(x) \le 1$, $u_o(x) \not\equiv 1$, from the point of view of dynamical systems. We describe the stability properties of the stationary solutions and show that solutions which start below unstable equilibria quench in finite time.

The stationary solutions are studied in Section 2. Section 3 contains the results on quenching.

2. Equilibria in dimension $N \ge 2$

In this section, we give a complete classification of the solutions of

$$(S) \quad \begin{cases} \Delta u = u^{-\beta} & \text{in } B_R = \{x \in \mathrm{IR}^N \, ; \, |x| < R\} \qquad (2.1) \\ u = 1 & \text{on } \partial B_R. \end{cases}$$

In addition we discuss their stability properties for the corresponding parabolic equation

$$u_t = \Delta u - u^{-\beta} \qquad (2.2)$$

subject to the same Dirichlet lateral boundary condition. Throughout this section it is assumed that $N \ge 2$, $\beta > 0$, $R > 0$.

We observe that all solutions of (S) must be radially symmetric. Indeed, by the Gidas – Ni – Nirenberg theorem [GNN] all solutions of (S) which are not radially symmetric have to intersect $u \equiv 1$ in B_R, which violates the maximum principle. Long before this symmetry result, the radially symmetric version of (S) was already studied in [JL], where by phase plane methods a curious dependence of the number of solutions on R, N

and β was shown. Thus the following theorem is due to [JL], although it is not formulated there in its full form.

Theorem 2.1 *Let* $N \geq 2$, $R > 0$, $\beta > 0$, $\gamma = \frac{1}{\beta+1}$, $L = \sqrt{2\gamma(N - 2 + 2\gamma)}$, *and* $D(N, \gamma) = (N - 2)^2 - 8(N - 2 + 2\gamma)(1 - \gamma)$.

(i) Suppose that $D(N, \gamma) \geq 0$. *Then (S) has no positive solution for* $R \geq L$, *and exactly one positive solution for* $R < L$.

(ii) Suppose that $D(N, \gamma) < 0$. *Then there exist numbers*

$$0 < r_1 < r_2 < \ldots \uparrow L,$$

and

$$R_1 > R_2 > \ldots \downarrow L,$$

depending on N *and* β, *such that when* $l(R)$ *denotes the number of positive radial solutions of (S),*

$$l(R) = \begin{cases} 1, & \text{if } 0 < R < r_1 \\ 2n, & \text{if } R = r_n \\ 2n + 1, & \text{if } r_n < R < r_{n+1} \\ \infty, & \text{if } R = L \\ 2n - 1, & \text{if } R = R_n \\ 2n, & \text{if } R_{n+1} < R < R_n \\ 0, & \text{if } R > R_1 \end{cases}$$

Because the stability analysis of these solutions requires the proof of this theorem as a preliminary, we shall indicate the main idea, which relies on the reduction of the second order non autonomous equation

$$U''(r) + \frac{N - 1}{r} U'(r) = U(r)^{-\beta}, \tag{2.3}$$

to a two–dimensional autonomous system (see also e.g. [BV], [H], [J], [G]). Here we do not use the transformation in [JL], but instead we set, following [H],

$$\tau = \log r; \quad \xi(\tau) = \frac{rU'(r)}{U(r)}; \quad \eta(\tau) = \frac{r^2}{U(r)^{\beta+1}}. \tag{2.4}$$

Then $\big(\xi(\tau), \eta(\tau)\big)$ is a solution of

$$(Q) \quad \begin{cases} \dfrac{d\xi}{d\tau} = (2 - N - \xi)\xi + \eta & (2.5) \\[2mm] \dfrac{d\eta}{d\tau} = \eta(2 - (1 + \beta)\xi). & (2.6) \end{cases}$$

Notice that (Q) is a *quadratic* system.

Now let $U = U_c(r)$ be the unique solution of (2.3) satisfying the initial conditions

$$U(0) = c > 0; \quad U'(0) = 0. \tag{2.7}$$

Then it is easily seen that (2.4) maps U to the unique orbit Γ of (Q) coming out of the origin along the eigenvector $\left(\begin{smallmatrix} 1 \\ N \end{smallmatrix}\right)$ of the eigenvalue 2 of the linearization of (Q) (see also [H]). The parametrization of Γ is determined by

$$\lim_{\tau \downarrow -\infty} \frac{\eta(\tau)}{e^{2\tau}} = \lim_{r \downarrow 0} \frac{1}{U(r)^{\beta+1}} = \frac{1}{c^{\beta+1}} \,. \tag{2.8}$$

This means that solutions $\big(\xi(\tau), \eta(\tau)\big)$ corresponding to larger values of c are running behind solutions corresponding to smaller values of c.

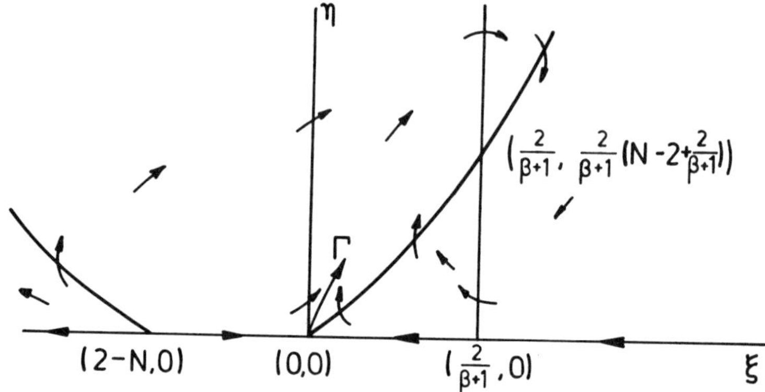

Figure 1. The phase plane for (Q) $(N > 2)$.

There are three critical points: $(2-N,0)$, $(0,0)$, $\big(\frac{2}{\beta+1}, \frac{2}{\beta+1}(N-2+\frac{2}{\beta+1})\big)$. The first one is not relevant here because of the positive invariance of the first quadrant. The second one has only the orbit Γ coming out into the first quadrant. The third critical point contains the singular solution

$$\varphi(r) = \frac{r^{2\gamma}}{\big(2\gamma(N - 2 + 2\gamma)\big)^{\gamma}} \,. \tag{2.9}$$

Because $u(R) = 1$ is equivalent to $\eta(\log R) = R^2$, the radial solutions of (S) are in a one to one correspondence to the intersections of Γ and the line $\{\eta = R^2\}$, where we must set $\tau = \log R$ at the intersection point. Thus the classification of the radial solutions of (S) depends on the global behaviour of Γ, which follows from the results in [JL]: the orbit Γ connects the critical points $(0,0)$ and $\big(2\gamma, 2\gamma(N - 2 + 2\gamma)\big)$, which implies

$$\lim_{c \downarrow 0} U_c(r) = \varphi(r) \qquad \text{for all } r \geq 0 \,. \tag{2.10}$$

The discriminant of the characteristic polynomial of the linearization around the critical point $\left(2\gamma, 2\gamma(N-2+2\gamma)\right)$ is exactly the number $D(N,\gamma)$ in Theorem 2.1. If it is negative Γ spirals clockwise around $(2\gamma, 2\gamma(N-2+2\gamma))$ an infinite number of times. Thus in Theorem 2.1(ii) R_n^2 is the η-value in the $(2n-1)$-th intersection of Γ and the line $\{\xi = 2\gamma\}$, and r_n^2 is the η-value in the $(2n)$-th intersection. On the other hand, if $D(N,\gamma) \geq 0$, then Γ approaches $(2\gamma, 2\gamma(N-2+2\gamma))$ monotonically, whence Theorem 2.1(i). In particular Γ lies below the line $\{\eta = 2\gamma(N-2+2\gamma)\}$, which means that U_c does not intersect the singular solution.

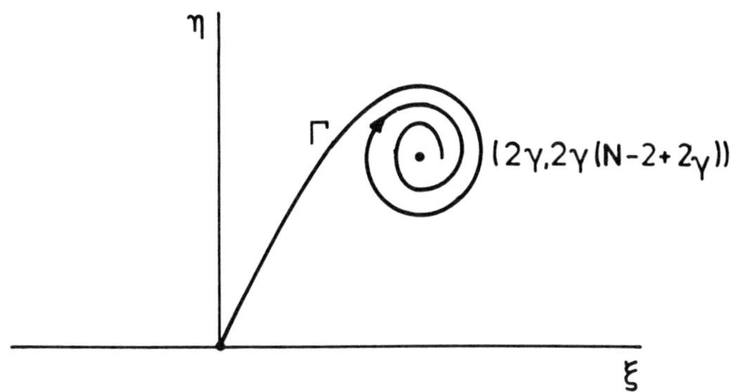

Figure 2. Γ if $D(N,\gamma) < 0$.

Remark. Computing $\frac{d\eta}{d\xi}$ along the line $l = \{\eta = N\xi\}$ we find that Γ lies below l. This implies that $\sqrt{\frac{2N}{\beta+1}} > R_1$.

Next we discuss the stability of the equilibria we found in Theorem 2.1. We assume that the reader is familiar with the notions of instability, asymptotic stability, and sub- and supersolution techniques as can be found in [A1,A2,S1,S2].

Theorem 2.2. *Suppose $R < L$ and $D(N,\gamma) \geq 0$. Then the positive solution of (S) is asymptotically stable.*

Proof. Let U be the positive solution of (S). The uniqueness and existence of U follows from Theorem 2.1(i). To establish the asymptotic stability of U it is sufficient to construct a supersolution strictly above and a subsolution strictly below U. The first one is easy, since $u \equiv 1$ is a supersolution. For the subsolution we consider $c < U(0)$, and the solution $U_c(r)$ of (2.3-2.7). We claim that U and U_c do not intersect. Indeed, the corresponding solutions $(\xi(\tau), \eta(\tau))$ and $(\xi_c(\tau), \eta_c(\tau))$ of (Q) both parametrize Γ, but $(\xi(\tau), \eta(\tau))$ runs behind $(\xi_c(\tau), \eta_c(\tau))$. Since Γ

is monotone (Fig. 3), it follows that $\eta_c(\tau) > \eta(\tau)$ for all τ. Consequently, $U_c(r) < U(r)$ for all $r \geq 0$. Thus U_c solves (2.1), and $U_c(R) < 1$, so it is a subsolution. ∎

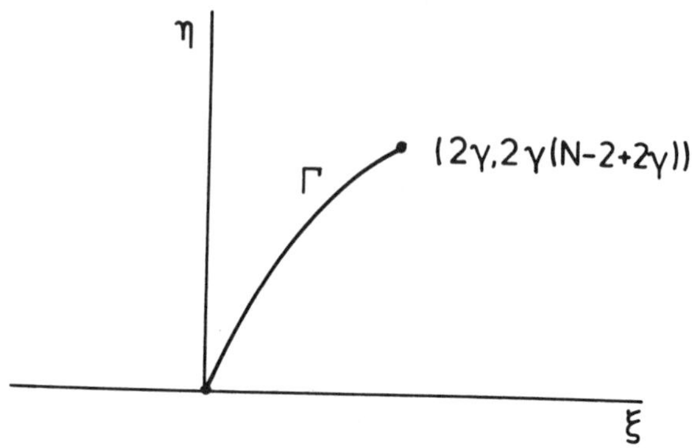

Figure 3. Γ if $D(N,\gamma) \geq 0$.

Proposition 2.3. *Suppose $D(N,\gamma) < 0$ and $R < R_1$. Then the solution of (S) corresponding to the first intersection of Γ and $\{\eta = R^2\}$ lies strictly above all the other solutions of (S) if they exist.*

Proof. Suppose U_c is the solution corresponding to the first intersection of Γ and $\{\eta = R^2\}$ and that U_{c*} is a solution corresponding to another intersection. Then, obviously, $\big(\xi_c(\tau), \eta_c(\tau)\big)$ runs behind $\big(\xi_{c*}(\tau), \eta_{c*}(\tau)\big)$. We consider all the different possibilities, the first three of which are sketched in Fig. 4. In all cases, we have $\eta_c(\log R) = \eta_{c*}(\log R) = R^2$, and $\xi_c(\log R) < \xi_{c*}(\log R)$. By (2.6) we have $\frac{\partial}{\partial \xi}\frac{d\eta}{d\tau} = -(1+\beta)\eta < 0$. It is easy to see that this implies that $\eta_c(\tau) < \eta_{c*}(\tau)$ for all $\tau < \log R$. Consequently, $U_c(r) > U_{c*}(r)$ for all $r \in [0, R)$. ∎

Proposition 2.4. *Suppose $D(N,\gamma) < 0$ and $r_1 < R < R_1$. Let U_{c_1} and U_{c_2} be two solutions of (S) not corresponding to the first intersection of Γ and $\{\eta = R^2\}$. Then U_{c_1} intersects U_{c_2} at least once on $[0, R)$.*

Proof. Again we consider all cases one of which is sketched in Fig. 5. Let $P_1 = \big(\xi_{c_1}(\log R), \eta_{c_1}(\log R)\big)$ and $P_2 = \big(\xi_{c_2}(\log R), \eta_{c_2}(\log R)\big)$ be the two intersections, and let $P = \big(\xi(\log R), \eta(\log R)\big)$ be the first intersection of Γ and $\{\eta = R^2\}$. Because $P \neq P_1$ and $P \neq P_2$, it is obvious that η_{c_1} and η_{c_2} must intersect in some $\tau = \tau^* < \log R$. Consequently, U_{c_1} and U_{c_2} intersect in $R^* = e^{\tau^*} < R$. ∎

Theorem 2.5. *Suppose $D(N,\gamma) < 0$ and $R < R_1$. Then the maximal solution U of (S) corresponding to the first intersection of Γ and $\{\eta = R^2\}$ is asymptotically stable.*

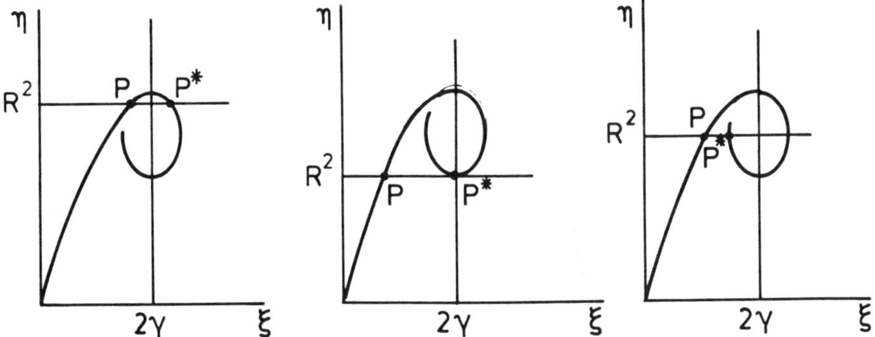

Fig. 4. The intersections P and P^* corresponding to solutions U_c and U_{c^*}

Proof. Since $u \equiv 1$ is a supersolution, it suffices to construct a subsolution strictly below U. Because Γ is monotone between $(0,0)$ and $(2\gamma, R_1^2)$, the proof is identical to the proof of Theorem 2.2, provided we choose $c < U(0)$ close enough to $U(0)$. Again, U_c is a subsolution, and obviously there is no other solution of (S) between U_c and U if c is close to $U(0)$. Hence U is stable. ∎

Theorem 2.6. *Suppose $D(N,\gamma) < 0$ and $r_1 \le R < R_1$. Then except for the maximal solution, every solution U of (S) is unstable from below and from above.*

Proof. We construct supersolutions arbitrarily close to U from below, and subsolutions arbitrarily close to U from above. Consider $c < U(0)$. Then $(\xi(\tau), \eta(\tau))$ runs behind $(\xi_c(\tau), \eta_c(\tau))$. Since η_c is not monotone on $(-\infty, \log R)$, it follows that η and η_c must intersect on $(-\infty, \log R)$. Suppose this happens for the first time in $\tau = \tau^* < \log R$. Let $R^* = e^{\tau^*}$, and define

$$\overline{U}_c(r) = \begin{cases} U_c(r), & \text{if } 0 \le r \le R^* \\ U(r), & \text{if } R^* < r \le R \end{cases} \tag{2.11}$$

Then $\overline{U}_c$ fails to be a classical supersolution in the sense of [A1,A2,S1,S2] because of the jump discontinuity of its first derivative at $r = R^*$. However since this jump is negative it easy to see that $\overline{U}_c$ is a supersolution for the parabolic problem. Obviously it lies below U and as $c \to U(0)$, $\overline{U}_c \to U$ uniformly. Thus U is unstable from below. In the same way one constructs subsolutions by considering $c > U(0)$, so that U is also unstable from above. ∎

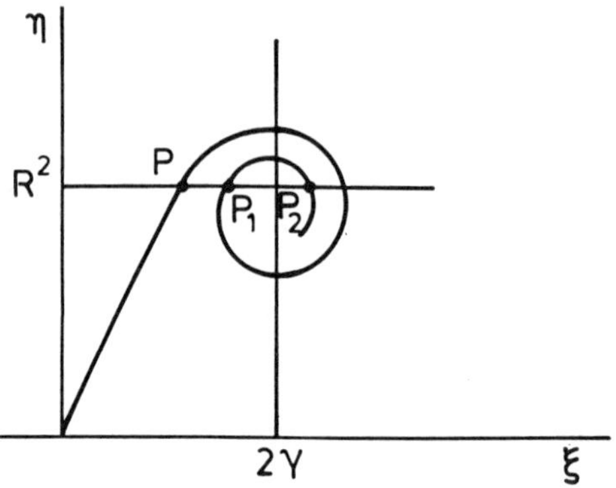

Figure 5. Three intersections of Γ and $\{\eta = R^2\}$.

It remains to describe the stability properties for $D(N,\gamma) < 0$ and $R = R_1$.

Theorem 2.7. *Suppose $D(N,\gamma) < 0$ and $R = R_1$. Then the unique positive solution is asymptotically stable from above, but unstable from below.*

Proof. Stability from above follows as in Theorem 2.2, and unstability from below as in Theorem 2.6. ∎

For later purposes we prove the following two propositions.

Proposition 2.8. *The function φ defined by (2.9) is the unique strictly positive solution of (2.4) on $(0,\infty)$, which satisfies the condition $\lim_{r\downarrow 0} U(r) = 0$.*

Proof. Let $\tilde{\varphi}$ be a strictly positive solution of (2.4) on $(0,\infty)$, $\lim_{r\downarrow 0}\tilde{\varphi}(r) = 0$, $\tilde{\varphi} \not\equiv \varphi$, and let $(\tilde{\xi},\tilde{\eta}) : \mathbb{R} \to \mathbb{R} \times \mathbb{R}$ be the transform of $\tilde{U} := \tilde{\varphi}$ given by (2.3). Because of (2.5) the positivity of $\tilde{U}'$ is equivalent to the positivity of $\tilde{\xi}$. Obviously $\tilde{U}'(r)$ must have positive values for r arbitrarily close to $r = 0$. Since the first quadrant is left invariant by (Q) this implies that it contains the orbit $\tilde{\Gamma}$ corresponding to $\tilde{U}$, and that $\tilde{\xi}(\tau) > 0$, $\tilde{\eta}(\tau) > 0$ for all $\tau \in \mathbb{R}$. By standard uniqueness results for ordinary differential equations $\tilde{\Gamma} \neq \Gamma$. The properties of the flow given by (Q) (cf. Fig. 1) then imply that $\lim_{\tau\to-\infty}\tilde{\xi}(\tau) = \lim_{\tau\to-\infty}\tilde{\eta}(\tau) = \infty$, and that $(\tilde{\xi}(\tau),\tilde{\eta}(\tau))$ lies below the parabola $(2 - N - \xi)\xi + \eta = 0$. Thus $\tilde{\eta}'(\tau)$ is at least of the order $\tilde{\eta}(\tau)^{\frac{3}{2}}$ as $\tau \to -\infty$, so that $\tilde{\eta}(\tau)$ blows up in finite time as τ decreases, contradiction. ∎

Proposition 2.9. *Suppose $D(N, \gamma) < 0$ and $R = L$. Then except for the maximal solution, every positive radial solution U of (S) intersects φ at least once in $(0, L)$.*

Proof. U corresponds to the n–th intersection of Γ and the line $\{\eta = L^2\}$ for some $n > 1$. Hence U crosses φ in $(0, L)$ exactly $n - 1$ times. $\blacksquare$

We conclude this section by examining $D(N, \gamma)$ as a function of $N \in [2, \infty)$ and $\gamma \in (0, 1)$. We have $D(2, \gamma) = 16\gamma(\gamma - 1) < 0$ for all $\gamma \in (0, 1)$ and obviously $D(N, \gamma) > 0$ for all $N \geq 10$ and all $\gamma > 0$. Only for $N \in (2, 10)$ we find a sign change of $D(N, \gamma)$ in $(0, 1)$ in $\gamma_o(N) = 1 - \frac{N}{4} + \frac{1}{2}\sqrt{N - 1}$. Thus

$$D(N, \frac{1}{\beta + 1}) \quad \begin{cases} > 0, & \text{if } 0 < \beta < K(N) \\ < 0, & \text{if } K(N) < \beta < \infty \end{cases},$$

where

$$K(N) = \frac{\nu}{2 - \nu}, \quad \nu := \frac{N}{2} - \sqrt{N - 1}.$$

(cf. (1.3) and (1.4)).

3. Quenching

If u is the solution of (P), then it is easy to see that u is radially symmetric, $u_r \geq 0$, $u_t \leq 0$.

The following result is a consequence of Theorem 2.1. Recall that R_o is the limiting radius for the global existence of u (see Section 1), L is defined in (1.2) and φ is defined by (2.9).

Theorem 3.1. *Assume that $R = R_o$ and put*

$$K(N) := \begin{cases} 0, & \text{if } N \leq 2 \\ \dfrac{\nu}{2 - \nu}, \ \text{where } \nu = \dfrac{N}{2} - \sqrt{N - 1}, & \text{if } 3 \leq N \leq 9 \\ \infty, & \text{if } N \geq 10. \end{cases}$$

Let u be the solution of (P). Then

(i) if $\beta > K(N)$, then $R_o > L$ and quenching in finite or infinite time cannot occur. Furthermore $u(\cdot, t)$ converges uniformly to the unique positive equilibrium as $t \to \infty$;

(ii) if $0 < \beta \leq K(N)$, then $R_o = L$ and u quenches in infinite time. More precisely, $u(\cdot, t) \to \varphi$ uniformly as $t \to \infty$.

Proof. For the one–dimensional case see [L2] or [L1] and the references there. For any $N \geq 1$, $\beta > 0$ it is known that

$$R_o = \sup\{R \,; \text{ there exists a strictly positive solution of (S)}\}$$

([AW], [AK]). If for $R = R_o$ a positive stationary solution v exists (obviously, $v < 1$ in Ω), then the maximum principle yields that $u(\cdot, t) > v$ in Ω for all $t \geq 0$ so the existence is global and quenching is impossible. By Theorem 2.1(ii) we know that $R_o = R_1$ for $\beta > K(N)$ and that a positive solution of (S) exists for $R = R_1$. By Theorem 2.7 it is stable from above. This proves (i).

If $0 < \beta \leq K(N)$ on the other hand then by Theorem 2.1(i) $R_o = L$. Also there is no positive stationary solution for $R = L$, but the singular stationary solution φ is present. Now we can use the arguments from the proof of Lemma 2 in [LM]. Suppose that u quenches in finite time. Then there is a $0 < T < \infty$ such that $u(0, t) \to 0$ as $t \to T$. According to the maximum principle $\psi := u - \varphi$ is positive for $t \in [0, T)$, $r \in [0, R)$. Since

$$(r^{N-1}\psi_r)_r = r^{N-1}(\psi_t + u^{-\beta} - \varphi^{-\beta}) \leq 0$$

and $r^{N-1}\psi_r = 0$ for $r = 0$, we see that $\psi_r(\cdot, t) < 0$ for any $t \in (0, T)$. This means that $\psi(\cdot, t)$ is maximized at $r = 0$, hence $u(\cdot, t) \to \varphi$ uniformly as $t \to T$. This is impossible in view of the strong maximum principle. To complete the proof of (ii) we show that $u(0, t) \to 0$ as $t \to \infty$. Recall that $u_t \leq 0$, $u_r \geq 0$. If $u(0, t)$ were bounded from below by a positive constant, then $u(\cdot, t)$ would have to approach a stationary solution – a contradiction with Theorem 2.1(i). By the same argument as above it follows that $u(\cdot, t) \to \varphi$ uniformly. $\blacksquare$

Now we turn to the quenching rate problem.

Theorem 3.2. *Let $N \geq 1$ and $\beta > 0$. If the solution u of (P) quenches in a finite time T, then*

$$\lim_{t \to T} u(x, t)(T - t)^{-\frac{1}{\beta+1}} = k := (\beta + 1)^{\frac{1}{\beta+1}}$$

uniformly for $|x| \leq C\sqrt{T - t}$, where C is an arbitrary positive constant.

In view of the results in [FH] and [G2] it suffices to give the proof for $\beta < 1$, $N = 1$ and $\beta \leq 1$, $N > 1$. We follow the strategy from [GK], which was modified for quenching problems in [G1], [G2]. We use the similarity variables

$$w(y, s) := u(r, t)(T - t)^{-\frac{1}{\beta+1}}, \quad y := r(T - t)^{-\frac{1}{2}}, \quad s := -\log(T - t).$$

Then w satisfies the equation

$$w_s = w_{yy} + \left(\frac{N - 1}{y} - \frac{y}{2}\right)w_y - w^{-\beta} + \frac{w}{\beta + 1} \tag{3.1}$$

in $W := \{(y, s)\,;\, 0 < y < Re^{s/2},\ s > -\log T\}$ with the boundary conditions

$$w_y(0, s) = 0, \quad w(Re^{s/2}, s) = e^{\frac{s}{\beta+1}}, \quad w(y, -\log T) = T^{-\frac{1}{\beta+1}}.$$

Our aim is to show that $w(y, s) \to k$ as $s \to \infty$.

Basic facts which allow us to proceed towards this aim are the following two ones: the existence of a $\delta > 0$ such that $w \geq \delta$ in W ([FK1, Thm. 1.2 b]) and the polynomial growth of $w(y, s)$ in y. The latter fact is stated in the next lemma.

Lemma 3.3.
(i) If $\beta < 1$, then $w(y, s) \leq C(y^{\frac{2}{\beta+1}} + 1)$ for some positive constant C.
(ii) If $\beta = 1$, then $w(y, s) \leq C(y^2 + 1)$ for some positive constant C.

Proof. (i) It is shown in [FK1, Lemma 2.2] that

$$P(x, t) := \frac{1}{2}|\nabla u|^2 - \frac{1}{1-\beta}u^{1-\beta} \leq 0 \quad \text{in } \Omega \times (0, T),$$

hence

$$\left(u^{\frac{\beta+1}{2}}\right)_r \leq C_1 := \frac{1+\beta}{\sqrt{2(1-\beta)}}.$$

Integrating the last inequality we obtain

$$u^{\frac{\beta+1}{2}}(r, t) \leq C_1 r + u^{\frac{\beta+1}{2}}(0, t).$$

From Theorem 1.2 a) in [FK1] we have

$$u(0, t) \leq \left[(1+\beta)(T-t)\right]^{\frac{1}{1+\beta}},$$

therefore

$$u(r, t) \leq C_2 r^{\frac{2}{\beta+1}} + C_3(T-t)^{\frac{1}{\beta+1}}.$$

If we multiply the last inequality by $(T-t)^{-\frac{1}{\beta+1}}$ and change the variables, we get the assertion.

(ii) The function u satisfies the equation

$$u_t = u_{rr} + \frac{N-1}{r}u_r - \frac{1}{u},$$

therefore $u_{rr} \leq \frac{1}{u}$, which yields $w_{yy} \leq \frac{1}{w}$. Now it is sufficient to recall that $w \geq \delta > 0$. $\blacksquare$

The polynomial growth in y together with the boundedness from below by a positive constant enable us to conclude (see [G1], [G2]) that $w(y, s) \to w_\infty(y) \geq \delta$, where w_∞ is a solution of the equation

$$w_{yy} + \left(\frac{N-1}{y} - \frac{y}{2}\right)w_y - w^{-\beta} + \frac{w}{\beta+1} = 0 \quad \text{for } y > 0 \qquad (3.2)$$

with the initial conditions

$$w_y(0) = 0, \quad w(0) = a \le k. \tag{3.3}$$

The second condition in (3.3) is fulfilled, because w_∞ must be a nondecreasing function (we have $u_r \ge 0$).

It is shown in [G2] that $\lim\limits_{y \to \infty} \dfrac{w'}{w}$ exists for any w which satisfies (3.2). Moreover, it is equal either to zero or to infinity and we call the corresponding solutions slow or fast solutions, respectively. Obviously, fast solutions grow at infinity faster than any power of y. Due to Lemma 3.3, only slow solutions are candidates for a limit element of $w(y, s)$ as $s \to \infty$.

Proof of Theorem 3.2. We proceed along the lines of Guo's proof in [G2]. thus, we only indicate the main steps.

(i) In the same manner as in [BPT, Lemmas 16 and 20] one shows that

$$g(y) := \frac{1}{\beta + 1} w - \frac{1}{2} y w_y \to 0 \qquad \text{as} \quad y \to \infty \tag{3.4}$$

for any nonconstant slow solution of (3.2).

(ii) Our next claim is, that g must have a zero for any slow solution of (3.2) which satisfies

$$w_y(0) = 0, \quad w(0) = a < k. \tag{3.5}$$

For the proof we could mimic the proof of Lemma 3.5 in [G2]. The only change is, that Case 1 in Guo's proof is ruled out by (3.4).

(iii) Since g has a zero, we can use the arguments from the proof of Lemma 3.6 in [G2] to show that any slow solution of (3.2-3.5) intersects φ at least twice, where φ is the singular solution from Section 2. It is also a singular solution of (3.2). Note that (2.9) makes sense for $N = 1$ if $\beta < 1$.

(iv) $w(\cdot, s)$ intersects φ exactly once for any $s > -\log T$ provided $R > L$. This follows from the maximum principle in the same way as in the proof of Lemma 4.9 in [G2]. Since we assume that u quenches in a finite time we consider only $R > L$.

By (iii) and (iv) it is impossible for $w(\cdot, s)$ to converge to a nonconstant slow solution. Therefore it must converge to k. This completes the proof. ∎

Theorem 3.4. *Suppose (1.4) holds and let $R \in [r_1, R_1]$, where r_1 and R_1 are as in Theorem 2.1(ii). Then u quenches in finite time provided $u_o \le U$, $u_o \not\equiv U$ and U is an arbitrary positive (radial) equilibrium which is unstable from below.*

Proof. Because of (1.4) we know that $D(N, \gamma) < 0$ in Theorem 2.1, so that case (ii) applies. It follows from the maximum principle that

$u(r, t_o; u_o) < U(r)$ for $t_o > 0$, so that $u(r, t_o; u_o) \leq \overline{U}_c(r)$ if $c < U(0)$ is close enough to $U(0)$ ($\overline{U}_c$ is the supersolution defined by (2.12)).

We show that $\overline{u}(r, t) := u(r, t; \overline{U}_c)$ quenches in finite time. Clearly, $\overline{u}$ cannot be bounded from below by a positive constant, because by Proposition 2.4 there is no positive equilibrium below U. We have only to rule out the possibility of infinite time quenching.

Assume that $\overline{u}$ quenches in infinite time. Since $\overline{u}_t \leq 0$, we know that there is a function $v = v(r) \geq 0$ with $v(0) = 0$, such that $\overline{u}(r, t) \downarrow v(r)$ as $t \to \infty$. Moreover, $v_r \geq 0$ because $\overline{u}_r > 0$ for $t > 0$. By the same arguments as in the proof of Lemma 2 in [LM] (see also the proof of Theorem 3.1) one shows that $v(r) > 0$ for $r > 0$, v is continuous on $[0, R]$, v satisfies (2.3) for $r > 0$. Thus $v \equiv \varphi$ because of Proposition 2.8 and since $v(R) = 1 = \varphi(R)$ this can only be if $R = L$. In view of Proposition 2.9 this is also impossible. ∎

Addendum. This paper was conceived after the conference took place.

REFERENCES

[AK] A. Acker & B. Kawohl, *Remarks on quenching*, Nonlinear Anal. TMA **13** (1989), 53–61.

[AW] A. Acker & W. Walter, *The quenching problem for nonlinear partial differential equations,* Springer Lecture Notes in Math. **564** (1976), 1–12.

[A1] H. Amann, *On the existence of positive solutions of nonlinear elliptic boundary value problems, Indiana Univ. Math. J.* **21** (1971), 125–146.

[A2] H. Amann, *Supersolutions, monotone iterations, and stability, J. Diff. Equ.* **21** (1976), 363–377.

[BPT] H. Brézis, L.A. Peletier & D. Terman, *A very singular solution of the heat equation with absorption,* Arch. Rat. Mech. Anal. **95** (1986), 185–209.

[BV] W.J. van den Broek & F. Verhulst, *A generalized Emden–Fowler equation,* Math. Meth. Appl. Sc. **4** (1982), 259–271.

[FH] M. Fila & J. Hulshof, *A note on the quenching rate,* to appear in Proc. A. M. S.

[FK1] M. Fila & B. Kawohl, *Asymptotic analysis of quenching problems,* to appear in *Rocky Mountain J. of Math.*

[FK2] M. Fila & B. Kawohl, *Is quenching in infinite time possible ?,* Quarterly of Appl. Math. **48**(1990), 531–534.

[GNN] B. Gidas, W.-M. Ni & L. Nirenberg, *Symmetry and related properties via the maximum principle, Comm. Math. Phys.* **68** (1979), 209–243.

[GK] Y. Giga & R.V. Kohn, *Asymptotically self–similar blow–up of semilinear heat equations*, Comm. Pure Appl. Math. **38** (1985), 297–319.

[G] R.J. Grundy, *Similarity solutions of the nonlinear diffusion equation*, Quarterly Appl. Math. **37** (1979), 259–280.

[G1] J.S. Guo, *On the quenching behavior of the solution of a semilinear parabolic equation*, J. Math. Anal. Appl. **151**(1990), 58–79.

[G2] J.S. Guo, *On the semilinear elliptic equation* $\Delta w - \frac{1}{2} y \nabla w + \lambda w - w^{-\beta} = 0$ *in* $\mathbb{R}^n$, IMA preprint #**531** (1989).

[H] J. Hulshof, *Similarity solutions of the porous medium equation with sign changes*, J. Math. Anal. Appl. **156** (1991).

[J] C.W. Jones, *On reducible nonlinear differential equations occurring in mechanics*, Proc. Roy. Soc. **A 217** (1953), 327–343.

[JL] D.D. Joseph & T.S. Lundgren, *Quasilinear Dirichlet problems driven by positive sources*, Arch. Rat. Mech. Anal. **49** (1973), 241–269.

[L1] H.A. Levine, *The phenomenon of quenching: a survey*, in: Trends in the Theory and Practice of Nonlinear Analysis, V. Lakshmikantham ed. North Holland (1985), 257–286.

[L2] H.A. Levine, *Quenching, nonquenching and beyond quenching for solutions of some parabolic equations*, Ann. Mat. Pura Appl. **155** (1989), 243–260.

[LM] H.A. Levine & J.T. Montgomery, *The quenching of solutions of some nonlinear parabolic equations*, SIAM J. Math. Anal. **11** (1980), 842–847.

[S1] D.H. Sattinger, *Monotone methods in nonlinear elliptic and parabolic boundary value problems*, Indiana Univ. Math. J. **21** (1972), 979–1000.

[S2] D.H. Sattinger, *Topics in Stability and Bifurcation Theory*, Springer Lecture Notes (1973).

Marek Fila	Josephus Hulshof	Pavol Quittner
Dept. of Math. Analysis	Mathematical Inst.	Inst. of Appl. Math.
Comenius Univ.	Leiden Univ.	Comenius Univ.
Mlynská dolina,	P.O.Box 9512,	Mlynská dolina
84215 Bratislava	2300 RA Leiden	84215 Bratislava
Czechoslovakia	The Netherlands	Czechoslovakia

Global Solutions for a Class
of Monge-Ampère Equations

BRUNO FRANCHI

Abstract

In this paper we prove, by shooting method, the existence of radially symmetric ground state solutions for a class of Monge-Ampère equations changing type.

In this note we shall prove some results concerning the existence of global (i.e. defined on all of $\mathbf{R}^n$) nonnegative solutions for a class of Monge-Ampère type equations. More precisely we are looking for sufficiently regular functions u such that

$$\text{(1)} \qquad \begin{cases} \det{(D^2u)} = (-1)^n g(|Du|^2)f(u) \text{ in } \mathbf{R}^n \\ u \geq 0, \quad \lim_{x \to \infty} u(x) = 0, \end{cases}$$

where $f : [0, \infty) \to \mathbf{R}$ is a continuous function such that

   (i) f is locally Lipschitz continuous on $(0, \infty)$;

   (ii)   there exist $\alpha > 0$ and $\gamma \in (\alpha, \infty]$ such that $f(u) < 0$ in $(0, \alpha)$, $f(0) = f(\alpha) = 0$ and $f(u) > 0$ in (α, γ). Moreover, if $\gamma < \infty$, then $f(\gamma) = 0$.

We shall assume that g is a continuosly differentiable function on $[0, \infty)$ such that

   (iii) $0 \leq tg'(t) \leq \dfrac{s}{2}g(t)$ for a suitable $s \geq 0$ and for any $t \geq 0$.

Moreover we shall suppose that $g(0) = 1$.

In what follows we shall call a solution of problem (1) a *ground state*. If $g \equiv 1$ the equation in (1) is the classical Monge-Ampère equation; if $g(p) = (1+p)^{(n+2)/2}$, the left hand side of (1) is the Gaussian curvature of the graph of the function u at the point $(x, u(x))$.

Unfortunately, in this paper we are not able to deal with the Gaussian curvature equation since a further condition concerning the growth at infinity of g will be required: we shall suppose that

$$\int_0^\infty \frac{\xi^n}{g(\xi^2)} \, d\xi = \infty.$$

---

[1] The author is partially supported by G.N.A.F.A. of C.N.R., Italy and Ministero dell'Università della Ricerca Scientifica e Tecnologica, Italy.

In fact, this condition is satisfied by $g(p) = (1+p)^{s/2}$ only if $0 \leq s \leq n+1$. Observe that the exponent $s = n+1$ is also critical for the Dirichlet problem in bounded convex regions of $\mathbf{R}^n$ (see [GT, Chapter 17]).

Some remarks about problem (1) are now in order.

The factor $(-1)^n$ at the right hand side cannot be avoided. In fact, suppose that $u = u(|x|)$ is a radial function. In this case the equation becomes

$$\frac{((u')^n)'}{g(|u'|^2)} = \frac{(-1)^n}{n} r^{n-1} f(u).$$

Since we are looking for solutions vanishing at infinity it is natural to deal with functions u which are decreasing and convex at infinity. Hence, taking into account that u is small at infinity and hence $f(u) < 0$ at infinity, the factor $(-1)^n$ is justified. Moreover, if Ω is a bounded convex open subset of $\mathbf{R}^n$, then P.L. Lions noted in[L] that strictly convex solutions of the problem

$$\begin{cases} \det (D^2 u) - (\lambda u)^n = 0 \text{ in } \Omega, \\ u = 0 \text{ on } \partial\Omega \end{cases}$$

satisfy properties analogous to those of the eigenvalues of the Laplace operator.

The equation in (1) is not of elliptic type since the function f changes sign at α ; we recall that very few results are known for Monge-Ampère equations changing of type also in bounded convex domains of the space: for recent results and bibliography, see [Z].

Existence (and uniqueness) results have been recently proved for ground states of the mean curvature equation: see e.g. [FLS 1], [FLS 2], [APS], [PS] and the papers quoted therein. In this case, the equation in (1) is replaced by

$$\text{div} \left(\frac{Du}{\sqrt{1 + |Du|^2}} \right) + f(u) = 0.$$

In this paper we shall restrict ourselves to radial ground states. In some sense, this choice is justified also a priori by the fact that for the mean curvature equation the ground states are necessarily symmetric under very natural hypotheses on f (see [FL]). Moreover, a radial symmetry result is proved for convex solutions of Monge-Ampère equation in a ball of $\mathbf{R}^n$ ([GNN] and [D]).

For radially symmetric functions $u = u(|x|) = u(r)$, problem (1) becomes

$$(2) \quad \begin{cases} \dfrac{((u')^n)'}{g(|u'|^2)} = \dfrac{(-1)^n}{n} r^{n-1} f(u) \quad \text{for } r > 0 \\ u \geq 0, \quad u'(0) = 0, \quad u(r) \to 0 \text{ if } r \to \infty, \end{cases}$$

where u is *regular* , i.e. u and $(u')^{n-1}$ are continuously differentiable on $[0,\infty)$.

In what follows we shall assume $n > 1$, since the more relevant cases of (2) with $n = 1$ are treated in [FLS 1].

Now put $t = r^{2n/(n+1)}$ and $a = (n-1)/2n, v(t) = u(r)$. The problem (2) becomes

$$(3) \begin{cases} \dfrac{((v')^{n-1}v')'}{g(t^{2a}|v'|^2)} + \dfrac{n-1}{2t} \dfrac{(v')^{n-1}v'}{g(t^{2a}|v'|^2)} + (-1)^{n-1}f(v) = 0, \text{ for } t > 0 \\ v \geq 0, \ v'(0) = 0, \ v(t) \to 0 \text{ if } t \to \infty, \end{cases}$$

where $g(p)$ and $f(u)$ denote the functions $g\left(\frac{2n}{n+1}p\right)$ and $\frac{1}{n^2}\left(\frac{n+1}{2n}\right)^{n+1} f(u)$ respectively.

We note that the condition $u'(0) = 0$ in (2) becomes $t^a v'(t) \to 0$ as $t \to 0+$ which could seem less restrective than $v'(0) = 0$ in (3). On the other hand, by de l'Hôpital's rule, it follows from (2) that u' is differentiable at $r = 0$ and we get

$$(-u''(0))^n = f(u(0)),$$

so that, again by de l'Hôpital's rule,

$$v'(t) = r^{-(n-1)/(n+1)} u'(r) \to 0 \text{ as } r \to 0+ .$$

Formally, the problem (3) is analogous to the problems considered in [FLS 1] but is not elliptic for arbitrary $n > 1$ since $\dfrac{d}{dp}(p^{n-1}p)$ has not a sign. Then it is natural to consider the following elliptic problem

$$(P) \begin{cases} \dfrac{(|v'|^{n-1}v')'}{g(t^{2a}|v'|^2)} + \dfrac{n-1}{2t} \dfrac{|v'|^{n-1}v'}{g(t^{2a}|v'|^2)} + f(v) = 0, \text{ for } r > 0 \\ v \geq 0, \ v'(0) = 0, \ v(t) \to 0 \text{ if } t \to \infty, \end{cases}$$

where v is regular, i.e. v and $|v'|^{n-1}v'$ are continuously differentiable on $[0,\infty)$.

In the sequel we shall prove that problem (P) has a regular solution v satisying $v' \leq 0$. Hence v is a solution of (2) and thus we obtain a solution of (1).

If $g \equiv 1$ the equation in (P) is the equation for a $(n+1)$-Laplace operator in $(n+1)/2$ space variables. In that case, existence (and uniqueness) results for problem (P) are proved in [FLS 1].

In the general case, the existence of a solution of (P) will be proved by the shooting method as in [BLP] and [FLS 1]. This method requires that solutions of the Cauchy problem associated with the differential equation depend continuously in some sense on the Cauchy data. Due to this fact we shall prove preliminarly the existence of a solution for regularized problems. To this end, we shall consider the following problems

$$(P_\varepsilon) \begin{cases} \dfrac{(\Omega_\varepsilon(v'))'}{g(t^{2a}|v'|^2)} + \dfrac{n-1}{2t} \dfrac{\Omega_\varepsilon(v')}{g(t^{2a}|v'|^2)} + f(v) = 0, \text{ for } r > 0 \\ v \geq 0, \; v'(0) = 0, \; v(t) \to 0 \text{ if } t \to \infty, \end{cases}$$

where $\Omega_\varepsilon(p) = (p^2 + \varepsilon^2)^{(n-1)/2} p$ and $0 \leq \varepsilon \leq \varepsilon_0$. If $\varepsilon > 0$, problems (P_ε) are elliptic regularisations of problem $(P) = (P_0)$.

The first step in order to study existence and properties of regular solutions of problems (P_ε) is the following fundamental identity:

(4)   for any $t_1, t_2 \in [0, \infty)$,

$$H(t_2, |v'(t_2)|) - H(t_1, |v'(t_1)|)$$
$$+ (n-1) \int_{t_1}^{t_2} D(\rho, |v'(\rho)|) \frac{d\rho}{\rho} = F(v(t_1)) - F(v(t_2)),$$

where

$$F(v) = \int_0^v f(s) \, ds,$$

$$H(t, p) = \int_0^p \frac{z \, d\Omega_\varepsilon(z)}{g(t^{2a} |\Omega_\varepsilon(z)|^{2/n})} = \int_0^{\Omega_\varepsilon(p)} \frac{z(\xi) \, d\xi}{g(t^{2a}\xi^{2/n})}$$

(here $z(\xi) = \Omega_\varepsilon^{-1}(\xi)$) and

$$(5) \quad D(\rho, p) = \frac{1}{2} \frac{\Omega_\varepsilon(p)\, p}{g(\rho^{2a}|\Omega_\varepsilon(p)|^{2/n})}$$
$$+ \frac{p}{n} \int_0^p \frac{g'(\rho^{2a}|\Omega_\varepsilon(\xi)|^{2/n})\, \rho^{2a}\, |\Omega_\varepsilon(\xi)|^{2/n} d\Omega_\varepsilon(\xi)}{g(\rho^{2a}|\Omega_\varepsilon(\xi)|^{2/n})^2}$$
$$- \frac{1}{n} \int_0^p dz \int_0^z \frac{g'(\rho^{2a}|\Omega_\varepsilon(\xi)|^{2/n})\, \rho^{2a}\, |\Omega_\varepsilon(\xi)|^{2/n} d\Omega_\varepsilon(\xi)}{g(\rho^{2a}|\Omega_\varepsilon(\xi)|^{2/n})^2}.$$

To prove (4) it is enough to differentiate both sides of the formula, keeping in mind that v and $\Omega_\varepsilon(v')$ are continuously differentiable functions on $[0, \infty)$.

The identity (4) is a good tool for the study of the problem (P_ε) since we are able to give precise estimates on the function D. In fact we have:

Lemma 1. *For any $\rho \geq 0$ and for any $p \geq 0$ we have:*

$$(6) \quad \frac{1}{2} \frac{\Omega_\varepsilon(p)\, p}{g(\rho^{2a}|\Omega_\varepsilon(p)|^{2/n})} \leq D(\rho, p) \leq \left(2^n + \frac{s}{n}\right) H(\rho, p).$$

Proof. Integrating by parts in (5) we get

$$D(\rho,p) = \frac{1}{2}\frac{\Omega_\varepsilon(p)\,p}{g(\rho^{2a}|\Omega_\varepsilon(p)|^{2/n})} + \frac{1}{n}\int_0^p \xi\frac{g'(\rho^{2a}|\Omega_\varepsilon(\xi)|^{2/n})\,\rho^{2a}\,|\Omega_\varepsilon(\xi)|^{2/n}d\Omega_\varepsilon(\xi)}{g(\rho^{2a}|\Omega_\varepsilon(\xi)|^{2/n})^2},$$

and the first inequality follows. On the other hand, keeping in mind that $0 \le g'(t) \le \frac{s}{2}tg(t)$, from (5) we obtain

$$D(\rho,p) \le \frac{1}{2}\frac{\Omega_\varepsilon(p)\ p}{g(\rho^{2a}|\Omega_\varepsilon(p)|^{2/n})} + \frac{1}{n}\int_0^p \frac{\xi d\Omega_\varepsilon(\xi)}{g(\rho^{2a}|\Omega_\varepsilon(\xi)|^{2/n})}.$$

Now

$$\frac{1}{2}\Omega_\varepsilon(p) \ge \Omega_\varepsilon(\frac{p}{2}) = \frac{p}{2}\left(\frac{p^2}{4}+\varepsilon^2\right)^{(n-1)/2} \ge \frac{1}{2^n}\Omega_\varepsilon(p),$$

and hence

$$
\begin{aligned}
H(\rho,p) &\ge \int_{p/2}^p \frac{z\,d\Omega_\varepsilon(z)}{g(\rho^{2a}\,|\Omega_\varepsilon(z)|^{2/n})}\\
&\ge \frac{p}{g(\rho^{2a}\,|\Omega_\varepsilon(p)|^{2/n})}\int_{p/2}^p d\Omega_\varepsilon(z)\\
&= \frac{p}{g(\rho^{2a}\,|\Omega_\varepsilon(p)|^{2/n})}\left(\Omega_\varepsilon(p) - \Omega_\varepsilon(p/2)\right)\\
&\ge \frac{1}{2^n}\frac{1}{2}\frac{\Omega_\varepsilon(p)\,p}{g(\rho^{2a}|\Omega_\varepsilon(p)|^{2/n})}
\end{aligned}
$$

and the second inequality follows.

Using the identity (4) we are able to prove some qualitive results for the solutions of our problem. The main result will be that ground states are monotonically decreasing functions, so that a solution of the problem (P) is in fact a solution of problem (3).

Lemma 2. *Let $\varepsilon \ge 0$. If v is a regular solution of the equation in (P_ε) in an interval I, and if $t_0 \in I$ is a critical point of v, then either*

$$v(t) \le v(t_0) \text{ for } t \in I, t \ge t_0$$

or

$$v(t) \ge v(t_0) \text{ for } t \in I, t \ge t_0.$$

Proof. The proof is similar to the proof of the same property in the quasilinear case in [FLS 1].

Lemma 3. *Let $\varepsilon \geq 0$. If v is a solution of the problem (P_ε), then $v'(t) \leq 0$ for $t \in [0, \infty)$. Moreover, if $t > 0$ and $v(t) > 0$, then $v'(t) > 0$.*

Proof. By Lemma 2, $v(t) \leq v(0)$ for $t \in [0, \infty)$, since v vanishes at the infinity. Suppose now that there exists $\bar{t} > 0$ such that $u'(\bar{t}) > 0$. Then there exists $\tau \in (0, \bar{t})$ such that $\min_{[0,\bar{t}]} v = v(\tau)$, and hence either $v(t) \leq v(\tau)$ or $v(t) \geq v(\tau)$ for $t \geq \tau$. The first case cannot occur since $v(\bar{t}) > v(\tau)$. In the second case, if $v(\tau) > 0$ we get a contradiction since $v(\infty) = 0$; on the other hand if $v(t)$ vanishes at $t = \tau$, by Lemma 2, $v(t) \equiv 0$ for $t > \tau$, and this contradicts again the fact that $v(\bar{t}) > v(\tau)$. The proof of the second assertion is analogous.

Lemma 4. *Let $\varepsilon \geq 0$. If v is a solution of the problem (P_ε), then $v'(t) \to 0$ as $t \to \infty$.*

Proof. If t is sufficiently large, then $f(v(t)) \leq 0$ and hence, the equation gives

$$(\Omega_\varepsilon(v'))' = -\frac{1}{t} \frac{n-1}{2} \Omega_\varepsilon(v') - g(t^{2a} |\Omega_\varepsilon(v')|^{2/n}) f(v(t)) \geq 0.$$

Hence $\Omega_\varepsilon(v')$ has a limit at infinity, which must be zero. Thus the assertion has been proved.

By Lemma 4, if v is a solution of the problem (P_ε), then $H(t, |v'(t)|) \to 0$ as $t \to \infty$. Thus, by (4), putting $t_1 = 0$, and $t_2 = \infty$, we get

$$(7) \qquad (n-1) \int_0^\infty D(\rho, |v'(\rho)|) \frac{d\rho}{\rho} = F(v(0)).$$

In particular identity (7) implies that a necessary condition for the existence of a solution of (P_ε) is $\{v > 0; F(v) > 0\} \neq \emptyset$.

Thus, as in [FLS 1] it is natural to put

$$\beta = \inf\{v > 0; F(v) > 0\}.$$

Obviously $\alpha < \beta$. *In what follows we shall assume that* $\beta \in (\alpha, \gamma)$.

The next result concerns existence, uniqueness and continuous dependence on the initial data of the solutions of the Cauchy problem associated with the equation in (P_ε) when $\varepsilon > 0$ and f is locally Lipschitz continuous on $[0, \infty)$. Without loss of generality, we may suppose that $f(v)$ is continued by zero for $v < 0$.

Theorem 5 *If $\varepsilon > 0$ and f is locally Lipschitz continuous on $[0, \infty)$, then the Cauchy problem*

$$(8) \begin{cases} \dfrac{(\Omega_\varepsilon(v'))'}{g(t^{2a}|v'|^2)} + \dfrac{n-1}{2t} \dfrac{\Omega_\varepsilon(v')}{g(t^{2a}|v'|^2)} + f(v) = 0 \text{ for } t > 0, \\ v(0) = \xi > 0, \quad v'(0) = 0 \end{cases}$$

has a unique solution v near $t = 0$ which can be continued on $[0, \infty)$. In addition, v and v' depend continuously on ξ on compact subsets of $[0, \infty)$.

Proof. Put $w = t^{(n-1)/2}\Omega_\varepsilon(v')$. The equation in (6) becomes

$$(9) \qquad \frac{w'}{g(|w|^{2/n})} = -t^{(n-1)/2}f(v)$$

and hence

$$G(w(t)) = -\int_0^t \sigma^{(n-1)/2}f(v(\sigma))\,d\sigma,$$

where

$$G(w) = \int_0^w \frac{d\sigma}{g(|\sigma|^{2/n})}.$$

The existence of a local solution $v(t)$ near $t = 0$ can be proved using a fixed point theorem for the operator

$$(11) \qquad T(v) = \xi - \int_0^t \Omega_\varepsilon^{-1}\left(r^{(1-n)/2}G^{-1}\int_0^t \sigma^{(n-1)/2}f(v(\sigma))\,d\sigma\right)d\tau.$$

Denote now by $[0, R_\xi)$ the domain of the maximal solution of (8), and put $R_0 = \sup\{t < R_\xi; v > 0 \text{ on } [0, t]\}$. If $R_0 < R_\xi$, then $v(R_0) = 0$ and $v'(R_0) < 0$ by standard uniqueness results. Hence v may be written explicitly after R_0 and we have $R_\xi = \infty$. Thus we may assume that $0 < v < v(0)$ in $(0, R_\xi)$.

Now by contradiction let us suppose that $R_\xi < \infty$. First, we note that v' is bounded on $[0, R_\xi)$. Indeed, by (4) and by Lemma 1, if $|v'(t)| > 1$, we have

$$\begin{aligned}
F(v(0)) - F(v(t)) &\geq \int_\varepsilon^{|v'(t)|} \frac{z\Omega_\varepsilon'(z)\,dz}{g(t^{2a}|\Omega_\varepsilon(z)|^{2/n})} \\
&\geq n\int_\varepsilon^{|v'(t)|} \frac{z\Omega_\varepsilon(z)\,dz}{g(t^{2a}2^{2a}z^2)} \\
&= (\text{putting } (2t)^a z = \xi)c_n\, t^{-a(n+1)}\int_{(2t)^a\varepsilon}^{(2t)^a|v'(t)|} \frac{\xi^n}{g(\xi^2)}\,d\xi \\
&\geq R_\xi^{-a(n+1)}\int_{(2t)^a\varepsilon}^{(2t)^a|v'(t)|} \frac{\xi^n}{g(\xi^2)}\,d\xi,
\end{aligned}$$

and hence $\sup_{[0,R_\xi)}|v'| < \infty$, since $\int_0^\infty \frac{\xi^n}{g(\xi^2)}\,d\xi = \infty$.

In particular $v(t)$ has a finite limit as $t \to R_\xi -$. Let us prove now that $v'(t)$ also has a (finite) limit as $t \to R_\xi -$. Let $(t_k)_{k\in\mathbf{N}}$ be any sequence in $(0, R_\xi)$ such that $t_k \to R_\xi$ and $v'(t_k) \to \ell$. By the boundedness of v' we can pass to the limit in (4), and we get

(12)

$$\int_0^{|\ell|} \frac{z\Omega_\varepsilon'(z)\,dz}{g(t^{2a}|\Omega_\varepsilon(z)|^{2/n})}$$

$$= F(v(0)) - F(v(R_\xi)) + (1-n)\int_0^{R_\varepsilon} D(\rho, |v'(\rho)|)\,\frac{d\rho}{\rho}.$$

Suppose now that there exist two sequences in $(0, R_\xi)$, say $(t_k')_{k\in\mathbf{N}}$ and $(t_k'')_{k\in\mathbf{N}}$ such that $t_k' \to R_\xi, t_k'' \to R_\xi, v'(t_k') \to \ell_1$ and $v'(t_k'') \to \ell_2$, with $\ell_1 \neq \ell_2$. By (12), $\ell_1 = -\ell_2 \neq 0$, so that there exists a sequence $(\tau_k)_{k\in\mathbf{N}}$ in $(0, R_\xi)$ such that $v'(\tau_k) = 0$ for every $k \in \mathbf{N}$, which contradicts (12). Hence $v'(t)$ has a finite limit as $t \to R_\xi-$, and the solution can be continued after R_ξ.

The remaining part of the assertion can be proved in a standard way keeping in mind the identity (11).

It is now possible to prove the existence of a ground state for the regularized equation if f is locally Lipschitz continuous on $[0, \infty)$.

Lemma 6. *If $\varepsilon > 0$, f is locally Lipschitz continuous on $[0, \infty)$ and $\gamma < \infty$, then problem (P_ε) has a solution v_ε such that $v_\varepsilon(0) \in [\beta, \gamma)$.*

Proof. Let ξ belong to $[\beta, \gamma)$. We will denote by $v = v(\xi, \cdot)$ the solution of the Cauchy problem (8). We note that, by (10), $w' < 0$ near zero and hence $v' < 0$ near zero. The following tree cases may occur:

(i) $v > 0$, $v' < 0$ in $[0, \infty)$;

(ii)   there exists $T_\xi > 0$ such that $v' < 0$ in $(0, T_\xi)$ and $v(T_\xi) = 0$. In this case, by the uniqueness of the Cauchy problem, $v'(T_\xi) < 0$;

(iii)   there exists $T_\xi > 0$ such that $v'(T_\xi) = 0, v' < 0$ in $(0, T_\xi)$. Observe that in this case $v(T_\xi) \neq \alpha$ by the uniqueness of the Cauchy problem. On the other hand $(\Omega_\varepsilon(v'))'(T_\xi) = -f(v(T_\xi))$, and hence $f(v(T_\xi)) \leq 0$, since $v' < 0$ before T_ξ. Thus $f(v(T_\xi)) < 0$ so that T_ξ is a minimum point. Then, by Lemma 1, $v(t) \geq v(T_\xi)$ for $t \geq 0$.

We argue as in [BLP] and [FLS 1]. Put

$$I_+ = \{\xi \in [\beta, \gamma); \inf v > 0\}$$

$$I_- = \{\xi \in [\beta, \gamma); v(R) = 0 \text{ for some } R > 0\}.$$

We shall prove that I_+ and I_- are open non empty subsets. Then a connectedness argument shows that there exists $\xi \in [\beta, \gamma)$ such that $\xi \notin I_{\pm}$. Hence (ii) and (iii) cannot hold so that v satisfies (i). On the other hand $v(\infty) = \inf\{v(t), t \geq 0\}$ cannot be a positive numer (since $\xi \notin I_+$) and hence v is a solution of the problem (P_ε).

I_- is open. The proof in the same as in [FLS 1].

I_+ is open. Let $\bar{v}$ be the solution of the Cauchy problem (8) such that $\bar{v}(0) = \xi_0 \in I_+$. If (iii) holds then the assertion follows easily from the continuous dependence of the solutions of (8). Thus we may suppose that (i) holds. Then $\bar{v}' < 0$ in $(0, \infty)$ and $\bar{v}(\infty) = \ell > 0$. Let us prove that $f(\ell) = 0$. By contradiction, suppose $f(\ell) \neq 0$. Since $\bar{v}' < 0$, then $\bar{v}(t) \to \ell+$ as $t \to \infty$ and hence $f(v(t))$ has a constant sign for t large. Then, by (9), $\bar{w}'$ has a constant sign for t large and hence $\bar{w}$ has a limit at infinity. Now, noting that the integral is divergent at infinity, we get

$$\lim_{t \to \infty} \frac{\int_0^t \tau^{(n-1)/2} f(\bar{v}(\tau)) g(|\bar{w}(\tau)|^{2/n}) \, d\tau}{\tau^{(n+1)/2}}$$
$$= \frac{2}{n+1} \lim_{t \to \infty} f(\bar{v}(t)) g(|\bar{w}(t)|^{2/n}).$$

Hence, if $f(\ell) \neq 0$, we have

$$\lim_{t \to \infty} |\Omega_\varepsilon(\bar{v}')| = \lim_{t \to \infty} t \frac{|\bar{w}(t)|}{t^{(n+1)/2}} = \infty.$$

But this implies that $|\bar{v}'(t)| \to \infty$ as $t \to \infty$, which is impossible, since $\bar{v}$ has a finite limit $\bar{v}(\infty)$ at infinity.

Thus $\ell = \alpha$, since $\bar{v}$ is decreasing and bounded away from zero. On the other hand $\bar{v}$ converges at infinity, and hence there exists a sequence $t_n \to \infty$ such that $|v'(t_n)| \leq 1$. By Lebesgue's theorem we get

$$(n - 1) \int_0^\infty D_\varepsilon(\rho, |\bar{v}'(\rho)|) \frac{d\rho}{\rho} = F(\bar{v}(0)) - F(\ell) < \infty.$$

Choose now $R > 0$ such that $(n-1) \int_R^\infty D(\rho, |\bar{v}'(\rho)|) \frac{d\rho}{\rho} < \frac{1}{4} |F(\alpha)|$ and let $\delta > 0$ be fixed such that, if $\xi \in [\beta, \gamma)$ and $|\xi - \xi_0| < \delta$, then $|F(\xi) - F(\xi_0)| < \frac{1}{4} |F(\alpha)|$ and the solution v of the corresponding Cauchy problem is positive on $[0, R]$. In addition we may choose δ such that

$$(n - 1) \left| \int_0^R (D_\varepsilon(\rho, |v'(\rho)|) - D_\varepsilon(\rho, |v'(\rho)|)) \frac{d\rho}{\rho} \right| < \frac{1}{4} |F(\alpha)|.$$

Indeed, by the continuous dependence of the solutions on Cauchy data, we need only to prove that the above integral is uniformly small near zero. To this end, we note that, by Lemma 1,

$$\frac{1}{\rho} D_\varepsilon(\rho, |v'(\rho)|) \leq \left(2^n + \frac{s}{n}\right) H_\varepsilon(\rho, |v'(\rho)|)$$
$$\leq \int_0^{|v'(\rho)|} z \, d\Omega_\varepsilon(z) \leq c_n \Omega_\varepsilon(|v'(\rho)|) (-v'(\rho))$$
$$= c_n(-v'(\rho)) \frac{|w(\rho)|}{\rho^{(n+1)/2}} = c_n'(-v'(\rho)) \frac{|w(r)|}{r^{(n-1)/2}},$$

for a suitable $r \in (0, \rho)$. Hence

$$\frac{1}{\rho} D_\varepsilon(\rho, |v'(\rho)|) \leq c_n' g(|w|^{2/n}) f(v(r)) (-v'(\rho)),$$

and the integral on $[0, \bar{\rho}]$ of the last term is uniformly small for ρ close to zero by the continuous dependence of v and v' on the Cauchy data. By our choice of δ, v is positive on $[0, R]$; on the other hand, if $t > R$, we get

$$F(\xi) - F(v(t)) \geq (n-1) \int_0^t D_\varepsilon(\rho, |v'(\rho)|) \frac{d\rho}{\rho}$$

$$(n-1) \int_0^R D_\varepsilon(\rho, |v'(\rho)|) \frac{d\rho}{\rho}$$

$$(n-1) \int_0^R D_\varepsilon(\rho, |\bar{v}'(\rho)|) \frac{d\rho}{\rho} - \frac{1}{4} |F(\alpha)| \geq$$

$$(n-1) \int_0^\infty D_\varepsilon(\rho, |\bar{v}'(\rho)|) \frac{d\rho}{\rho} - \frac{1}{2} |F(\alpha)|$$

$$= F(\xi_0) - \frac{1}{2} F(\alpha) > F(\xi) + \frac{1}{4} |F(\alpha)|,$$

and hence $F(v(t)) > 0$ for any $t \in [0, \infty)$.

This proves that $\xi \in I_+$ if ξ is close to ξ_0, and hence I_+ is open.

I_+ *is not empty.* As in [FLS 1] we can prove that $\beta \in I+$.

I_- *is not empty.* By contradiction, suppose that $I_- = \emptyset$ and let $\xi \in [\beta, \gamma)$ be any shooting point. If we put $R = \sup\{t > 0$ such that $v' < 0$ on $(0, t)\}$, then either $R < \infty, m = v(R) > 0$ and $v'(R) = 0$, or $R = \infty$ and $m = \lim_{t \to \infty} v(t) \geq 0$. First, let us prove that $f(m) \leq 0$. Indeed, if $R < \infty$ and $f(m) > 0$, by the equation v' is decreasing at R, contradicting the fact that $v' < 0$ in a left neighbourhood of R. If $R = \infty$, then, using the method which we used to prove that I_+ is open, we can show that $f(m) = 0$. Since $f(m) \leq 0$, in particular $F(m) \leq 0$. Let now $\bar{\beta} \in (\beta, \gamma)$ be fixed. If the shooting point $\xi > \bar{\beta}$ then, since $m \in [0, \alpha]$, there exists $R_\xi \in (0, R)$ such that $v(R_\xi) = \bar{\beta}$. Since the solutions depend continuously on the Cauchy data, $R_\xi \to \infty$ as $\xi \to \gamma$. From the identuty (4) with $t_1 = R_\xi$ and $t_2 = R$, we get:

$$-\int_0^{|v'(R_\xi)|} \frac{z \, d\Omega_\varepsilon(z)}{g(R_\xi^{2a} |\Omega_\varepsilon(z)|^{2/n})}$$

$$+ (n-1) \int_{R_\xi}^R D_\varepsilon(\rho, |v'(\rho)|) \frac{d\rho}{\rho} = F(\bar{\beta}) - F(m).$$

The above formula holds also in the case $R = \infty$, since it is possible to find a divergent sequence $(t_n)_{n\in\mathbb{N}}$ such that $|v'(t_n)| \leq 1$ Now

$$F(\beta) < (n-1) \int_{R_\xi}^{R} D_\varepsilon(\rho, |v'(\rho)|) \frac{d\rho}{\rho} + F(m)$$

$$\leq (n-1) \int_{R_\xi}^{R} D_\varepsilon(\rho, |v'(\rho)|) \frac{d\rho}{\rho}.$$

Now, by Lemma 1,

$$D_\varepsilon(\rho, |v'(\rho)|) \leq c_n H_\varepsilon(\rho, |v'(\rho)|) \leq c'_n |v'(\rho)| (1 + H_\varepsilon(\rho, |v'(\rho)|))$$
$$\leq \text{ (by (4))} \quad c'_n (1 + F(\gamma) + |F(\alpha)|) |v'(\rho)| = c_2 |v'(\rho)|,$$

where c_2 is independent of ε and ξ.

Then

$$F(\bar{\beta}) \leq c_2(n-1) \frac{1}{R_\xi} (v(R_\xi) - v(R))$$

$$\leq \frac{c_2(n-1)(\bar{\beta} - m)}{R_\xi},$$

and hence, if $\xi \to \gamma -$ we obtain $F(\bar{\beta}) \leq 0$ which in turn is impossible, since $\bar{\beta} \in (\beta, \gamma)$.

Thus, the proof of the Lemma is completed.

We are now able to prove our existence theorem.

Theorem 7. *Let f be a continuous function on $[0, \infty)$ which is Lipschitz continuous on $(0, \infty)$. Suppose*

    (i)   *there exist $\alpha > 0$, $\gamma \in (\alpha, \infty)$ such that $f(u)< 0$ in $(0, \alpha)$, $f(u) > 0$ in (α, γ) and $f(0)= f(\alpha)= f(\gamma)= 0$;*

    (ii) *there exists $\beta = \inf\{v > 0; F(v) > 0\} \in (\alpha, \gamma)$;*

    (iii) $\int_0^\infty \dfrac{\xi^n}{g(\xi^2)} d\xi = \infty.$

Then, the problem (P) has a regular solution.

Proof. Denote by v_ε the solution of the problem P_ε where we replaced the function f by the smooth function

$$f_\varepsilon(v) = \int f(v - \varepsilon + \varepsilon t)\omega(t)\, dt,$$

where ω is a usual mollifier. Since we may always suppose, without loss of generality, that $f(v) < 0$ on (γ, ∞), then f_ε still satisfies (i) and (ii) for suitable $\alpha_\varepsilon, \beta_\varepsilon, \gamma_\varepsilon$. In addition

$$F_\varepsilon(v_\varepsilon(0)) - F_\varepsilon(v_\varepsilon(t)) \leq F_\varepsilon(\gamma\varepsilon) + |F_\varepsilon(\alpha_\varepsilon)| \leq M^*,$$

208 BRUNO FRANCHI

where M^* does not depend on ε.

Let $T > 0$ be fixed. If $t \in [0, T]$ and $|v'_\varepsilon(t)| > 1$, then by (4) we get:

$$M^* \geq H_\varepsilon(t, |v'_\varepsilon(t)|) \geq n \int_0^{|v'_\varepsilon(t)|} \frac{z^n}{g(T^{2a}|\Omega_1(z)|^{2/n})}\, dz$$

$$\geq n \int_1^{|v'_\varepsilon(t)|} \frac{z^n}{g(T^{2a}|\Omega_1(z)|^{2/n})}\, dz$$

$$\geq n \int_1^{|v'_\varepsilon(t)|} \frac{z^n}{g((2T)^{2a}z^2)}\, dz$$

$$= (2T)^{a(n+1)} \int_{(2T)^{2a}}^{(2T)^{2a}|v'_\varepsilon(t)|} \frac{\xi^n}{g(\xi^2)}\, d\xi.$$

Hence, by (iii), there exists $M_T > 0$ such that

$$(13) \qquad |v'_\varepsilon(t)| \leq M_T \quad \text{on } [0, T] \text{ if } \varepsilon \leq \varepsilon_0 \leq 1.$$

Then there exists a continuous function v such that for any compact subset $K \subset [0, \infty), \sup_K |v_j - v| \to 0$ as $j \to \infty$, where $v_j = v_{\varepsilon_j}$ is a suitable subsequence.

With obvious meaning of the symbols, by Lebesgue's theorem we get:

$$G(w_j) = G(t^{(n-1)/2}\Omega_j(v'_j)) = -\int_0^t \sigma^{(n-1)/2} f_j(v_j)\, d\sigma$$

$$\to -\int_0^t \sigma^{(n-1)/2} f(v)\, d\sigma$$

as $j \to \infty$. If $G(\infty) = \infty$, this implies that

$$w_j(t) \to G^{-1}\left(-\int_0^t \sigma^{(n-1)/2} f(v(\sigma))\, d\sigma\right)$$

$$= \text{(by definition) } w(t)$$

for any $t > 0$. On the other hand, if $G(\infty) < \infty$, then

$$0 \leq \int_0^t \sigma^{(n-1)/2} f_j(v_j(\sigma))\, d\sigma = G(|w_j|) \leq G\left(T^{(n-1)/2}\Omega_1(M_T)\right)$$

$$< G(\infty)$$

and we can argue as above.

Obviously the function w is continuously differentiable. Moreover, if $t > 0$, then

$$(14) \qquad -v'_j(t) \to t^{(1-n)/2n}(-w(t))^{1/n} = -\tilde{v}(t)$$

as $j \to \infty$. Indeed, if $0 < t \le T$, keeping in mind (13), we have:

$$|v_j'|^n < -v_j'(v_j'^2 + \varepsilon_j^2)^{(n-1)/2} \le |v_j'|^n + C_{T,n}\varepsilon_j.$$

Hence

$$-t^{(n-1)/2}w_j(t) - C_{T,n}\varepsilon_j \le |v_j'|^n \le -t^{(n-1)/2}w_j(t).$$

Obviously, the function $\tilde{v}$ is continuous. Moreover, $\tilde{v}(t) \to 0$ as $t \to 0+$, since

$$\lim_{t\to 0+} (\tilde{v}(t))^n = (-1)^{n+1} \lim_{t\to 0+} \frac{w(t)}{t^{(n-1)/2}}$$

$$= \frac{2}{n-1} \lim_{t\to 0+} \frac{(G^{-1})'\left(-\int_0^t \sigma^{(n-1)/2}f(v)\,d\sigma\right) t^{(n-1)/2}f(v(t))}{t^{(n-3)/2}}$$

$$= \frac{2(-1)^{n+1}}{n-1} \lim_{t\to 0+} t(G^{-1})'\left(\int_0^t \sigma^{(n-1)/2}f(v)\,d\sigma\right) f(v(t)) = 0$$

since $(G^{-1})'(0) = 1$. Thus v is differentiable in $[0,\infty)$ and $v' = \tilde{v}$. In addition $v' \le 0$ and

$$w(t) = t^{(n-1)/2}|v'|^{n-1}v' = G^{-1}\left(-\int_0^t \sigma^{(n-1)/2}f(v(\sigma))\,d\sigma\right).$$

The above identity shows that v is regular, $v'(0) = 0$, and v satisfies the equation in (P). Moreover, v has a limit $\ell \in [0,\gamma]$ at infinity, since $v' \le 0$. To complete our proof we only need to show that $\ell = 0$. First, let us prove that $\ell \ne \gamma$ (i.e. $v \not\equiv \gamma$). By contradiction, suppose $v \equiv \gamma$; if $T > 0$ is fixed, we have:

$$0 = (n-1)\int_0^\infty D(\rho, |v'(\rho)|)\,\frac{d\rho}{\rho} \ge (n-1)\int_0^T D(\rho, |v'(\rho)|)\,\frac{d\rho}{\rho}$$

$$= (n-1)\int_0^\infty \left(D(\rho, |v'(\rho)|) - D_j(\rho, |v_j'(\rho)|)\right)\frac{d\rho}{\rho}$$

$$+ (n-1)\int_0^\infty D_j(\rho, |v_j'(\rho)|)\,\frac{d\rho}{\rho} - (n-1)\int_T^\infty D_j(\rho, |v_j'(\rho)|)\,\frac{d\rho}{\rho}$$

$$= I_1 + I_2 + I_3.$$

Now $I_2 = F_j(v_j(0)) > \dfrac{2}{3}F(\gamma)$ if j is large enough. On the other hand,

$$|I_3| \le \frac{n-1}{T}\int_T^\infty D_j(\rho, |v_j'(\rho)|)\,d\rho$$

$$\le \frac{c_n}{T}\int_T^\infty (1 + H_j(\rho, |v_j(\rho)|))(-v_j'(\rho))\,d\rho$$

$$\le \text{(by (4))}\ \frac{C_n(1+M*)}{T}(v_j(T) - v_j(\infty))$$

$$\le \frac{C_n(1+M*)}{T}\gamma < \frac{F(\gamma)}{4}$$

if T is large enough.

Finally, if T is fixed as above and j is large enough, then

$$|I_1| < \frac{F(\gamma)}{4}.$$

In fact, we need only to note that the integrals are uniformly small near zero (as we did in the proof of Lemma 6) and the assertion follows by Lebesgue's theorem and the continuous dependence on the Cauchy data.

Combining the previous estimates, we have

$$0 \geq \frac{F(\gamma)}{6} > 0$$

and hence we get a contradiction, showing that $v \not\equiv \gamma$. In addition, the same argument shows that

$$F(v(0)) = (n-1) \int_0^\infty D(\rho, |v'(\rho)|) \frac{d\rho}{\rho}.$$

On the other hand, since v has a finite limit at infinity, we can repeat previous argument showing that we can pass to the limit in the identity (4). Thus, we get:

$$F(v(0)) - F(\ell) = (n-1) \int_0^\infty D(\rho, |v'(\rho)|) \frac{d\rho}{\rho}$$

and hence $F(\ell) = 0$. We shall show that $\ell = 0$. By contradiction, suppose $\ell = \beta$ (since $0 \leq v \leq \gamma$, either $\ell = 0$ or $\ell = \beta$). In this case the function $w(t) = t^{(n-1)/2}|v'(t)|^{n-1}v'(t)$ is negative and strictly decreasing in a neighbourhood of infinity. Hence $w(t) \to w_\infty \in [-\infty, 0)$ as $t \to \infty$. Thus there exists $c_0 > 0$ such that

$$-v'(t) \geq c_0 t^{(1-n)/2n}$$

for large t. This is impossible, since $\dfrac{n-1}{2n} < \dfrac{1}{2}$ and v converges at infinity. Thus $\ell = 0$ and the assertion is completely proved.

The above existence result concerns the case $\gamma < \infty$. An analogous result can be proved in a similar way when $\gamma = \infty$ under suitable restrictions on the growth of f at infinity.

Theorem 8. *Suppose $s \in [0, n+1)$. Let f be a continuous function on $[0, \infty)$ which is Lipschitz continuous on $(0, \infty)$. Suppose*

    (i)   *$f(0) = 0, f(u) < 0$ on $(0, \alpha), f(u) > 0$ on $(\alpha, \infty), f(\alpha) = 0$;*

    (ii)   *there exists $\beta = \inf\{v > 0; F(v) > 0\} \in (\alpha, \infty)$;*

    (iii) *$\liminf_{\xi \to \infty} \dfrac{F(\xi)}{\xi^{n+1-s}} = 0$.*

Then, the problem (P) has a regular solution.

Proof. First we note that, by the assumption (iii) on g, $\int_0^\infty \frac{\xi}{g(\xi^2)}\,d\xi = \infty$, since $s < n+1$. In the present case, we may repeat the arguments of the proof of Lemma 7 except of the last step, where we showed that I_- is not empty. Let us sketch how it is possible to modify the proof at this point, following the ideas used in [FLS 1]. With the notations we introduced in the proof of Lemma 7, let us prove first that $\limsup_{\xi\to\infty} R_\xi = \infty$. Let $\tau \in [0, R_\xi]$ be such that $|v'(\tau)| = \max_{[0,R_\xi]} |v'|$. We have:

$$0 < \xi - \bar\beta = v(0) - v(R_\xi) \le |v'(R_\xi)|.$$

Thus the identity (4) in $[0,\tau]$ gives

$$\int_0^{(\xi-\bar\beta)/R_\xi} \frac{\Omega'_\varepsilon(z)\,z\,dz}{g(R_\xi^{2a}|\Omega_\varepsilon(z)|^{2/n})} \le F(\xi).$$

Suppose now $\xi - \bar\beta \ge 2R_\xi$. Keeping in mind that

$$\Omega'_\varepsilon(z) \ge nz^{n-1} \text{ for } z > 0 \text{ and } g(\theta q) \le \theta^{s/2} g(q) \text{ for } \theta > 1 \text{ and } q \ge 0,$$

we have:

$$F(\xi) \ge c(n,s)\frac{1}{2g(R_\xi^{2a})}\left(\left(\frac{\xi - \bar\beta}{R_\xi}\right)^{n+1-s} - 1\right).$$

Hence

$$g(R_\xi^{2a})R_\xi^{n+1-s}$$
$$\ge \min\left\{\frac{c(n,s)\,\xi^{n+1-s}}{2\,F(\xi)}\left(1 - \frac{\bar\beta}{\xi}\right)^{n+1-s}, \frac{\xi - \bar\beta}{2}\right\}$$

and the assertion follows.

Let now $R_0 \in [R_\xi, R)$ be such that

$$H(R_0, |v'(R_0)|) = \max_{[R_\xi, R]} H(t, |v'(t)|).$$

The identity (4) in $[R_0, R]$ gives

$$H(R_0,|v'(R_0)|)$$
$$= F(m) - F(v(R_0) + (n-1)\int_{R_0}^R D(\rho, |v'(\rho)|)\frac{d\rho}{\rho}$$
$$\le \sup_{[0,\beta]} |F| + (n-1)\int_{R_0}^R D(\rho, |v'(\rho)|)\frac{d\rho}{\rho}$$
$$\le \text{(by Lemma 1)} \sup_{[0,\beta]} |F| + c_1\int_{R_0}^R H(\rho, |v'(\rho)|)\frac{d\rho}{\rho}$$
$$\le \sup_{[0,\beta]} |F| + \frac{c_1}{R_\xi}\int_{R_0}^R (1 + H(\rho, |v'(\rho)|))\,d\rho$$
$$\le \sup_{[0,\beta]} |F| + \frac{c_1}{R_\xi}\bar\beta(1 + H(R_0, |v'(R_0)|)).$$

[GT] D. Gilbarg and N.S. Trudinger, *Elliptic Partial Differential Equations of Second Order*, Second Edition, Springer-Verlag, Berlin, 1983.

[L] P.L. Lions, *Two Remarks on the Monge-Ampère Equations*, Ann. Mat. Pura Appl. **142** (1985), 263–275.

[PS] L.A. Peletier and J. Serrin, *Ground States for the Prescribed Mean Curvature Equation*, Proc. Amer. Math. Soc.. **100** (1987), 694–700.

[Z] C. Zuily, *Existence locale de solutions* C^∞ *pour des équations de Monge-Ampère changeant de type*, Comm. Partial Differential Equations **14** (1989), 691–697.

Dipartimento Matematico dell'Università
University of Torino
Via Carlo Alberto 10
10123 Torino, Italy

The Structure of Solutions near an Extinction Point in a Semilinear Heat Equation with Strong Absorption: A Formal Approach

V. A. GALAKTIONOV

M. A. HERRERO, and J. J. L. VELÁZQUEZ

Abstract

We consider nonnegative solutions of the semilinear parabolic equation

$$u_t - u_{xx} + u^p = 0, \quad -\infty < x < +\infty, \quad t > 0, \quad 0 < p < 1,$$

which vanish at the extinction point $x = 0$ at a time $t = T$. By means of formal methods, we derive a family of asymptotic expansions for solutions and interface curves (these last separating the regions where $u = 0$ and $u > 0$), as (x, t) approaches $(0, T)$.

1. Introduction

In this paper we shall consider nonnegative solutions of the equation

$$(1.1) \qquad u_t - u_{xx} + u^p = 0, \quad -\infty < x < +\infty, \quad t > 0,$$

where

$$0 < p < 1,$$

such that they vanish in a finite time, in the sense that

$$u(x, t) \equiv 0 \quad \text{for} \quad t \geq T \quad \text{and some} \quad T < +\infty.$$

The existence of the so-called extinction time $T_E = \inf\{t : u(x, t) \equiv 0\}$ is well known for a variety of initial and boundary value problems corresponding to (1.1). It is also known that solutions to (1.1) may develop interfaces or fronts, separating the regions where $u > 0$ and $u = 0$, even when they are initially absent; cf. for instance [K], [BF], [EK], [FH].

Let us introduce some notation and assumptions. We say that x_o is an extinction point for a solution $u(x, t)$ of (1.1) if there exist sequences $\{x_n\} \to x_o$, $\{t_n\} \to T_E$ such that $u(x_n, t_n) > 0$ for any n. It has been recently shown that, under fairly general assumptions on initial values $u_0(x)$, there is only a finite number of extinction points for the solution of the Cauchy problem corresponding to (1.1) (cf. [CMM]). We shall accordingly consider isolated extinction points, and assume the following hypothesis.

(1.3)   Let x_o be an extinction point corresponding to an extinction time T_E. Then

$$\lim_{t \to T_E} (T_E - t)^{-\frac{1}{1-p}} u(x,t) = (1-p)^{\frac{1}{1-p}}$$

uniformly on sets $|x - x_o| \le C (T_E - t)^{1/2}$ for any $C > 0$.

Assumption (1.3) is certainly satisfied by the explicit solution

$$(1.4) \qquad\qquad u_M(x,t) = \left(M^{1-p} - (1-p)t\right)_+^{\frac{1}{1-p}},$$

where $(s)_+ = \max(s, 0)$, and $M > 0$ is an arbitrary fixed constant. Moreover, in [FH] conditions are given which ensure the occurrence of (1.3) in Cauchy-Dirichlet problems associated to (1.1).

We shall set for simplicity $x_o = 0$ and $T_E = T$ henceforth. In this paper we want to describe the possible behaviors of the solutions and interfaces as (x,t) approaches $(0, T)$. To this end, we shall use a formal approach, consisting in the use of matched asymptotic expansions. Although our procedure is not a rigorous one, we believe that it provides us with a rational insight into the problem under consideration.

We now describe briefly our conclusions. Let us consider a solution $u(x,t)$ of (1.1) which vanishes at a time $t = T$ and has the origin as an estinction point. Set now

$$y = x(T - t)^{-1/2}, \quad \tau = -\log(T - t),$$

$$u(x,t) = (T - t)^{\frac{1}{1-p}} \Phi(y, \tau), \quad G = \Phi^{1-p}.$$

Then (1.3) can be rephrased as stating that $G(y, \tau) \to (1 - p)$ as $\tau \to \infty$, uniformly on sets $\{|y| \le C\}$, $C > 0$ fixed. Let us also write

$$G(y, \tau) = (1 - p) + \Psi(y, \tau).$$

In view of our foregoing analysis, and up to the first order, we expect three kinds of possible behaviors for $\Psi(y, \tau)$,

(i) $\Psi(y, \tau) = 0,$

(ii) $\Psi(y, \tau) \approx -\dfrac{(1 - p)^2 (4\pi)^{1/4}}{2^{1/2} p} \cdot \dfrac{H_2(y)}{\tau} \quad$ as $\quad \tau \to \infty,$

(iii) $\Psi(y, \tau) \approx B \, e^{(1 - \frac{n}{2})\tau} H_n(y) \quad$ as $\quad \tau \to \infty, \quad n = 3, 4, \ldots,$

where $H_n(y) = c_n \widetilde{H}_n(y/2)$, $c_n = (2^{n/2}(4\pi)^{1/4}(n!)^{1/2})^{-1}$, $\widetilde{H}_n(s)$ is the standard Hermite polynomial of nth order, and B is a free constant. In particular, (ii) can be rewritten in the old variables x and t, to give

$$(1.5) \quad u(x,t) \approx (T-t)^{\frac{1}{1-p}} \left[(1-p) - \frac{(1-p)^2}{4p} \cdot \frac{x^2}{(T-t)|\log(T-t)|} \right]_+^{\frac{1}{1-p}}$$

$$\text{as} \quad t \to T,$$

where the above expression holds in regions where $|x|^2 \leq C(T-t)|\log(T-t)|$ for any $C > 0$ such that the quantity between the braces stays positive. We shall show below (cf. Appendix B) that there exist indeed solutions $u(x,t)$ which are bounded below by the right-hand side of (1.5) as $t \to T$.

On the other hand, we say that $x_1(t)$, $x_2(t)$, are interface curves corresponding to the extinction point $x = 0$ if $-\infty < x_1(t) < x_2(t) < +\infty$ for any $t \leq T$, $u(x,t) > 0$ if $x_1(t) < x < x_2(t)$, $u(x,t) = 0$ for $x > x_2(t)$ and $x < x_1(t)$, and $\lim_{t \to T} x_1(t) = \lim_{t \to T} x_2(t) = 0$. The onset of interfaces for (1.1) has been discussed in [FH] and [CMM]. We do not expect interfaces to appear in case (i). (Actually, we conject that the only solution which behaves in this way is precisely that in (1.4)). In cases (ii) and (iii) we derive the following expansions for the interfaces

(ii)
$$|x_{2,i}(t)| = \left[\frac{4p}{1-p} \right]^{1/2} (T-t)^{1/2}|\log(T-t)|^{1/2}$$

$$+ \theta_1 \frac{\log(|\log(T-t)|)(T-t)^{1/2}}{|\log(T-t)|^{1/2}} + \cdots,$$

(iii)
$$|x_{n,i}(t)| = \theta_2 (T-t)^{1/n} + \sum_{\ell=n+1}^{2(n-1)-1} \theta_\ell (T-t)^{\ell/2-1+1/n}$$

$$+ \theta_{2(n-1)}(T-t)^{1-1/n}|\log(T-t)| + \cdots, \quad n = 3, 4, \ldots$$

as $t \to T$, where $i = 1, 2$, θ_1 is a given constant, and θ_2, θ_ℓ ($\ell = n + 1, \ldots, 2(n-1)+1, \quad n = 3, 4 \ldots$) are some real constants. Moreover (iii) is to be understood in the sense that all the θ_ℓ may vanish, and only the first nonzero coefficient in the family $\{\theta_\ell, \theta_{2(n-1)}\}$ is to be retained.

2. The Results

2.1. *Preliminaries*

As described in Section 1, we begin by changing variables in (1.1) as follows

(2.1a) $$y = x(T-t)^{-1/2}, \quad \tau = -\log(T-t),$$

(2.1b) $$u(x,t) = (T-t)^{\frac{1}{1-p}}\Phi(y,\tau),$$

(2.1c) $$G = \Phi^{1-p},$$

so that G solves

(2.2) $$G_\tau = G_{yy} - \frac{1}{2}yG_y + G + \frac{p}{1-p}\cdot\frac{(G_y)^2}{G} - (1-p) \quad \text{in} \quad \Sigma_T,$$

where $\Sigma_T = \{(y,\tau) : -\infty < y < +\infty, -\log T < \tau < +\infty\}$. We shall also use the following notation

$$L^2_\omega(\ ) = \left\{ f \in L^2_{\text{loc}}(\ ) : \int_{-\infty}^{+\infty} |f(y)|^2 e^{-y^2/4}\,dy < +\infty \right\},$$

$$\langle f, g \rangle = \int_{-\infty}^{+\infty} f(y)g(y)e^{-y^2/4}\,dy, \quad \|f\| = \langle f, f \rangle^{1/2}.$$

As suggested by (2.2), an important role will be played by the linear operator

(2.3) $$A\phi(y) = \phi''(y) - \frac{1}{2}y\phi'(y) + \phi(y)$$

with domain $D(A) = \{f \in H^2_{\text{loc}}(\ ) : \phi, \phi' \text{ and } \phi'' \text{ belong to } L^2_\omega(\ )\}$. This operator is self-adjoint in $L^2_\omega(\ )$ with spectrum $\sigma(A)$ consisting of the eigenvalues $\{1 - \frac{n}{2} : n = 0, 1, 2, \ldots\}$. The eigenfunction corresponding to the nth eigenvalue is the modified Hermite polynomial $H_n(y)$, given by

(2.4) $$H_n(y) = c_n \widetilde{H}_n(y/2), \quad \|H_n\| = 1,$$

where $c_n = (2^{n/2}(4\pi)^{1/4}(n!)^{1/2})^{-1}$ and $\widetilde{H}_n(s)$ is the standard n^{th} Hermite polynomial. We shall repeatedly use the integrals

(2.5) $$A_{n,m,\ell} = \int_{-\infty}^{+\infty} H_n(y)H_m(y)H_\ell(y)e^{-y^2/4}\,dy.$$

Clearly, $A_{n,m,\ell}$ is invariant under permutation of its indexes. The numerical value of the $A_{n,m,\ell}$ has been obtained in [GHV]. For the reader's convenience, however, we shall repeat the corresponding computation in Appendix A below, where it is shown that

(2.6) $\quad A_{n,m,\ell} \neq 0 \quad$ if and only if $n + m + \ell$ is even and $n \leq m + \ell$, $m \leq n + \ell$, $\ell \leq m + n$. In such case, there holds

$$A_{n,m,\ell} = (4\pi)^{-1/4}(n!)^{1/2}(m!)^{1/2}(\ell!)^{1/2}$$
$$\left[\left[\frac{m+n-\ell}{2}\right]!\left[\frac{n+\ell-m}{2}\right]!\left[\frac{m+\ell-n}{2}\right]!\right]^{-1}$$

2.2 First order asymptotics near an extinction point

We first note that rewriting (1.3) in terms of the function $G(y,\tau)$ given in (2.1), we have

$$G(y,\tau) \to (1-p) \quad \text{uniformly in compact sets } \{|y| \le C\} \text{ as } \tau \to \infty.$$

We now linearize around $(1-p)$, which is a stationary solution of (2.2). We thus set

$$(2.7) \qquad G(y,\tau) = (1-p) + \Psi(y,\tau)$$

so that $\Psi(y,\tau)$ solves

$$(2.8) \qquad \Psi_\tau = \Psi_{yy} - \frac{1}{2}y\Psi_y + \Psi + \frac{p}{1-p}\cdot\frac{(\Psi_y)^2}{((1-p)+\Psi)}.$$

Assume now that

$$(2.9) \qquad \Psi(y,\tau) = \sum_{k=0}^{\infty} a_k(\tau)H_k(y).$$

Recalling the distribution of eigenvalues of operator A in (2.3), we obtain for the coefficients $a_\ell(\tau)$

$$(2.10)$$
$$\dot{a}_\ell(\tau) = \left(1 - \frac{\ell}{2}\right)a_\ell(\tau) + \frac{p}{1-p}\left\langle H_\ell(y), \frac{\left[\sum_{k=0}^{\infty} a_k(\tau)H_k'(y)\right]^2}{\left[(1-p)+\sum_{k=1}^{\infty} a_k(\tau)H_k(y)\right]}\right\rangle$$
$$\equiv \left(1 - \frac{\ell}{2}\right)a_\ell(\tau) + f_\ell(\Psi), \quad \ell = 0,1,2,\ldots.$$

Notice that $f_\ell(\Psi)$ is formally a second order term. By (2.4) and well-known properties of Hermite polynomials

$$(2.11) \qquad H_k'(y) = \left(\frac{k}{2}\right)^{1/2} H_{k-1}(y) \quad \text{for} \quad k = 1,2,\ldots$$

so that for any nonnegative integer ℓ, we obtain that whenever Ψ is small enough

$$\left\langle H_\ell(y), \frac{[\sum_{k=0}^\infty a_k(\tau)H_k'(y)]^2}{[(1-p) + \sum_{k=1}^\infty a_k(\tau)H_k(y)]} \right\rangle$$

$$\approx \frac{1}{1-p}\left\langle H_\ell(y), \left[\sum_{k=1}^\infty a_k(\tau)H_k'(y)\right]^2 \right\rangle$$

$$= \frac{1}{2(1-p)}\left\langle H_\ell(y), \left[\sum_{k=1}^\infty k^{1/2}a_k(\tau)H_{k-1}'(y)\right]^2 \right\rangle$$

$$= \frac{1}{2(1-p)} \sum_{k,n=1}^\infty (kn)^{1/2}a_k a_n A_{\ell,k-1,n-1},$$

where $A_{\ell,k-1,n-1}$ is given by (2.5). Substituting this in (2.10), we get

$$(2.12) \qquad \dot{a}_\ell(\tau) = \left(1 - \frac{\ell}{2}\right) a_\ell(\tau)$$

$$+ \frac{p}{2(1-p)^2} \sum_{k,n=1}^\infty (kn)^{1/2}a_k a_n A_{\ell,k-1,n-1} + \ldots .$$

By analogy with standard ODE results, we may consider the possible asymptotic behaviors of $\Psi(y,\tau)$ as $\tau \to \infty$ as corresponding to such situations in which one of the modes $a_k(\tau)H_k(y)$ predominates in (2.9). To begin with, the two positive eigenvalues 1 and 1/2 yield unstable behavior for the solution $\Psi = 0$. Since we are supposing here that (1.3) holds, we must assume that a_0, a_1 are negligible in the first approximation. When $n = 3, 4, \ldots$ (2.12) is approximately a linear equation with solutions

$$(2.13a) \qquad a_n(\tau) \approx \mu e^{\left(1 - \frac{n}{2}\right)\tau} \quad \text{as} \quad \tau \to \infty$$

for arbitrary real μ. On the other hand, for such solutions where a_2 predominates, we have

$$\dot{a}_2(\tau) \approx \frac{p}{2(1-p)^2} \sum_{k,n=1}^\infty (kn)^{1/2}a_k a_n A_{2,k-1,n-1} = \frac{a_2^2 p}{(1-p)^2}A_{2,1,1} + \ldots ,$$

whence

$$(2.13b) \qquad a_2(\tau) \approx -\frac{(1-p)^2(4\pi)^{1/4}}{2^{1/2}p} \cdot \frac{1}{\tau} \quad \text{as} \quad \tau \to \infty.$$

Summing up these results we obtain that, in the first approximation, the following asymptotic behaviors are possible

$$(2.14) \quad \begin{cases} \text{(i)} \quad \Psi(y,\tau) = 0, \\[2mm] \text{(ii)} \quad \Psi_2(y,\tau) \approx -\dfrac{(1-p)^2(4\pi)^{1/4}}{2^{1/2}p} \cdot \dfrac{1}{\tau} H_2(y) \\[2mm] \qquad\qquad\qquad\qquad\qquad \text{as} \quad \tau \to \infty, \\[2mm] \text{(iii)} \quad \Psi_n(y,\tau) \approx k\, e^{\left(1-\frac{n}{2}\right)\tau} H_n(y) \quad \text{as} \quad \tau \to \infty, \\[2mm] \qquad\qquad\qquad\qquad\qquad\qquad n = 3,4,\ldots. \end{cases}$$

We shall refer henceforth to Ψ_2 and Ψ_n $(n = 3, 4, \ldots)$ as second and third type solutions respectively.

2.3 Higher-order asymptotics at the inner region

We now proceed to derive a higher-order expansion for second and third type solutions. To this end, we set $\delta_{j,k} = 1$ if $j = k$, $\delta_{j,k} = 0$ otherwise and write

$$(2.15) \qquad a_n(\tau) = -\frac{d_p}{\tau}\delta_{n,2} + \omega_n(\tau), \quad |\omega_n(\tau)| << \frac{1}{\tau} \quad \text{as} \quad \tau \to \infty,$$

where $d_p = \frac{(1-p)^2(4\pi)^{1/4}}{2^{1/2}p}$ (cf. (2.14)). Trying (2.15) in (2.12), it follows that for $n \neq 2$ dominant terms read

$$\dot{\omega}_n(\tau) = \left(1 - \frac{n}{2}\right)\omega_n(\tau) + \frac{pd_p^2}{(1-p)^2\tau^2}A_{n,1,1} + o\left(\frac{1}{\tau^2}\right) \quad \text{as} \quad \tau \to \infty.$$

By (2.6), $A_{n,1,1} \neq 0$ if and only if $n = 0, 2$. Since $A_{0,1,1} = (4\pi)^{-1/4}$, we obtain

$$\dot{\omega}_o(\tau) = \omega_o(\tau) + \frac{(4\pi)^{1/4}(1-p)^2}{2p} \cdot \frac{1}{\tau^2} + o\left(\frac{1}{\tau^2}\right) \quad \text{as} \quad \tau \to \infty.$$

As $|\omega_o(\tau)| << \frac{1}{\tau}$ for large τ, we then deduce that

$$\omega_o(\tau) \approx -\frac{(4\pi)^{1/4}(1-p)^2}{2p}\int_\tau^\infty \frac{e^{(\tau-s)}}{s^2}\,ds,$$

whence

$$(2.16) \qquad \omega_o(\tau) \approx -\frac{(4\pi)^{1/4}(1-p)^2}{2p}\frac{1}{\tau^2} \quad \text{as} \quad \tau \to \infty.$$

As to $\omega_2(\tau)$, we need to take into account third order terms in (2.10). Since

$$\frac{(\Psi_y)^2}{(1-p)+\Psi} \approx \frac{(\Psi_y)^2}{1-p}\left[1 - \frac{\Psi}{1-p} + O\left(|\Psi|^2\right)\right],$$

we obtain from (2.10)

$$\dot{a}_n = \left(1 - \frac{n}{2}\right) a_n + \frac{p}{(1-p)^2} \left\langle H_n, (\Psi_y)^2 \right\rangle$$
$$- \frac{p}{(1-p)^3} \left\langle H_n, \Psi (\Psi_y)^2 \right\rangle + O\left(\|\Psi\|^4\right).$$

To estimate the term $\langle H_n, \Psi(\Psi_y)^2 \rangle$, we notice that, since by (2.14) and (2.11), $\Psi(y, \tau) \approx -\frac{d_p}{\tau} H_2(y)$, $\Psi_y(y, \tau) \approx -\frac{d_p}{\tau} H_1(y)$ as $\tau \to \infty$, there holds

$$\left\langle H_2, \Psi (\Psi_y)^2 \right\rangle \approx -\frac{d_p^3}{\tau^3} \left\langle H_2, H_2 (H_1)^2 \right\rangle = -\frac{d_p^3}{\tau^3} \langle H_2 H_1, H_2 H_1 \rangle$$
$$= -\frac{d_p^3}{\tau^3} \sum_{\ell=0}^{\infty} \langle H_2 H_1, H_\ell \rangle \langle H_\ell, H_2 H_1 \rangle = -\frac{d_p^3}{\tau^3} \sum_{\ell=0}^{\infty} A_{2,1,\ell}^2.$$

By (2.6), we have $A_{2,1,1} = 2^{1/2}(4\pi)^{-1/4}$, $A_{2,1,3} = 3^{1/2}(4\pi)^{-1/4}$, so that

$$-\frac{p}{(1-p)^3} \left\langle H_2, \Psi (\Psi_y)^2 \right\rangle \approx \frac{5(4\pi)^{1/4}(1-p)^3}{2^{3/2}p^2} \cdot \frac{1}{\tau^3} \quad \text{as} \quad \tau \to \infty,$$

and therefore the equations for $\omega_2(\tau)$ and $\omega_4(\tau)$ are

(2.17a)

$$\dot{\omega}_2(\tau) = \frac{p}{2(1-p)^2} \left[-2\frac{d_p}{\tau} \sum_{k=1}^{\infty} (2k)^{1/2} A_{2,1,k-1} \, \omega_k(\tau) + O\left(\omega_k \omega_j\right) \right]$$
$$- \frac{5(4\pi)^{1/4}(1-p)^3}{2^{3/2}p^2} \cdot \frac{1}{\tau^3} + \dots \quad \text{as} \quad \tau \to \infty,$$

(2.17b)

$$\dot{\omega}_4(\tau) = -\omega_4(\tau) + O\left(\frac{1}{\tau^3}\right) \quad \text{as} \quad \tau \to \infty.$$

By (2.6), the only nonzero coefficients $A_{n,\ell,k}$ in (2.17a) are $A_{2,1,1}$ and $A_{2,1,3}$, this last appearing as a coefficient to ω_4. An analysis of (2.17) reveals that $\omega_k \omega_j = o(\omega_k)$, and $\omega_4(\tau) = o(\frac{1}{\tau^{3-\epsilon}})$, for some $\epsilon > 0$ small enough, as $\tau \to \infty$. We are thus led to

(2.18)

$$\dot{\omega}_2(\tau) = \frac{2p d_p A_{2,1,1}}{(1-p)^2} \cdot \frac{\omega_2(\tau)}{\tau} - \frac{5(4\pi)^{1/4}(1-p)^3}{2^{3/2}p^2} \cdot \frac{1}{\tau^3} + \dots$$
$$= -\frac{2}{\tau} \omega_2(\tau) - \frac{5(4\pi)^{1/4}(1-p)^3}{2^{3/2}p^2} \cdot \frac{1}{\tau^3} + \dots \quad \text{as} \quad \tau \to \infty.$$

Integrating (2.18) we get

$$\omega_2(\tau) = \frac{\alpha}{\tau^2} - \frac{5(4\pi)^{1/4}(1-p)^3}{2^{3/2}p^2} \cdot \frac{\log \tau}{\tau^2} + \dots \quad \text{as} \quad \tau \to \infty,$$

where α is an arbitrary real constant (depending on initial data). Recalling (2.16), we thus obtain the following asymptotic expansion for $\Psi_2(y,\tau)$ in (2.14)

(2.19a)

$$\Psi_2(y,\tau) \approx -\frac{(1-p)^2(4\pi)^{1/4}}{2^{1/2}p} \cdot \frac{H_2(y)}{\tau} + \frac{\alpha}{\tau^2}H_2(y)$$
$$-\frac{5(4\pi)^{1/4}(1-p)^3}{2^{3/2}p^2} \cdot \frac{\log\tau}{\tau^2}H_2(y) - \frac{(4\pi)^{1/4}(1-p)^2}{2p} \cdot \frac{H_0(y)}{\tau^2} + \cdots,$$

where, in view of (2.4) and standard results

$$H_o(y) = (4\pi)^{-1/4}, \quad H_2(y) = \left(2^{3/2}(4\pi)^{1/4}\right)^{-1}\left(y^2-2\right).$$

We next obtain the corresponding expansions for the higher modes in (2.14). We set

(2.20)
$$\Psi_n(y,\tau) = C\, e^{\left(1-\frac{n}{2}\right)\tau}H_n(y) + \omega(y,\tau)$$
$$= C\, e^{\left(1-\frac{n}{2}\right)\tau}H_n(y) + \sum_{k=0}^{\infty}\omega_k(\tau)H_k(y).$$

Recalling (2.13a), we obtain

$$a_\ell(\tau) = C\, e^{\left(1-\frac{n}{2}\right)\tau}\delta_{\ell,n} + \omega_\ell(\tau),$$

whence

(2.21)
$$\dot\omega_\ell(\tau) \approx \left(1-\frac{\ell}{2}\right)\omega_\ell(\tau) + \frac{p}{2(1-p)^2}\sum_{k,m=1}^{\infty}(km)^{\frac{1}{2}}A_{\ell,k-1,m-1}C^2 e^{2\left(1-\frac{n}{2}\right)\tau}\delta_{k,n}\delta_{m,n}$$
$$= \left(1-\frac{\ell}{2}\right)\omega_\ell(\tau) + \frac{npA_{\ell,n-1,n-1}}{2(1-p)^2}C^2 e^{2\left(1-\frac{n}{2}\right)\tau}.$$

By (2.6), $A_{\ell,n-1,n-1} = 0$ if $\ell > 2(n-1)$. The asymptotic solution to (2.21) is

(2.22)
$$\begin{cases} \alpha_\ell e^{\left(1-\frac{\ell}{2}\right)\tau} + \frac{pC^2 n}{(1-p)^2}\frac{A_{\ell,n-1,n-1}}{\ell-2(n-1)}e^{2\left(1-\frac{n}{2}\right)\tau} + o\left(e^{2\left(1-\frac{n}{2}\right)\tau}\right) \\ \qquad\qquad\qquad\qquad\qquad\qquad \text{if } \ell \neq 2(n-1), \\[2mm] \alpha_{2(n-1)}e^{(2-n)\tau} + \frac{pC^2 n}{2(1-p)^2}A_{\ell,n-1,n-1}\tau e^{(2-n)\tau} + o\left(e^{2\left(1-\frac{n}{2}\right)\tau}\right) \\ \qquad\qquad\qquad\qquad\qquad\qquad \text{if } \ell = 2(n-1), \end{cases}$$

where C is any real constant, and $\alpha_\ell = 0$ if $\ell = 0, 1, 2, \ldots, n$ (because otherwise (2.14), (iii) would not hold), α_ℓ being an arbitrary constant for $\ell = n + 1, \ldots, 2(n - 1)$. Putting together (2.20) - (2.22), we arrive at

(2.23)

$$\Psi_n(y, \tau) \approx C\, e^{\left(1 - \frac{n}{2}\right)\tau} H_n(y)$$

$$+ \frac{pn}{(1 - p)^2} C^2 e^{2\left(1 - \frac{n}{2}\right)\tau} \sum_{\ell=0}^{n} \frac{A_{\ell, n-1, n-1}}{(\ell - 2(n - 1))} H_\ell(y)$$

$$+ \sum_{\ell=n+1}^{2(n-1)-1} \left(\alpha_\ell e^{\left(1 - \frac{\ell}{2}\right)\tau} H_\ell(y) + \frac{pnC^2}{(1 - p)^2} \frac{A_{\ell, n-1, n-1}}{\ell - 2(n - 1)} e^2 \left(1 - \frac{n}{2}\right) \tau H_\ell(y) \right)$$

$$+ \left(\alpha_{2(n-1)} e^{(2-n)\tau} + \frac{pnC^2 A_{2(n-1), n-1, n-1}}{2(1 - p)^2} \tau e^{(2-n)\tau} \right) H_{2(n-1)}(y).$$

2.4 *Higher-order asymptotics for second type solutions: The intermediate region*

A close look at (2.19) reveals that this expansion is no longer valid when

$$\frac{|y|^2}{\tau} \approx 1 \quad \text{or equivalently,} \quad |x|(T - t)^{-1/2} \approx |\log(T - t)|^{1/2}.$$

This suggests that another expansion has to be tried when $|y|^2 \approx \tau$. We then introduce the new variable

$$\xi = \frac{y}{\sqrt{\tau}},$$

which transforms (2.2) into

$$(2.24a) \quad G_\tau = \frac{1}{\tau} G_{\xi\xi} + \frac{1}{2}\left(\frac{1}{\tau} - 1\right) \xi G_\xi + G + \frac{1}{\tau} \frac{p}{1 - p} \frac{(G_\xi)^2}{G} - (1 - p).$$

Taking into account (2.7), it follows that (2.19a) can be written in the new variables as

$$G(\xi, \tau) = (1 - p) - \frac{(1 - p)^2}{4p} \xi^2 - \frac{5(1 - p)^3}{8p^2} \cdot \frac{\log \tau}{\tau} \xi^2 + \frac{(1 - p)^2}{2p} \frac{1}{\tau}$$

$$+ \frac{\beta}{\tau} \xi^2 + \frac{5(1 - p)^3}{4p^2} \frac{\log \tau}{\tau^2} - \frac{2\beta}{\tau^2} - \frac{(1 - p)^2}{2p} \frac{1}{\tau^2} + \ldots \quad \text{as} \quad \tau \to \infty$$

for some real constant β. This suggests trying the following expansion for the new intermediate region

$$(2.25) \quad G(\xi, \tau) = G_o(\xi) + \frac{\log \tau}{\tau} G_1(\xi) + \frac{1}{\tau} G_2(\xi) + \ldots \quad \text{as} \quad \tau \to \infty.$$

Substituting (2.25) into (2.24) yields

$$(2.26a) \qquad -\xi\frac{G_o'(\xi)}{2} + G_o(\xi) = 1 - p,$$

$$(2.26b) \qquad -\xi\frac{G_1'(\xi)}{2} + G_1(\xi) = 0,$$

$$(2.26c) \qquad G_o''(\xi) - \xi\frac{G_2'(\xi)}{2} + \frac{1}{2}\xi G_o'(\xi) + G_2(\xi)$$
$$+ \frac{p}{1-p}\frac{(G_o'(\xi))^2}{G_o(\xi)} = 0.$$

Solving (2.26), we obtain

$$(2.27a) \qquad G_o(\xi) = (1-p) + c_o\xi^2, \quad G_1(\xi) = c_1\xi^2.$$

Matching (2.25) and (2.24b) yields then

$$(2.27b) \qquad c_o = -\frac{(1-p)^2}{4p}, \quad c_1 = -\frac{5(1-p)^3}{8p^2},$$

whereas integrating (2.26c) gives now

$$(2.28) \quad G_2(\xi) = c_3\xi^2 + \frac{(1-p)^2}{2p} - \frac{(1-p)^2}{4p}\xi^2 \log\left((1-p) - \frac{(1-p)}{4p}\xi^2\right).$$

Substituting (2.27) and (2.28) in (2.25), we obtain an expression for $G(\xi, \tau)$, valid for $\xi \approx 1$ which matches with (2.19a) as $\xi \downarrow 0$, namely

$$(2.29)$$
$$G(\xi, \tau) = \left((1-p) - \frac{(1-p)^2}{4p}\xi^2\right) - \frac{5(1-p)^3}{8p^2}\frac{\log\tau}{\tau}\xi^2$$
$$+ \frac{1}{\tau}\left(c_3\xi^2 + \frac{(1-p)^2}{2p} - \frac{(1-p)^2}{4p}\xi^2 \log\left((1-p) - \frac{(1-p)^2}{4p}\xi^2\right)\right) + \ldots$$

as $\tau \to \infty$, where c_3 is a free real constant.

2.5 *Higher-order asymptotics for second type solutions: The behavior near the free boundary*

One readily sees that expansion (2.29) cannot be valid when $\xi^2 \approx \frac{4p}{1-p}$. Actually, a dimensional analysis in equation (2.24a) indicates that the diffusion term becomes then very important, and a new boundary layer appears. Set $\xi_o^2 = \frac{4p}{1-p}$; we shall analyze the region near $\xi_o = -\left[\frac{4p}{1-p}\right]^{1/2}$, the case corresponding to $\xi_o = \left[\frac{4p}{1-p}\right]^{1/2}$ being similar. A dimensional balance between the terms $\frac{1}{\tau}G_{\xi\xi}$, $\frac{\xi G_\xi}{2}$ and $(1-p)$ in (2.24) when $\xi \approx \xi_o$, suggests selecting new variables as follows

$$\chi = \tau\left(\xi + \xi_o\right),$$
$$G(\xi,\tau) = \frac{1}{\tau}H(\chi,\tau).$$

Equation (2.24) is then transformed into

(2.30)
$$-\frac{1}{\tau^2}H(\chi,\tau) + \frac{\chi}{\tau^2}H_\chi(\chi,\tau) + \frac{1}{\tau}H_\tau(\chi,\tau) = H_{\chi\chi}(\chi,\tau)$$
$$+ \frac{(\xi_o - \chi/\tau)}{2}H_\chi - \frac{1}{2\tau}\left(\xi_o - \frac{\chi}{\tau}\right)H_\chi + \frac{1}{\tau}H + \frac{p}{1-p}\frac{(H_\chi)^2}{H} - (1-p).$$

We now expand $H(\chi,\tau)$ as follows

(2.31) $$H(\chi,\tau) = H_o(\chi) + \beta_1(\tau)H_1(\chi) + \dots \quad \text{as} \quad \tau \to \infty.$$

Substitution of (2.31) into (2.30) gives

(2.32a) $$(H_o)_{\xi\xi} + \frac{\xi_o}{2}(H_o)_\chi + \frac{p}{1-p}\frac{\left((H_o)_\chi\right)^2}{H_o} - (1-p) = 0.$$

We add to (2.32a) the following conditions

(2.32b) $$H_o(0) = H_o'(0) = 0, \quad H_o(\chi) > 0 \quad \text{for} \quad \chi > 0$$

to obtain that

(2.33) $$H_o(\chi) \approx \frac{(1-p)^2}{2(1+p)}\chi^2 \quad \text{as} \quad \chi \to 0,$$
$$H_o(\chi) \approx \frac{2(1-p)\chi}{\xi_o} - \frac{4p}{\xi_o^2}\log\chi + K \quad \text{as} \quad \chi \to \infty,$$

where K is a real constant. Notice that, for fixed $\chi > 0$, we have

$$(1-p) - \frac{(1-p)^2}{4p}\xi^2 = \frac{(1-p)^2}{4p}\left(\xi_o^2 - \xi^2\right) \approx \frac{1}{\tau}\frac{(1-p)^2}{2p}\xi_o\chi \quad \text{as} \quad \tau \to \infty,$$

so that

(2.34)
$$G(\xi,\tau) \approx \frac{1}{\tau}\left\{\frac{(1-p)^2}{2p}\xi_o\chi - \frac{5(1-p)^3}{8p^2}\log\tau\xi_o \right.$$
$$+ c_3\xi_o^2 + \frac{(1-p)^2}{2p} + \frac{(1-p)^2}{4p}\xi_o^2\log\tau$$
$$\left. - \frac{(1-p)^2}{4p}\xi_o^2\log\left(\frac{(1-p)^2}{2p}\xi_o\right) - \frac{(1-p)^2}{4p}\xi_o^2\log\chi\right\} + \dots$$

as $\tau \to \infty$. If we now try to match up to the first order (2.34) and (2.31) (with $H_o(\chi)$ satisfying (2.33)), we see that this in general is not possible unless we introduce a shift in the free boundary, $\chi_o(\tau)$, so that we replace $H_o(\chi)$ in (2.31) by $H_o(\chi - \chi_o(\tau))$. By (2.33)

$$H\left(\chi - \chi_o(\tau), \tau\right) \approx \frac{2(1-p)}{\xi_o}\chi - \frac{4p}{\xi_o^2}\log\chi + \left(K - \frac{2(1-p)}{\xi_o}\chi_o(\tau)\right)$$

$$\text{as} \quad \tau \to \infty.$$

Then a standard matching procedure yields

$$\frac{(1-p)^2}{2p}\xi_o = \frac{2(1-p)}{\xi_o}, \quad \frac{(1-p)^2}{4p}\xi_o^2 = \frac{4p}{\xi_o^2},$$

which are automatically satisfied by our choice of ξ_o, and

$$-\frac{5(1-p)^3}{8p^2}\xi_o^2\log\tau + \frac{(1-p)^2}{4p}\xi_o^2\log\tau \approx -\frac{2(1-p)}{\xi_o}\chi_o(\tau),$$

whence

(2.35) $$\chi_o(\tau) \approx -\theta_1 \log\tau \quad \text{as} \quad \tau \to \infty,$$

for some constant θ_1 depending on p, and the free boundary lies at

$$\xi + \xi_o = \frac{\chi_o(\tau)}{\tau}.$$

We thus obtain an asymptotic expansion for the free boundary, namely

(2.36)
$$|x_i(t)| = \left[\frac{4p}{1-p}\right]^{1/2}(T-t)^{1/2}|\log(T-t)|^{1/2}$$
$$+ \theta_1\frac{\log(|\log(T-t)|)}{|\log(T-t)|^{1/2}}(T-t)^{1/2} + \dots \quad \text{as} \quad t \to T.$$

2.6 *Higher-order asymptotics for third type solutions*

Going back to formula (2.23), we realize that such expansion is no longer valid when $|y| \approx e^{(1/2-1/n)\tau}$. This suggests introducing as a new variable

(2.37) $$\xi = y\, e^{-(1/2-1/n)\tau} = x(T-t)^{-1/n}.$$

228   GALAKTIONOV, HERRERO, AND VELÁZQUEZ

Substituting (2.37) in (2.23), we obtain for each of the terms appearing there

$$e^{(1-\frac{n}{2})\tau} H_n(y) = c_n \left\{ \xi^n - \frac{n(n-1)}{2}\xi^{n-2}e^{(-1+\frac{2}{n})\tau} + \ldots \right\},$$

$$e^{2(1-\frac{n}{2})\tau} \sum_{\ell=0}^{n} \frac{A_{\ell,n-1,n-1}}{(\ell - 2(n-1))} H_\ell(y) = \frac{A_{n,n-1,n-1}}{2-n} c_n \xi^n e^{(1-\frac{n}{2})\tau} + \ldots,$$

$$\sum_{\ell=n+1}^{2(n-1)-1} \alpha_\ell H_\ell(y)e^{(1-\frac{\ell}{2})\tau} = \sum_{\ell=n+1}^{2(n-1)-1} \alpha_\ell c_\ell \xi^\ell e^{(1-\frac{\ell}{n})\tau} + \ldots,$$

$$\sum_{\ell=n+1}^{2(n-1)-1} \frac{pH_0 nC^2}{(1-p)^2}\frac{A_{\ell,n-1,n-1}}{(\ell - 2(n-1))} H_\ell(y)e^{2(1-\frac{n}{2})\tau}$$

$$= -\frac{C^2 pn c_{2(n-1)} - 1}{(1-p)^2} A_{2(n-1)-1,n-1,n-1}\xi^{2(n-1)-1}e^{(-\frac{3}{2}+\frac{3}{n})\tau} + \ldots,$$

$$e^{(2-n)\tau} H_{2(n-1)}(y) = c_{2(n-1)}\xi^{2(n-1)}e^{(-1+\frac{2}{n})\tau},$$

where $n > 2$, and only terms of order higher than or equal to $e^{(-1+\frac{2}{n})\tau}$ have been retained; here c_j is the coefficient of ξ^j in the corresponding modified Hermite polynomial $H_j(\xi)$. Setting now $G(\xi,\tau) \equiv G(y(\xi),\tau)$, we obtain that

(2.38)

$$G(\xi,\tau) = (1-p) + Cc_n\xi^n - \frac{Cn(n-1)}{2}c_n\xi^{n-2}e^{(\frac{2}{n}-1)\tau}$$

$$+ \sum_{\ell=n+1}^{2(n-1)} \alpha_\ell c_\ell \xi^\ell e^{(1-\frac{\ell}{n})\tau}$$

$$+ \frac{pn A_{2(n-1),n-1,n-1}}{2(1-p)^2}C^2 c_{2(n-1)}\xi^{2(n-1)}\tau e^{(\frac{2}{n}-1)\tau} + \ldots$$

at the inner region, whereas taking into account (2.37), we see that $G(\xi,\tau)$ satisfies

(2.39) $$G_\tau = e^{(\frac{2}{n}-1)\tau}G_{\xi\xi} - \frac{\xi G_\xi}{n} + G + \frac{p}{1-p}e^{(\frac{2}{n}-1)\tau}\frac{(G_\xi)^2}{G} - (1-p).$$

Motivated by (2.38), we try in (2.39)

$$G(\xi,\tau) = G_o(\xi) + \sum_{\ell=n+1}^{2(n-1)} G_\ell(\xi)e^{(1-\frac{\ell}{n})\tau} + Q(\xi)\tau e^{(\frac{2}{n}-1)\tau} + \ldots$$

to get

$$(2.40a) \qquad -\frac{1}{n}\xi\,(G_o)_\xi + G_o = 1 - p,$$

$$(2.40b) \qquad \left(1 - \frac{\ell}{n}\right) G_\ell = -\frac{1}{n}\xi(G_\ell)_\xi + G_\ell,$$

$$(2.40c) \qquad \left(\frac{2}{n} - 1\right) Q = -\frac{1}{n}\xi(Q)_\xi + Q,$$

$$(2.40d) \qquad Q + \left(\frac{2}{n} - 1\right) G_{2(n-1)} = (G_o)_{\xi\xi}$$

$$-\frac{1}{n}\xi(G_{2(n-1)})_\xi + \frac{p}{1-p}\frac{\left((G_o)_\xi\right)^2}{G_o}.$$

Integrating (2.40a) - (2.40c) and comparing with (2.38), we obtain

$$G_o(\xi) = (1 - p) + A_o\xi^n, \quad A_o = Cc_n,$$

$$G_\ell(\xi) = A_\ell\xi^\ell, \quad A_\ell = \alpha_\ell c_\ell, \quad \ell = n+1,\dots,2(n-1)-1,$$

$$Q(\xi) = R\xi^{2(n-1)}, \quad R = \frac{pnA_{2(n-1),n-1,n-1}}{2(1-p)^2}C^2 c_{2(n-1)},$$

$$\frac{1}{n}\xi\,(G_{2(n-1)})_\xi - \frac{2(n-1)}{n}G_{2(n-1)}$$

$$= Cn(n-1)c_n\xi^{n-2} + \frac{p}{1-p}\frac{n^2 A_o^2\xi^{2(n-1)}}{\left((1-p) + A_o\xi^n\right)} - R\xi^{2(n-1)}$$

$$= Cn(n-1)c_n\xi^{n-2} - \frac{RnA_o\xi^{2(n-1)+n}}{(1-p) + A_o\xi^n},$$

whence

$$G_{2(n-1)}(\xi) = S\xi^{2(n-1)} - Cn(n-1)c_n\xi^{n-2}$$

$$- RnA_o\xi^{2(n-1)}\int_0^\xi \frac{s^{n-1}}{(1-p) + A_o s^n}\,ds,$$

for some real constant S. Summing all these results up, and expanding the integral above, we get

(2.41)

$$G(\xi,\tau) = ((1-p) + c_n C\xi^n) + \sum_{\ell=n+1}^{2(n-1)} \alpha_\ell c_\ell \xi^\ell e^{(1-\frac{\ell}{2})\tau}$$

$$+ \frac{pn A_{2(n-1),n-1,n-1}}{2(1-p)^2} C^2 c_{2(n-1)} \xi^{2(n-1)} \tau e^{(\frac{2}{n}-1)\tau}$$

$$+ e^{(\frac{2}{n}-1)\tau} \left\{ S\xi^{2(n-1)} - Cn(n-1)c_n\xi^{n-2} \right.$$

$$\left. - \frac{pn A_{2(n-1),n-1,n-1}}{2(1-p)^2} C^2 c_{2(n-1)} \log\left((1-p) + Cc_n\xi^n\right)\xi^{2(n-1)} \right\} + \cdots .$$

In (2.41), the sign of C and the parity of n play an important role in the existence of solutions with interfaces. In particular, interfaces do exist if $C < 0$, in which case $(1-p)+c_n C\xi^n$ has always a root; one or two interfaces may appear if n is odd or even respectively. No root exists at all if $C > 0$ and n is even.

We shall consider here the case where interfaces exist, by analogy with the study already done for second type solutions. We therefore suppose that $C < 0$, and examine the situation where there is an interface at $\xi > 0$. Setting $C = -\alpha$, $\alpha > 0$, we see that the first term in (2.41) vanishes at $\xi_o = \left(\frac{1-p}{c_n \alpha}\right)^{1/n} > 0$. To deal with the boundary layer appearing near the interface, we introduce new variables as follows

$$\chi = e^{(1-\frac{2}{n})\tau} (\xi_o - \xi),$$

$$G(\xi,\tau) = e^{(\frac{2}{n}-1)\tau} H(\chi,\tau)$$

to get

$$\left(\frac{2}{n} - 1\right) e^{(\frac{2}{n}-1)\tau} H + \left(1 - \frac{2}{n}\right) e^{(\frac{2}{n}-1)\tau} \chi H_\chi + e^{(\frac{2}{n}-1)\tau} H_\tau$$

$$= H_{\chi\chi} + \frac{1}{n}\left(\xi_o - \chi e^{(\frac{2}{n}-1)\tau}\right) H_\chi + e^{(\frac{2}{n}-1)\tau} H + \frac{p}{1-p} \frac{(H_\chi)^2}{H} - (1-p).$$

Now we try

(2.42) $$H(\chi,\tau) = H_o\left(\chi - \chi_o(\tau)\right) + \sigma(\tau) H_1(\chi) + \cdots \quad \text{as} \quad \tau \to \infty$$

to obtain to the first order

$$(H_o)_{\chi\chi} + \frac{\xi_o}{n}(H_o)_\chi + \frac{p}{1-p}\frac{\left((H_o)_\chi\right)^2}{H_o} - (1-p) = 0.$$

Assuming $H_o(0) = H_o'(0) = 0$, $H_o(\chi) > 0$ for $\chi > 0$, we obtain as in the case of second type solutions

$$H_o(\chi) = \frac{n(1-p)}{\xi_o}\chi - \frac{n^2 p}{\xi_o^2}\log\chi + \left(K - \frac{n(1-p)\chi_o}{\xi_o}\right) + \cdots$$

for some real constant $K = K(p, n)$. We now write $G(\xi, \tau)$ in (2.41) in the new variables. Note that then the first term reads

$$(1-p) - c_n\alpha\xi^n = c_n\alpha\left(\xi_o^n - \xi^n\right)$$
$$= c_n\alpha\left(\xi_o^n - \left(\xi_o - \chi e^{\left(\frac{2}{n}-1\right)\tau}\right)^n\right)$$
$$= c_n n\alpha\xi_o^{n-1}\chi e^{\left(\frac{2}{n}-1\right)\tau} + \cdots$$

and a matching of first order terms in (2.41), (2.42) yields in particular

$$\frac{n(1-p)}{\xi_o} = nc_n\alpha\xi_o^{n-1},$$
$$\frac{n^2 p}{\xi_o^2} = \frac{pnA_{2(n-1),n-1,n-1}}{2(1-p)^2}c_{2(n-1)}\alpha^2\xi_o^{2(n-1)}.$$

Since

$$\frac{pnA_{2(n-1),n-1,n-1}}{2(1-p)^2}c_{2(n-1)} = \frac{pn}{(1-p)^2}nc_n^2\alpha^2,$$

these two conditions are indeed satisfied by our choice of ξ_o. We also obtain for the interface

$$(2.43) \qquad \xi = \xi_o - e^{\left(\frac{2}{n}-1\right)\tau}\chi_o(\tau)$$
$$= \xi_o - \frac{\xi_o}{n(1-p)}\left(\sum_{\ell=n+1}^{2(n-1)}\alpha_\ell c_\ell\xi_o^\ell e^{\left(1-\frac{\ell}{2}\right)\tau}\right.$$
$$\left. + \frac{pnc_n^2\alpha^2}{(1-p)^2}\xi_o^{2(n-1)}\tau e^{\left(\frac{2}{n}-1\right)\tau} + \cdots\right),$$

where in the above expansion any of the coefficients α_ℓ may be equal to zero, so that (2.43) is to be interpreted as a formula where only the first nonzero term is to be retained. Back to our original variables, we have thus obtained the following expression for the interface

$$(2.44)$$
$$x(t) \approx \left(\frac{1-p}{c_n\alpha}\right)^{1/n}(T-t)^{1/n} - \sum_{\ell=n+1}^{2(n-1)}a_\ell(T-t)^{\ell/2-1+1/n}$$
$$- b(T-t)^{1-1/n}|\log(T-t)|$$

we arrive at

$$(A5) \qquad I_{n,m,\ell} = 2^n \sqrt{\pi} m! \ell! \sum_{k=0}^{n} \frac{\binom{n}{k} 2^{m-k}}{(m-k)!} \delta_{2k, m+n-\ell}.$$

Therefore $I_{n,m,\ell} \neq 0$ if $m + n - \ell$ is an even integer, $0 \leq m + n - \ell \leq 2n$, whence $m+n+\ell$ has to be even and $\ell \leq m+n$, $m \leq \ell+n$. Then $k = \frac{m+n-\ell}{2}$, and (A5) gives

$$(A6) \qquad I_{n,m,\ell} = \frac{2^n \sqrt{\pi} m! \ell! 2^{m - \frac{m+n-\ell}{2}}}{\left(m - \frac{m+n-\ell}{2}\right)!} \binom{n}{\frac{m+n-\ell}{2}}$$

$$= 2^{\frac{n+m+\ell}{2}} \sqrt{\pi} \frac{n! m! \ell!}{\left(\frac{m+\ell-n}{2}\right)! \left(\frac{m+n-\ell}{2}\right)! \left(\frac{n+\ell-m}{2}\right)!},$$

having obtained (A6), we turn our attention to the integral

$$A_{n,m,\ell} = \int_{-\infty}^{+\infty} \widetilde{H}_n(x) \widetilde{H}_m(x) \widetilde{H}_\ell(x) \exp\left(-\frac{x^2}{4}\right) \, dx,$$

where $\widetilde{H}_n(x) = c_n H_n(\frac{x}{2})$, $c_n = (2^{n/2}(4\pi)^{1/4}(n!)^{1/2})^{-1}$, and we just notice that

$$A_{n,m,\ell} = 2 c_n c_m c_\ell I_{n,m,\ell},$$

whence the result.

Appendix B

We now show that there exist nonnegative initial functions

$$(B1) \qquad u(x,0) = \Phi(x), \quad -\infty < x < +\infty$$

such that the solution of the Cauchy problem (1.1), (B1) satisfies the inequality

$$(B2) \quad u(x,t) \geq (T-t)^{\frac{1}{1-p}} \left((1-p) - \frac{(1-p)^2}{4p} \cdot \frac{x^2}{(T-t)|\log(T-t)|} \right)_+^{\frac{1}{1-p}}$$
$$(1 + o(1))$$

as $t \to T$ (cf. (1.5)), in any compact set $\{0 \leq |\eta| \leq \eta_*\}$, where

$$\eta \equiv \frac{x}{(T-t)^{1/2} |\log(T-t)|^{1/2}}$$

and η_* is a fixed constant such that $\eta_* \in (0, \eta_o)$, $\eta_o = 2(\frac{p}{1-p})^{1/2}$.

In order to prove this lower estimate we use a modification of a method due to Friedman and McLeod [FM]. We suppose that

$$\Phi = \Phi(r) \geq 0 \ (r = |x|), \ \Phi \in C^2, \ \Phi(0) > 0, \ \Phi' \leq 0, \ \Phi''(0) < 0,$$

Φ has compact support and

$$0 \leq \Phi(r) \leq m \quad \text{for} \quad r \geq 0,$$

where $m \in (0,1)$ is some constant. By standard results $u(r,t) \in C^\infty$ in $\{u > 0\}$. Consider the function

$$J(r,t) = u_r(r,t) + rF(u(r,t)),$$

where F is a smooth nonnegative increasing function. Then as in [FM], [FH] we deduce by the Maximum Principle that

(B3) $$J(r,t) \geq 0 \quad \text{for} \quad r \geq 0, \quad t \in (0,T)$$

provided that

(B4) $$pF(u)u^{p-1} - F'(u)u^p + 2F(u)F'(u) \geq 0 \quad \text{in} \quad (0,m],$$

(B5) $$F''(u) \leq 0 \quad \text{in} \quad (0,m],$$

(B6) $$J(r,0) \equiv \Phi_r + rF(\Phi) \geq 0 \quad \text{for} \quad r \geq 0.$$

It is readily seen that function $F(u)$ given by

(B7) $$F(u) = \frac{u^p}{2p|\log u|} \quad \text{in} \quad (0,m], \quad F(0) = 0$$

satisfies (B4) and (B5) for small enough $m > 0$. An example of initial function satisfying (B6) is

$$\Phi(r) = A \left(a^2 - r^2\right)_+^{\frac{2}{1-p}} \quad \text{for} \quad r \geq 0$$

if

$$Aa^{\frac{4}{1-p}} < 1, \quad A^{\frac{1-p}{2}} \leq \frac{(1-p)^2 e}{16p}.$$

Integrating inequality (B3) with $F(u)$ given in (B7) we get

(B8) $$u(r,t) \geq R^{-1} \left\{ \left[R(u(0,t)) - \frac{(1-p)r^2}{4p} \right]_+ \right\},$$

where R^{-1} is the inverse function of $R(u) = u^{1-p}|\log u|$. Using now the well known estimate (see for instance [FH])

$$u(0,t) \geq [(1-p)(T-t)]^{\frac{1}{1-p}} \quad \text{in} \quad [0,T),$$

(B2) follows from (B8).

Remark. The function given in (B7) is the minimal solution (with respect to equivalence as $u \to 0$) of inequality (B4). Hence (B2) is the best possible lower estimate which can be obtained by this method.

Acknowledgments. This work was partially done during a visit of the first author to Universidad Complutense in Madrid; he is very thankful to this institution for its hospitality. The first author has been partially supported by OIVTA AN USSR Grant "Mathematical Modelling in Nonlinear Phenomena. Synergetics" as well as by EEC Contract SC1-0019-C. The second and third authors have been partially supported by CICYT Research Grant PB86-01120C0202, as well as by EEC Contract SC1-0019-C.

REFERENCES

[BF] H. Brezis and A. Friedman, *Estimates on the support of solutions of parabolic variational inequalities*, Illinois J. Math.,**20**(1976), 82–98.

[CMM] X. Chen, H. Matano and M. Mimura, *Finite-point extinction and continuity of interfaces in a nonlinear diffusion equation with strong absorption*, to appear.

[EK] L. C. Evans and B. F. Knerr, *Instantaneous shrinking of the support of nonnegative solutions to certain nonlinear parabolic equations and variational inequalities*, Illinois J. Math. **23** (1979), 153–166.

[FH] A. Friedman and M. A. Herrero, *Extinction properties of semilinear heat equations with strong absorption*, J. Math. Anal. and Appl. **124** (1987), 530–546.

[FM] A. Friedman and J. B. McLeod, *Blow-up of positive solutions of semilinear heat equations*, Indiana Univ. Math. J. **34** (1985), 425–447.

[GHV] V. A. Galaktionov, M. A. Herrero and J. J. L. Velázquez, *The space structure near a single point blow-up for semilinear heat equations: A formal approach*, to appear.

[K] A. S. Kalashnikov, *The propagation of disturbances in problems of nonlinear heat conduction with absorption* USSR Comp. Math. Phys. **14** (1974), 70–85.

V. A. Galaktionov
Keldysh Institute
of Applied Mathematics
Academy of Sciences USSR
Miusskaya sq. 4
125047 Moscow, USSR

M. A. Herrero and J. J. L. Velázquez
Departamento de Matemática Aplicada
Facultad de Matemáticas
Universidad Complutense
28040 Madrid, Spain

On a conjecture by Hagan and Brenner

D. HILHORST and H.J. HILHORST

1. Introduction

In this note we consider the problem

$$P^0 \quad \begin{cases} u_t = (D(u)u_x)_x & \text{for } (x,t) \in Q := (0,1) \times \mathbb{R}^+ \\ u(0,t) = u(1,t) = 0 & \text{for } t > 0 \\ u(x,0) = 1 & \text{for } x \in (0,1), \end{cases}$$

where D is a smooth function such that $D > 0$ and $D' \leq 0$ on $[0,1]$. We denote the unique solution of Problem P^0 by u. We furthermore consider for $i = 0, 1$ the problems

$$P_i^0 \quad \begin{cases} u_t = D(i)u_{xx} & \text{for } (x,t) \in Q \\ u(0,t) = u(1,t) = 0 & \text{for } t > 0 \\ u(x,0) = 1 & \text{for } x \in (0,1). \end{cases}$$

Inspired by a conjecture of Hagan and Brenner [2] in a higher dimensional context we show that

$$u \leq u_1 \text{ in } Q \tag{1.1}$$

and

$$\int_0^1 u_0(x,t)dx \leq \int_0^1 u(x,t)dx \text{ for all } t > 0. \tag{1.2}$$

The proof of (1.1) is based on a simple comparison argument. This inequality is in fact a special case of results due to Benilan and Diaz [1].

The proof of (1.2) is based on integrating in space the equation for $v = u - u_0$ and on sign considerations for the inhomogeneous term in this equation. In fact we deduce from the maximum principle that

$$\int_x^{1-x} u_0(s,t)ds \leq \int_x^{1-x} u(s,t)ds$$

for all $x \in [0, 1/2)$ and all $t > 0$. Essential for the proof are the properties of the initial function being concave and symmetric about $x = 1/2$.

After completing this note Benilan informed us that he has proved (1.2) by a method based on semi-group theory.

Acknowledgment. The authors are grateful to Professor L.A. Peletier for suggesting this problem to them and for an inspiring discussion.

2. Some qualitative properties of the solutions and the inequality $u \leq u_1$

First we remark that since u, u_0 and u_1 are symmetric with respect to the line $x = 1/2$, u satisfies the problem

$$P \quad \begin{cases} u_t = (D(u)u_x)_x & (x,t) \in \tilde{Q} := (0,1/2) \times \mathbb{R}^+ \\ u(0,t) = 0 \quad u_x(1/2,t) = 0 \quad t > 0 \\ u(x,0) = 1 & x \in (0,1/2) \end{cases}$$

and similarly the functions u_i, $i = 0,1$ satisfy the problems

$$P_i \quad \begin{cases} u_t = D(i)u_{xx} & (x,t) \in \tilde{Q} \\ u(0,t) = 0 \quad u_x(1/2,t) = 0 \quad t > 0 \\ u(x,0) = 1 & x \in (0,1/2) \end{cases}$$

In order to study these problems it is useful to approximate the function D and the initial function 1 by sequences of smooth functions D_n and φ_n such that

$$D_n \geq \inf_{s \in [0,1]} D(s), \quad D'_n \leq 0 \text{ on } [0,1], \quad D'_n(0) = 0 \text{ and } D_n \to D \text{ uniformly on}$$
$[0,1]$ as $n \to \infty$;

$0 \leq \varphi_n \leq 1$, $\varphi_n(0) = \varphi_n(1) = 0$, $\varphi_n(1/2 + x) = \varphi_n(1/2 - x)$ for all $x \in [0,1/2]$, $\varphi'_n(x) \geq 0$ for $x \in [0,1/2]$, $\varphi''_n(x) \leq 0$ for all $x \in [0,1]$, $\varphi''_n(0) = \varphi''_n(1) = 0$, $\varphi_n \to 1$ as $n \to \infty$ uniformly on compact subsets of $(0,1)$.

In what follows we denote by P_n^0 and P_n the problems corresponding to P^0 and P respectively where D is replaced by D_n and with initial function φ_n and we denote by $P_{i,n}^0$ and $P_{i,n}$, $i=0,1$, the problems corresponding to P_i^0 and P_i with $D(i)$ replaced by $D_n(i)$ and initial function φ_n.

It is standard [3] that the problems P_n^0 and $P_{i,n}^0$, $i = 0,1$, each have a unique classical solution which we shall denote by u_n and u_{in}, $i = 0,1$, respectively, and that u_n (resp. $u_{in}, i = 0,1$) converges to u (resp. $u_i, i = 0,1$) when $n \to \infty$.

Lemma 2.1. *The functions u_n, u_{0n} and u_{1n} satisfy*

(i) $\qquad\qquad\qquad 0 \leq u_n, u_{0n}, u_{1n} \leq 1 \text{ in } Q;$

(ii)
$$u_{nx}, u_{0nx}, u_{1nx} \geq 0 \quad \text{in } (0, 1/2) \times \mathbb{R}^+,$$
$$u_{nx}, u_{0nx}, u_{1nx} \leq 0 \quad \text{in } (1/2, 1) \times \mathbb{R}^+;$$

(iii)
$$u_{0nxx}, u_{1nxx} \leq 0 \quad \text{in } Q.$$

Proof. (i) This follows from the standard maximum principle [4]. (ii) We prove the property for u_n for $x \in (0, 1/2)$. The corresponding results for u_{0n} and u_{1n} and for $x \in (1/2, 1)$ can be shown in a similar way. Since u_n attains its minimum 0 at the boundary $x = 0$, it follows from the strong maximum principle that $u_{nx}(0, t) > 0$ for all $t > 0$. Let $q = u_{nx}$. Then q satisfies

$$\begin{cases} q_t = D_n(u_n)q_{xx} + 3D_n'(u_n)qq_x + D_n''(u_n)q^3 & (x,t) \in \tilde{Q} \\ q(0,t) > 0 \qquad q(1/2, t) = 0 \qquad t > 0 \\ q(x, 0) \geq 0 \qquad x \in (0, 1/2) \end{cases}$$

which again by maximum principle implies that $q > 0$ in D. (iii) Set $p = u_{inxx}$. Then p satisfies

$$\begin{cases} p_t = D_n(i)p_{xx} & (x,t) \in Q \\ p(0,t) = p(1,t) = 0 & t > 0 \\ p(x, 0) \leq 0 & x \in (0, 1) \end{cases}$$

so that $p \leq 0$ in Q.

Next we show an inequality which is a special case of results of Benilan and Diaz [1]. Because it can be proven so easily in our case, we give the proof below.

Theorem 2.2. *The functions u and u_1 are such that*

$$u \leq u_1 \text{ in } Q.$$

Proof. We show below that $u_n \leq u_{1n}$; set $w = u_n - u_{1n}$. Then w satisfies

$$\begin{aligned} w_t &= \{D_n(w + u_{1n})w_x\}_x + \{(D_n(u_n) - D_n(1))u_{1nx}\}_x \\ &= \{D_n(w + u_{1n})w_x\}_x + (D_n(u_n) - D_n(1))u_{1nxx} \\ &\quad + D_n'(u_n)u_{nx}u_{1nx} \end{aligned}$$

so that

$$\begin{cases} \{D_n(w + u_{1n})w_x\}_x - w_t \geq 0 & (x,t) \in Q \\ w(0,t) = w(1,t) = 0 & t > 0 \\ w(x, 0) = 0 & x \in (0, 1) \end{cases}$$

which implies that $w \leq 0$.

3. Comparison between the integrals of u and u_0

The purpose of this section is to show that

$$\int_0^1 \{u(x,t) - u_0(x,t)\}dx \geq 0 \qquad (3.1)$$

for all $t > 0$. We define

$$v = u_n - u_{0n}.$$

Then v satisfies

$$v_t = \{D_n(u_n)u_{nx} - D_n(0)u_{0nx}\}_x$$
$$= \{D_n(v + u_{0n})v_x\}_x + \{(D_n(v + u_{0n}) - D_n(0))u_{0nx}\}_x$$

so that we rewrite the problem for v in the form

$$\begin{cases} v_t = \{A(x,t)v_x\}_x + B(x,t) & (x,t) \in \tilde{Q} \\ v(0,t) = 0 \quad v_x(1/2,t) = 0 & t > 0 \\ v(x,0) = 0 \quad x \in (0,1/2) \end{cases}$$

with

$$A(x,t) = D_n(v + u_{0n}) \qquad (x,t) \in \tilde{Q}$$

and

$$B(x,t) = \{(D_n(v + u_{0n}) - D_n(0))u_{0nx}\}_x \qquad (x,t) \in \tilde{Q}.$$

Proving (3.1) amounts to showing that

$$\int_0^{1/2} v(x,t)dx \geq 0 \qquad (3.2)$$

which is a consequence of the following theorem.

Theorem 3.1. *Let* $z(x,t) = \int_x^{1/2} v(s,t)ds$. *Then*

$$z \geq 0 \text{ in } \tilde{Q}.$$

In particular inequalities (3.1) and (3.2) are satisfied.

Proof. The function z satisfies the problem

$$\begin{cases} z_t = A(x,t)z_{xx} + \int_x^{1/2} B(r,t)dr & (x,t) \in \tilde{Q} \\ z_x(0,t) = 0 \quad z(1/2,t) = 0 & t > 0 \\ z(x,0) = 0. \end{cases}$$

Since

$$\int_{x}^{1/2} B(r,t)dr = -\{D_n((v+u_{0n})(x,t)) - D_n(0)\}u_{0nx}(x,t) \geq 0,$$

the result of Theorem 3.1 follows from the maximum principle.

REFERENCES

[1] BENILAN Ph. and J.I. DIAZ, Comparison of solutions of nonlinear evolution problems with different nonlinear terms, *Israel J. of Math. 42,* (1982) 241-257.

[2] HAGAN P.S. and D. BRENNER, Sorption limits with increasing diffusion coefficients, *SIAM J. Applied Math. 48,* (1988), 917-920.

[3] LADYZENSKAJA O.A., V.A. SOLONNIKOV and N.N. URAL'CEVA, Linear and Quasilinear Equations of Parabolic Type, Translations of Mathematical Monographs, vol. 23, Amer. Math. Soc., Providence R.I., 1968.

[4] PROTTER M.H. and H.F. WEINBERGER, Maximum Principles in Differential Equations, Prentice Hall, Englewood Cliffs, N.J., 1967.

D. Hilhorst

Lab. d'Analyse Numérique

CNRS et Université Paris-Sud

Bâtiment 425

91405 Orsay, France

H.J. Hilhorst

Lab. de Phys. Théorique et Hautes Energies

Université Paris-Sud

Bâtiment 211

91405 Orsay, France

A Nonlinear Diffusion-Absorption Equation with Unbounded Initial Data

S. KAMIN[1], L.A. PELETIER[2] and J. L. VAZQUEZ[2]

1. Introduction

We consider the questions of existence, uniqueness and interface behaviour for nonnegative solutions of the Cauchy Problem

$$\text{(I)} \begin{cases} u_t = (u^m)_{xx} - u^p & \text{in} \quad S = \mathbf{R} \times \mathbf{R}^+ \\ u(x,0) = u_0(x) & \text{on} \quad \mathbf{R} \end{cases} \tag{1.1}$$

$$\tag{1.2}$$

where u_0 is a continuous and nonnegative function which can be *unbounded* on $\mathbf{R}$, in the exponent range

$$1 < p < m .$$

It is well-known that a growth condition on the initial data is necessary and sufficient for the existence of nonnegative solutions of the usual examples of parabolic equations like the heat equation [F], where the condition is of the type of quadratic exponential growth, or the porous media equation $u_t = \Delta u^m$ with $m > 1$ [A], where the condition says $u_0 = O(|x|^{2/(m-1)})$. In these examples uniqueness is guaranteed in the class of nonnegative and continuous solutions [F], [DK].

In the case of Problem (I) it has been known for some time [K] that for existence the asymptotic behaviour

$$u_0(x) \sim M(1 + |x|)^{2/(m-p)} \qquad \text{as} \qquad x \to \pm\infty \tag{1.3}$$

is critical. In this note we shall show that there is a range of *non-uniqueness* for initial data with critical growth in our problem, due to the interaction between diffusion and absorption as we explain below. Two explicit solutions are helpful in visualizing this interaction and will play a big role in the subsequent theory. They are the *stationary solution*

$$U_1(x) = c_0|x|^\gamma , \tag{1.4}$$

[1]Supported by grant No. 06-0361-0682 from the United States-Israel BSF.
[2]Supported by EEC Contract SC1-0019-C(TT).

where

$$\gamma = 2/(m - p)$$

is the *critical exponent* and

$$c_0 = \left\{ \frac{(m-p)^2}{2m(m+p)} \right\}^{\frac{1}{m-p}} , \tag{1.5}$$

and the *flat solution*

$$U_2(t) = A^* t^{-\alpha} , \quad \text{with} \quad \alpha = \frac{1}{p-1} \quad \text{and} \quad A^* = \alpha^\alpha . \tag{1.6}$$

We observe that for $t = 0$ the solution U_2 takes on infinite initial values so that every reasonable solution of the Cauchy Problem (I) obeying the Maximum Principle should satisfy $u(x,t) \leq A^* t^{-\alpha}$, being therefore bounded for all positive times. This is not the case of solution U_1. Hence the problem that we want to investigate.

Our results can be very conveniently exemplified by taking into consideration the family of initial functions

$$u_0(x) = \varphi(x) = \begin{cases} cx^\lambda & x \geq 0, \\ 0 & x < 0, \end{cases} \tag{1.7}$$

where c and λ are positive numbers. We shall prove the following results.

    I. If $\lambda < \gamma$, then Problem (I) with $u_0 = \varphi$ has a unique solution.

    II. If $\lambda > \gamma$, then Problem (I) with $u_0 = \varphi$ has no solution.

    III. If $\lambda = \gamma$ the results depend strongly on the value of c. We have three different situations. Thus,

(a) suppose $c \leq c_0$ and let $\beta = \alpha/\gamma$. Then Problem (I) with $u_0 = \varphi$ has a self-similar solution of the form

$$W(x,t) = \begin{cases} t^{-\alpha} f(\eta), & \eta = xt^\beta & x \geq 0 \\ 0 & & x < 0 \end{cases} \tag{1.8}$$

where f is a strictly increasing function such that

$$f(0) = 0 , \quad f(\infty) = A^* .$$

Thus, W is bounded above by the flat solution $U_2(t)$. Moreover, there exists $c_* \leq c_0$ such that for $c < c_*$ W is the unique solution of Problem (I) satisfying a growth condition of the form

$$u(x,t) \leq c(k^2 + |x|^2)^{\gamma/2} . \tag{1.9}$$

(b) Now, for $c = c_0$ (IIIa) yields an example of non-uniqueness because the stationary solution

$$\overline{u}(x) = \begin{cases} c_0 x^\gamma & x \geq 0 \\ 0 & x < 0 \end{cases} \tag{1.10}$$

is also a solution of Problem (I), different from W because it is not bounded above like W. While $\overline{u}$, as an equilibrium state, is invariant in time, W can be viewed as a curve $[0, \infty) \to C(\mathbf{R})$ joining this equilibrium state ($W(0) = \overline{u}$) to the stable level $u = 0$. This latter property is essentially a consequence of the fact that the ODE $u_t = -u^p$ has a solution with initial data $u(0) = \infty$, namely $U_2(t)$, which will serve as an absolute upper bound for all 'reasonably' behaved solutions, which we will identify later on with the class of *minimal* solutions of Problem (I).

(c) There exists a number $c_1 \geq c_0$ such that Problem (I) with $u_0 = \varphi$ has no solution for $c > c_1$.

Because the support of φ is bounded from below, the solution of Problem (I) with $u_0 = \varphi$, whenever it exists, has an interface. We denote it by $x = \zeta(t)$; plainly $\zeta(t) \to 0$ as $t \to 0$. Let T denote the waiting time:

$$T = \sup\{t \geq 0 : \zeta(s) = 0 \quad \text{for} \quad 0 \leq s \leq t\} . \tag{1.11}$$

For the self-similar solutions W and the stationary solution $\overline{u}$, $\zeta(t) = 0$ for all $t \geq 0$ so that $T = \infty$. We shall show, however, that for the solutions in Case (I), when $\lambda < \gamma$, the waiting time is finite. Thus, the growth rate x^γ near zero is also important in determining whether the waiting time is finite or infinite.

The results shown above for the class of functions φ can be extended to general initial functions u_0. Thus, we prove in particular the following results:

(1) Suppose that

$$\liminf |x|^{-\gamma} u_0(x) > c_1$$

as either $x \to -\infty$ or $x \to +\infty$ (or both). Then Problem (I) has no solution. On the contrary, for u_0 satisfying (1.9) with $c \leq c_0$, Problem (I) has at least one solution.

(2) Suppose that

$$\limsup_{|x| \to \infty} |x|^{-\gamma} u_0(x) < c_* .$$

Then there exists a unique solution of Problem (I) satisfying the same condition as $|x| \to \infty$ for all $t > 0$.

The plan of the paper is as follows. In Section 2 we establish the basic existence and uniqueness results and introduce the class of minimal solutions which will prove to be useful in cases of non-uniqueness. Section 3 discusses the existence of selfsimilar solutions for initial data of type (1.7). Section 4 contains results about existence of waiting times of the three possible types: zero, finite positive and infinite, depending on the initial data. In Section 5 we use a waiting time result in conjunction with the comparison results proved in Section 2 to show the above mentioned non-existence result. Finally, several possible extensions are briefly commented upon in Section 6.

In view of the above results the growth constants $0 < c_* \leq c_0 \leq c_1$ have a relevance in understanding the existence and uniqueness theory. It has been recently proved [MPV] that the best constant c_1 is strictly larger than c_0. However, our estimated value of c_* is less than c_0 and we do not know at this time whether the constant c_* in the uniqueness result can be taken equal to c_0 or not.

2. Existence and Uniqueness. Minimal solutions

In this section, we define the notion of solution of Problem (I), of minimal solution of the same problem and prove existence and uniqueness results.

Definition. By a *solution* of Problem (I) we shall mean a function $u : \overline{S} \to [0, \infty)$ such that

    (i) u is continuous in $\overline{S}$

    (ii) For any $T > 0$ and any $a > 0$,

$$\int_{-a}^{a} u(T)\psi(T)\, dx - \int_{0}^{T} \int_{-a}^{a} (u\psi_t + u^m \psi_{xx} - u^p \psi)\, dx dt$$
$$= \int_{-a}^{a} u_0 \psi(0)\, dx - \int_{0}^{T} u^m \psi_x \big|_{-a}^{a}\, dt \tag{2.1}$$

for every $\psi \in C^{2.1}(\overline{Q})$, where $Q = Q_{a,T} = (-a, a) \times (0, T)$, such that $\psi \geq 0$ in $\overline{Q}$ and $\psi = 0$ on $\{-a, a\} \times [0, T]$. If (2.1) is satisfied with $\leq$ instead of $=$ we say that u is a *subsolution*, if with $\geq$ a *supersolution*.

We begin with an existence theorem essentially contained in [K] to which we add the observation that it produces minimal solutions.

Theorem 2.1. *Suppose that* $\Phi(x,t) \geq 0$ *is a supersolution of Problem (I), and let* $u_0 \in C(\mathbf{R})$ *be bounded above by* $\Phi(x,0)$. *Then there exists a solution* $u(x,t)$ *of Problem (I) with initial data* u_0, *which is minimal in the set of solutions of Problem (I), such that*

$$0 \leq u(x,t) \leq \Phi(x,t) \qquad in \ \ S\,. \tag{2.2}$$

It also satisfies

$$u(x,t) \leq A^* t^{-\alpha} \qquad in \quad S. \tag{2.3}$$

Moreover, if u_0 is positive in $\mathbf{R}$ then u is smooth and positive in S.

Proof. For $n = 1, 2, \ldots$ we consider the weak solution u_n to equation (1.1) with initial data

$$u(x,0) = u_{0n}(x) \tag{2.4}$$

which are smooth and nonnegative in $(-n, n)$, vanish for $x = \pm n$ and approximate u_0 as $n \to \infty$ from below: $u_{0,n}(x) \leq u_{0,n+1}(x) \leq u_0(x)$ for every $x \in \mathbf{R}$, [K]. On the lateral boundaries of the cylinders $Q_n = (-n, n) \times \mathbf{R}^+$ we set

$$u_n(\pm n, t) = 0 \ . \tag{2.5}$$

From the maximum principle applied in Q_n, we conclude that

$$u_n(x,t) \leq \Phi(x,t) \qquad in \quad \overline{Q}_n \ . \tag{2.6}$$

Moreover, the solutions u_n are uniformly continuous in $\overline{Q}_n$ (we use (2.6) to go down to $t = 0$). We may thus pass to the (monotone) limit and set $u = \lim u_n$, which is a solution of (1.1) and satisfies the initial condition.

The maximum principle applied in Q_n also implies that for every n there exists a constant $\varepsilon_n > 0$ such that

$$u_n(x,t) \leq A^*(t + \varepsilon)^{-\alpha} \qquad in \quad Q_n \tag{2.7}$$

for every $\varepsilon < \varepsilon_n$, since the second member is also a solution of (1.1), and its data on the parabolic boundary are larger than those of u_n. Hence, in the limit we have $u(x,t) \leq A^* t^{-\alpha}$.

Finally, if $\tilde{u}(x,t) \geq 0$ is another solution of Problem (I), we again have

$$u_n(x,t) \leq \tilde{u}(x,t) \qquad in \quad Q_n$$

and hence $u(x,t) \leq \tilde{u}(x,t)$. This shows that u is minimal.

Now for the positivity and smoothness we recall, see [BNP, K], that at any point where the initial data are positive the solutions will be positive for all times, and this happens uniformly in n since the proof can be based on comparing with an appropriate explicit subsolution. Hence u will be positive. Since under those conditions the equation ceases to be degenerate, standard theory gives the C^∞ smoothness. $\square$

Remarks. 1) The observation that positive initial data produce smooth and positive solutions applies not only to minimal solutions but to *any* solutions of Problem (I).

2) Since the functions in the approximating sequence $\{u_n\}$ are bounded above by the absolute supersolution $U(t) = A^* t^{-\alpha}$ the conclusion that the limit $u = \lim u_n$ exists and is a solution of equation (1.1) follows for all nonnegative initial data u_0, even in $L^1_{loc}(\mathbf{R})$. What is not guaranteed in general is the fact that this solution takes on the initial data, i.e. that the data are *admissible*. Examples of non-admissible initial data will be presented in Section 3.

3) Estimate (2.3), which is essentially the bound provided by the ODE $u_t = -u^p$, implies that there is an absolute rate with which all admissible data are attracted towards the nullstate via the corresponding minimal solution. Moreover, all minimal solutions become bounded for $t \geq \tau > 0$.

As an immediate consequence of Theorem 2.1 we have a non-uniqueness result.

Example. Corresponding to initial data $u_0 = U_1(x) \equiv c_0 |x|^\gamma$ there are at least two solutions, namely the stationary solution $u(x, t) = U_1(x)$ and the minimal solution which decays in time like $O(t^{-\alpha})$. This example will be pursued further in Section 3.

The occurrence of non-uniqueness gives interest to the study of special classes of solutions with nice properties. This is the case with the class of minimal solutions. Thus, we have

Corollary 2.2. *Let u and v be two minimal solutions of Problem (I) with initial data u_0 and v_0 such that $u_0 \leq v_0$. Then for every $t > 0$ we have $u(t) \leq v(t)$.*

Proof. The function v is a supersolution for Problem (I) with initial data u_0. $\square$

Corollary 2.3. *Assume that*

$$0 \leq u_0(x) \leq c(k^2 + |x|^2)^{\gamma/2} \qquad on \quad \mathbf{R} \tag{2.8}$$

for some $0 < c \leq c_0$ and $k \geq 0$. Then Problem (I) admits a solution u and

$$u(x, t) \leq c(k^2 + |x|^2)^{\gamma/2}.$$

Proof. The right hand side of (2.8) is a supersolution of (1.1). $\square$

In view of this situation it will be interesting to characterize the minimal solutions by simple direct criteria, like growth conditions. This purpose is pursued in the rest of this section.

We first introduce a new constant

$$c_* = \left\{ \frac{p(m-p)^2}{2m^2(3m-p)} \right\}^{\frac{1}{m-p}}, \tag{2.9}$$

We observe that $0 < c_* < c_0$ for $0 < p < m$ with $c_* = c_0 = 0$ for $p = m$. This constant plays an important role in our technique of proving uniqueness and comparison results. However, we do not know if it is optimal in that respect.

We now introduce the growth condition which is basic for our uniqueness results. A solution u of Problem (I) will be said to belong to the class $\mathcal{B}$ if there are constants $M < c_*$ and $k \geq 0$ such that for every $t > 0$

$$u(x,t) \leq M(k^2 + |x|^2)^{\gamma/2} . \tag{2.10}$$

We will also say that u is a $\mathcal{B}$-solution. A similar definition applies to sub- and supersolutions.

We next state a basic technical result.

Lemma 2.4. *Assume that u is a $\mathcal{B}$-subsolution and v is a $\mathcal{B}$-supersolution of (1.1) with initial data u_0 and v_0. Assume moreover that one of them is positive in S and that $(u_0 - v_0)_+ \equiv \max\{u_0 - v_0, 0\}$ is integrable. Then*

$$\int_{\mathbf{R}} (u(t) - v(t))_+ \, dx \leq \int_{\mathbf{R}} (u_0 - v_0)_+ \, dx \tag{2.11}$$

for every $t > 0$.

Remark. Clearly, the same applies if we replace $(u - v)_+$ by $(u - v)_- = (v - u)_+$. Adding both inequalities we get that

$$\|u(t) - v(t)\|_1 \leq \|u_0 - v_0\|_1 \tag{2.12}$$

for every $t > 0$. In particular, if $u_0 = v_0$ then $u = v$, i.e. we have uniqueness under the present assumptions. (2.12) is known as the contraction property in L^1, while (2.11) is sometimes given the name of T-contraction. It implies the comparison principle.

Proof of Lemma 2.4. In outline, we shall follow the uniqueness proof of Kersner for bounded solutions of Problem (I) [Ke]. Because many arguments are similar, we shall not always give the complete details.

Thus, let u and v have initial values u_0 and v_0 such that $(u_0 - v_0)_+ \in L^1(\mathbf{R})$. Moreover, let $M < c_*$ and $k > 0$ be constants such that (2.10) is satisfied by $u(t)$ and $v(t)$ for every $t \geq 0$. Finally, either u or v is positive and smooth. Then it will be enough to prove that for every $t > 0$ and every $\chi \in C_0^\infty(\mathbf{R})$, such that $0 \leq \chi \leq 1$ we have

$$\int_{\mathbf{R}} (u(t) - v(t)) \chi \, dx \leq \|(u_0 - v_0)_+\|_1 . \tag{2.13}$$

Subtracting the integral identity (2.1) for v from the one for u we obtain

$$\int_{-a}^{a} \{u(T) - v(T)\}\, \psi(T)\, dx - \int_{0}^{T} \int_{-a}^{a} (u - v)(\psi_t + A\psi_{xx} - C\psi)$$

$$\leq -\int_{0}^{T} (u^m - v^m)\psi_x\big|_{-a}^{a}\, dt + \int \{u_0 - v_0\}\psi(0)\, dx, \qquad (2.14)$$

where

$$\begin{cases} A = \dfrac{u^m - v^m}{u - v}\,, \quad C = \dfrac{u^p - v^p}{u - v} & \text{if} \quad u \neq v \\[2mm] A = mu^{m-1}\,, \quad C = pu^{p-1} & \text{if} \quad u = v\,. \end{cases}$$

Now, for a and $T > 0$ let $Q = Q_{a,T} = (-a, a) \times (0, T)$. We observe that if we put $\delta_1 = \inf\{u(x,t) : (x,t) \in Q\}$, $\delta_2 = \inf\{v(x,t) : (x,t) \in Q_a\}$ and $\delta = \max\{\delta_1, \delta_2\}$, then by our hypotheses $\delta > 0$. Moreover, $A > \delta^{m-1}$ and $C > \delta^{p-1}$ in $\overline{Q}$.

On $\overline{Q}_{a,T}$ we uniformly approximate A and C by decreasing sequences $\{A_j\}$ and $\{C_j\}$ of smooth functions and we write (2.14) as follows:

$$\int_{-a}^{a} \{u(T) - v(T)\}\psi(T)\, dx \leq -\int_{0}^{T} (u^m - v^m)\psi_x\big|_{-a}^{a}\, dt$$

$$+ \int_{0}^{T} \int_{-a}^{a} (u - v)\,(A - A_j)\psi_{xx} - \int_{0}^{T} \int_{-a}^{a} (u - v)(C - C_j)\psi \qquad (2.15)$$

$$+ \int_{0}^{T} \int_{-a}^{a} (u - v)(\psi_t + A_j\psi_{xx} - C_j\psi) + \sup(\psi(0))\|(u_0 - v_0)_+\|_1\,.$$

Let χ be as above with $\operatorname{supp}(\chi) \in (-a, a)$. We then choose for ψ the solution of the initial-boundary value problem

$$\begin{cases} \psi_t + A_j\psi_{xx} - C_j\psi = 0 & \text{in} \quad Q \\[1mm] \psi(x, T) = \chi(x)\,, \quad \psi(\pm a, t) = 0\,. \end{cases}$$

By classical theory such a solution exists, is unique and

$$0 < \psi(x, t) \leq \max\{\chi(x) : -a \leq x \leq a\} \leq 1 \qquad \text{in} \quad Q\,.$$

In addition, we have

Lemma 2.5 *Let μ be a positive number and let M satisfy*

$$M < M_\mu \equiv \left(\frac{p}{m\mu(\mu + 1)}\right)^{1/(m-p)}. \qquad (2.16)$$

Then there exist positive constants $K_1(\mu, k)$, $K_2(\mu, k)$ and $K_3(\mu, k, a)$ such that

(a) $\psi(x, t) \leq K_1(k + |x|)^{-\mu} \quad in \quad \overline{S}$
(b) $|\psi_x(\pm a, t)| \leq K_2(k + a)^{-\mu} \quad for \quad 0 \leq t \leq T,$
(c) $\int_{0}^{T} \int_{-a}^{a} (\psi_{xx})^2 \leq K_3.$

Proof. (a) Suppose that supp $\chi \subset [-b, b]$ with $b > 0$. Then we consider the comparison function (we may assume $k > 0$)

$$z(x,t) = \left(\frac{k+b}{k+|x|} \right)^{\mu} ,$$

Plainly, on $\mathrm{supp}(\chi)$ $|x| < b$ and

$$z(x,T) \geq 1 \qquad \text{if} \qquad x \in \mathrm{supp}(\chi) ,$$

so that $z \geq \chi$, which equals ψ at $t = T$. Along the lateral boundary of Q, we have $z > 0 = \psi$.

Let $Q_+ = (0, a) \times (0, T)$. In Q_+ we have

$$\begin{aligned}
\mathcal{L}(z) &\equiv A_j z_{xx} + z_t - C_j z \\
&\leq z A_j \left\{ \frac{\mu(\mu+1)}{(k+x)^2} - \frac{C_j}{A_j} \right\} \\
&= z\, A_j (k+x)^{-2} \Phi(x) ,
\end{aligned}$$

where

$$\Phi(x) = \mu(\mu+1) - \frac{C_j}{A_j}(k+x)^2 .$$

Because of the growth condition on u and v, we have

$$\frac{C_j}{A_j} \geq \frac{p}{m} M_j^{p-m}(k^2 + |x|^2)^{-1} ,$$

where $M_j \to M$ as $j \to \infty$ and we have used the fact that $\gamma(m - p) = 2$. It follows that

$$\Phi \leq \mu(\mu+1) - \frac{p}{m} M_j^{p-m} . \tag{2.17}$$

and so we obtain the supersolution condition $\mathcal{L}z \leq 0$ if

$$M_j^{m-p} \leq \frac{p}{m\mu(\mu+1)} .$$

This condition will be satisfied for j sufficiently large if

$$M^{m-p} < \frac{p}{m\mu(\mu+1)} . \tag{2.18}$$

Since $z(x,t) \geq \psi(x,t)$ on the parabolic boundary of Q_+ it follows from the Maximum Principle that

$$\psi(x,t) \leq z(x,t) = (k+b)^{\mu}(k+|x|)^{-\mu} , \qquad \text{on} \qquad \overline{Q}_+ \tag{2.19}$$

and the same argument and conclusion apply to $Q_- = (-a, 0) \times (0, T)$, so that (2.19) holds in $Q_{a,T}$.

For the proof of Parts (b) and (c) we can refer to [Ke]. $\quad\square$

Returning to (2.15) we obtain

$$\int_{-a}^{a} \{u(T) - v(T)\}\, \chi \leq I_1 + I_2 + I_3 + \|(u_0 - v_0)_+\|_1 \,, \tag{2.20}$$

where the terms I_i represent bounds for the integrals in the second member of (2.15) and we take into account that the last integrand vanishes identically. By Lemma 2.5(b),

$$I_1 = \left| \int_0^T (u^m - v^m)\psi_x \big|_{-a}^{a} \right| = O\left(a^{m\gamma - \mu}\right) \qquad \text{as} \quad a \to \infty$$

and so, if we choose $\mu > m\gamma$ then $I_1 \to 0$ as $a \to \infty$. Next,

$$\begin{aligned}
I_2 &\leq \sup_{Q_a} |u - v| \cdot \sup_{Q_a} |A - A_j| \sqrt{2aT} \left(\int_0^T \int_{-a}^{a} (\psi_{xx})^2 \right)^{1/2} \\
&\leq K(a) \sup_{Q_a} |A - A_j|
\end{aligned}$$

which tends to 0 as $j \to \infty$ when a is kept fixed. Similarly,

$$I_3 \leq 2aT \sup_{Q_a} |u - v| \cdot \sup |C - C_j|$$

also tends to 0 as $j \to \infty$ for fixed a. Thus, if we let successively $j \to \infty$ and then $a \to \infty$, the right hand side of (2.20) vanishes and (2.13) results with $t = T$.

It remains to verify whether we can always choose $\mu > m\gamma$ since the choice of μ affects – through Lemma 2.5 – the constant M in the upper bound (2.10). However, since we have $c_* = M_{m\gamma}$, because by assumption $M < c_*$, and finally since M_μ decreases with increasing μ, it is possible to choose a $\mu > m\gamma$ so that

$$M < M_\mu < M_{m\gamma} \,.$$

This completes the proof. $\quad\square$

Our next result is

Theorem 2.6. *Solutions of Problem (I) in class $\mathcal{B}$ are uniquely determined by their initial data. Therefore, they are minimal solutions.*

Proof. Assume that we have two $\mathcal{B}$-solutions u and v corresponding to the same initial data f and let us prove that $u \leq v$ in S. If one of them is positive then the conclusion follows from the Remark after Lemma

2.4. If this is not true, then we consider for $\varepsilon > 0$ initial data f_ε with the following properties: f_ε is smooth, positive, larger than f, smaller than $M(k^2 + |x|^2)^{\gamma/2}$, where $M < c_*$ is chosen so that (2.10) is satisfied also by $u(t)$ and $v(t)$ for all $t \geq 0$. Finally,

$$\int_{\mathbf{R}} |f_\varepsilon - f| < \varepsilon. \qquad (2.21)$$

Now we apply formula (2.12) to the pairs (u, u_ε) and (v, u_ε) to obtain

$$\|u(t) - u_\varepsilon(t)\|_1 \leq \varepsilon, \qquad \|v(t) - u_\varepsilon(t)\|_1 \leq \varepsilon \qquad (2.22)$$

Since $\varepsilon > 0$ is arbitrary we conclude that $u = v$ in S.

Finally, if u is a $\mathcal{B}$-solution of Problem (I) with data u_0 and v is the minimal solution with the same data, clearly $v \leq u$, hence v is also a $\mathcal{B}$-solution and we conclude that $v = u$, i.e. u is minimal. $\quad\square$

Theorem 2.6 together with Corollary 2.2 give the following comparison result.

Corollary 2.7. *Let u be a $\mathcal{B}$-subsolution and v an arbitrary supersolution of Problem (I) corresponding to initial values u_0 and v_0. Then*

$$u_0 \leq v_0 \Rightarrow u(t) \leq v(t) \qquad in \qquad \overline{S} \quad for\ every\ t > 0. \qquad (2.23)$$

In order to finish our characterization of minimal solutions we introduce the concept of limit solution. We say that u is a *limit solution* if it is the increasing limit of a sequence of $\mathcal{B}$-solutions.

Theorem 2.8. *The concepts of limit solution and minimal solution coincide.*

Proof. Assume that u is a limit solution with initial data u_0 which is the increasing limit of a sequence of $\mathcal{B}$-solutions u_n and let v be any other solution such that $u_0 \leq v_0$. Then, by Corollary 2.7 we have $u_n \leq v$ for every n, hence in the limit $u \leq v$.

For the converse statement we assume that u is a minimal solution with initial data u_0. If u is not a $\mathcal{B}$-solution we take the sequence of bounded solutions u_n with initial data $u_{0,n} = \min\{n, u_0\}$. Since each u_n is in $\mathcal{B}$ and is therefore minimal, we have $u_n \leq u$ (Corollary 2.2). In the limit of the nondecreasing sequence u_n we obtain as in the proof of Theorem 2.1 a solution $\tilde{u} \leq u$. By the minimality assumption $\tilde{u} = u$. This completes the proof. $\quad\square$

3. Similarity Solutions

In this section we construct a minimal solution $W(x,t)$ for Problem (I) with the special initial function $\varphi = \varphi_c(x)$ given by

$$\varphi_c(x) = \begin{cases} cx^\gamma & x \geq 0 \\ 0 & x < 0, \end{cases} \tag{3.1}$$

where $c > 0$ and $\gamma = 2/(m-p)$. By the results of the previous section if we approximate the initial data by bounded functions

$$U_{0n}(x) = \min\{\varphi(x), n\} \tag{3.2}$$

and let u_n be the bounded solution of Problem (I) with initial value $u_n(x,0) = U_{0n}(x)$, we will have:

(1) $n_1 < n_2 \Rightarrow u_{n_1} \leq u_{n_2}$ in S,
(2) For every $n > 0$ we have $u_n(x,t) \leq A^* t^{-\alpha}$ $(A^* = \alpha^\alpha)$.

As in Theorem 2.1 the family of functions $\{u_n(x,t)\}$ is increasing with n and uniformly bounded above for any $(x,t) \in S$, hence there exists the limit

$$W(x,t) = \lim_{n \to \infty} u_n(x,t) . \tag{3.3}$$

Moreover, if $c \leq c_0$ we can use $c_0|x|^\gamma$ as a supersolution to conclude as in Theorem 2.1 that the initial data u_0 are taken, thus obtaining a minimal solution of the initial-value problem. We have proved

Lemma 3.1 *Suppose $c \leq c_0$. Then the function $W = W_c$ defined by (3.3) is a solution of Problem (I) with initial function $\varphi_c(x)$. Besides*

$$W_c(x,t) \leq c_0|x|^\gamma . \tag{3.4}$$

In the next two lemmas we give a characterization of $W(x,t)$.

Lemma 3.2. *The function W has the selfsimilar form*

$$W(x,t) = t^{-\alpha} f(\eta) , \qquad \eta = xt^\beta , \tag{3.5}$$

with α and β given in the Introduction.

Proof. We need to check that the transformation $u \to T_k u$ given for $k > 0$ by

$$(T_k u)(x,t) = k^{-\gamma} u(kx, k^{-1/\beta}t) \tag{3.6}$$

transforms solution of (1.1) again into solutions of (1.1). It clearly follows that it must transform minimal solutions into minimal solutions. Now we observe that our initial data φ are invariant under the transformation. In fact

$$k^{-\gamma}\varphi(kx) = \varphi(x).$$

The uniqueness of minimal solutions implies then that for every $(x,t) \in S$ and every $k > 0$ we have $(\mathcal{T}_k W)(x,t) = W(x,t)$, i.e.

$$W(x,t) = k^{-\gamma}W(kx, k^{-1/\beta}t).$$

or, if we set $t = 1$, write $k^{-1/\beta} = s$ and $kx = y$:

$$W(s^\beta y, 1) = s^\alpha W(y,s).$$

Replacing y and s by x and t again, we arrive at the desired expression for W:

$$W(x,t) = t^{-\alpha}f(\eta),$$

where $\eta = xt^\beta$ and $f(\eta) = W(\eta, 1)$. $\square$

Lemma 3.3. *Suppose $c \leq c_0$ and let*

$$W(x,t) = t^{-\alpha}f(\eta)$$

be the solution of Problem (I) with $u_0 = \varphi$ constructed in Lemma 3.1. Then f is a solution of the boundary value problem:

$$(f^m)'' - \beta\eta f' + \alpha f - f^p = 0 \qquad 0 < \eta < \infty \tag{3.7}$$

$$f'(\eta) > 0 \qquad 0 < \eta < \infty \tag{3.8}$$

$$\lim_{\eta \to 0} \eta^{-\gamma}f(\eta) = c, \qquad \lim_{\eta \to \infty} f(\eta) = A^* \tag{3.9a,b}$$

Proof. It follows from the construction of W that $W(x,t) > 0$ if $x > 0$ and $t \geq 0$. Thus

$$f(\eta) > 0 \qquad 0 < \eta < \infty. \tag{3.10}$$

By standard regularity theory, W must be a classical solution of (1.1). Hence, f is a smooth function. When we substitute (3.6) into (1.1) we obtain (3.7).

For every $n > 0$, φ_n is nondecreasing and hence so is $u_n(\cdot, t)$ for every $t \geq 0$. Therefore $W_x \geq 0$, which implies that

$$f'(\eta) \geq 0 \qquad 0 < \eta < \infty. \tag{3.11}$$

To prove strict inequality, we first observe that because

$$W(x,t) \leq U_2(t),$$

we have

$$f(\eta) < A^*. \tag{3.12}$$

Note that equality would mean $W = U$, which is clearly impossible.

Suppose now that at some $\eta_0 > 0$, $f'(\eta_0) = 0$. Then by the equation (3.7) and the bounds (3.10) and (3.12) on f, it follows that $f''(\eta_0) < 0$ and so that $f' < 0$ in a right neighbourhood of η_0. This contradicts (3.11) and thus proves (3.8).

To prove (3.9a), we use the initial condition: $\lim_{t \to 0} W(x,t) = cx^\gamma$ for $x > 0$, namely

$$\lim_{t \to 0} x^{-\gamma} t^{-\alpha} f(\eta) = c \qquad \eta = xt^\beta \ .$$

Since for $x > 0$ fixed, $t \to 0$ implies $\eta \to 0$, and $x^{-\gamma} t^{-\alpha} = \eta^{-\gamma}$, the limit follows.

Finally, for proving (3.9b) we use the behaviour of solutions of equation (1.1) as $t \to \infty$ established in [BNP]. For any $n > 0$ it follows from [BNP] that for $x > 0$ fixed,

$$t^\alpha u_n(x,t) \to A^* \qquad \text{as} \qquad t \to \infty \ .$$

Hence, because $u_n \le W \le A^* t^{-\alpha}$, we have $t^\alpha W(x,t) \to A^*$ as $t \to \infty$, namely

$$f(\eta) \to A^* \qquad \text{as} \qquad \eta \to \infty \ .$$

This completes the proof. $\square$

Corollary 3.4. *The ODE problem (3.7)-(3.9) has a solution, if $c \le c_0$.*

Summarizing we have proved the following existence theorem.

Theorem 3.5. *Suppose $u_0 = \varphi$, where φ is given by (3.1). Then for every $c \in (0, c_0]$, Problem (I) has a minimal solution W which has the form*

$$W(x,t) = t^{-\alpha} f(\eta), \qquad \eta = xt^\beta \tag{3.13}$$

where $\alpha = 1/(p-1)$, $\beta = \alpha/\gamma$ and f satisfies (3.7)-(3.9).

To end this study we show that W is a monotone function in both variables. The fact that it is nondecreasing in x for every fixed $t > 0$ has been already observed. Monotonicity in time is shown next

Proposition 3.6. *For every positive t and x we have $W_t < 0$. Therefore, the function $f(\eta)/\eta^\gamma$ is decreasing in $(0, \infty)$.*

Proof. It consists of applying the Maximum Principle to the equation satisfied by $z = W_t$ in $R = (0, \infty) \times (0, \infty)$:

$$z_t = (mu^{m-1} z)_{xx} - pu^{p-1} z \tag{3.14}$$

and observing that $z(0,t) = 0$ and that

$$z(x,0) = W_t(x,0) = (c^m x^{\gamma m})_{xx} - (cx^\gamma)^p \tag{3.15}$$

is negative for $c < c_0$ and 0 for $c = c_0$. To justify the application of the Maximum Principle to an unbounded function we can consider approximations u_n constructed in this section for which $(\partial_t u_n)(x,0) = (u_n^m)_{xx}(x,0) - u_n(x,0)^p \le 0$.

For the last statement we observe that $W_t = \beta t^{-(\alpha+1)}(\eta f'(\eta) - \gamma f)$.

$\square$

Remark. The construction of this selfsimilar solution which is bounded for $t \geq \tau > 0$ and we call *Leiden solution* was the starting example in our investigation of the question of non-uniqueness for Problem (I).

4. Waiting Time

In this section we shall assume that u_0 is a continuous and nonnegative function such that

$$\text{supp}(u_0) = [0, x_0] \tag{4.1}$$

for some $x_0 \in (0, \infty)$. The interface starting at $x = 0$ is described by the function

$$\zeta(t) = \inf\{x \in \mathbf{R} : u(x, t) > 0\}.$$

Its waiting time is defined as $T = \sup\{t \geq 0 : \zeta(t) = 0\}$. Because of the relatively strong absorption term $(1 < p < m)$, we now have three possibilities for T: $T = 0$, $0 < T < \infty$ and $T = \infty$, the latter being absent in the case of weak absorption $(p \geq m)$ [BKP].

In this section we shall investigate the influence of the *local* behaviour of u_0 near $x = 0$ on T. We give two conditions, one ensuring that $T = 0$ and the other that $T < \infty$.

Theorem 4.1. *Suppose u_0 satisfies (4.1) and for some $a > 0$,*

$$x^{-1}u_0^{m-1}(x) \to \frac{m-1}{m}a \qquad as \qquad x \downarrow 0. \tag{4.2}$$

Then $\lim_{t \downarrow 0} t^{-1}\zeta(t) = -a$. Therefore the waiting time is zero.

Proof. We shall compare the solution $u(x, t)$ of Problem (I) in the vicinity of the origin $(0, 0)$ with two travelling wave solutions of (1.1) with speeds close to $-a$.

Let V_a be a travelling wave solution of (1.1) on S with wave speed $-a$, i.e. a solution of the form

$$V_a(x, t) = q_a(x + at) \qquad x \in \mathbf{R}, \, t \geq 0.$$

In [HV] such a solution was found to exist, be unique (as a travelling wave with fixed speed), and have the properties

$$q_a(\xi) \begin{cases} > 0 & \xi > 0, \\ 0 & \xi \leq 0, \end{cases} \tag{4.3}$$

$$q_a'(\xi) > 0 \quad \text{for all} \quad \xi > 0, \tag{4.4}$$

$$\lim_{\xi \downarrow 0}(q_a^{m-1}(\xi))' = \frac{m-1}{m}a. \tag{4.5}$$

Moreover, the family (q_a) is monotone increasing with a.

Choose $\varepsilon > 0$ and small. We shall use the travelling wave solutions

$$z_{\pm}(x,t) = V_{a(1\pm\varepsilon)}(x,t)$$

as comparison functions. Note that by (4.5), $z_+ < z_-$ near the origin. By assumption (4.2) there exists a point $x_\varepsilon > 0$ such that

$$z_+(x,0) < u_0(x) < z_-(x,0) \quad \text{for} \quad 0 < x \leq x_\varepsilon$$

and hence, by continuity, there exists a time $\tau_\varepsilon > 0$ such that

$$z_+(x_\varepsilon,t) < u(x,t) < z_-(x_\varepsilon,t) \quad \text{for} \quad 0 \leq t \leq \tau_\varepsilon \,.$$

We now apply the Maximum Principle in the rectangle $R = [-1, x_\varepsilon] \times [0, \tau_\varepsilon]$. Because $z_+ \leq u \leq z_-$ on the parabolic boundary, it follows that

$$z_+(x,t) \leq u(x,t) \leq z_-(x,t) \qquad \text{on} \qquad \mathbf{R}\,,$$

and hence

$$-a(1+\varepsilon)t \leq \zeta(t) \leq -a(1-\varepsilon)t \qquad \text{for} \qquad 0 \leq t \leq \tau_\varepsilon \,.$$

Therefore, the waiting time is zero, and $\zeta(t)/t$ tends to $-a$ as $t \downarrow 0$.

Remark. If we weaken (4.2) to

$$\liminf_{x \downarrow 0} x^{-1} u^{m-1}(x) > 0 \,,$$

the second part of Theorem 4.1 remains true.

Remark. The condition derived in Theorem 4.1 ensuring that $T = 0$ is the same as the one valid for the porous media equation without absorption [Kn].

Whereas in Theorem 4.1 the absorption term turned out to have no affect, in the next theorem, which gives a condition ensuring that $T < \infty$, the absorption term must necessarily play its part, since without it, we would always have $T < \infty$. It is interesting to see how here the growth $|x|^\gamma$ as $x \to 0$, which was so important for the existence and uniqueness theory, plays again a critical rôle.

Theorem 4.2. *There exists a number $c_1 > 0$ such that whenever u_0 satisfies (4.1) and*

$$\liminf_{x \to 0^+} x^{-\gamma} u_0(x) = c \,. \tag{4.6}$$

with $c > c_1$, then the waiting time is finite.

Proof. We shall prove Theorem 4.2 by constructing a solution v_k of Problem (I), with $v_k(x,0)$ suitably chosen so that (i) $u \geq v_k$ in S and (ii) $v_k(0, t_0) > 0$ for some $t_0 > 0$.

To begin with, we introduce the function

$$\theta(x) = \begin{cases} \{x(1-x)\}^{1/(m-1)} & 0 < x < 1, \\ 0 & \text{elsewhere}. \end{cases} \tag{4.7}$$

Let z be the solution with initial data θ. Then, because $x^{-1}\theta^{m-1}(x) \to 1$ as $x \to 0+$, it follows from Theorem 4.1 that there exists a point $(-2\delta, \tau)$, $\delta > 0$, $\tau > 0$ such that $z(-2\delta, \tau) > 0$. Now write

$$v(x,t) = z(x - \delta, t).$$

Then,

$$\operatorname{supp} v(\cdot, 0) = [\delta, 1 + \delta]$$

and

$$\operatorname{supp} v(\cdot, \tau) \supset [-\delta, 1 + \delta]. \tag{4.8}$$

Assume now that c is so large that $v(x,0) \le \frac{1}{2}cx^\gamma$ for $x > 0$. We introduce the rescaled solutions

$$v_k(x,t) = k^{-\gamma}v(kx, k^{-1/\beta}t).$$

It is clear that v_k is again a solution of (1.1), that $v_k(x,0) \le \frac{1}{2}cx^\gamma$ and that $\operatorname{supp} v_k(\cdot, 0) = [\delta/k, (1+\delta)/k]$. Moreover,

$$\left[\frac{-\delta}{k}, \frac{1+\delta}{k}\right] \subset \operatorname{supp} v(\cdot, \tau k^{1/\beta}), \tag{4.9}$$

Assume now that u_0 satisfies (4.6) with c as above. Then, if k is large enough we will have

$$u_0(x) \ge v_k(x, 0) \qquad \text{for } x \in \mathbf{R},$$

and therefore

$$u(0, \tau k^{1/\beta}) \ge v_k(0, \tau k^{1/\beta}) > 0.$$

This completes the proof. $\square$

Let us consider next the existence of solutions with an infinite waiting time. The simplest example of such a solution is the stationary solution

$$u(x,t) = c_0(x_+)^\gamma.$$

Global comparison with this solution allows one to obtain the following general result.

Theorem 4.3. *Let $u_0(x) \le c_0(x_+)^\gamma$ for every $x \in \mathbf{R}$ with $u_0(x) > 0$ for $x > 0$. Then Problem (I) admits a minimal solution whose left interface is $\zeta(t) = 0$ for all $t > 0$.*

The proof is an easy consequence of the construction in Theorem 2.1.

To end this section we supply an example of a solution with a *finite* and *nonzero waiting time*.

Example. We know that the solution $u_1(x,t)$ with initial data $u_1(x,0) = \theta(x)$ given by (4.7) has an interface with zero waiting time, therefore, for some small $\delta > 0$ there exists $t_1 > 0$ such that its left interface $\zeta_1(t) < -\delta$ if $t \geq t_1$.

Now let $u(x,t)$ be the solution of (I) with continuous initial data $u_0(x) \geq \theta(x)$ such that $u_0(x) = 0$ for $x \leq -\delta$ and $x > 1$ and $0 < u_0(x) < c_0(x+\delta)^\gamma$ for $-\delta < x \leq 0$.

By comparison $u(x,t) \geq u_1(x,t)$ so that the left interface $\zeta(t)$ of u, which starts at $x = -\delta$, is moving at least for $t > t_1$. On the other hand, by local comparison with $u_2(x,t) = c_0(x+\delta)^\gamma_+$ we conclude that $\zeta(t) = -\delta$ for some interval $0 < t < t_2$. This means that $0 < t_2 \leq T \leq t_1 < \infty$.

5. Nonexistence

To begin with, we consider solutions of Problem (I) with an initial function of the form

$$\varphi(x) = \begin{cases} cx^\gamma & x > 0 \\ 0 & x \leq 0 \end{cases} \tag{5.1}$$

where c is some positive number.

Lemma 5.1. *Suppose u_0 is given by (5.1). Then if $c > c_1$, where c_1 is defined in Theorem 4.2, Problem (I) has no solution.*

Proof. Suppose Problem (I) does have a solution u. To construct a minimal solution with the same initial data we approximate these initial data from below

$$\hat{u}_{0,n} = \begin{cases} \varphi(x) & \text{if} \quad \varphi(x) < n \\ 0 & \text{if} \quad \varphi(x) \geq n \end{cases} \tag{5.2}$$

Let $\hat{u}_n$ be the corresponding solution. Since $\hat{u}_0$ has compact support, so does $\hat{u}(t)$ for every $t > 0$. By the results of Section 2 $u \geq \hat{u}_n$ hence

$$\hat{u}(x,t) = \lim_{n \to \infty} \hat{u}_n(x,t) \ . \tag{5.3}$$

will be the minimal solution of the same problem, $\hat{u} \leq u$. Proceeding as in Section 3, we find that $\hat{u}$ can be written in the form

$$\hat{u} = W(x,t) = t^{-\alpha} f(xt^\beta) \ , \tag{5.4}$$

where f has the properties

$$f(\eta_1) \leq f(\eta_2) \qquad \text{if} \qquad \eta_1 < \eta_2 \tag{5.5.a}$$

$$f(\eta) \to A^* \qquad \text{as} \qquad \eta \to \infty \ . \tag{5.5.b}$$

Fix $n > 0$ and let $\zeta_n(t)$ be the interface corresponding to the solution $\hat{u}_n$. Because $c > c_1$, it follows from Theorem 4.2 that eventually $\zeta_n(t) < 0$ and so that there exists a point (x_0, t_0) such that $x_0 < 0$ and $u(x_0, t_0) > 0$. According to [BNP],

$$t^\alpha u_n(x_0, t) \to A^* \qquad \text{as} \qquad t \to \infty \ .$$

Because $\hat{u}_n \leq W \leq A^*$, this means that

$$t^\alpha W(x_0, t) \to A^* \qquad \text{as} \qquad t \to \infty$$

or, in view of (5.4), that

$$f(x_0 t^\beta) \to A^* \qquad \text{as} \qquad t \to \infty \ . \tag{5.6}$$

Recall that $x_0 < 0$. So $t \to \infty$ implies $\eta \to -\infty$, whence we can rewrite (5.6) as

$$f(\eta) \to A^* \qquad \text{as} \qquad \eta \to -\infty \ .$$

In view of the monotonicity of f (5.5.a) and its limit at $\eta \to +\infty$ (5.5.b), we conclude that

$$f(\eta) \equiv A^* \qquad \text{for all} \qquad \eta \in \mathbf{R} \ ,$$

and so, that

$$W(x, t) = A^* t^{-\alpha} \ .$$

Thus, in view of the inequality $u \geq \hat{u}$, the hypothetical solution u coincides with the flat solution:

$$u(x, t) = A^* t^{-\alpha} \qquad \text{in} \qquad S \ .$$

In particular, it cannot satisfy the initial condition. $\quad\square$

In the next theorem, we use Lemma 5.1 to obtain a general nonexistence theorem.

Theorem 5.2. *Suppose u_0 has the property*

$$\liminf |x|^{-\gamma} u_0(x) > c_1 \tag{5.7}$$

as $x \to -\infty$ or $x \to \infty$. Then Problem (I) has no solution.

Proof. Suppose that (5.7) holds for $x \to \infty$, and write

$$\liminf_{x \to \infty} x^{-\gamma} u_0(x) = c_1 + 2\varepsilon \ .$$

Then there exists a number $\xi \geq 0$ such that

$$|x|^{-\gamma} u_0(x) \geq c_1 + \varepsilon \qquad \text{for} \qquad x \geq \xi ,$$

and hence

$$u_0(x) \geq v_0(x) \equiv \begin{cases} (c_1 + \varepsilon)(x - \xi)^\gamma & x \geq \xi \\ 0 & x < \xi . \end{cases}$$

Now, we have proved in Lemma 5.1 that there exists no solution of Problem (I) with initial data v_0. By virtue of Theorem 2.1 this implies that there can be no solution of (1.1) with initial data u_0. In fact, every approximation by an increasing family of bounded solutions must give as its limit the flat solution. $\quad\square$

6. Final comments

The discussion of existence, uniqueness and interface behaviour performed in the preceding sections relies on the interplay between two phenomena, diffusion and absorption, which in the range of exponents considered in this paper imply on the one hand the existence of nontrivial equilibium solutions, and on the other hand uniform decay to the zero level for all solutions in a certain class. In this respect, our results can be easily extended to a number of related models. To quote

(i) The corresponding equation in several space variables

$$u_t = \Delta u^m - u^p \qquad \text{in } S = \mathbf{R}^N \times (0, \infty) \tag{6.1}$$

for the same range of exponents $1 < p < m$. Clearly, there will be a number of properties specific to several dimensions, but for instance the theory of Section 2 applies with only the natural changes.

(ii) The version with *p-Laplacian*-diffusion, i.e.

$$u_t = (|u_x|^{p-2} u_x)_x - u^q$$

where now $1 < q < p - 1$, or its N-dimensional counterpart.

(iii) Equations with variable coefficients as long as the two main ingredients are conserved. For instance, we may consider absorption terms of the form $-c(x)u^p$, with the same restriction on p if $c(x)$ is a continuous function which satisfies $0 \leq k \leq c(x) \leq K$. However, for unbounded $c(x)$ the critical growth exponent γ has to be modified.

REFERENCES

[A] D. G. Aronson, *The Porous Medium Equation*, in Nonlinear Diffusion Problems, Lecture Notes in Maths. **1224**. C.I.M.E. Series, Springer Verlag, Berlin, 1986..

[BKP] M. Bertsch, R. Kersner, L. A. Peletier, *Positivity versus localization in degenerate diffusion equations*, Nonlin. Anal. TMA **9** (1985), 987-10098.

[BNP] M. Bertsch, T. Nanbu, L. A. Peletier, *Decay of solutions of a degenerate nonlinear diffusion equation*, Nonlin. Anal. TMA **6** (1982), 539-554.

[DK] B. E. J. Dahlberg, C. E. Kenig, *Non-negative solutions of the porous medium equation*, Comm. P. D. E. **9** (1984), 409-437.

[F] A. Friedman, Partial Differential Equations of Parabolic Type, Prentice Hall, Englewood Cliffs, N. J., 1962..

[HV] M. A. Herrero, J. L. Vazquez, *Thermal waves in absorbing media*, J. Diff. Eq. 74 **2** (1988), 218-233.

[K] A. S. Kalashnikov, *The effect of absorption of heat propagation in a medium in which thermal conductivity depends on temperature*, Zh. vychisl. math. i math. phys. **16** (1976).

[Ke] R. Kersner, *Degenerate parabolic equations with general nonlinearities*, Nonlin. Anal., TMA **4** (1980), 1043-1062.

[Kn] B. F. Knerr, *The porous medium equation in one dimension*, Trans. Amer. Math. Soc. **234** (1977), 381-415.

[MPV] J. B. McLeod, L. A. Peletier, J. L. Vazquez, *Solutions of a nonlinear ODE appearing in the theory of diffusion with absorption*, Differential and Integral Equations 4 (1991), 1-14.

S. Kamin
Raymond & Beverly Sackler
Faculty of Exact Sciences
Tel Aviv University
Ramat Aviv, Israel

L.A. Peletier
Mathematical Institute
University of Leiden
PB 9512, 2300 RA Leiden
The Netherlands

J. L. Vazquez
División de Matemáticas
Universidad Autónoma de Madrid
Cantoblanco, 28049 Madrid, Spain

A Free Boundary Problem Arising in Plasma Physics

HANS G. KAPER and MAN KAM KWONG[1]

Abstract

This article is concerned with free boundary problems for the differential equation

$$u'' + (1/x)u' + f(u) = 0, \ x > 0.$$

In particular, it addresses the questions of existence and uniqueness of a finite point P and a solution u satisfying the conditions

$$u'(0) = 0; \ u(x) > 0, \ 0 < x < P; \ u(P) = u'(P) = 0.$$

Among the admissible nonlinearities are functions of the type $f(u) = u^p - u^q$ for certain combinations (p, q), where $0 \leq q < p \leq 1$, including the special case $f(u) = \sqrt{u} - 1$, which has been proposed in plasma physics as a simple model for Tokamak equilibria with magnetic islands.

1 Introduction

In this article we study free boundary problems for the differential equation

$$u'' + (1/x)u' + f(u) = 0, \ x > 0. \tag{1}$$

We prove that, for a broad class of functions f, there exists a finite point P and a solution u of (1) satisfying the conditions

$$u'(0) = 0; \ u(x) > 0, 0 < x < P; \ u(P) = u'(P) = 0. \tag{2}$$

The class of admissible nonlinearities includes functions of the type $f(u) = u^p - u^q$ for certain combinations (p, q), where $0 \leq q < p \leq 1$. The special case $f(u) = \sqrt{u} - 1$, proposed recently by Miller, Faber, and White [1] in plasma physics as a very simple model of Tokamak equilibria with magnetic islands, provided the motivation for the present investigation.

---

[1]This work was supported by the Applied Mathematical Sciences subprogram of the Office of Energy Research, U. S. Department of Energy, under Contract W-31-109-Eng-38.

Let F be the integral

$$F(u) = -\int_0^u f(s)\, ds, \quad u > 0. \tag{3}$$

Throughout this article we assume that f satisfies the following conditions.

[F1] f is continuous on $[0, \infty)$ and locally Lipschitz on $(0, \infty)$.

[F2] $F(u) > 0$ for $0 < u < \beta$ and $f(u) > 0$ for $u \geq \beta$, for some $\beta > 0$.

[F3] $\lim_{\epsilon \downarrow 0} \int_\epsilon^\beta F^{-1/2}(u)\, du$ exists and is finite.

[F4] $\liminf_{u \to \infty} f(u) > 0$.

[F5] $u \mapsto \dfrac{f(u)}{u - \beta}$ is nonincreasing for $u > \beta$.

Condition **[F1]** imposes reasonable continuity requirements on f; the Lipschitz continuity guarantees the existence and uniqueness of solutions of initial value problems whenever the initial data are positive.

Conditions **[F2]** and **[F3]** are necessary for the existence of a solution of the free boundary problem. Condition **[F2]** requires, first of all, that f be more negative than positive near the origin and, furthermore, that f eventually become positive. These properties certainly hold if f has exactly one zero, provided $f(u)$ is negative near $u = 0$ and positive for large u. Condition **[F3]** is satisfied, for instance, if $f(u)$ behaves like $-u^q$ $(0 < q < 1)$ as $u \to 0$, or if $f(0) \neq 0$. If $f(u) = u^2 - u$, the free boundary problem does not have a solution, which indicates that **[F3]** is close to being necessary and sufficient. Condition **[F3]** was first introduced by Peletier and Serrin [2].

Condition **[F4]** guarantees that the nonlinear term $f(u)$ retains sufficient strength to bend the solution of (1) down toward the x-axis as the initial height at $x = 0$ increases.

Condition **[F5]** expresses the fact that f grows sublinearly beyond β. The condition is sufficient for the existence of a solution; however, it is not necessary (see [7]).

In Section 2 we prove the existence of a solution of the free boundary problem. The proof is based on the use of the shooting method. In Section 3 we briefly discuss the uniqueness of the solution of the free boundary problem. The uniqueness is essentially an immediate consequence of a theorem of Peletier and Serrin [2], as extended by the authors in [3] or, more specifically, in [4]

The results of this article have been extended to free boundary problems for the more general equation $u'' + g(x)u' + f(u) = 0$; see [7].

2   Existence

We use the shooting method, replacing the boundary problem by an
initial value problem,

$$u'' + (1/x)u' + f(u) = 0, \ x > 0, \tag{4}$$

$$u(0) = A, \ u'(0) = 0. \tag{5}$$

Here, A is a positive constant. We denote a solution of this initial value
problem by u_A. According to the classical theory of initial value problems,
u_A exists in a neighborhood of the origin $x = 0$. We consider u_A only as
long as it is strictly positive or until it reaches its first zero.

If, for some $A > 0$, the solution u_A of (4), (5) is such that $u_A(x) > 0$
for $0 \leq x < P_A$ and $u_A(P_A) = u'(P_A) = 0$ for some finite point P_A, then
(P_A, u_A) is a solution of the free boundary problem (1), (2).

With any solution u_A of (4), (5) we associate a function E_A by the
expression

$$E_A(x) = \frac{1}{2}[u'_A(x)]^2 - F(u_A(x)), \ x \geq 0. \tag{6}$$

Since u_A satisfies (4), $E'_A(x) = -(1/x)[u'_A(x)]^2 \leq 0$, so E_A is nonincreasing
along a solution curve.

Lemma 1 *Every solution u_A of (4), (5) is bounded.*

Proof. Since $F(u_A(x)) \geq -E_A(x) \geq -E_A(0) = F(A)$ for all $x \geq 0$, F is
bounded along any solution curve. According to Condition [F4], the lim
inf of f at infinity is (strictly) positive, so $F(u_A)$ remains bounded if and
only if u_A is bounded. ∎

It follows from Lemma 1 that either any solution u_A of (4), (5) extends (as
a positive-valued function) to the entire half-line $[0, \infty)$ or there is a finite
point P_A where u_A reaches the x-axis. (In the latter case, we consider only
the interval $[0, P_A]$.)

Let $\mathcal{O}$ denote the set of values $A > 0$ for which u_A extends to $[0, \infty)$.
Clearly, only starting values A that do not belong to $\mathcal{O}$ can lead to solutions
of the free boundary problem.

Lemma 2 *If $0 < A < \beta$, then $A \in \mathcal{O}$.*

Proof. If $0 < A < \beta$, $E_A(0) = -F(A) < 0$. Because E_A is nonincreasing,
$E_A(x) < 0$ for all $x \geq 0$. On the other hand, it follows from the definition (6) that $E_A(x) \geq 0$ whenever $u_A(x) = 0$. Hence, if $0 < A < \beta$, u_A
cannot reach the x-axis, so it must be the case that u_A extends to $[0, \infty)$.
∎

Because of Lemma 2, only starting values $A \geq \beta$ can lead to solutions of
the free boundary problem.

Suppose that $\mathcal{P}(x)$ is bounded for some $x_b > x_a$: $u_A(x_b) \leq C$ for all $A > 0$, say. We show that this supposition leads to a contradiction.

Take any $A > 0$. Without loss of generality we may assume that $u_A'(x_b) < 0$. Then either $u_A'(x) < 0$ for all $x \geq x_b$ or there is a finite point x_c, where u_A ceases to be monotone. In the former case, we put $x_c = \infty$. We apply Hadamard's inequality to the function y, defined by the expression $y(x) = u_A(x) - u_A(x_b)$ on the interval (x_b, x_c). As $\|y\| \leq C$, $\|y'\| = \|u_A'\|$, and $\|y''\| \leq (1/x_b)\|u_A'\| + K$, where $K = \max\{|f(u)| : 0 \leq u \leq C\}$, we have

$$\|u_A'\|^2 \leq 2C\left(\frac{1}{x_b}\|u_A'\| + K\right). \tag{9}$$

This quadratic inequality shows that $\|u_A'\|$ is bounded by a constant independent of A. In particular, $|u_A'(x_b)| \leq D$ for some constant D.

Given that both u_A and u_A' are bounded at x_b by constants that are independent of A, it is not difficult to show that u_A is bounded at x_a. If $L = \max\{-f(u) : u \geq 0\}$ (notice that L is finite), then

$$u_A(x) \leq v(x), \ x_a \leq x \leq x_b, \tag{10}$$

where v is the solution of the (backwards) initial value problem

$$v'' + (1/x_a)v' - L = 0, \ x < x_b; \ v(x_b) = C, \ v'(x_b) = -D. \tag{11}$$

Since v is independent of A, it follows in particular that $u_A(x_a)$ is bounded by a constant that does not depend on A. But this contradicts our earlier finding that $\mathcal{P}(x)$ is unbounded at x_a. It must therefore be the case that $\mathcal{P}(x)$ is unbounded for all $x > x_a$.

Since we already know that $\mathcal{P}(x)$ is unbounded for $0 \leq x \leq x_a$, we conclude that $\mathcal{P}(x)$ is unbounded for all x, as claimed.

Step 2. Let $\mu = \frac{1}{2}\liminf_{u\to\infty} f(u)$; because of Condition [F4], μ is positive. Let γ be fixed, such that $f(u) \geq \mu$ for all $u \geq \gamma$; without loss of generality we may assume that $\gamma \geq \beta$.

Let ϵ be an arbitrarily small positive constant, and let $x_0 = 1/\epsilon$.

Since the set $\mathcal{P}(x) = \{u_A(x) : A > 0\}$ is unbounded for all x, there is an $A > 0$ for which $u_A(x_0) = \mu/\epsilon^2$. Let A be so determined, and let x_1 be the point where u_A reaches the value γ. We claim that we can make $|u_A'(x_1)|$ arbitrarily large by choosing ϵ sufficiently small.

Because u_A is monotone, at least until it reaches the value β, it is certainly monotone on $[x_0, x_1]$; we denote the inverse function by $u \mapsto x(u)$ and define R by the expression

$$R(u) = [u_A'(x(u))]^2, \ u_A(x_0) \geq u \geq \gamma. \tag{12}$$

Notice that R is strictly positive on its interval of definition, because u_A is decreasing on $[x_0, x_1]$. This function satisfies the initial value problem

$$\frac{dR}{du} = 2\left(\frac{1}{x}\sqrt{R(u)} - f(u)\right), \ u < u_A(x_0); \ R(u_A(x_0)) = (u_A'(x_0))^2. \tag{13}$$

We compare R with the solution r of the initial value problem,

$$\frac{dr}{du} = 2\left(\epsilon\sqrt{r(u)} - \mu\right), \; u < u_A(x_0); \; r(u_A(x_0)) = 0, \qquad (14)$$

From the theory of differential inequalities we conclude that $R(u) \geq r(u)$ for all $u_A(x_0) \geq u \geq \gamma$. In particular, $R(\gamma) \geq r(\gamma)$.

The solution of (14) is most easily found by going back to the associated second-order problem,

$$u'' + \epsilon u' + \mu = 0, \; x > 0, \; u(0) = \mu/\epsilon^2, \; u'(0) = 0. \qquad (15)$$

The solution u is monotone; and, if $x \mapsto x(u)$ denotes the inverse function, then r, defined by $r(u) = (u'(x(u)))^2$, satisfies (14). (Recall that $u_A(x_0) = \mu/\epsilon^2$.) Thus we find that

$$u(x) = (\mu/\epsilon^2)(1 - \epsilon x - (1 - e^{-\epsilon x})), \qquad (16)$$

$$u'(x) = -(\mu/\epsilon)(1 - e^{-\epsilon x}), \qquad (17)$$

and therefore,

$$r(u) = (\mu/\epsilon)^2(1 - e^{-\epsilon x(u)})^2. \qquad (18)$$

In particular,

$$r(\gamma) = (\mu/\epsilon)^2(1 - e^{-\epsilon x(\gamma)})^2, \qquad (19)$$

where $x(\gamma)$ is determined by the equation

$$(\mu/\epsilon^2)(1 - \epsilon x(\gamma) - (1 - e^{-\epsilon x(\gamma)})) = \gamma. \qquad (20)$$

Since γ and μ are both fixed, it follows that $1 - \epsilon x(\gamma) - (1 - e^{-\epsilon x(\gamma)}) = O(\epsilon^2)$ and, hence, $x(\gamma) = O(\epsilon^{-1})$ as $\epsilon \to 0$. Therefore, $r(\gamma) = O(\epsilon^{-2})$ as $\epsilon \to 0$.

Since $R(\gamma) \geq r(\gamma)$, we conclude that $R(\gamma) = O(\epsilon^{-2})$ as $\epsilon \to 0$. This proves the claim that we can make $|u'_A(x_1)|$ arbitrarily large by taking ϵ sufficiently small.

Step 3. We now show that the conclusion of Step 2 leads to a contradiction by applying Hadamard's inequality beyond x_1.

Either $u'_A(x) < 0$ for all $x \geq x_1$ or there is a finite point x_2, where u_A ceases to be monotone. In the former case, we put $x_2 = \infty$. The function y, defined on (x_1, x_2) by the expression $y(x) = u_A(x) - u_A(x_1)$, satisfies the norm estimates $\|y\| \leq \gamma$, $\|y'\| = \|u'_A\|$, and $\|y''\| \leq \epsilon\|u'_A\| + L$, where $L = \max\{-f(u) : u \geq 0\}$, so

$$\|u'_A\|^2 \leq 2\gamma(\epsilon\|u'_A\| + L). \qquad (21)$$

It follows that $\|u'_A\|$ is bounded if ϵ is sufficiently small; in particular, $|u'_A(x_1)|$ is bounded.

The conclusions of Step 2 and Step 3 are clearly incompatible, so the supposition underlying the arguments of the proof, namely that $\mathcal{O}$ is all of $(0, \infty)$, is untenable. We conclude that $\mathcal{O}$ is indeed a proper subset of $(0, \infty)$, as asserted. ∎

Now we consider the complement of the set $\mathcal{O}$. Let $\mathcal{C}$ be the set of initial heights A for which u_A crosses the x-axis at some finite point P_A with a (strictly) negative slope, $u'_A(P_A) < 0$.

Lemma 8 *The sets $\mathcal{C}$ and $\overline{\mathcal{O}}$, the closure of $\mathcal{O}$, are disjoint.*

Proof. The sets $\mathcal{C}$ and $\mathcal{O}$ are disjoint by definition, so it suffices to show that a point of $\mathcal{C}$ cannot be a boundary point of $\mathcal{O}$.

Let A be a point in $\mathcal{C}$. Given any small positive number ϵ, we can find a point x_0 to the left of the point P_A, but close enough that $u_A(x_0) = \epsilon$ and $|u'_A(x_0)| > \frac{1}{2}|u'_A(P_A)|$. Suppose that A is a boundary point of $\mathcal{O}$. Since the solution depends continuously on the initial height, there is a point $B \in \mathcal{O}$ sufficiently close to A that $u_B(x_0) < 2\epsilon$ and $|u'_B(x_0)| > \frac{1}{2}|u'_A(P_A)|$. Now, u_B does not reach the x-axis, so either u_B is monotone on $[x_0, \infty)$, or there is a first point x_1 beyond x_0 where u_B ceases to be monotone. In the former case, we set $x_1 = \infty$. We apply Hadamard's inequality to the function $y(x) = u_B(x_0) - u_B(x)$ on (x_0, x_1). Since $\|y\| < 2\epsilon$, $\|y'\| = \|u'_B\|$, and $\|y''\| \leq (1/x_0)\|u'_B\| + N$, where $N = \max\{|f(u)| : 0 < u < 2\epsilon\}$, we obtain

$$\|u'_B\|^2 \leq 4\epsilon((1/x_0)\|u'_B\| + N). \tag{22}$$

This quadratic inequality implies that $\|u'_B\|$ is bounded. But then we can make the expression in the right member of (22) and, hence, $\|u'_B\|$ arbitrarily small by decreasing ϵ. But now we have a contradiction, since $|u'_B(x_0)|$ is at least equal to $\frac{1}{2}|u'_A(P_A)|$. ∎

We are now ready to prove the existence result.

Theorem 1 *If f satisfies the conditions [F1] − [F5], then there exists a solution of the free boundary problem (1), (2).*

Proof. The set $\mathcal{O}$ is open and does not coincide with the entire ray $(0, \infty)$. Hence, it must have a finite boundary point A, which is neither in $\mathcal{O}$ nor in $\mathcal{C}$ (Lemma 8). The solution u_A of (4), (5) that starts with the initial value $u_A(0) = A$ reaches the x-axis at P_A with a horizontal slope, $u'_A(P_A) = 0$. Thus, u_A defines a solution of the free boundary problem, with $P = P_A$. ∎

3   Uniqueness

Uniqueness results for boundary value problems for the Emden-Fowler equation are hard to establish. The most comprehensive result for equations with a sublinear (for large u) term $f(u)$ is due to Peletier and Serrin

[2]. These authors were concerned with ground states, i.e., nonnegative, nontrivial solutions of the equation

$$u'' + \frac{n-1}{x}u' + f(u) = 0, \ x > 0, \tag{23}$$

for which $\lim_{x\to\infty} u(x) = 0$. They established the uniqueness of a ground state when $n \leq 2$ under Conditions [F1], [F2], and [F5] and when $n > 2$ under the additional condition that f is negative near the origin. These results of Peletier and Serrin were subsequently improved by the authors [3], who showed that Conditions [F1], [F2], and [F5] suffice for *all* values of n ($n \geq \frac{3}{2}$). It turns out that the same proof works for solution of free boundary problems of the type studied in the present article. We state the result and omit the proof.

Theorem 2 *If f satisfies the conditions* [F1] $-$ [F5]*, then the free boundary problem (1), (2) has one and only one solution.*

REFERENCES

[1] G. Miller, V. Faber, and A. B. White, Jr., "Finding plasma equilibria with magnetic islands", J. Comp. Phys. **79** (1988), 417-435.

[2] L. A. Peletier and J. Serrin, "Uniqueness of non-negative solutions of semilinear equations in $\mathbf{R}^n$", J. Diff. Eq. **61** (1986), 380-397.

[3] H. G. Kaper and Man Kam Kwong, "Uniqueness of non-negative solutions of semilinear elliptic equations", pp. 1-17 in *Nonlinear Diffusion Equations and Their Equilibrium States* II, Wei-Ming Ni, L. A. Peletier, and J. Serrin (eds.), MSRI Conf. Proc., Springer-Verlag, New York, 1989.

[4] H. G. Kaper and Man Kam Kwong, "Ground states of semilinear diffusion equations", in *Proc. Int'l Conf. on Diff. Equations and Applications*, F. Kappel and W. Schappacher (eds.), Retzhof, June 18-24, 1989 (to be published).

[5] J. Hadamard, "Sur le module maximum d'une fonction et de ses dérivées", C. R. Acad. Sci. Paris **42** (1914), 68-72.

[6] Man Kam Kwong and A. Zettl, "Remarks on best constants for norm inequalities among powers of an operator", J. Approx. Theory **26** (1979), 249-258.

[7] H. G. Kaper and Man Kam Kwong, "Free boundary problems for Emden-Fowler equations", Diff. and Int. Eq. **3** (1990), 353-362.

Hans G. Kaper and Man Kam Kwong
Mathematics and Computer Science Division
Argonne National Laboratory
Argonne, IL

Remarks on Quenching, Blow Up
and Dead Cores

BERNHARD KAWOHL

Introduction

The experience in the investigation of nonlinear diffusion equations suggests
a) that many qualitative effects which are known for linear equations can
also be expected for solutions of nonlinear equations, and b) that there can
be specific nonlinear phenomena. In my lecture I will attempt to survey
some aspects of b). There is the traditional trinity of subjects, in this case
quenching, blow up and dead cores. In all of those three types of equation
the diffusion term becomes negligable in a certain sense, and the behaviour
of the solution is essentially controlled by an ordinary differential equation.

Several participants of this conference have contributed to the study of
these equations, namely C. Bandle, C. Brauner, M.K. Kwong, H. Levine,
B. McLeod, J. Serrin, J. Spruck, J.L. Vazquez and F. Weissler. There
are important contributions of other colleagues as well, which I shall try
to mention along the way. Last but not least I shall advertise some of the
results which I had the pleasure of discovering with my coauthors A. Acker,
L. Peletier, M. Fila and G. Dziuk.

1. Mean Curvature Flow and Quenching

On a recent visit to Heidelberg S. Angenent gave a lecture on curve
shortening flow. Imagine a closed curve in $I\!\!R^2$ which moves with velocity
proportional to its curvature in the firection of the interior normal. There
are many beautiful results on this geometric problem, which are surveyed
in a lucid style by B. White [W] in a recent issue of the Mathematical
Intelligencer. One of these results states that a closed curve will not separate into two curves under this flow. This result is no longer true if one
generalizes the problem to higher space dimension.

So suppose that a twodimensional smooth closed surface moves in $I\!\!R^3$
with velocity proportional to its mean curvature. If the initial surface is
convex, then G. Huisken has shown in [H] that it stays convex for positive
times and shrinks to a point in a strong topology. In general, however,
a single surface can split into two separate ones. To see this imagine a
dumbbell domain, i.e. an initial surface which consists essentially of two
large disjoint spheres and of a thin pipe or handle connecting them. We
expect the spheres to shrink slower than the pipe under this flow; and

it has in fact been shown by M. Grayson in [G] that the dumbbell will experience a "necking","pinching", or "quenching" phenomenon. These abstract considerations are confirmed by numerical experiments, e.g. by G. Dziuk [D] or by J.A. Sethian [S], and I tried to illustrate them in Figure 1 below.

The differential equation describing the flow can be described as follows. Let x be the axial coordinate and let t denote time and $u(t,x)$ the radial distance of the surface to its axis of rotation. Then the normal velocity of the surface is $-1\sqrt{1+u_x^2}\, u_t \to -\dfrac{u_t}{\sqrt{1+u_x^2}}$. Notice that u_t is the velocity in radial direction. The principle curvatures of the surface are $\dfrac{-u_{xx}}{(1+u_x^2)^{3/2}}$ in axial direction and $\dfrac{1}{u\,\sqrt{1+u_x^2}}$ in the other direction. Therefore u satisfies

$$u_t - \frac{u_{xx}}{1+u_x^2} = -\frac{1}{u} \qquad \text{where } u > 0. \tag{1.1}$$

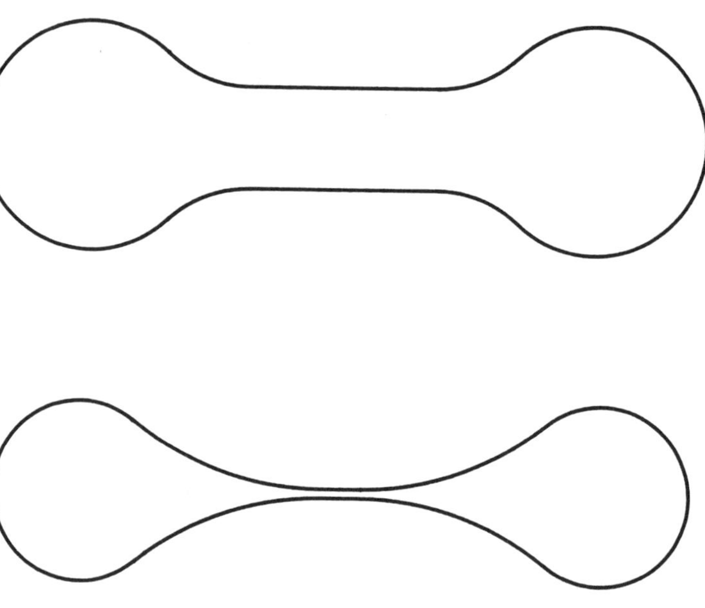

Figure 1. Mean curvature flow of a dumbbell

Some of the problems that have puzzled differential geometers are the questions

i) when quenching occurs, i.e. when $u \to 0$ in some points?

ii) if quenching occurs in isolated points x or on entire intervals. Numerical experiments have not offered a conclusive answer to this question. and

iii) what happens after quenching?

Question iii) has recently been treated in papers of L.C. Evans and J. Spruck [ES], see also [CGG], by introducing a notion of a weaker solution. For mathematical convenience let us consider equation (1.1) on $(0, \infty) \times (0, a)$ under Dirichlet and Neumann boundary conditions

$$u_x(t, 0) = 0, \qquad u(t, a) = b \qquad \text{for } t > 0, \tag{1.2}$$

and under initial data

$$u(0, x) = u_0(x) \qquad \text{for } x \in (0, a). \tag{1.3}$$

Moreover we assume without further mentioning that our initial data are positive and compatible with the boundary condition.

Theorem 1. [DK] *Let u be a solution of* (1.1),(1.2),(1.3).

i1) If a is large in the sense that $2a > \pi b$. Then u will quench in finite time.

i2) If a is small in the sense that the equation $b = \frac{1}{\alpha} cosh(\alpha x)$ has a solution, then u need not quench in finite or infinite time.

ii) Suppose that u quenches in finite time T and that u_0 satisfies the following assumptions, which are explained in Remark 2 below:

$$u_{0t} \leq 0, \quad u_{0x} \geq 0 \quad \text{and} \quad u_{0xt} \geq 0 \quad \text{in } (0, a), \tag{1.4}$$

then $u_{xt} \geq 0$ in $(0, T) \times (0, a)$ and u quenches only in the origin, i.e. at $(T, 0)$.

Remark 2. Condition (1.4) is a shorthand for stating that

$$v_0 := \frac{u_{0xx}}{1 + u_{0x}^2} - \frac{1}{u_0} \geq 0, \qquad u_{0x} \geq 0 \qquad \text{and} \qquad v_{0x} \geq 0.$$

Remark 3. The assumption in statement i1) can be weakened to the assumption that the equation $b = \frac{1}{\alpha} cosh(\alpha x)$ has no solution and that u_0 satisfies (1.4). In that sense the result i2) is optimal. See [DK] for details.

Proof. i1) Suppose a solution exists and is positive for every finite time. We multiply (1.1) by u and integrate with respect to x to obtain

$$\frac{\partial}{\partial t} \int_0^a \frac{u^2}{2} \, dx = \int_0^a (arctan\ u_x)_x u \, dx - a$$

$$= -\int_0^a u_x\ arctan\ u_x \, dx + b\ arctan\ u_x(a) - a \cdot$$

$$\leq \frac{b\pi}{2} - a$$

Therefore

$$\|u(t)\|_{L^2(0,a)}$$

becomes negative in finite time, a contradiction.

To prove i2), we observe that the Ljapunov functional associated with (1.1) is given by the surface area

$$V(t) = 2\pi \int_0^a u\sqrt{1 + u_x^2} \, dx,$$

since along a solution of (1.1), (1.2) we have

$$\frac{dV}{dt} = -2\pi \int_0^a \frac{u}{\sqrt{1 + u_x^2}} \, u_t^2 \, dx.$$

But the minimal surfaces of revolution are catenoids and they are described by hyperbolic cosines. If we pick those as initial data, we have a positive stationary solution. Incidentally, functionals such as V have recently been investigated by Bemelmans and Dierkes in detail [BD].

For the proof of iii) we set $w := u_{xt}$ and derive a differential inequality for w by differentiating (1.1) with respect to x and t:

$$w_t - c_1(t, x)w_{xx} + c_2(t, x)w_x + c_3(t, x)w = -c_4(t, x)u_t u_x, \qquad (1.5)$$

where

$$c_1(t, x) = \frac{1}{1 + u_x^2}$$

$$c_2(t, x) = \frac{4u_x u_{xx}}{(1 + u_x^2)^2}$$

$$c_3(t, x) = \left(\frac{2u_{xxx}u_x + 2u_{xx}^2}{(1 + u_x^2)^2} - 8\frac{u_{xx}^2 u_x^2}{(1 + u_x^2)^3} - \frac{1}{u^2} \right)$$

$$c_4(t, x) = \frac{2}{u^3}$$

are locally bounded coefficients. If we can show that the right hand side in (1.5) is nonnegative, then by the maximum principle w will attain any negative minimum on the parabolic boundary. But on the initial part of the parabolic boundary we have $w = 0$ by assumption, on the left end we have $w = 0$ because of the Neumann boundary condition, and on the right end we have $w \geq 0$ provided we can prove that $u_t \leq 0$ near the right end, if we observe that $u_t = 0$ due to the Dirichlet condition.

Therefore it remains to show that $u_t \leq 0$ and $u_x \geq 0$. But these two inequalities can be derived from the maximum principle in a similar fashion by differentiation of (1.1). We refer to [DK] for details.   Q.E.D

The idea for the proof of single point quenching (Theorem 1 ii)) was taken from a paper of A. Acker and myself, where we considered among other equations the following model problem

$$u_t - \Delta u = -\frac{1}{u^\alpha} \qquad \text{with } \alpha > 0 \quad \text{in } \{t > 0\} \times B_R(0) \qquad (1.6)$$

$$u = 1 \qquad \text{on the parabolic boundary} \qquad (1.7)$$

For this problem we showed single point quenching, but I shall not repeat the result here. Instead I want to indicate another new result from [AK]. For the case of one space dimension and $\alpha = 1$ Theorem 4 was formulated by H. Kawarada; his original proof however contained a gap that has been noticed by several people [AK,CK].

Theorem 4[AK] *Suppose that the solution of* (1.6), (1.7) *quenches, i.e. goes to zero at some point* $(T, 0)$. *Then* $u_t \to -\infty$ *as* $u \to 0$.

Proof. The auxiliary function $u^\alpha u_t$ stays away from zero in the quenching point. This is again a consequence of the maximum principle, because $u^\alpha u_t$ satisfies a suitable differential inequality.   Q.E.D.

Remark 5. In the proof of Theorem 4 we have seen that there is a positive constant c such that

$$u^\alpha u_t \leq -c \qquad (1.8)$$

near the quenching point $(T, 0)$. Integration of (1.8) from $t < T$ to T yields

$$u(t, 0) \geq \tilde{c}(T - t)^{1/(1+\alpha)} \qquad (1.9)$$

for $t < T$ and t close to T. This gives us a onesided bound on the rate at which u_t tends to $-\infty$. To see an inequality that complements (1.9), namely

$$u(t, 0) \leq \tilde{C}(T - t)^{1/(1+\alpha)}, \qquad (1.10)$$

consider the center of the ball. Here u attains its minimum and so $u_t(t, 0) \geq -u^{-\alpha}(t, 0)$. Upon integration this implies (1.10).

Remark 6. Inequalities (1.10) and essentially (1.9) have been extended from a radial domain to a convex domain by K. Deng and H.A. Levine [DL]. They showed in our terms that there exists a constant c such that

$$J := u_t - \tilde{c}u^{-\alpha} \leq 0, \qquad (1.11)$$

from which (1.9) follows.

In order to derive (1.11) they first showed that quenching points stay in a compact subset of the spatial domain Ω and adapted the moving plane method as in a paper of A. Friedman and B. McLeod [FM] on blow up equations, see also [FK1].

Remark 5 contained a proof for the time asymptotics of solutions near the quenching point. Let me summarize this and results on the space asymptotics in the following Theorem.

Theorem 7 [FK1] *Let u be a solution of* (1.6), (1.7) *and suppose that u quenches at time T.*
   i) Then $u(t,0) \sim (T-t)^{1/(1+\alpha)}$.
   ii) If $0 < \alpha < 1$ then $u_r(T,0) = 0$.
   iii) If $1 < \alpha$ and if $u_0(|x|) = 1 - \varepsilon|x|^2$ then $u_r(T,0) = \infty$.

Proof. For details of the proof I refer to [FK1]. The main idea for the proof of i) was mentioned above in Remark 5. To prove ii) one introduces the so-called P-function which is defined by $P = \frac{1}{2}|\nabla u|^2 - \frac{1}{1-\alpha}u^{1-\alpha}$. Then one derives a suitable differential inequality for P and uses the maximum principle to conclude that $P \le 0$. Therefore $|\nabla u|$ at the quenching time is bounded from above by a power of u. Finally one integrates this inequality. The P-function has been used by many people, e.g. by D. Phillips in [P]. A good reference book on various applications of the P-function is the book of R. Sperb [Sp]. To prove iii) one follows an idea from [FM] and introduces the auxiliary function $J = r^{n-1}u_r - \varepsilon r^n u^{-\gamma}$ with $\gamma < \alpha$. After playing with derivatives of J one discovers that $J \ge 0$, again due to the maximum principle.Q.E.D.

Remark 8. The assumption on u_0 in Theorem 7iii) appears to be of a technical nature. Our proof shows that for very strong absorption, i.e. for $\alpha > 1$ the function u develops a cusp in space, while for $\alpha \in (0,1)$ it remains differentiable. I believe that for $\alpha = 1$ the solution will develop a corner. At present this appears to be an open problem.

Remark 9. Another more or less open problem is whether quenching is possible in infinite time. There are results on blow up equations which state that either u blows up in finite time or u stays globally bounded, but u does not blow up in infinte time, see e.g. [F1,Gi,NST] and the references therein. Marek Fila and I have obtained some sufficient criteria to rule out quenching in infinite time for problem (1.6), (1.7), see [FK2]. These criteria on α depend on the space dimension N. For a related problem where quenching is caused by a nonlinear boundary condition, the answer is known to depend on the space dimension as well. Without going into details let me simply refer to [LL,FK2].

2. Blow up

The purpose of the present and the last section is to clarify the relation between blow up and dead core problems. It is not a mere coincidence that "while extinction and blow up are quite opposite facts, they are amenable to analysis by similar techniques." [FH].

Consider the equation

$$u_t - \Delta u = u^p \tag{2.1}$$

for $p > 1$ and under Dirichlet boundary conditions $u = 0$. It is well understood, that solutions of such equations can blow up in finite time, depending on the initial data. There are results on single point blow up and so on, and I apologize for not listing all the remarkable contributions here. Instead let me just mention the works of F. Weissler or V. Galaktionov and refer to the bibliography of this lecture as a starting point for those who want to learn more about it.

Recently M. Chipot and F. Weissler [CW] have studied the question if blow up can be prevented by adding a sink term $-|\nabla u|^q$ to the right hand side of (2.1). They found that for $q \leq 2p/(p+1)$ blow up is still possible. I tried to see what happens if the nonlinearity u^p is replaced by the popular exponential e^u. Formally e^u is a polynomial of infinite degree, so the upper bound for q should be $q = 2$ in this case. But as it turns out the case $q = 2$ is very special, even if we stick to the nonlinearity u^p.

Theorem 10 [KP] *Suppose that u solves the problem*

$$u_t - \Delta u = u^p - |\nabla u|^2 \qquad \text{for } t > 0, \ x \in \Omega, \tag{2.2}$$

$$u(t,x) = 0 \qquad \text{for } t > 0, \ x \in \partial\Omega, \tag{2.3}$$

$$u(0,x) = u_0(x) \geq 0 \qquad \text{for } x \in \Omega. \tag{2.4}$$

i) If $p > 2$ there can be blow up.
ii) If $1 < p \leq 2$ there is no blow up.

Proof. Set $v := \exp(-u)$. Then v satisfies the differential equation

$$v_t - \Delta v = -h(v) \tag{2.4}$$

with $h(v) = v(-\log v)^p$ and under conditions

$$v(t,x) = 1 \quad \text{on the boundary} \tag{2.5}$$

and

$$v(0,x) = v_0(x) \quad \text{initially.} \tag{2.6}$$

Notice that the question whether u blows up is now transformed into the question whether v can reach zero somewhere in Ω in finite time. Problems of this nature are called dead core problems, and for the remainder of the proof I refer to Theorem 13 in the next section. Q.E.D.

Remark 11. There is a heuristic argument for Theorem 10. The reaction term u^p which can cause blow up interacts with the damping term

$|\nabla u|^2$. Theorem 10 tells us that the order of those two terms decides which term wins. Further results on this and other equations can be found in [KP,F2,BE].

Remark 12. It is instructive to study the equation

$$u_t - \Delta u = u^p + |\nabla u|^2 \qquad \text{for } t > 0,\ x \in \Omega, \tag{2.7}$$

as well. Notice that it differs from (2.2) by the plus sign in front of the gradient term. Let us consider (2.7), (2.3), (2.4) for the case that Ω is a ball and that u_0 is radially decreasing. Then the reaction term u^p is maximal where u is maximal and the gradient term is minimal (it vanishes in the origin). So the reaction term will enhance single point blow up. The gradient term, on the other hand is large "close to the boundary" and this will have the effect, that the profile of u tends to be flattened out at the top. I tried to illustrate this in Figures 2 and 3, where the "strength" of those two effects is visualized by the length of vectors. So the gradient term will counteract single point blow up.

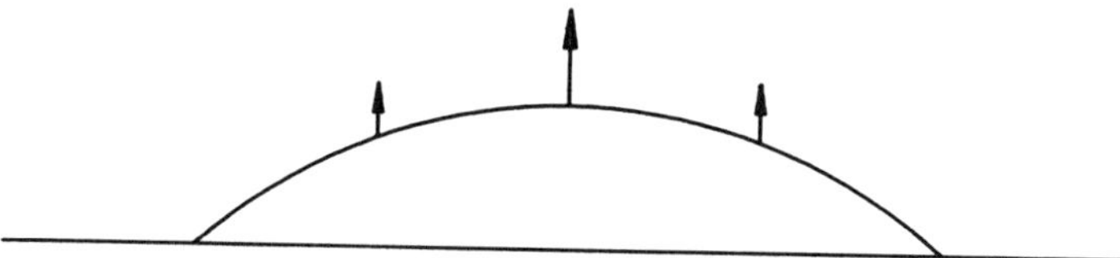

Figure 2. The effect of u^p on u_0

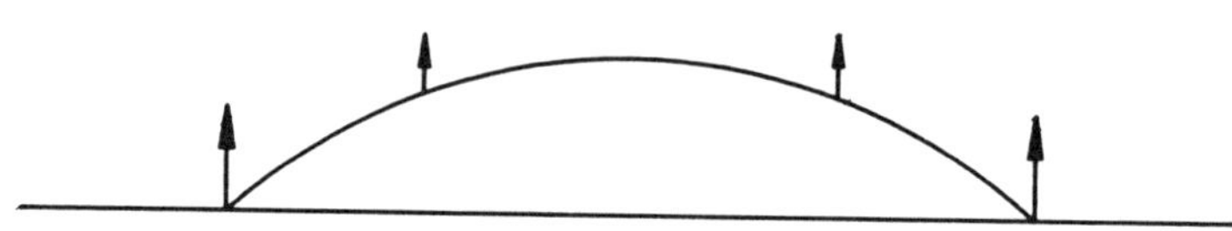

Figure 3. The effect of $|\nabla u|^2$ on u_0

As in Theorem 10 one can ask which term "wins" in the sense that we have either single point blow up or global blow up. Global blow up is the effect that for every $x \in \Omega$ the solution $u(t, x)$ becomes unbounded as t approaches the blow up time. For $p > 2$ one has single point blow up, for $p < 2$ one can prove global blow up, and for $p = 2$ one has at least regional blow up, i.e. blow up on a subset of Ω with positive measure. The proofs of these results are contained in [FM,L], once one uses the transformation $v = \exp(u) - 1$ and derives the transformed equation

$$v_t - \Delta v = (1 + v)(\log(1 + v))^p \qquad \text{for } t > 0, \ x \in \Omega.$$

3. Dead cores

In the last section of my lecture I shall study the differential equation

$$v_t - \Delta v = -h(v) \tag{3.1}$$

subject to

$$v(t, x) = 1 \quad \text{on the boundary} \tag{3.2}$$

and

$$v(0, x) = v_0(x) \quad \text{initially.} \tag{3.3}$$

In particular I shall concentrate on the question whether positive initial data v_0 can lead to a solution that becomes zero somewhere in Ω in finite time. Contrary to Section 1 we assume that the nonlinearity h on the right hand side of (3.1) is at least continuous at zero. For the case that h behaves essentially like u^α with $\alpha \in (0, 1)$ for positive u close to zero there is a result of C.Bandle, R.Sperb and I.Stakgold [BSS] which states that the solution can become zero (it develops a dead core) as long as the domain Ω is large enough. If Ω is too small, then u can asymptotically tend to a positive equilibrium state. This phenomenon is in accordance with Theorem 1i) of the present lecture. Our aim is to get a dead core result for a more general class of nonlinearities h, which (hopefully) includes the one in (2.2). To this end we introduce the notation

$$H(s) = \int_0^s h(\tau) \, d\tau.$$

Theorem 13 [KP]. *Suppose that u solves (3.1), (3.2), (3.3) and that $h(0) = 0$, $h(s) > 0$ for $s \in (0, 1)$ and that h is nondecreasing in a neighborhood $[0, \omega)$ of the origin.*

i) If $\int_0^1 (H(s))^{-1/2} \, ds < \infty$, there can be a dead core.

ii) If $\int_0^1 (H(s))^{-1/2} \, ds = \infty$, there is no dead core.

Proof. Let me just outline the main ideas. The proof goes by a comparison argument. There are two equations which are "simpler" than (3.1), namely the ordinary differential equation

$$dv/dt = -h(v) \quad \text{for } t > 0 \tag{3.4}$$

subject to initial condition (3.3) and the elliptic equation

$$\Delta v = h(v) \quad \text{in } \Omega \tag{3.5}$$

subject to (3.2). Our assumptions on h include the possibility that $h(1) = 0$, and in this case the constant function $v \equiv 1$ is a solution of (3.1), (3.2), (3.3). This situation has to be dealt with, but it does not cause essential difficulties.

The behaviour of solutions to Problem (3.5), (3.2) is decisive for the occurence of a dead core. In fact if a solution of (3.5), (3.2) is somewhere zero, then we expect a solution of (3.1), (3.2), (3.3) to develop a dead core. And if solutions of (3.5) (3.2) are everywhere positive, then so are the solutions of (3.1), (3.2), (3.3). Of course these statements require proofs, but they can be found in [KP].

Why and how does the condition on H enter into the proof of Theorem 13? To understand this consider the one-dimensional case of (3.5), (3.2) and suppose that there is a stationary solution with dead core. Then the associated P-function $P(x) := (1/2)v_x^2 - H(v)$ is constant in Ω, and since there is a dead core P must vanish everywhere, i.e.

$$\frac{dv}{dx} = (2\,H(v))^{1/2}. \tag{3.6}$$

Now let us exchange the independent and dependent variable and interpret x as a function of v. Then (3.6) is formally rewritten as $dx/dv = (2\,H(v))^{-1/2}$ and we realize (see Figure 4) that the quantity $d := \int_0^1 (2\,H(s))^{-1/2}\,ds$ tells us the distance which it takes v to start at zero and reach height 1.

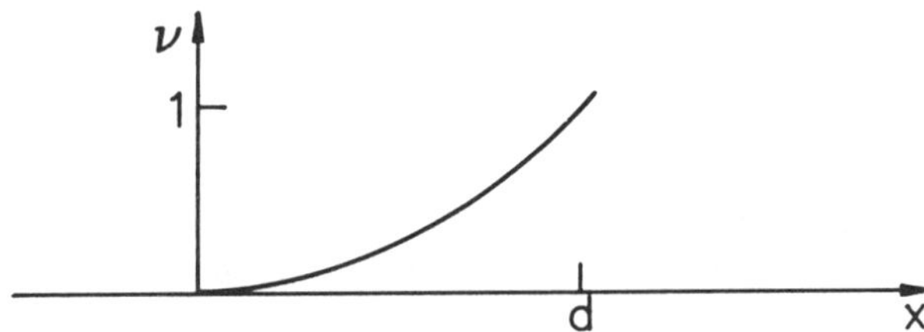

Figure 4. Solution of (3.5) with dead core

If $d = \infty$, this will never happen, otherwise it can happen for sufficiently large one-dimensional domains Ω. Once we have the result for the one-dimensional case, we can derive it for radial domains in arbitrary dimensions, and – by comparison of general domains Ω with balls – for any bounded open Ω in higher dimensions.   Q.E.D.

Remark 14. That the condition on H is decisive for the occurence of zeroes, has been noted in a number of papers for various related equations. I refer to [DH,PS,V].

REFERENCES

[AK] A.Acker, B.Kawohl, "Remarks on quenching", Nonlinear Analysis, TMA **13** 1989, 53–61.

[BE] J..Bebernes, D.Eberly, "Characterization of blow up for a semilinear equation with convection term", Quart. J. Mech. Appl. Math. **42** (1989) 447–456.

[BD] J.Bemelmans, U.Dierkes, "On a singular variational integral with linear growth. I: existence and regularity of minimizers", Preprint 3, SFB 256, Bonn 1987

[BSS] C.Bandle, R.Sperb, I.Stakgold "Diffusion and reaction with monotone kinetics", Nonlinear Anal. TMA **8** !984) 321–333.

[CGG] Y.G.Chen, Y.Giga, S.Goto, "Uniqueness and existence of viscosity solutions of generalized mean curvature flow equations", Proc. Japan Acad. Ser. A **65** 1989, 207–210.

[CK] C.Y.Chen, M.K.Kwong, "Quenching phenomena for singular nonlinear parabolic equations", Nonlinear Analysis, TMA **12** 1988, 1377–1383.

[CW] M.Chipot, F.Weissler, "Some blow up results for a nonlinear parabolic equation with a gradient term", SIAM J. Math. Anal. **20** (1989) 886–907.

[DH] J.I.Diaz, J.Hernandez, "On the existence of a free boundary for a class of reaction diffusion systems", SIAM J. Math. Anal. **5** 1984, 670–685.

[DL] K.Deng, H.Levine, "On the blow up of u_t at quenching", Proc. Amer. Math. Soc. **106** 1989, 1049–1056.

[D] G.Dziuk, "An algorithm for evolutionary surfaces", report no. 5, SFB 256, Bonn 1989.

[DK] G.Dziuk, B.Kawohl, "On rotationally symmetric mean curvature flow", J. Differ. Equations, to appear.

[ES] L.C.Evans, J.Spruck, "Motion of level sets by mean curvature I and II", Preprints 37 and 38, G.A.N.G., Amherst 1989

[F1] M.Fila, "Boundedness of global solutions of nonlinear diffusion equations", Preprint 497, SFB123, Heidelberg 1988

[F2] M.Fila, "Remarks on blow up for a nonlinear parabolic equation with a gradient term", Preprint 502, SFB 123, Heidelberg 1988

[FK1] M.Fila, B.Kawohl, "Asymptotic analysis of quenching problems", Rocky Mt. Math. J., to appear.

[FK2] M.Fila, B.Kawohl, "Is quenching in infinite time possible?", Quarterly of Appl. Math., to appear.

[FH] A.Friedman, M.Herrero, "Extinction properties of semilinear heat equations with strong absorption, J. Math. Anal. Appl. **124** 1987, 530-546.

[FM] A.Friedman, B.McLeod, "Blow up of positive solutions of semilinear heat equations", Indiana Univ. Math. J. **34** 1985, 425–447.

[Gi] Y.Giga, "A bound for global solutions of semilinear heat equations",
      Commun. Math. Phys. **103** 1986, 415–421.

[G]   M.Grayson, "A short note of the evolution of surfaces via mean cur-
      vature", Duke Math J. **58** 1989, 555–558.

[H]   G.Huisken, "Flow by mean curvature of convex surfaces into spheres",
      J. Diff. Geom. **20** 1984, 237–266.

[K]   H.Kawarada, "On solutions of initial boundary problem for $u_t = u_{xx}+$
      $1/(1-u)$", Publ. RIMS Kyoto Univ. **10** 1975 729–736.

[LL] H.A.Levine, G.M.Lieberman, "Quenching of solutions of parabolic equa-
      tions with nonlinear boundary conditions in several dimensions", J.
      Reine Angew. Math. **345** 1983, 23–38.

[NST] W.M.Ni, P.E.Sacks, J.Tavantzis, "On the asymptotic behavior of so-
      lutions of certain quasilinear equations of parabolic type", J. Differ.
      Equations **54** 1984, 97–120.

[PS] L.A.Peletier, J.Serrin "Uniqueness of non-negative solutions of semi-
      linear equations in $I\!R^N$", J. Differ. Equations **61** 1986, 380–397.

[P]   D.Phillips, "Existence of solutions of quenching problems", Applic.
      Anal. **24** 253–264.

[S]   J.A.Sethian, "The collapse of a dumbbell moving under its mean cur-
      vatureg, manuscript, 1989.

[Sp] R.Sperb, *Maximum principles and their applications*. New York, Aca-
      demic Press 1981

[V]   J.L.Vazquez, "A strong maximum principle for some quasilinear ellip-
      tic equations", Appl. Math. Optim. **12** 1984, 191–202.

[W]   B.White "Some recent developments in differential geometry", Math-
      ematical Intelligencer **11** 1989, 41–47.

Bernhard Kawohl
SBF 123, Universität Heidelberg
Im Neuenheimer Feld 294
D 6900 Heidelberg
West Germany

Bifurcation at boundary points
of the continuous spectrum

TASSILO KÜPPER* and CHARLES A. STUART

1. Introduction

Nonlinear ordinary or partial differential equations on unbounded domains have been studied intensively with respect to bifurcation from the lowest point of the continuous spectrum [1,2,9] . Results have been obtained mainly by variational methods .

In this paper we report on results concerning bifurcation at arbitrary boundary points of the continuous spectrum . This study was motivated by nonlinear perturbations of Hill's equation . For simplicity we restrict our attention here to the following class of problems which captures all the essential features ; more general nonlinearities have been treated in [7,8] :

$$-u''(x) + q(x)u(x) \pm a(x) \mid u(x) \mid^{\sigma} u(x) = \lambda u(x) \quad , $$
$$x \in I\!R, u, u'' \in L^2(I\!R) \tag{1.1$\pm$}$$

where

$$q \in L^{\infty}(I\!R) \quad \text{with} \quad q(x+1) \equiv q(x) \geq q_0 > 0 \quad \text{a.e.} \tag{1.2}$$

and

$$\sigma > 0 \quad , \quad a \in L^{\infty}(I\!R), a(x) \geq 0 \quad \text{a.e.} \;, \; \lim_{|x| \to \infty} a(x) = 0 \quad . \tag{1.3}$$

Then $u \equiv 0$ is a (trivial) solution for (1.1$\pm$) for all $\lambda \in I\!R$. Nontrivial solutions can bifurcate from $u \equiv 0$ only at values of λ belonging to the spectrum of the linerarized problem.

---

*Support for the participation at the stimulating conference is gratefully acknowledged.

The problem linearized at $u \equiv 0$ is Hill's equation

$$Sh := -h'' + qh = \lambda h$$
$$h \in D_S = H^2(\mathbb{R}) \quad .$$

$$(1.4)$$

The spectrum of S is purely continuous and consists of at most countably many closed intervals $[\mu_i, \lambda_i](i = 1, 2, ...)$ which we call spectral blocks [3] . Successive spectral blocks are seperated by an open interval called a gap in the spectrum .

While bifurcation can only occur within the spectrum results concerning linear perturbations of Hill's equation tell us that nontrivial solutions can only be expected " over the gaps " !

Theorem 1.1.[8] *Assume that $(\lambda, u) \in \mathbb{R} \times H^1(\mathbb{R})$ is a weak solution of (1.1±) with $u \neq 0$. Then $\lambda \notin \bigcup_i(\mu_i, \lambda_i)$ if $\sigma \geq 2$ or if $0 < \sigma < 2$ and $\int_{-\infty}^{+\infty} a(x)^{2/(2-\sigma)}dx < \infty$.*

The proof is based on a result of Hinton-Shaw [6] concerning linear perturbations

$$-h'' + qh + ph = \lambda h$$

of Hill's equation ; for our application we set $p = \pm r \mid u \mid^\sigma$.

We will derive sufficient conditions such that boundary points of $\sigma(S)$ are infact bifurcation points ; indeed we will show that for equation(1.1+) there is no bifurcation at lower endpoints μ_i of spectral blocks and that there is bifurcation at upper endpoints λ_i into the gaps . Similar results hold for equation (1.1-) .

These results are obtained as a special case of a general operator theoretical approach which allows other applications such as partial differential equations (see [7]) . Further applications illustrating the quality of the abstract bifurcation results are given at the end of a class of explicity solvable equations involving a multiplication operator as the linear part; this example shows that the abstract bifurcation results are sharp.

2. Abstract Setting

We consider nonlinear eigenvalue problems of the form

$$Su \pm N(u) = \lambda u \quad .$$

$$(2.1\pm)$$

Concerning the linear part S we assume that
(H0) $S: D(S) \subseteq H \longmapsto H$ is a positive selfadjoint operator in a real
   Hilbert space H with norm $\| \cdot \|$ and scalar product $\langle \cdot, \cdot \rangle$.

The spectrum of S is denoted by $\sigma(S)$. Then $T := S^{1/2}$ is defined and let H_1 denote the Hilbert space obtained by equipping the domain $D(S^{1/2})$ of T with the scalar product

$$\langle u, v \rangle_1 := \langle u, v \rangle + \langle Tu, Tv \rangle \quad \forall u, v \in H_1 \quad .$$

Identifying H with its dual H^* we have

$$H_1 \subseteq H = H^* \subseteq H_1^* \quad .$$

The injections are continuous and $T : H_1 \mapsto H$ is a bounded linear operator and its dual $T' : H^* = H \mapsto H_1^*$ is also linear and bounded . The operator $T'T : H_1 \mapsto H_1^*$ is a bounded operator that extends S in the sense $D(S) = \{u \in H_1 \mid T'Tu \in H\}$ and $T'Tu = Su$ for all $u \in D(S)$.

The nonlinearity N is assumed to be a gradient operator satisfying the following hypotheses :

(H1)$\exists \varphi \in C^2(H_1, \mathbb{R}), \varphi(0) = 0, \varphi'(0) = 0, \varphi$ even ,convex

(H2)$N = \varphi'$; i.e. $\langle N(u), v \rangle = \varphi'(u)v \ (\forall u, v \in H_1)$

(H3) $N : H_1 \longmapsto H_1^*$ is compact

(H4) $\exists_{1 < \eta \leq \gamma} \forall_{u \in H_1} \quad 0 \leq \eta \langle N(u), u \rangle \leq \langle N'(u)u, u \rangle \leq \gamma \langle N(u), u \rangle$

(H5) $\exists_{C, \delta > 0, \nu \in (0,1)} \forall_{u \in H_1, \|u\|_1 < \delta} \forall_{z \in H_1} \quad \langle N'(u)z, z \rangle \leq C \varphi(u)^\nu \parallel z \parallel_1^2$

We now consider the equation $(2.1\pm)$ in its generalized form

$$T'Tu \pm N(u) = \lambda u \quad (\lambda, u) \in \mathbb{R} \times H_1 \qquad (2.2\pm)$$

and solutions of $(2.2\pm)$ are weak solutions of $(2.1\pm)$.

We want to study bifurcation at boundary points of $\sigma(S)$. For that reason we recall that a spectral block of S is an interval $J = [l, r] \subset \sigma(S)$ such that for some $\varepsilon > 0$, $(l - \varepsilon, r + \varepsilon) \cap \sigma(S) = [l, r]$.

We call $\lambda \in \mathbb{R}$ a gap-bifurcation point if there exist solutions (λ_n, u_n) of $(2.2\pm)$ such that $\lambda_n \in \rho(S) = \mathbb{R} \backslash \sigma(S), u_n \neq 0$ and $\lambda_n \to \lambda, \parallel u_n \parallel_1 \to 0$ $(n \to \infty)$.

The following theorem tells at which boundary points of the spectrum we may expect gap-bifurcation .

Theorem 2.1. *Assume that (H0),...,(H5) hold and let $J = [l, r]$ denote a spectral block with $l < r$. Then*
    (i) l is not a gap-bifurcation point for $(2.2+)$.
    (ii) r is not a gap-bifurcation point for $(2.2-)$.

Remark 2.2. For a proof we refer to [7 ,Thm. 3.1] ; we note however that the proof of (ii) in [7] requires as an additional hypothesis an estimate for φ . The details for the improved version will be given in [5] .

To prove gap-bifurcation at r for (2.2+) and at l for (2.2-) we used the spectral block $J = [l, r]$ to define a generalized Lyapunov-Schmidt-reduction; for details see [7] .

Let $P_J : H \mapsto H$ denote the orthogonal projection onto the part of H associated with the spectral block J of S . Then

$$H = P_J H \oplus (I - P_J)H$$

and $H_J := P_J H$ is a closed subspace of H_1 . Using $v := P_J u, w = (I - P_J)u$ one obtains that

$$T'Tu \pm N(u) = \lambda u$$

is equivalent to

$$0 = T'Tv \pm P'_J N(v + w) - \lambda v \tag{2.3$\pm$}$$

$$0 = T'Tw \pm (I - P'_J)N(v + w) - \lambda w =: F(\lambda, v, w) \tag{2.4$\pm$}$$

As λ can be chosen close to $[l, r]$ it follows that

$$\frac{\partial F}{\partial w}(\lambda, 0, 0) = (T'T - \lambda I)|_{H_1 \cap (I - P_J)H}$$

has a bounded inverse . Applying the Implicit Function Theorem to (2.4$\pm$) one obtains that the solutions of (2.4$\pm$) close to $(\lambda, 0, 0)$ are described by a relation $w = g(\lambda, v)$ with a suitable function g . Hence w can be eliminated and (2.2$\pm$) is locally equivalent to the <u>Reduction Problem</u> for $v \in H_J$, v small

$$(T'T - \lambda I)v \pm P'_J N(v + g(\lambda, v)) = 0 \quad . \tag{2.5$\pm$}$$

It corresponds to the so-called bifurcation equation but in contrast to bifurcation at eigenvalues of finite multiplicity it is still an infinite-dimensional problem .

It is further possible to eliminate $\lambda = q(v)$. Setting $G(v) = g(q(v), v)$ one is lead to a formal equation in v

$$T'Tv \pm P'_J N(v + G(v)) = q(v)v$$

whose solutions are characterized as stationary points of the functional

$$F_\pm(v) := \frac{1}{2} \| Tv \|^2 + \frac{1}{2} \| TG(v) \|^2 \pm \varphi(v + G(v))$$

restricted to the manifold

$$M_c := \{v \in H_J \mid \; \| \, v + G(v) \, \| = c\} \quad .$$

Through the reduction process we are now in a similar situation as in the case of bifurcation from the lowest point of the continuous spectrum . The existence of stationary points can be guaranteed with the aid of test functions in M_c . As M_c and the functional $F_\pm$ are not given explicitly due to the implicit definition of G we need to rely on estimates of G obtained from the implicit function theorem .

The following theorem is a combination of Thms. 5.3 and 5.4 of [7] :

Theorem 2.3. *Assume that (H1),...,(H5) hold and let $J = [l, r]$ denote a spectral block for S . Assume that there exists a sequence of test functions $\{z_n\} \in H_J$ with $\| \, z_n \, \| = 1$ and $\varphi(z_n) \neq 0$. Set $\mu = l$ resp. $\mu = r$. If for* $\delta = [(\sigma + 1)(\sigma + 2) - 2]/[(\sigma + 1)(\sigma + 2) - (\gamma + 1)]$

$$\lim_{n \to \infty} \frac{\| \, T z_n \, \|^2 - \mu}{\varphi(z_n)^\delta} = 0 \quad , \tag{2.6}$$

then μ is a gap-bifurcation point for (2.2-) resp. (2.2+) .

Remark 2.4.
(i) Sometimes it is enough to verify that (2.6) holds for some $\delta > 1$. This has been established if

$$\text{either} \quad (a) \quad \text{we treat only (2.2+)}$$
$$\text{or} \quad (b) \quad N \text{ is a homogeneous function .}$$
$$[\text{By (H4) the degree will be } \gamma = \eta \,]$$

(ii) As stated , Theorem 2.3 is a slight improvement of the results in [7] This modification as well as those mentioned in remark (i) will justified in [5] .

3. Nonlinear pertubations of Hill's equations

We sketch the application of the abstract theorem to nonlinear perturbations of Hill's equation ; for simplicity we restrict our attention to the homogeneous case (1.1±) so that bifurcation at lower resp. upper ends of a spectral block occurs depending on the sign of the nonlinearity.

We set $H = L^2(\mathbb{R}), S$ as in (1.4), $T = S^{1/2}$ and $N(u) := a(x) \mid u \mid^\sigma u.$ The potential of N is given by

$$\varphi(u) = \left(\int_{-\infty}^{\infty} a(x) \mid u \mid^{\sigma+2} (x) dx \right) /(\sigma + 2) \quad .$$

Applying the improved version of Thm.2.3 we obtain

Theorem 3.1. *Assume $\sigma > 0$, $a \in C^0(\mathbb{R})$, $a \not\equiv 0$ and let $[l, r]$ denote a spectral block for S . Then l resp. r is a gap-bifurcation point for (1.1-) resp. (1.1+) if*

$$\text{either} \quad (i) \quad 0 < \sigma < 2$$

$$\text{or} \quad (ii) \quad \exists_{A>0, \tau \in (0,1)} \quad a(x) \geq A(1+ \mid x \mid)^{-\tau} \quad a.e. \text{ on } \mathbb{R} \quad \text{and}$$

$$0 < \sigma < 2(2 - \tau) \quad .$$

The proof requires the verification of (H1),...,(H5) (see [8]) and the choice of suitable test functions in H_J . For details we refer to [8]; the interesting part is the construction of test functions . In [8] test functions have been constructed as eigenpackets of the operator S . They are built up by a family of bounded solutions $u(x, \mu)(\mu \in J)$ of the differential equation

$$-u''(x, \mu) + q(x)u(x, \mu) = \mu u(x, \mu) \quad . \tag{3.1}$$

It has then to be checked that

$$\Phi(\lambda)(x) := \int_l^\lambda u(x, \mu)d\mu \quad \in L^2(\mathbb{R})$$

satisfies the properties of an eigenpacket . The functions $\Phi(\lambda)$ are used to define the test functions .

Another device of test functions which can be used for our purposes as well has been derived by Heinz [4] :

The advantage of Heinz' approach is that it does not involve the construction of test functions in H_J , but simply in H_1 . In application to partial differential equations , such as the periodic Schroedinger equation which is analogue of (3.1) in higher dimensions , we find it easier to follow Heinz' construction [4] .

4. Nonlinear Perturbations of a multiplication operator

The application of the abstract theorems and the construction of suitable test functions is illustrated by the following simple class of problems in the Hilbert space

$$H := L^2(0, 1) \quad \text{with the usual norm} \quad \parallel u \parallel := \int_0^1 u^2(x)dx \quad .$$

The linear part S is given as a multiplication operator :

$$Su(x) := f(x)u(x)$$

where $f \in L^\infty(0,1)$. Then $S : H \mapsto H$ is a bounded selfadjoint operator with spectrum $\sigma(S) = \overline{R(f)}$. To generate a gap in the spectrum we assume that f is of the form

$$f(x) = \begin{bmatrix} f_1(x) & \text{for} & 0 \le x < 1/2 \\ f_2(x) & \text{for} & 1/2 < x \le 1 \end{bmatrix}$$

where

$$f_1 \in C^1[0, 1/2] \quad , f_2 \in C^1[1/2, 1] \quad ,$$
$$f(0) = 0 \quad , \text{f strictly increasing and}$$
$$f_1(1/2) < f_2(1/2) \quad .$$

Then $\sigma(S) = [0, f_1(1/2)] \cup [f_2(1/2), f_2(1)]$.
Further $Tu(x) = \sqrt{f(t)}u(t)$ and $H_1 = H$.
The nonlinearity N is given by

$$N(u) := A^*(\mid Au \mid^\sigma Au)$$

where

$$(Au)(x) := \int_0^1 g(x,y)u(y)dy$$

with $g \in C^0[0,1]^2$ so that $A : H \mapsto H$ is a compact linear operator and Au is continuous on $[0,1]$. In particular , $\parallel Au \parallel_\infty \le \parallel g \parallel_\infty \parallel u \parallel$ and N is well-defined in H .

Then $N \in C^1(H,H)$ is a gradient operator with potential

$$\varphi(u) := \frac{1}{\sigma+2} \int_0^1 \mid Au \mid^{\sigma+2} (x)dx$$

and

$$N'(u)v = (\sigma+1)A^*(\mid Au \mid^\sigma Av) \quad .$$

In addition we assume that $g(x, 1/2) \not\equiv 0$ on $[0,1]$.

Clearly, N satisfies (H1),(H2);(H3) and (H4) holds with $\eta = \gamma = \sigma+1$. Finally (H5) is fulfilled with $\nu = \sigma/(\sigma+2)$ because of :

$$\langle N'(u)z, z \rangle = (\sigma+1)(A^*(\mid Au \mid^\sigma Az), z)$$
$$= (\sigma+1)(\mid Au \mid^\sigma Az, Az)$$
$$= (\sigma+1)\int_0^1 \mid Au \mid^\sigma \mid Az \mid^2 dx$$
$$\le (\sigma+1)\left(\int_0^1 \mid Au \mid^{\sigma+2} dx\right)^{\sigma/(\sigma+2)} \left(\int_0^1 \mid Az \mid^{\sigma+2} dx\right)^{2/(\sigma+2)}$$
$$\le (\sigma+1)(\sigma+2)^\nu \varphi^\nu(u) \left(\int_0^1 \left(\int_0^1 g(x,y)^2 dy\right)^{(\sigma+2)/2}\right)^{\sigma+2} \parallel z \parallel^2$$

Since N is homogenous we need only discuss the case

$$Su - N(u) = \lambda u \quad ,$$

because the equation with the other sign can be transformed to this form. Set $l = f_2(1/2)$ and $r = f_2(1)$. Assume that for some $\alpha \geq 1$ and $h \in C^0[1/2, 1]$ with $h(1/2) > 0$:

$$f_2(x) = l + (x - 1/2)^\alpha h(x) \quad .$$

Application of the improved version of Thm.2.3 gives : There is gap-bifurcation at l if $\sigma < 2(\alpha - 1)$.

Remark.

(i) In the special case $g(x, y) \equiv 1$ the nonlinear eigenvalue problem can be solved explicitly . There exists a unique $\lambda^* \in (f_1(1/2); f_2(1/2))$ such that for $\lambda \in (-\infty, 0) \cup (\lambda^*, f_2(1/2))$ a solution of (4.1) is given by

$$u_\lambda(x) = \frac{(f(x) - \lambda)^{-1}}{\left(\int_0^1 (f(y) - \lambda)^{-1} dy \right)^{(\sigma+1)/\sigma}} \quad .$$

Bifurcation occurs at $l = f_2(1/2)$ iff $0 < \sigma < 2(\alpha - 1)$.

(ii) As another illustration take g as the Green's function to a regular differential operator L of the order m together with boundary conditions $B_i = 0$ $(i = 1, ..., m)$. Setting $Au = v$ and $w = A(\mid v \mid^\sigma v)$ leads to the singular system

$$\begin{aligned}
Lv &= w/(f(t) - \lambda) \\
Lw &= \mid v \mid^\sigma v \\
B_i[v] &= B_i[w] = 0 \quad (i = 1, ..., m) \quad .
\end{aligned}$$

Proof. 1) We have to construct a suitable sequence of test functions z_n . For $J = [l, r]$ we obtain

$$H_J = \{h \in H \mid h(x) = 0 \quad \text{a.e. in } (0, 1/2)\}$$

Set $z_n(x) := \sqrt{n}$ $(1/2 \leq x \leq \varepsilon_n := 1/2 + 1/n)$ and $z_n(x) = 0$ else . Then
(i)$z_n \in H_J$ and $\parallel z_n \parallel = 1$.
(ii)$\parallel Tz_n \parallel^2 = \int_{1/2}^{\varepsilon_n} f(x) z_n^2(x) dx$, hence

$$0 \leq \parallel Tz_n \parallel^2 - l \leq \int_{1/2}^{\varepsilon_n} (x - 1/2)^\alpha h(x) dx = n^{-\alpha} \int_0^1 t^\alpha h(1/2 + t/n) dt \quad .$$

(iii)

$$\varphi(z_n) = \int_0^1 | \int_0^1 g(x,y)z_n(y)dy |^{\sigma+2} dx/(\sigma+2)$$

$$= n^{(\sigma+2)/2} \int_0^1 | \int_{1/2}^{\varepsilon_n} g(x,y)dy |^{\sigma+2} dx/(\sigma+2)$$

$$= n^{(\sigma+2)/2} \left[\frac{1}{n} \int_0^1 | g(x,1/2) |^{\sigma+2} dx + O(1/n^2) \right]^{\sigma+2} /(\sigma+2)$$

$$\geq Cn^{-(\sigma+2)/2} \quad \text{for some } C > 0 \quad .$$

Hence

$$0 \leq \frac{\| Tz_n \|^2 - l}{\varphi(z_n)^\delta}$$

$$\leq \frac{n^{-\alpha} \int_0^1 t^\alpha h(1/2 + t/n)dt}{C^\delta n^{(-(\sigma+2)/2)\delta}}$$

$$\leq \tilde{C} n^{-\alpha+((\sigma+2)/2)\delta}$$

and bifurcation occurs if $2(\alpha - 1) > \sigma$.

2) The special case $g \equiv 1$ corresponds to the equation

$$f(x)u(x) - | \int_0^1 u(y)dy |^\sigma \int_0^1 u(y)dy = \lambda u(x) \quad . \tag{4.1}$$

Since $N(u)$ is constant for all $u \in H$, u must satisfy

$$u(x) = \frac{c}{f(x) - \lambda}$$

for some constant c . The value of λ must be such that $d_\lambda := (f(x)-\lambda)^{-1} \in H$ and c must be chosen so that (4.1) is satisfied . Hence

$$c = | \int_0^1 d_\lambda(y)dy |^\sigma \int_0^1 d_\lambda(y)dy \, | c |^\sigma c$$

i.e.

$$| c |^\sigma = \left(| \int_0^1 d_\lambda(y)dy |^\sigma \int_0^1 d_\lambda(y)dy \right)^{-1}$$

provided $\int_0^1 d_\lambda(y)dy \neq 0$. Thus for equation (4.1) we have solutions at λ iff

$$a) \quad \int_0^1 (f(x) - \lambda)^{-2}dx < \infty \quad \text{and}$$

$$b) \quad \int_0^1 (f(x) - \lambda)^{-1}dx > 0 \quad .$$

A Comparison Result and Elliptic Equations Involving Subcritical Exponents

MAN KAM KWONG

Abstract

It is well-known that good bounds for solutions of nonlinear differential equations are difficult to obtain. In this paper, we establish a theorem comparing non-negative solutions (having identical initial values) of the equations $u''(t)+q(t)u^p(t)+r(t)u(t) = 0$ and $v''(t)+k(t)q(t)v^p(t)+r(t)u(t) = 0$, respectively. If $q(t), r(t) \geq 0$, $k(t) \geq 1$, $k(t)$ is non-decreasing, and the first equation satisfies a certain uniqueness criterion, our result asserts that $u(t) \geq v(t)$. Both the uniqueness assumption on the equation and the monotonicity requirement on $k(t)$ are necessary. A particular case of this theorem plays a central role in a recent paper of Atkinson and Peletier in the study of asymptotic behavior of nonlinear elliptic equations involving a critical exponent. A simple corollary of our result provides information on the same type of equations with subcritical exponents.

1. Introduction

The celebrated Sturm comparison theorem is a useful tool for obtaining bounds for linear second-order ordinary differential equations. Suppose that $u(t)$ and $v(t)$ are, respe, non-negative solutions of the equations

$$u''(t) + q(t)u(t) = 0, \tag{1.1}$$

and

$$v''(t) + Q(t)v(t) = 0, \tag{1.2}$$

on (a, b), and they satisfy the same initial conditions

$$u(a) = v(a) \geq 0, \quad u'(a) = v'(a). \tag{1.3}$$

I am grateful to Professor Atkinson and Professor Peletier for drawing my attention to their result. The present work is a direct response to their suggestion to investigate whether a more general comparison theorem is possible. In this paper the following main theorem is established. The concept of the *"uniqueness condition"* will be made precise in the next section. A reflection (replacing t by $-t$) has been executed, so that instead of requiring $u(t)$ and $v(t)$ to agree at some terminal "point", namely ∞, we make them agree at an initial point $t = a$.

Main Theorem *Let $u(t)$ be a positive solution of*

$$u''(t) + q(t)u^p(t) + r(t)u(t) = 0, \quad t \in (a,b) \tag{1.20}$$

where $q(t), r(t) \geq 0$, and $p > 1$. Furthermore, suppose that (1.20) satisfies a "uniqueness condition" for boundary value problems on subintervals of (a,b). Let $k(t) \geq 1$ be any increasing function of t. Then the solution $v(t)$ of the equation

$$v''(t) + k(t)q(t)v^p(t) + r(t)v(t) = 0, \tag{1.21}$$

satisfying the same initial conditions as those of $u(t)$ at $t = a$, is smaller than $u(t)$, before $v(t)$ vanishes for the first time.

This theorem includes the lemma of Atkinson and Peletier. The approach adopted is completely different from theirs. There are two major extensions. First, a linear term is included, and more general coefficients for the nonlinear term other than $1/t^k$ are allowed. Even in the particular case when this coefficient is a power of t, this power does not have to be tied to one special exponent of u as in the lemma in [2]. Second, the two solutions, $u(t)$ and $v(t)$, can be compared starting from any initial point $t = a$, finite or not.

The exponent $(2k-3)$ that appears in (1.18) is the well-known (Sobolev) critical exponent for the Emden-Fowler equation. The dynamical behavior of the solutions changes radically as the exponent increases from the subcritical to the supercritical case. There is therefore much interest in attempting to extend the work of Atkinson and Peletier to include both the subcritical and supercritical cases. Our main theorem confirms that in these noncritical cases a lemma analogous to that of Atkinson and Peletier holds.

For equations with non-critical exponents, the solutions satisfying the asymptotic condition (1.19) (or even solutions of the critical exponent case satisfying other initial conditions) no longer have a simple closed form. As the proof of Atkinson and Peletier relies implicitly on such formulas, there seems to be no easy way to extend it to such cases.

In this paper, the main theorem is deduced from a special case, in which the function $k(t)$ is a constant. Our method makes extensive use of the Sturm comparison theorem and is closed related to a method first used by Coffman [6,7] to obtain uniqueness results for boundary value problems. His ideas have been successfully applied by Ni [11], Ni and Nussbaum [13], McLeod and Serrin [10], and Kwong [8]. For a survey of the method and known results, see the survey articles [9,12].

In Section 3 we show how the lemma of Atkinson and Peletier and its generalizations to non-critical exponents follow from the preceding theorem. An application to nonlinear elliptic equations involving subcritical exponents gives results analogous to those of Brezis and Nirenberg for the critical exponent case. These results have been obtained previously using the variational approach, see for example, Ambrosetti and Rabinowitz [1]. We expect that our main theorem will also play an important role in the detailed analysis of the structure of the solution space in the supercritical exponent case.

2. Main Results

We are interested in comparing the positive solutions of the Emden-Fowler equation

$$u''(t) + q(t)u^p(t) + r(t)u(t) = 0, \quad t \in (a, b) \tag{2.1}$$

with those of a similar one having a larger coefficient. We assume that $q(t)$ and $r(t)$ are piecewise continuous and

$$q(t) \geq 0, \ r(t) \geq 0, \ p > 1. \tag{2.2}$$

The interval (a, b) can be compact or otherwise, i.e., $-\infty \leq a < b \leq \infty$. However, for most of our discussion, it is assumed to be compact.

We may assume without loss of generality that $q(t)$ is not identically zero in any right neighborhood of the left endpoint $t = a$. In the contrary case, we can simply bypass such a neighborhood by shifting the left endpoint over it without affecting the validity of our main result. The requirement that $p > 1$ puts the equation (2.1) in the superlinear category. For such equations it is known that if the solution has either a sufficiently large initial height or a sufficiently large initial slope, it must have a zero close to the initial point a. This is one of the important facts used in the shooting method; see, for example, Bandle and Kwong [3].

Proof. The proof consists of a continuity argument making use of elementary topological properties of the plane. We do not insist on absolute rigor while presenting the proof. Theoretical details can be easily filled in.

Let us first look at the case in which B is a zero of one of the solutions, say $u_1(B) = 0$. Then $u_2(B) > 0$ because by (U) there cannot be two solutions that have the same boundary condition at B. At $t = a$, either $u_2(a) > u_1(a)$ or $u_2(a) < u_1(a)$. We use a shooting method argument to show that the first case is vacuous. By keeping the ratio $u'(a)/u(a) = s$, and increasing $u(a)$ starting from $u_2(a)$, we can shoot out various solutions. In case the ratio $u_1(a) = u_2(a) = 0$ is zero, the solutions $u(t)$ starts out with $u(a) = 0$, but with progressively increasing initial slope $u'(a)$. In the following we will not point out this modification explicitly. The value $u(B)$ depends continuously on the initial height. By (U), $u(B)$ cannot vanish, so it must remain positive for all initial height $u(a)$. On the other hand, superlinearity implies that for $u(a)$ large enough, the solution must have a zero in (a, B). Pick the first initial height at which this happens. Because $u(B) > 0$, the solution can be tangent to the t-axis only at this zero, but this is impossible. It follows that the first case, $u_2(a) > u_1(a)$ is empty, as claimed.

In the second case, the two solutions must intersect. Suppose they do so more than once. We shoot out solutions as before but with progressively decreasing initial height. Since $u(B)$ remains positive and two solutions of (2.1) cannot be tangential at any point, the number of points of intersection of $u(t)$ with $u_1(t)$ has to be a constant; in particular, it is greater than one. It is geometrically obvious that as $u(a)$ decreases towards 0, the first intersection point of $u(t)$ with $u_1(t)$ approaches the endpoint B. In other words, if $u(a)$ is sufficiently small, all the intersections occur within a very small right neighborhood of B. Let $W(t) = u(t) - u_1(t)$. Then $W(t)$ changes sign (oscillates) more than once in this neighborhood. The function satisfies the second-order "linear" differential equation

$$W''(t) + \left[\frac{q(t)[u^p(t) - u_1^p(t)]}{u(t) - u_1(t)} + r(t)\right] W(t) = 0. \qquad (2.19)$$

Observe that the "coefficient" in this equation, the expression enclosed in large square brackets, is a bounded function in this neighborhood, and it is well known that solutions of such equations cannot oscillate in an arbitrarily small interval. We therefore have a contradiction.

Now suppose that $B = b$ and that both $u_1(b)$ and $u_2(b)$ are positive. We may assume that $u_1(a) < u_2(a)$. We first consider the case where $u_1(b) \leq u_2(b)$. If the two solutions do not intersect, we have nothing to prove. So suppose they do. As a consequence, part of the graph of $u_2(t)$ lies below that of $u_1(t)$. We will show that there are two other solutions that satisfy the same boundary conditions as those of $u_1(t)$, thus contradicting (U). If $u_1(b) = u_2(b)$, the first solution is simply $u_2(t)$. In the contrary case, the first solution is obtained by shooting with initial height above $u_2(a)$. As pointed out before, if $u(a)$ is large enough, the solution must have a zero near the left endpoint. So as $u(a)$ increases from $u_2(a)$, the other end of the curve $u(b)$ must eventually come down and pass through $u_1(b)$, giving the first solution. Next, since both $u_1(t)$ and $u_2(t)$ are bounded away from zero, we see that if we shoot with sufficiently small initial height, the solution will remain small and so will not cross either function. As we gradually increase the initial height, there must be a first time when the solution $u(t)$ intersects one of these given solutions. By the choice of this critical case, the graphs of $u_1(t)$ and $u_2(t)$ must lie entirely above that of $u(t)$. This rules out the possibility that $u(t)$ coincides with $u_1(t)$ since, by assumption, part of the graph of $u_2(t)$ lies below that of the $u_1(t)$. If a point of intersection of $u(t)$ with the other two functions is an interior point of the interval (a, b), then $u(t)$ must be tangential to the solution that it intersects at this point. This contradicts the uniqueness of initial value problems. So the only possibility left is that $u(t)$ intersects $u_1(t)$ at b, giving the second solution we need.

Now consider the case $u_1(b) > u_2(b)$. If the two solutions intersect more than once, they must do so at least three times (recall that solutions cannot be tangent to each other). Let us shoot out solutions as before with increasing initial heights starting from $u_2(a)$, and follow their terminal values $u(b)$. We know that eventually $u(b)$ must hit the t-axis, but before it does so, it may or may not pass through the point $u_1(b)$. Suppose it does. At the moment when this first happens, we are back to the previous case. Note that the number of times $u(t)$ intersects $u_1(t)$ remains a constant during this continuous deformation of $u(t)$ (as $u(a)$ is increased). Suppose that $u(b)$ does not pass through $u_1(b)$ before it hits the t-axis. At the moment when $u(b)$ first becomes zero, we have the very first situation at the beginning of the proof. In all cases, we have a contradiction, and so the proof is complete. ∎

We are now ready to show that the converse of Lemma 1 holds.

LEMMA 4 *Suppose that (2.1) satisfies (U) with respect to s. Let $U(t)$ be a positive solution of (2.1) in (a, b) such that $U'(a)/U(a) = s$. Then for any $\lambda > 1$, the solution $v(t; \lambda)$ of (2.3) and (2.4) satisfies*

$$v(t; \lambda) \leq U(t) \tag{2.20}$$

for all t before the first zero of $v(t)$.

Proof. As in the proof of Lemma 1, we define $u(t;\alpha) = \alpha v(t;\lambda)$, $\alpha = \lambda^{1/(p-1)}$, by scaling the function $v(t;\lambda)$. Let B be the first zero of $v(t;\lambda)$ or b if $v(t;\lambda)$ does not vanish. If $u(t;\alpha)$ intersects $U(t)$ in $(a, B]$, it does so at a unique point, by Lemma 3. Denote this point by C. If the two graphs do not intersect, we take $C = B$. Since $\lambda > 1$, the scaling is a stretching and $u(a;\alpha) > U(a)$. It follows that in (C, B), $v(t;\lambda) < u(t;\alpha) < U(t)$. It remains to show that $v(t;\lambda) < U(t)$ in (a, C) as well.

Let $z(t) = U(t) - v(t;\lambda)$. It satisfies the differential equation

$$z''(t) + q(t)\left[\frac{U^p(t) - v^p(t;\lambda)}{U(t) - v(t;\lambda)}\right] z(t) + r(t)z(t) = (\lambda - 1)q(t)v(t,\lambda). \quad (2.21)$$

It also satisfies the initial conditions

$$z(a) = z'(a) = 0. \quad (2.22)$$

Note that the righthand side of equation (2.21) is positive in (a, C). Using the variation of parameter formula it is easy to see that in some neighborhood of the endpoint $t = a$, $z(t)$ is positive. If positivity prevails throughout (a, C), the proof is complete. So suppose the contrary, and let $D < C$ be the first point at which $z(D) = 0$. We compare equation (2.21) with the one satisfied by $w(t) = u(t;\alpha) - U(t)$,

$$w''(t) + q(t)\left[\frac{u^p(t;\alpha) - U^p(t)}{u(t;\alpha) - U(t)}\right] w(t) + r(t)w(t) = 0. \quad (2.23)$$

In (a, D), $v(t;\lambda) \leq U(t) \leq u(t;\alpha)$. This implies that the expression inside the square brackets in (2.23) is larger than the corresponding expression in (2.21). Rewriting (2.21) in homogeneous form:

$$z''(t) + q(t)\left[\frac{U^p(t) - v^p(t;\lambda)}{U(t) - v(t;\lambda)} - \frac{(\lambda - 1)q(t)v(t,\lambda)}{z(t)}\right] z(t) + r(t)z(t) = 0, \quad (2.24)$$

we see that it has a coefficient smaller than that of (2.23). Hence $w(t)$ oscillates faster than $z(t)$ in $[a, D]$. This contradicts the fact that $w(t)$ has no zero in (a, D) whereas $z(t)$ vanishes at both endpoints a and D. The proof of the lemma is now complete. $\blacksquare$

So far we have been considering only a finite left endpoint $a \geq -\infty$. When $a = -\infty$, some technical details have to be added. Since we are comparing only essentially positive solutions, we can include only those equations (2.1) that admit a nonoscillatory solution near $-\infty$. Because of the concavity of $u(t)$, $u'(t)$ approaches a finite non-negative limit as $t \to -\infty$. The value $\lim_{t \to -\infty} u(t)$ may or may not be bounded. In any case, the boundary conditions (2.4) have to be interpreted as asymptotic relations. We can get around this difficulty by requiring that the solution v of (2.3), or of (2.25) below, can be approximated by solutions satisfying (2.4) for finite values of a as we let $a \to -\infty$.

THEOREM 1 *Let $u(t)$ be a positive solution of (2.1) in (a, b), $-\infty < a < b \leq \infty$, where $q(t) \geq 0$, and $p > 1$, and suppose that condition (U) with respect to $s = u'(a)/u(a)$ is satisfied. Let $k(t) \geq 1$ be any increasing function of t. Then the solution $v(t)$ of the equation*

$$v''(t) + k(t)q(t)v^p(t) + r(t)v(t) = 0, \tag{2.25}$$

such that

$$v(a) = u(a), \quad v'(a) = u'(a), \tag{2.26}$$

satisfies

$$v(t) \leq u(t), \quad t \in (a, B), \tag{2.27}$$

where B is the first zero of $v(t)$ or b if $v(t)$ does not vanish.

Now if $u(t)$ is a positive solution of (2.1) on $(-\infty, b)$, and if the solution $v_a(t)$ of (2.25) satisfying (2.26) for some finite $a < b$ converges uniformly to a solution $v(t)$ of (2.25) as $a \to -\infty$, then (2.27) holds in $(-\infty, B)$.

Proof. The result for an unbounded interval follows from that for compact intervals using a continuity argument. It also suffices to assume that $k(t)$ is a step function. The general case is obtained by taking limits. The proof of the theorem in the reduced case is done most easily by induction on the number of steps of $k(t)$. By leveling the last step of $k(t)$ with the previous one, we obtain a function $k_n(t)$ with one less step. The original $k(t)$ can easily be recovered as the product of $k_n(t)$ and a two-step function. Let $v_n(t)$ be the solution of the equation of the form (2.25) with $k(t)$ replaced by $k_n(t)$. By applying Lemma 4 to the interval of the last step of $k(t)$, we see that $v(t) \leq v_n(t)$. By the induction hypothesis, $v_n(t) \leq u(t)$. So the conclusion of the theorem follows. ∎

We remark that Theorem 1 is no longer true if the monotonicity hypothesis on $k(t)$ is simply removed. Numerical experimentation quickly yields the following example:

$$k(t) = \begin{cases} 2 & t \leq 1 \\ 1 & t > 1 \end{cases}, \tag{2.28}$$

$$u''(t) + u^3(t) = 0, \quad t \in (0,4), \tag{2.29}$$

$$v''(t) + k(t)v^3(t) = 0, \quad t \in (0,4), \tag{2.30}$$

$$u(0) = v(0) = 0, \quad u'(0) = v'(0) = 0.5. \tag{2.31}$$

The solution $v(t)$ intersects $u(t)$ at approximately $t = 3.143$ and remains larger than $u(t)$ after that.

3. The Atkinson and Peletier Lemma and Subcritical Equations

The following simple consequence of Theorem 1 covers the lemma of Atkinson and Peletier. To see this, we first have to make a reflection, replacing t by $-t$, and then choose $a = -\infty$, $q(t) = t^{-k}$ and $r(t) = 0$. Their condition (1.16) on the function $f(u)$ is restated in an equivalent form which makes the proof more transparent. A differentiation shows that (1.16) holds if and only if $f(u)/u^{2k-3}$ is non-increasing.

THEOREM 2 *Let $p > 1$ and $f(u)$ be a C^1 function for $u > 0$ satisfying the condition*

$$\frac{f(u)}{u^p} \text{ is a non-increasing function of } u. \tag{3.1}$$

Suppose that (2.1) satisfies the uniqueness condition (U) and $u(t)$ is a positive solution of (2.1) in (a,b) such that

$$u'(a) \leq 0. \tag{3.2}$$

Then the solution of the initial value problem

$$v''(t) + q(t)f(v(t)) + r(t)v(t) = 0, \tag{3.3}$$

$$v(a) = u(a), \quad v'(a) = u'(a), \tag{3.4}$$

satisfies the inequality

$$v(t) \leq u(t), \tag{3.5}$$

before $v(t)$ changes sign.

Proof. The boundary condition (3.2) implies that $v(t)$ is decreasing in t in the interval (a, B). By (3.1), the function $f(v(t))/v^p(t)$ is therefore an increasing function of t in (a, B). Rewriting (3.3) as

$$v''(t) + \frac{f(v(t))}{v^p(t)}q(t)v^p(t) + r(t)u(t) = 0, \tag{3.6}$$

we bring it into the form of (2.25), with $k(t) - f(v(t))/v^p(t)$. Theorem 1 now applies. ∎

As pointed out in [2], all functions of the form $f(u) = u^p + \lambda u^q$, or more generally $f(u) = \sum \lambda_i u^{q_i}$, with $0 < q, q_i < p, \lambda, \lambda_i > 0$ satisfy (3.1).

Let Ω denote the unit ball in R^n. Brezis and Nirenberg in [5] studied the nonlinear eigenvalue problem

$$\Delta u + u^p + \lambda u^q = 0 \quad \text{in } \Omega, \tag{3.7}$$

with Dirichlet boundary condition

$$u = 0 \quad \text{on } \partial\Omega, \tag{3.8}$$

where $1 < q < p = \dfrac{n+2}{n-2}$. They obtained, using variational techniques, necessary and sufficient conditions on the value of λ for the existence of a solution. What is most interesting is that the necessary and sufficient range depends on the value of q as well as on n. Two or three distinct cases can be distinguished according to whether $n \geq 4$ or $n = 3$. For all values of n, the cutoff value of q for the first case is $n/(n-2)$, whereas for $n = 3$, the value $q = 1$ is in a separate category by itself.

An alternative approach using ordinary differential equation methods was adopted in [2] to reconfirm and to refine the results of Brezis and Nirenberg. It is known that any solution of the eigenvalue problem must be radially symmetric. Thus we are really dealing with an ordinary differential equation. A scaling in the independent variable further changes the problem to the equivalent one of studying the location of the first zero of the solutions of the initial value problem

$$u''(t) + \frac{n-1}{t}u'(t) + u^p(t) + u^q(t) = 0, \tag{3.9}$$

$$u(0) = \alpha, \quad u'(0) = 0. \tag{3.10}$$

As the initial height α varies from 0 to ∞, the first zero is tracked; the range of possible locations is related to the range of possible eigenvalues of (3.7).

We remark that the method presented below works without change for the more general equation

$$\Delta u + u^p + \sum_{i=1}^{N} c_i u_i^q + \lambda u^q = 0 \quad \text{in } \Omega, \tag{3.11}$$

where $c_i > 0$, and $q < q_i < p$, and the equivalent scaled equation

$$u''(t) + \frac{n-1}{t}u'(t) + u^p(t) + \sum_{i=1}^{N} C_i u_i^q + u^q(t) = 0. \qquad (3.12)$$

It is natural to ask what the corresponding result of Brezis and Nirenberg is in cases where the main exponent p is non-critical. The variational approach has been successfully applied to the subcritical case to answer this question. See, for example, the work of Ambrosetti and Rabinowitz [1]. The shooting method of Atkinson and Peletier provides an alternative. In fact, in the case of p being subcritical, the extension of their key lemma as given by Theorem 2 is all we need. As our next result shows, the corresponding Brezis and Nirenberg result is simpler, comprising always two cases, $q = 1$ and $q > 1$.

THEOREM 3 *Suppose* $1 \leq q < p < \dfrac{n+2}{n-2}$. *If* $q > 1$, *the set of the first zero of all solutions of* (3.9) *(more generally,* (3.9′)*) and* (3.10) *is* $(0, \infty)$. *Equivalently, the eigenvalue problem* (3.7) *(more generally,* (3.7′)*) and* (3.8) *has a solution for all* $\lambda > 0$.

If $q = 1$, *the set of the first zero of all solutions of* (3.9) *(more generally,* (3.9′)*) and* (3.10) *is* $(0, T)$ *where* T *is the first zero of the solution of the linear initial value problem*

$$\phi''(t) + \frac{n-1}{t}\phi'(t) + \phi(t) = 0, \qquad (3.13)$$

$$\phi(0) = 1, \quad \phi'(0) = 0. \qquad (3.14)$$

Equivalently, the eigenvalue problem (3.7) *(or* (3.7′)*) and* (3.8) *has a solution for all* $\lambda \in (0, T^2)$.

Proof. To emphasize the fact that the solution to (3.9) and (3.10) depends on the initial shooting height α, we write it as $u(t; \alpha)$. Its first zero is therefore also a function of α; we denote it by $b(\alpha)$. Let us first show that in all cases

$$\lim_{\alpha \to \infty} b(\alpha) = 0. \qquad (3.15)$$

Let $U(t; \alpha)$ denote the solution of the initial value problem

$$U''(t; \alpha) + \frac{n-1}{t}U'(t; \alpha) + U^p(t; \alpha) = 0, \qquad (3.16)$$

$$U(0) = \alpha, \quad U'(0) = 0, \tag{3.17}$$

and $B(\alpha)$ the first zero of $U(t; \alpha)$. Using the well-known Emden transform, we can rewrite (3.9) and (3.16) in a form similar to (1.13), for which we can apply Theorem 2 to conclude that

$$u(t; \alpha) \le U(t, \alpha) \quad \text{for all } t \in (0, b(\alpha)). \tag{3.18}$$

As a consequence,

$$b(\alpha) \le B(\alpha). \tag{3.19}$$

Incidentally this establishes the fact that all solutions of (3.9) and (3.10) must have a finite zero; in other words $b(\alpha) < \infty$. Now (3.15) follows if we can show that

$$\lim_{\alpha \to \infty} B(\alpha) = 0. \tag{3.20}$$

This is a well-known fact, since $U(t; \alpha)$ can be obtained from the special case $U(t; 1)$ by scaling, namely,

$$U(t; \alpha) = \alpha U(\alpha^{(p-1)/2} t; 1), \tag{3.21}$$

and

$$B(\alpha) = \frac{B(1)}{\alpha^{(p-1)/2}}. \tag{3.22}$$

Next let us show that for $q > 1$,

$$\lim_{\alpha \to 0} b(\alpha) = \infty. \tag{3.23}$$

We exploit the method of scaling again. Define

$$v(t; \alpha) = \frac{1}{\alpha} u\left(\frac{t}{\alpha^{(q-1)/2}}; \alpha\right). \tag{3.24}$$

Then $v(t; \alpha)$ satisfies the initial value problem

$$v''(t; \alpha) + \frac{n-1}{t} v'(t; \alpha) + \alpha^{p-q} v^p(t; \alpha) + v^q(t; \alpha) = 0, \tag{3.25}$$

$$v(0) = 1, \quad v'(0) = 0. \tag{3.26}$$

Note that as $\alpha \to 0$, the coefficient of the term v^p in (3.25) goes to zero. Thus by continuity, $v(t; \alpha)$ converges uniformly to $U(t; 1)$ in any finite interval. The first zero of $v(t; \alpha)$ therefore approaches $B(1)$. It follows that $b(\alpha)$, being the first zero of $v(t; \alpha)$ divided by $\alpha^{(q-1)/2}$, approaches ∞ as $\alpha \to 0$.

By (3.15) and (3.23), the set of $b(\alpha)$ contains arbitrarily large and arbitrarily small positive values. By connectedness, the set must therefore be $(0, \infty)$.

Now let us look at the case $q = 1$. In view of (3.15), it remain to show that $b(\alpha)$ is always less than T, but can be arbitrarily close to T. Writing (3.9) in the form of a "linear" equation:

$$u''(t) + \frac{n-1}{t}u'(t) + \left[u^{p-1}(t) + 1\right] u(t) = 0, \qquad (3.27)$$

we see that it oscillates more than (3.16), since the coefficient of the last term in (3.27) is larger than the corresponding coefficient of the last term in (3.16). Hence the first zero of $u(t; \alpha)$ is strictly less than the first zero of $U(t)$; in other words, $b(\alpha) < T$. On the other hand, $u(t; \alpha) \leq \alpha$ in $(0, b(\alpha))$ so that if α is very small, the first term of the expression inside the square brackets in (3.27) is very small, say less than some $\epsilon > 0$. Hence it oscillates slower than the equation

$$w''(t) + \frac{n-1}{t}w'(t) + \left[\epsilon + 1\right] w(t) = 0, \qquad (3.28)$$

implying that $b(\alpha)$ is larger than the first zero of $w(t)$ (which is assumed to satisfy the initial conditions $w(0) = 1$ and $w'(0) = 0$). Since the first zero of $w(t)$ tends to T as $\epsilon \to 0$, so does $b(\alpha)$. This completes the proof of the theorem. ∎

After establishing existence, the next natural question to ask is how many solutions there are for each given λ or $b(\alpha)$. Results of Ni [11] and Ni and Nussbaum [13] imply uniqueness when $1 \leq q < p < \dfrac{n}{n-2}$. The situation is more complicated for larger values of p. So far most of the knowledge is derived from numerical computation. The number of solutions can be read from the graph plotting α against λ or $b(\alpha)$, the so-called bifurcation diagram. Atkinson and Peletier [2] proved that when $2 < n < 4$, p is critical, and $1 < q < (6-n)/(n-2)$, there are at least two solutions for each large $b(\alpha)$. The bifurcation curve looks like one branch of a hyperbola with the horizontal axis as one of its asymptotes, and the other end of the curve runs off to infinity at the top right-hand corner. Theorem 3 shows that if p is decreased from the critical value, this end of the curve will approach the vertical axis $b = 0$ instead. Hence if p is sufficiently close to the critical value, the bifurcation curve will maintain the hyperbolic shape of the curve for the critical exponent case within a bounded region, but the upper portion of the curve will, for large α, be bent back towards the vertical axis. This creates (at least) a doublefold curve. It follows that there are values of $b(\alpha)$ or λ for which there are at least three distinct solutions.

A similar phenomenon was observed by Ni and Nussbaum [13] when p is supercritical and q is subcritical. As p is further reduced, numerical evidence indicates that the doublefold curve gradually unfolds; below a certain threshold, depending on q, the bifurcation curve becomes strictly monotone, and uniqueness for the boundary value problems for all $b(\alpha)$ is regained. It will be interesting to see whether these facts can be verified theoretically and the threshold value can be determined.

Acknowledgment

This work was supported by the Applied Mathematical Sciences subprogram of the Office of Energy Research, U.S. Department of Energy, under Contract W-31-109-Eng-38.

REFERENCES

[1] Ambrosetti, A. and Rabinowitz, P. H., *Dual variational methods in critical point theory and applications*, J. Funct. Anal., 14 (1973), 349-381.

[2] Atkinson, F. V., and Peletier, L. A., *Emden-Fowler equations involving critical exponents*, Nonlinear Analysis, 10 (1986), 755-776.

[3] Bandle, C., and Kwong, Man Kam, *Semilinear elliptic problems in annular domains*, ZAMP (J. of Applied Math. and Phy.), 40 (1989), 245-257.

[4] Bellman, R., *Stability Theory of Differential Equations*, McGraw-Hill, New York, 1953.

[5] Brezis, H., and Nirenberg, L., *Positive solutions of nonlinear elliptic equations involving critical Sobolev exponents*, Comm. Pure Appl. Math., 36 (1983), 437-477.

[6] Coffman, C. V., *On the positive solutions of boundary value problems for a class of nonlinear differential equations*, J. Diff. Eq., 3 (1967), 92-111.

[7] Coffman, C. V., *Uniqueness of the ground state solution for $\Delta u - u + u^3 = 0$ and a variational characterization of other solutions*, Arch. Rational Mech. Analysis, 46 (1972), 81-95.

[8] Kwong, Man Kam, *Uniqueness of positive solutions of* $\Delta u - u + u^p = 0$ *in* R^n, Arch. Rational Mech. Anal., 105 (1089), 243-266.

[9] Kwong, Man Kam, *On the Kolodner-Coffman method for the uniqueness problem of Emden-Fowler BVP*, Preprint MCS-P44-0189, Mathematics and Computer Science Division, Argonne National Laboratory, 1989.

[10] McLeod, K., and Serrin, J., *Uniqueness of positive radial solutions of* $\Delta u + f(u) = 0$ *in* R^n, Arch. Rational Mech. Anal., 99 (1087), 115-145.

[11] Ni, W. M., *Uniqueness of solutions of nonlinear Dirichlet problems*, J. Diff. Eq., 50 (1983), 289-304.

[12] Ni, W. M., *Uniqueness, nonuniqueness and related questions of nonlinear elliptic and parabolic equations*, Nonlinear Functional Analysis and Its Applications, Part 2 (Berkeley, California, 1983), 219-228.

[13] Ni, W. M., and Nussbaum, R., *Uniqueness and nonuniqueness for positive radial solutions of* $\Delta u + f(u, r) = 0$, Comm. Pure and Appl. Math., 38 (1985), 69-108.

[14] Wong, James S. W., *On the generalized Emden-Fowler equation*, SIAM Review, 17 (1975), 339-360.

Advances in Quenching

HOWARD A. LEVINE

Introduction

In this paper we shall survey the literature on the so called quenching problem since 1985, when the last survey on the subject appeared [18]. We shall also present some open problems which have arisen in consequence of the results of the recent literature.

Let us first recall the old result of Kawarada [16]. He considered the initial-boundary value problem (in our notation).

$$
\begin{aligned}
& u_t = u_{xx} + \epsilon/(1-u) \qquad 0 < x < 1, \qquad t > 0, \qquad \epsilon > 0 \\
\text{(K)} \quad & u(0,t) = u(1,t) = 0 \hspace{4.5cm} t > 0 \\
& u(x,0) = u_0(x) \equiv 0 \hspace{2.8cm} 0 \le x \le 1
\end{aligned}
$$

For (K) he proved the following:

(a) If $\epsilon > 8$, there is $T(\epsilon) < \infty$ such that $\displaystyle\lim_{t \to T^-} u(\tfrac{1}{2},t) = 1$,

(b) Whenever (a) holds, $\displaystyle\lim_{t \to T^-} \max_x u_t(x,t) = +\infty$

He says that whenever (a), (b) occur, u "quenches". Several authors have noted that the proof of (b) appears to be incomplete [1,8]. We shall say more about this question later.

The literature prior to 1985 is mostly concerned with various extensions of Kawarada's result to other classes of problems, both parabolic and hyperbolic. These problems fall into four classes.

I. Multidimensional Parabolic Problems

Ia. Singular nonlinearity in the equation. A typical problem of this type would be

$$
\begin{aligned}
u_t &= \Delta_N u + \varphi(u) && \text{in} \quad \Omega \times (0,T) \\
u &= 0 && \text{on} \quad \partial\Omega \times (0,T) \\
u(x,0) &= u_0(x) \ge 0 \quad, \quad u_0 < 1 && \text{on} \quad \Omega
\end{aligned}
$$

where $\Omega \subset R^N$ is bounded, $\varphi(u) \sim \epsilon(1-u)^{-\beta}$ for some $\beta > 0$ when u is near 1, and $\epsilon > 0$. (Δ_N is the N dimensional Laplace operator.)

Supported in part by NSF Grant DMS - 8822788.

Ib.   Singular nonlinearity in a boundary condition. An example of such a problem is

$$
\begin{aligned}
u_t &= \Delta_N u && \text{in} \quad \Omega \times (0, T) \\
u &= 0 && \text{on} \quad \sigma \times (0, T) \\
\frac{\partial u}{\partial n} &= \varphi(u) && \text{on} \quad \Sigma \times (0, T) \\
u(x, 0) &= u_0(x),
\end{aligned}
$$

where $\partial \Omega = \sigma \cup \overline{\Sigma}$, $\sigma \cap \Sigma = \emptyset$ and where σ, Σ are open, smooth submanifolds of $\partial \Omega$ (which is likewise assumed to be at least piecewise smooth).

II. Hyperbolic problems

IIa.   Singular nonlinearity in the equation. Here we have in mind, for example, problems of the form

$$
\begin{aligned}
u_{tt} &= \Delta_N u + \varphi(u) && \text{in} \quad \Omega \times (0, T) \\
u(x, t) &= 0 && \text{on} \quad \partial \Omega \times (0, T) \\
u(x, 0) &= u_0(x) < 1 && \text{on} \quad \Omega \\
u_t(x, 0) &= v_0(x)
\end{aligned}
$$

where φ, Ω, Δ_N are as above.

IIb.   Singular nonlinearity in a boundary condition. This is the hyperbolic analog of Ib above, viz:

$$
\begin{aligned}
u_{tt} &= \Delta_N u && \text{in} \quad \Omega \times (0, T) \\
u &= 0 && \text{in} \quad \sigma \times (0, T) \\
\frac{\partial u}{\partial n} &= \epsilon \varphi(u) && \text{on} \quad \Sigma \times (0, T) \\
u(x, 0) &= u(x) && \text{on} \quad \bar{\Omega}, u_0 < 1 \quad \text{on} \quad \Sigma
\end{aligned}
$$

Here σ, Σ are as in IIb.

Problems under the Ia heading were considered by Acker and Walter, Levine and Montgomery and by Walter, while those under Ib were investigated by Levine (in one dimension) and later by Levine and Lieberman in several dimensions. Problems under the heading IIa were studied (in one dimension) by Chang and Levine. Except for a paper in which $-\Delta_N$ is replaced by $(-\Delta_N)^p$ for p sufficiently large (Levine and Smiley), there is no literature on these problems in several dimensions prior to 1985. Results for problems under the heading IIb are even rarer. Prior to 1985, there is only a single paper of Levine, and that is restricted to one space dimension.

As remarked earlier, all of these papers are surveyed in [16].

The plan of this survey is as follows:

A. We survey the recent literature which in extensions of problems considered under Ia,b and IIa,b above.

B. We discuss more fully the question of the blow up of the derivatives of the solution near quenching points. Some of this discussion overlaps some of the discussion of Kawohl elsewhere in this volume.

C. We consider known results for solutions of (Ia) beyond quenching.

D. We give a fairly complete discussion of the long-time behavior of the solution of the one dimensional problem.

E. We discuss some interesting miscellaneous results.

As our discussion proceeds, we will mention open problems under each of the headings **A- E**.

A. Extensions of (Ia,b) and (IIa,b).

Ia. In two recent papers, Chan, Chen, Kaper, and Kwong [5,6] undertook the study of the following IBVP $b < 1$:

$$u_t = u_{xx} + \frac{b}{x} u_x + \epsilon \varphi(u) \quad 0 < x < 1, \quad t > 0.$$

(A-Ia)
$$u(1, t) = 0$$
$$u(0, t) = 0 \qquad\qquad\qquad t > 0$$
$$u(x, 0) \equiv 0 \qquad\qquad 0 < x < 1$$

If $b \geq 1$, this problem together with the boundary condition $u_x(0, t) = 0$ in lieu of the Dirichlet condition at $x = 0$ is simply a radial problem in $b + 1$ "space" dimensions. For the special case $\varphi(u) = 1/(1 - u)$ they give an upper bound on $\epsilon*$, the critical value with the property that if $\epsilon > \epsilon*$; there is always quenching while if $\epsilon < \epsilon*$, there is no quenching. They also characterize the existence of $\epsilon*$ in terms of the existence of a numeral, nonnegative stationary solution, something that had been done earlier for $b = 0$ by Levine and Montgomery and is higher dimensions by Acker and Kawohl. From the numerical results of [6], we see that $\epsilon*$ is a decreasing function of b and that the quenching point (for $\epsilon > \epsilon*$) moves to the left as b increases. There is an easy intuitive explanation for the second of these observations. If we view $-bu_x/x$ as a convection term, then for $b > 0$, transport by convection is to the left while if $b < 0$, transport by convection is to the right. Thus, as b increases, we expect the quenching point to move to the left. We would also expect that this term to make a larger contribution to u_t near $x = 0$ than near $x = 1$ so that quenching should be inhibited for negative b and encouraged for positive b. Proofs would be desirable.

Ib. Recently S. R. Park [19] examined the following problem:

$$\text{(A-Ib)} \quad \begin{aligned} u_t &= u_{xx} + \delta u_x u & \delta \text{ real,} \quad 0 < x < 1, \quad t > 0 \\ u(0,t) &= 0 & t > 0 \\ u_x(1,t) &= \epsilon(1 - u(1,t))^{-\beta} & \epsilon > 0, & \quad t > 0 \\ u(x,0) &= u_0(x) & u_0(1) < 1 \end{aligned}$$

He was interested in the effect of convection on quenching. He gave a complete (numerical) description of the set of stationary solutions of this problem and showed that when $\delta > 0$ the convective term inhibits quenching while when $\delta < 0$ it aids quenching. He also showed that when quenching occurs (necessarily at $x = 1$), $u_t(1,t)$ also blows up in finite time.

Park also studied (A-Ib) when the boundary condition was replaced by $u_x(1,t) = \epsilon u^{-\beta}(1,t)$. The results are of some theoretical interest because one can have blow up in <u>infinite</u> time for such problems.

IIa. Smith [24], examined the following (hyperbolic) problem:

$$\text{(A-IIa)} \quad \begin{aligned} u_{tt} &= \Delta_n u + \epsilon\varphi(u) & \text{if} \quad (x,t)\epsilon\Omega \times (0,T) \\ u &= 0 & \text{on} \quad \partial\Omega \times (0,T) \\ u(x,0) &= u_0(x) \, (<1) & \text{on} \quad \Omega \\ u_t(x,0) &= v_0(x) & \text{on} \quad \Omega \end{aligned}$$

where $n = 2,3,4\ldots$. He established the following interesting result: Let $\epsilon > 0$ be such that the above problem has a stationary solution $f(x)$ say. Then for any integer $k > 0$, and any number $T_0 > 0$ there exists $u_0\epsilon C^k(\bar{P})$ with $u_0 = 0$ in $\partial\Omega$ and $u_0 < 1$ in $\bar{\Omega}$ such with

$$\int_\Omega |\nabla u_0 - f|^2 dx < \delta$$

such that the corresponding solution of the time dependent solution with $u_0 \equiv 0$ quenches in finite time $T < T_0$. This result is comparable to a similar result of Sternberg [25] where $\varphi(u)$ was replaced by $|u|^{p-1}u$.

He also extended the local existence (in time) result of Chang and Levine [22] to dimensions $N = 2,3$ but only for sufficiently small $\epsilon > 0$.

He established (numerically) the existence of a number $\epsilon_N > 0$ such that if $\epsilon < \epsilon_N$, the solution of (A-IIa) with $u_0 = v_0 = 0$ is global. This result was established in one dimension [22] but the proof relies upon the continuous imbedding of $H_0^1(\Omega)$ into $L^\infty(\Omega)$ in one dimension, a result that is false in higher dimensions. It can also be shown that for each $N > 1$, there is $\epsilon_* > 0$ such that if $\epsilon > \epsilon_*$, solutions of A-IIa quenches in finite time. We record some of his results here. (Table 1)

Table 1

(Values of ϵ_N, ϵ_* for $\varphi(u) = 1/(1 - u)$ [24].)

N	ϵ_N	ϵ_*
1	0.341	0.383
2	1.017	1.309
3	1.520	2.139
7	2.563	6.000

The solutions are plotted in Figures 2, 3 for cases (a), (c).

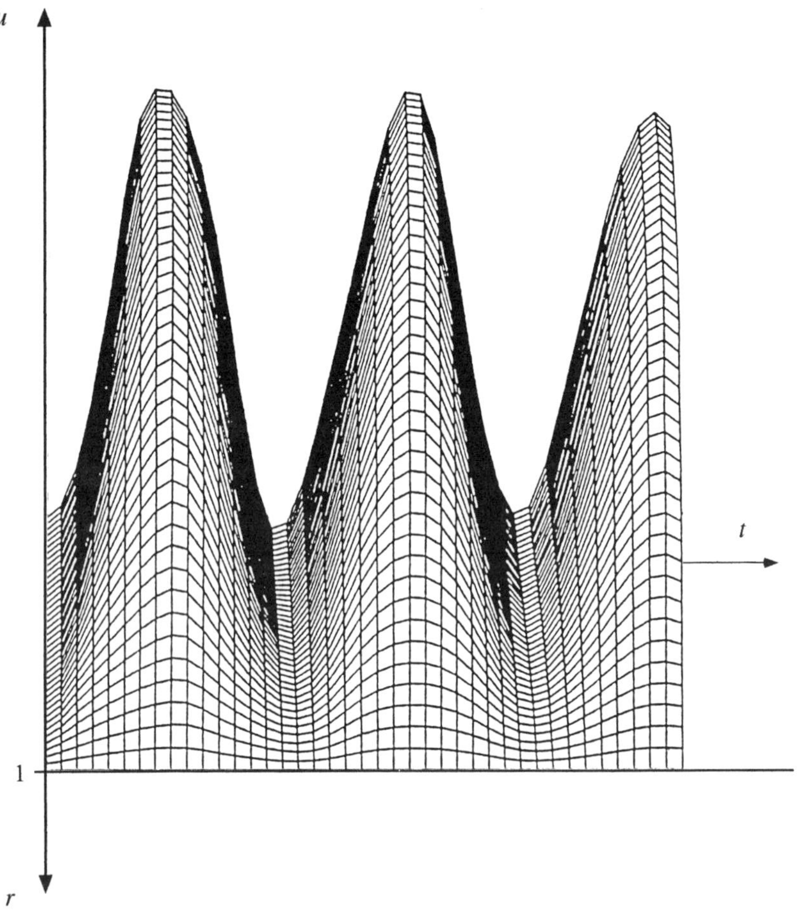

Figure 1. A solution of (A-IIa) with $\epsilon < \epsilon_N$ and $u_0 = v_0 = 0$.

These results suggest the existence of a critical $\bar{\epsilon}(N)$ such that (with $u_0 = v_0 \equiv 0$)

(a) $\epsilon < \bar{\epsilon}(N)$ $\implies u \le 1 - \delta$ for all $t > 0$ and some $\delta < 1$.

(b) $\epsilon = \bar{\epsilon}(N)$ $\implies u$ quenches in infinite time.

(c) $\epsilon > \bar{\epsilon}(N)$ $\implies u$ quenches in finite time.

(d) $\epsilon(N)$ increases with N.

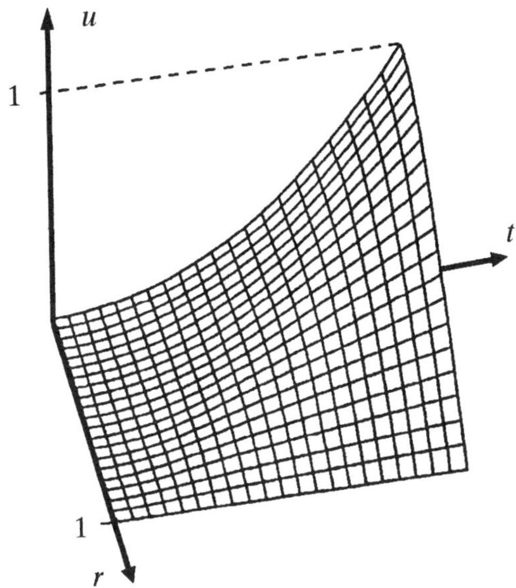

Figure 2. A solution of (A-IIa) with $\epsilon > \epsilon_N$ and $u_0 = v_0 \equiv 0$.

IIb. In [23] M. Rammaha undertook the study of the following IBVP:

$$\begin{aligned}
u_{tt} &= \Delta_N u && \text{in} && \Omega \times (0, T) \\
u(x, t) &= 0 && \text{on} && \sigma \times (0, T) \\
\frac{\partial u}{\partial n}(x, t) &= \epsilon\varphi(u) && \text{on} && \Sigma \times (0, T) \\
u(x, 0) &= u_0(x) && \text{on} && \bar{\Omega} \times (0, T)
\end{aligned}$$

(A-IIb)

with $u_0 < 1$ on Σ, and where Ω, σ, Σ are as above.

Rammaha's local (in time) existence result is somewhat better than Smith's in that the restriction that ϵ be small is not required. However, the restriction to $N = 2$ or $N = 3$ is forced on him as well. He established the following result:

(a) If $\int_0^1 \varphi(u)du < \infty$ and $\epsilon \geq\geq 0$, u quenches in finite time on Σ.

(b) If $\int_0^1 \varphi(u)du = \infty$ and $\epsilon \geq\geq 0$, u quenches in finite or infinite time.

Here $\epsilon \geq\geq 0$ means

$$\epsilon > \lambda_1 \sup_{[0,1]} \ (u/\varphi(u))$$

where λ_1 is the smallest eigenvalue of the associated Steklov problem.

Finally, by considering an annular region, Ω,

$$\Omega = \{x \epsilon R^2 | \ 1 < |x| < L\}$$

with $u = 0$ on $|x| = 1$ and $u_r = \varphi(u(L, t))$ on $r = L$, he shows that in 2 and 3 dimensions (A-IIb) will have global solutions if ϵ is small.

Such a result is also possible for the problem (A-IIa) in an annulus. However even for the disk, there is no known global existence result for (A-IIa). (The results for the annulus parallel those of Levine [27], in one dimension.)

B. Blow up of Derivatives. It is clear from the differential equation Kawarada considered that if u quenches, then at least one of u_t or u_{xx} must blow up in a finite time also. But which one of them blows up? Perhaps both do.

In [11], Kawarada asserted that $u_t(\frac{1}{2}, \cdot)$ blows up at time T if $u(\frac{1}{2}, \cdot)$ reaches one in finite time T. However, Chan and Kwong [8] have observed that his argument is incomplete.

We turn therefore to a discussion of the blow up of the derivatives of solutions at quenching for problems (Ia,b, IIa,b).

For Problem Ia in one space dimension, Chan and Kwong [8] showed that if u quenches (in finite time) and if

$$\int_0^1 \varphi(u)du = \infty,$$

then u_t blows up in finite time. For $\varphi(u) = \epsilon(1 - u)^{-\beta}$ this means that $\beta \geq 1$. They considered the case of homogeneous initial and boundary values.

Independently, Acker and Kawohl, [1], considered (Ia) when $\Omega = \{x \epsilon R^N | \ |x| \leq 1\}$ with nonhomogeneous initial values. They confined their discussion to radical solutions. Specifically, they showed that if

$$u_r(r, 0) \leq 0, \quad u_t(r, 0) \geq 0, \quad u_{rt}(r, 0) \leq 0$$

and if u quenches, then it does so only at $r = 0$ and u_t blows up there.

Recently Deng and Levine, [9] have shown that if Ω is *convex* and $u_t(x,0) \geq 0$, then the set of quenching points is a compact subset of Ω (quenching cannot occur at the boundary) and u_t blows up at every quenching point. The results of [1,8] are based on variants of the maximum principle while those of [9] combine the maximum principle with arguments of Friedman and McLeod [13].

Deng and Levine also considered the third boundary value problem:

$$\frac{\partial u}{\partial t} = \Delta_N u + \epsilon \varphi(u) \quad \text{in} \quad \Omega \times (0, T)$$

$$\frac{\partial u}{\partial n} + \beta(x, t)u = 0 \qquad \text{on} \quad \partial\Omega \times (0, T)$$

$$u(x, 0) = u_0(x) < 1 \qquad \text{on} \quad \Omega$$

They showed that when u quenches, u_t blows up. (Ω need not be convex.) However, the possibility of quenching occurring on the boundary is not excluded. An example where it occurs on the boundary would be nice.

Until recently, nothing was known about the behavior of the spacial derivatives, $\vec{\nabla}_x u$, Δu at quenching. Elsewhere in this volume, Kawohl discusses some recent results he and Fila have obtained along these lines. Briefly, from their results, one has, if u quenches at time T at $r = 0$ (for Ω a ball and radial solutions, say) then $u_r(0, T) = 0$ for $0 < \beta < 1$ while $u_r(0, T) = \infty$ if $\beta > 1$. The case $\beta = 1$ is open. Their arguments involve the study of the "P-function", i.e.

$$P(x, t) = \frac{1}{2}|\nabla_x u|^2 + \epsilon \Phi(u(x, t))$$

near a quenching point ($\Phi(0) = 0, \Phi'(u) = \varphi(u)$) in the case $0 < x < 1$ and the study of a function of the form

$$J = r^{n-1} u_r + c(r) F(u)$$

for a clever choice of c(r) and F(u) when $\beta > 1$. (The latter argument is based on that of [13].)

There appears to be nothing known thus far about the behavior of Δu near a quenching point.

The best asymptotic results of which the author is aware are due to Guo [15] who has shown that if x_0 is a quenching point and T the quenching time, then for $\varphi(u) = \epsilon(1 - u)^{-\beta}$

$$\lim_{t \nearrow T^-} (1 - u(x, t))(T - t)^{-1/(\beta+1)} = [(\beta + 1)\epsilon]^{1/(\beta+1)}$$

if $N = 1$ and $\beta \geq 3$. (More recently, Fila and Hulsof, [10], have shown this result to hold for $\beta \geq 1$). This result should hold for $\beta > 0$. The

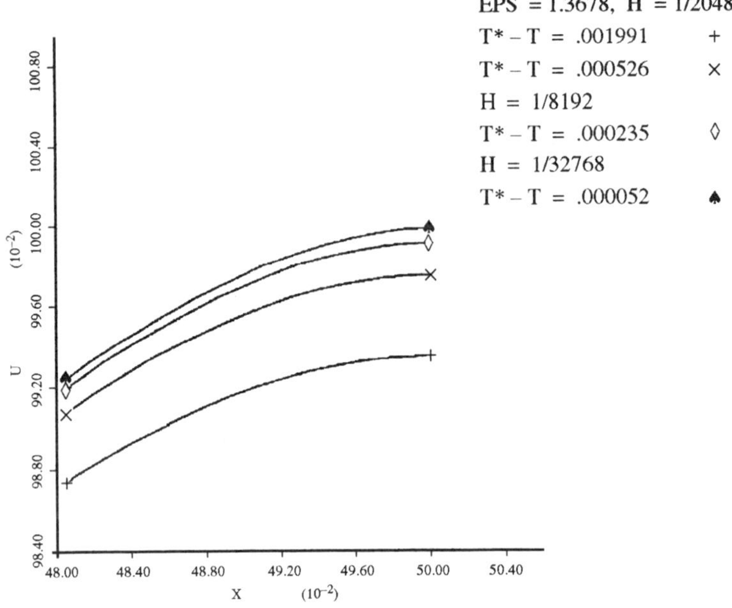

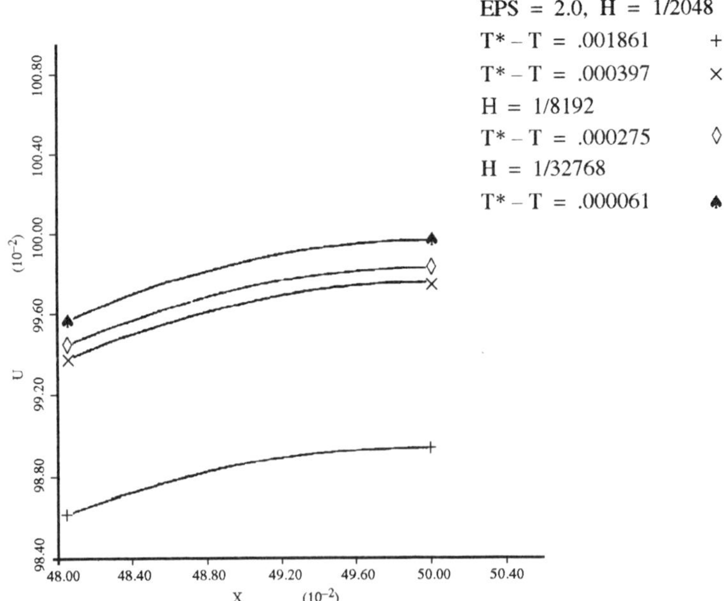

Figure 3.(a-b). Profiles of u for two values of ϵ corresponding to $T^* > 0.5$.

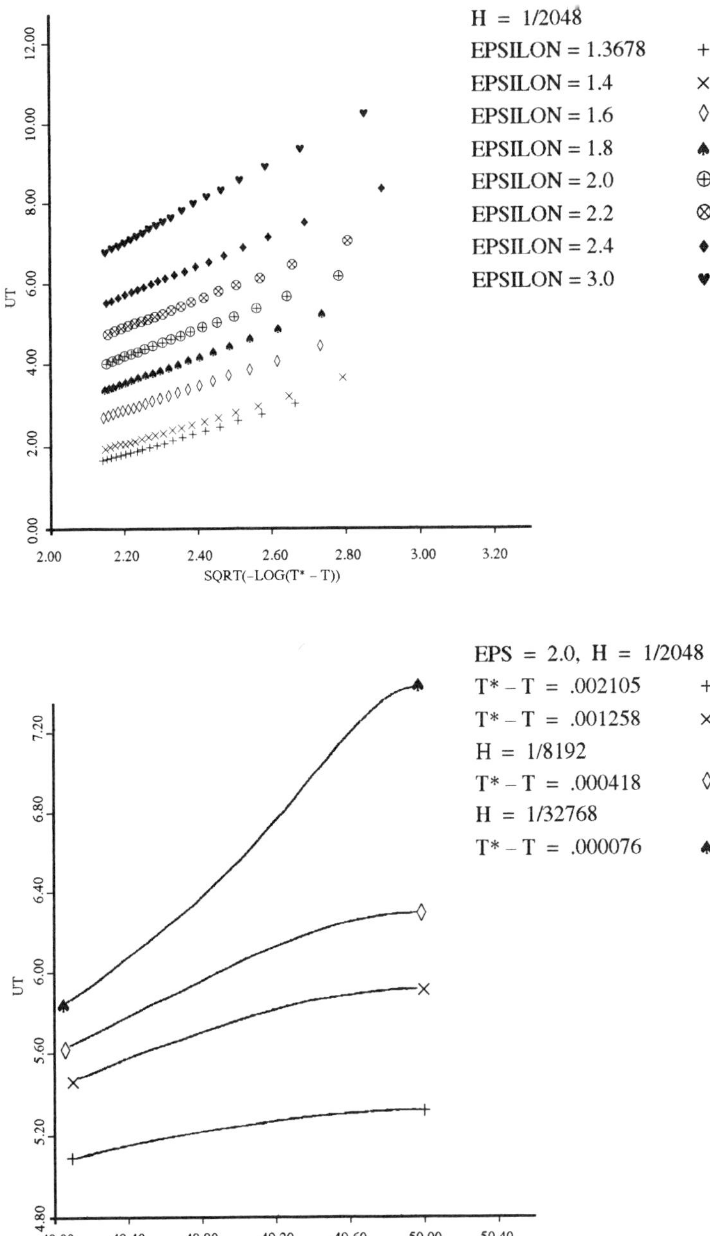

Figure 4.(a-b). Graphs showing the blow-up of $u_t(0.5, t)$.

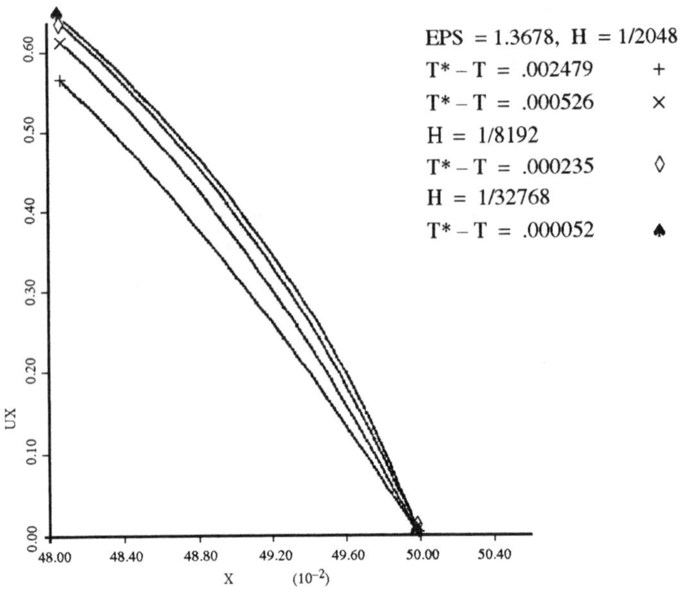

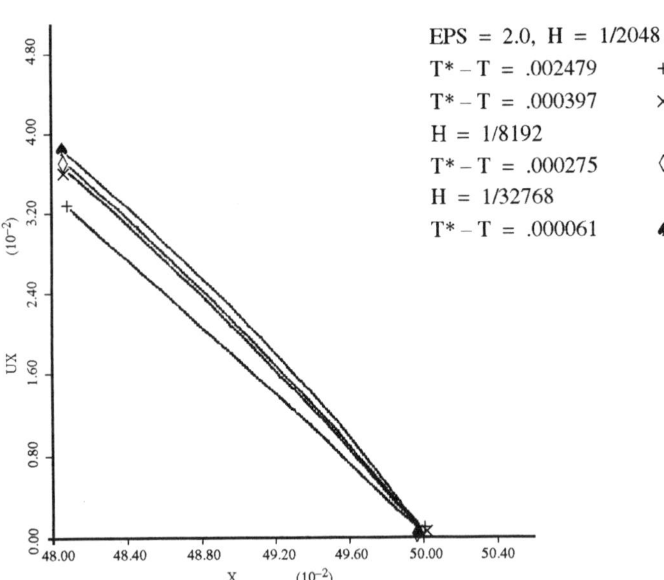

Figure 5.(a-b). Graphs showing that $u_x(x, t)$ does not blow up for x near .05, t near T in two late quenching cases.

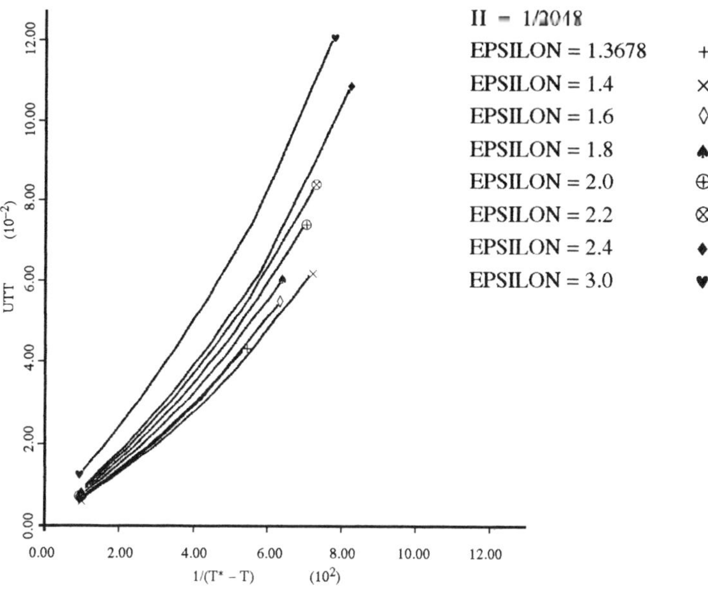

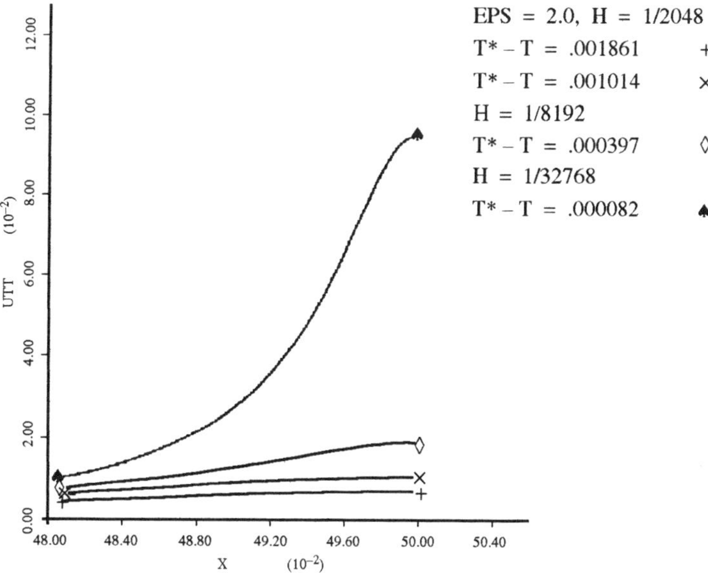

Figure 6.(a-b). Graphs showing the blow-up of $u_{tt}(0.5, t)$. (In 6(a), 20 points were plotted for each value of ϵ.)

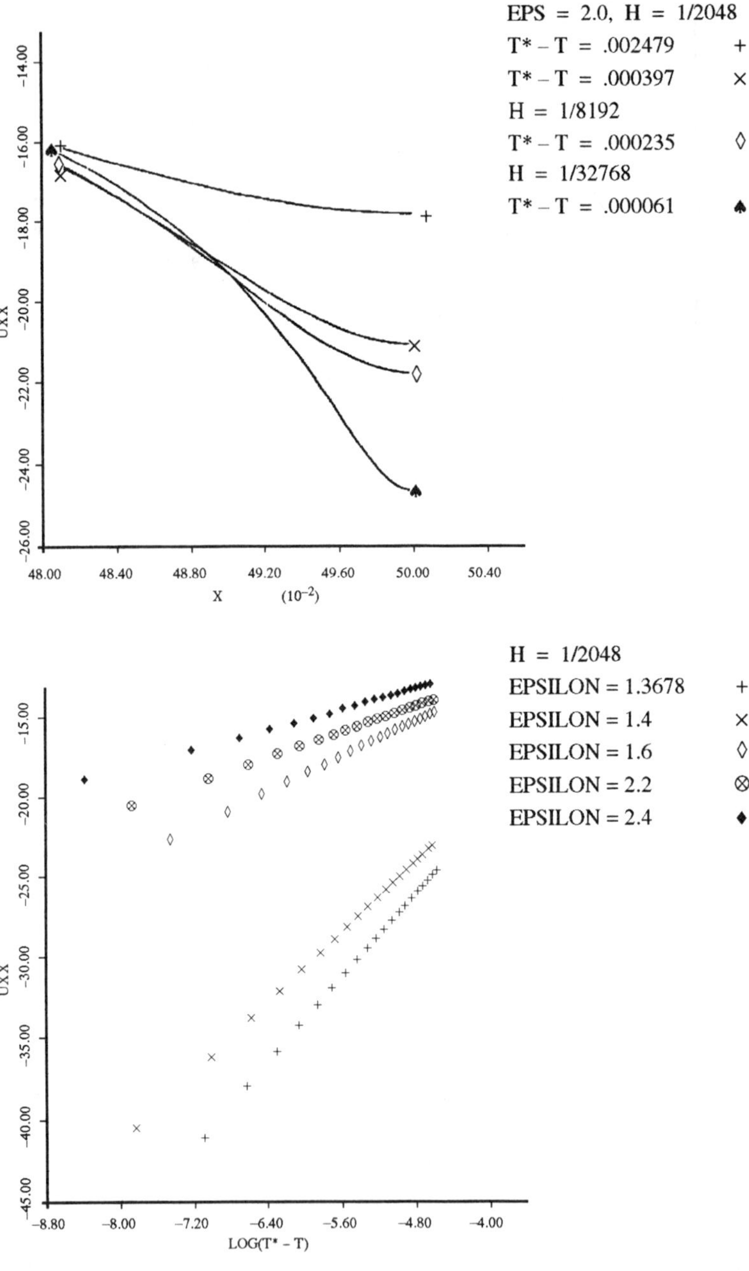

Figure 7.(a-b). Graphs showing the blow-up of $u_{xx}(0.5, t)$.

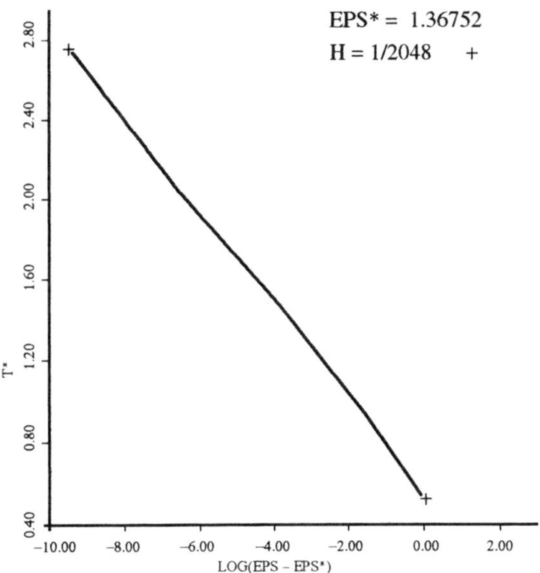

Figure 8. Graph of T^* vs. $\ln(\epsilon - \epsilon^*)$ for an assumed value of ϵ^*.

C. Beyond Quenching. In [17,18] we proposed to study the behavior
of solutions of

$$
\begin{aligned}
u_t &= \Delta u + \epsilon(1-u)^{-\beta} && \text{in} && \Omega \times (0,T) \\
\text{(C- Ia)} \qquad u(x,t) &= 0 && \text{on} && \partial\Omega \times (0,T) \\
u(x,0) &= u_0(x) && \text{on} && \Omega
\end{aligned}
$$

"beyond quenching". There have been, to the author's knowledge, two
independent attempts to do this, which we shall briefly discuss here, [2,20].

The idea in both of these papers is to replace $\epsilon(1-u)^{-\beta}$ by some kind of
nonsingular nonlinearity and define a solution of (C-Ia) beyond the quenching time in an appropriate weak sense. Here $0 < \beta < 1$.

For example, Bandle and Brauner consider, in our notation, the regularized problem:

$$
\begin{aligned}
\text{(C-Ia)} \qquad u_t^\delta &= \Delta u^\delta + \frac{\epsilon(1-u^\delta)}{\delta + (1-u^\delta)^{\beta+1}} && \text{in} && \Omega \times (0,\infty) \\
u^\delta(x,t) &= 0 && \text{on} && \partial\Omega \times (0,\infty) \\
u^\delta(x,0) &= 0 && \text{on} && \Omega
\end{aligned}
$$

for $\delta > 0$. They show that $u^\delta \in (0,1)$ for $t > 0$ and is global. Moreover
$u^\delta(x,t)$ increases as t increases and as δ <u>decreases</u>. They show that $u^\delta \to u$
in the sense that

$$
\begin{aligned}
u^\delta &\to u && \text{in } L^P\left(\Omega \times (0,T)\right) && \forall\, T > 0 \ \ \text{and } 1 \le p < \infty, \\
u^\delta &\to u && \text{in } L^\infty((0,T); H_0^1(\Omega)) && \text{in the weak* topology,} \\
u^\delta &\to u && \text{in } C^0((0,T); L^2(\Omega)) && \text{strongly.}
\end{aligned}
$$

and

$$
u_t^\delta \to u_t \quad \text{in} L^2(\Omega \times (0,T)) \qquad\qquad \text{weakly for all} \quad T > 0
$$

The limit solution satisfies, in the weak sense,

$$
\begin{aligned}
\text{(C-Ia)}_w \qquad \frac{\partial u}{\partial t} &= \Delta u + \frac{\epsilon\,\chi(u<1)}{(1-u)^\beta} + \nu && \text{in} && \Omega \times (0,\infty) \\
u &= 0 && \text{on} && \partial\Omega \times (0,\infty) \\
u &= 0 && \text{on} && \Omega \times \{0\}
\end{aligned}
$$

where χ is the indicator function of $(-\infty, 1)$ and ν is a measure supported in the complement of S where

$$S = \bigcup_{0 < t < \infty} (\Omega - \Omega(t))$$

and

$$\Omega(t) = \{x|\ u(x, t) = 1\}$$

This u is minimal, in the sense that if U is any other solution of $(\text{C-Ia})_w$, then $u \leq U$.

Phillips, [20], on the other hand, starts with the problem (again in our notation)

$$
\begin{aligned}
(\text{C-Ia})_P \qquad u_t &= \Delta u + \epsilon(1 - u)^{-\beta}\chi(u < 1) && \text{in} && \Omega \times (0, \infty) \\
u(x, t) &= \psi(x, t) && \text{on} && \partial\Omega \times (0, \infty) \\
u(x, 0) &= \psi(x, 0) && \text{on} && \bar{\Omega}
\end{aligned}
$$

where $\psi(x, t) \leq 1 - \delta_0$. He also shows that this problem has a continuous weak solution if $\partial\Omega \in C^{2+\alpha}$, $\psi \in C^{2+\alpha}$. He uses the same regularization of $\epsilon(1 - u)^{-\beta}$ as Bandle and Brauner.

In both articles, no claim is made as to the uniqueness of the weak solution. Nor is there any assertion of the agreement of the constructed solution with those constructed by alternate regularizations, such as

$$\Phi_p(\beta, u) = \epsilon[\beta + (1 - u)^{\beta + p}]^{-1} \cdot (1 - u)^p$$

for $p > 0$ or by

$$\Phi_\delta(u) = \epsilon \begin{cases} \delta^{-\beta} & 1 \geq u \geq \delta \\ (1 - u)^{-\beta} & u \leq 1 - \delta. \end{cases}$$

This lack of uniqueness makes it difficult to study the solution beyond quenching. However, we can make some further conjectures in this direction. This we do in the next section.

D. Dynamical structure of (Ia). In this section we look more closely at Phillips' problem but in one dimension: We consider

$$
\begin{aligned}
u_t &= u_{xx} + \epsilon(1 - u)^{-\beta}\chi(u < 1) & 0 < x < 1, \quad t > 0 \\
(\text{D-Ia}) \qquad u(0, t) &= u(1, t) = 0 & t > 0 \\
u(x, 0) &= u_0(x), \qquad (u_0 < 1) & 0 \leq x \leq 1
\end{aligned}
$$

We shall assume that u_0 is continuous and only that $\beta > 0$.

Definition. A distribution f is a stationary solution of (D-Ia) if (with $\varphi(u) = \epsilon(1-u)^{-\beta}\chi(u < 1)$),

$$\text{(D-Ia)}_S \qquad\qquad 0 = \int_0^1 [\psi''(x)f(x) + \chi(x)\varphi(f(x))]dx$$

for all $\psi\epsilon\, C_0^\infty\,(0,1)$.

In [17] we showed the following:

I. If $\beta \geq 1$, then every stationary solution is classical and satisfies $f < 1$ on $[0,1]$. Also, there exists exactly one $\epsilon(\beta) > 0$ such that:

(a) If $\epsilon > \epsilon(\beta)$ there are no stationary solutions.
(b) If $\epsilon = \epsilon(\beta)$ there is exactly one stationary solution.
(c) If $0 < \epsilon < \epsilon(\beta)$ there are exactly two stationary solutions.

In case (c), we always have $f_-(x,\epsilon) < f_+(x,\epsilon)$ on $(0,1)$ (ordering of solutions).

II. If $0 < \beta < 1$, there are numbers $\epsilon_0(\beta)$, $\epsilon(\beta)$ with $0 < \epsilon_0(\beta) < \epsilon(\beta)$ such that

(a) There are no classical stationary solutions if $\epsilon > \epsilon(\beta)$.
(b) There is exactly one classical stationary solution if $\epsilon = \epsilon(\beta)$ or $0 < \epsilon < \epsilon_0(\beta)$.
(c) There are precisely two classical stationary solutions if $\epsilon_0(\beta) < \epsilon < \epsilon(\beta)$.
(d) If $\epsilon \geq \epsilon_0(\beta)$, there is exactly one singular stationary solution $f_s(x,t)$ of class $C^{1+\gamma}(0,1)$ where $\gamma = (1-\beta)/(1+\beta)$. It is given by

$$f_s(x,t) = \begin{cases} 1 - (1-x/a)^{1+\gamma} & 0 \leq x < a \\ 1 & a \leq x \leq \frac{1}{2} \\ f_s(1-x,a) & \frac{1}{2} \leq x \leq 1 \end{cases}$$

where $a^2 = 2(1-\beta)/[(1+\beta)^2\epsilon]$ and where $\epsilon_0(\beta) = 8(1-\beta)(1+\beta)^{-2}$

Moreover, wherever any of f_-, f_+, f_s are defined, we have

$$f_-(x,\epsilon) < f_+(x,\epsilon) < f_s(x,\epsilon)$$

on $(0,1)$. Solutions diagrams are given in Figures 7a,b.

We have the following results for the time dependent problem. (All limits are pointwise unless otherwise stated.)

I. $\beta \geq 1$.

(i) If $\epsilon > \epsilon(\beta)$, then for all $u_0 < 1$, u quenches in finite time.

(ii) Let $\epsilon = \epsilon(\beta)$: (a) If $u_0 \leq f\,(x, \epsilon(\beta)) = f_+ = f_-$, then $\lim\limits_{t \to \infty} u(x,t) = f(x, \epsilon(\beta))$. (b) for all $\delta \geq 0$, there is $u_0 \geq f(\cdot, \epsilon(\beta))$, $\|u_0 - f(\cdot, \epsilon(\beta))\|_{L^\infty} < \delta$ such that $u(x, t; u_0)$ quenches.

(iii) Let $\epsilon < \epsilon(\beta)$. (a) If $u_0(x) < f_+(x, t)$ on $(0, 1)$ and $u_0 \epsilon C^1$, then $\lim\limits_{t \to +\infty} u(x, t) = f_-(x, \epsilon)$. (b) For every $\delta > 0$, there is $u_0 < 1$ with $\|u_0 - f_+\|_{L^\infty} < \delta, u_0 \geq f_+$ such that u quenches.

II. $0 < \beta < 1$.

(i) If $\epsilon > \epsilon_0$ then (i), (ii), (iii) of the above hold according as $\epsilon > \epsilon(\beta)$, $\epsilon = \epsilon(\beta)$, $\epsilon < \epsilon(\beta)$.

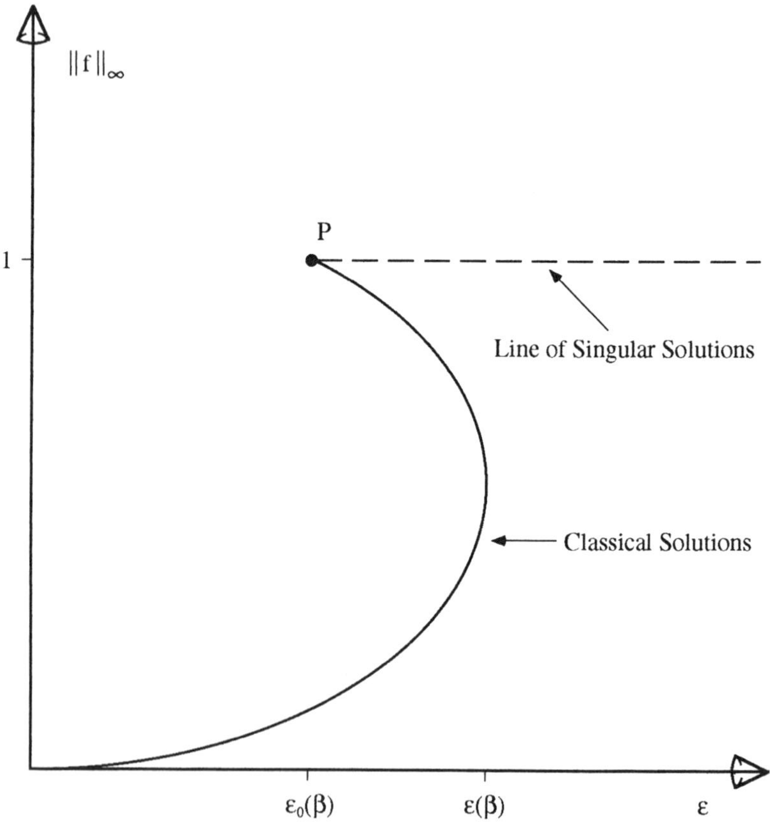

Figure 9a. Bifurcation diagram for (D-Ia)$_S$ with $0 < \beta < 1$.

$$\lim_{\beta \to 0^+} \epsilon_0(\beta) = \lim_{\beta \to 0^+} \epsilon(\beta) = 8.$$

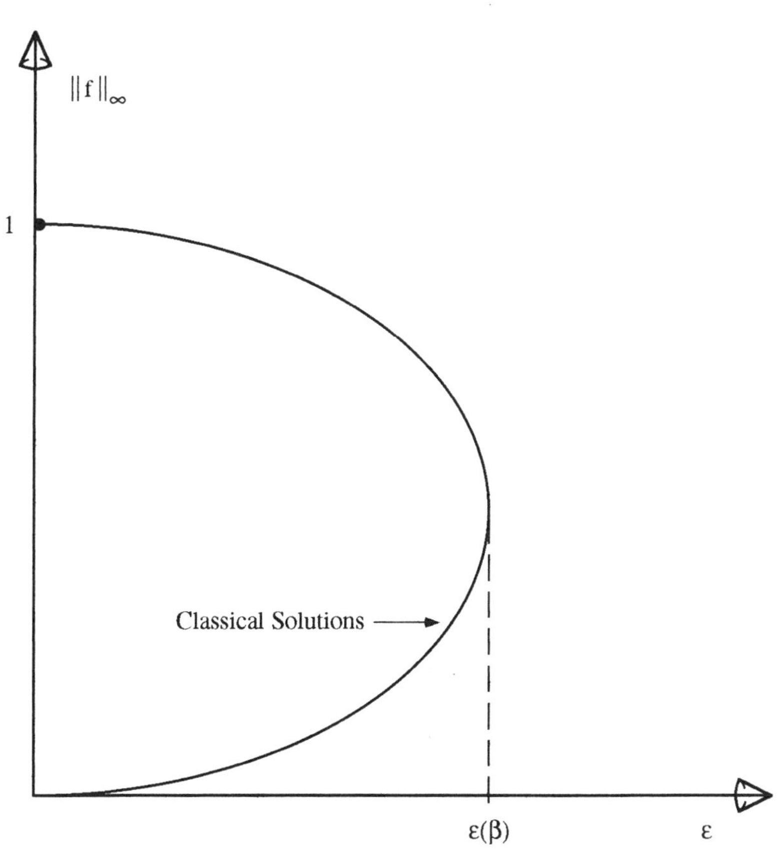

Figure 9b. Bifurcation diagram for (D-Ia)$_S$ with $\beta \geq 1$.

(ii) Let $\epsilon = \epsilon_0$ (a) If $u_0 < f_s(x, \epsilon_0), u_0 \epsilon C^1$, then u is classical and $\lim_{t \to \infty} u(x, t) = f_{-(x, \epsilon_0)}$. (b) There exist smooth initial values $u_0 < 1$ such that u quenches in finite time. (These are necessarily larger than f_s somewhere on $(0, 1)$).

(iii) Let $0 < \epsilon < \epsilon_0$. (a) If $u_0(x) < f_s(x, \epsilon_0)$, u cannot quench and $u(x, t) \to f(x, \epsilon)$ as $t \to \infty$. (b) There are smooth initial values $u_0 < 1$, for which u quenches (in finite time).

The above results suggest several interesting problems, some of which we shall discuss later. However, it is more natural to present a few of these problems here.

Sacks has raised the following problem: Suppose u_0 is of class C^1 and both $(u_0 - f_+)^- > 0$ and $(u_0 - f_+)^+ > 0$ on open subsets of $(0, 1)$. Then what is the nature of the long time behavior of $u(x, t : u_0)$?

The above results suggest that quenching in infinite time may not be possible, although they do not rule out this possibility. Recently Fila and Kawohl [11] made a more systematic study of this issue. They showed that if $N = 1$ or $N = 2$ and $\beta > 1$ or if $N = 3$ and $\beta > 3$, then infinite time quenching is not possible.

These observations lead us to make the following **conjectures**:

1. If $\beta > 1$ and quenching occurs, then we have "complete quenching" in finite time. "Complete quenching" is defined in the same way as the notion of "complete blow up" of Baras and Cohen [3].

2. If $\beta = 1$ and quenching occurs, then we have complete quenching, but in infinite time, i.e. $\lim_{t \to +\infty} u(x, t) = 1$ for all $x \epsilon (0, 1)$.

3. Suppose $0 < \beta < 1$. If quenching occurs and $\epsilon \geq \epsilon_0(\beta)$, then the limit solution is $f_s(x, \epsilon)$ and the onset of quenching occurred in finite time if $\epsilon > \epsilon_0(\beta)$ and in infinite time if $\epsilon = \epsilon_0(\beta)$. If quenching occurs with $\epsilon < \epsilon_0(\beta)$ then it occurs in finite time and complete quenching takes place in infinite time.[1]

In particular, we conjecture that for all $\beta > 0$, when $N = 1$ quenching in infinite time is not possible except where $0 < \beta < 1$ and $\epsilon = \epsilon_0(\beta)$. This statement is already true for all $\beta > 0$, $N = 1, 2, \ldots$ provided ϵ is sufficiently large $(\epsilon > \epsilon_0(\beta))$.

E. Miscellaneous results and other open problems.

In one dimension, Guo, [15], showed that if $u_0 < 1$, $\beta > 1$, then the set

$$Q = \{x | u(x, \cdot) \text{ quenches}\} = Q(u_0)$$

---

[1]Recently, Fila, Levine and Váquez have shown that for $0 < \beta < 1$, and $\epsilon < \epsilon_0$, then the classical stationary solution is stable. In particular, even when quenching occurs, the solution of (D-Ia) becomes classical in finite time and tends to the unique stationary solution of the problem. Thus, the third conjecture is not correct. However, the first two conjectures are still open.

is always finite. Recently, Giga and Kohn [27], have given an example of initial values for the problem

$$u_t = u_{xx} + |u|^{p-1}u \quad 0 < x < 1, \quad t > 0$$

(B) $$u(0,t) = u(x,t) = 0 \qquad\qquad t > 0$$

$$u(x,0) = u_0(x) \qquad\qquad 0 < x < 1$$

so that finite time blow up occurred at **exactly** two points ([27]). Question: For problem (Ia) (in one dimension) given any integer $m > 0$, are these initial values $u_0(x;m)$ such that quenching takes place **precisely** at m points. Can we specify the points?

The result of Guo is clearly false in the annulus. However, the question then arises; given u_0, what is the Hausdorf dimension of $Q(u_0)$ assuming $Q \neq 0$? Moreover, is Guo's result true if $0 < \beta \leq 1$? For radial problems in regions, we claim $H(Q) \geq N - 1$ where $H(Q) = $ Hausdorf dimension of Q. For radial problems in the annulus, this is clearly true.

The blow up points for (B), like the quenching points, always lie in a compact subset of $(0,1)$. Recently, [12], Floater examined the problem

$$x^q u_t = u_{xx} + u^p \quad 0 < x < 1, \quad t > 0$$

(Fℓ) $$u(0,t) = u(1,t) = 0 \qquad\qquad t > 0$$

$$u(x,0) = u_0(x) \geq 0 \quad 0 \leq x \leq 1$$

where $q > 0$, $p > 1$ and u_0 is continuous on $(0,1)$ with $u_0(0) = u_0(1) = 0$. It is well known [29] that for u_0 so large that

$$\frac{1}{2}\int_0^{'}(u_0')^2 - \frac{1}{p+1}\int_0^{'}u_0^{p+1}(x)dx < 0,$$

solutions of $(F\ell)$ are not global.

Floater shows: If $(u_0(x)/x)' \leq 0$, $1 < p \leq q + 1$ and if finite time blow up occurs, then it occurs at $x = 0$. Numerical evidence suggests that it occurs in the interior of $(0,1)$ if $p > q + 1$. We would like to know what happens if u^p is replaced by $(1 - u)^{-\beta}$. That is under what conditions on β will the quenching points for

(QFℓ) $$x^q u_t = u_{xx} + \epsilon(1-u)^{-\beta} \quad 0 < x < 1, \quad t > 0$$

$$u(0,t) = u(1,t) = 0 \qquad\qquad t > 0$$

$$u(x,0) = u_0(x) \quad (<1) \qquad 0 < x < 1$$

occur at $x = 0$?

Moreover, for either (Fℓ) or (QFℓ) could the addition of a convective term on the right hand side of the equation (with the right sign so that drift is to the right) prevent blowup (or quenching) at $x = 0$?

As we remarked earlier, for Dirichlet data, quenching cannot occur on the boundary if Ω is convex. However, if Ω is not convex, can one construct an Ω for which quenching does occur on the boundary? For example, can quenching occur on the boundary at a reentrant corner?

Finally, we might consider the whole spectrum of quenching problems for other classes of problem. Dziuk and Kawohl, [26] have already done this in one dimension for the first initial-boundary value problem:

$$u_t = u_{xx}/(1 + u_x^2) + 1/(1 - u).$$

One is then led to consider problems of the form

$$u_t = \vec{\nabla} \cdot (f(|\nabla u|)\vec{\nabla} u) + \epsilon(1 - u)^{-\beta}.$$

If $f(s) = (1 + s^2)^{-\frac{1}{2}}$, the first nonlinear operator is the mean curvature operator. If $f(s) = s^{p-1}$, then we are dealing with the p Laplacian.

One of our students is investigating quenching problems of the form

$$u_t = \Delta(u^{m+1}) + \epsilon(1 - u)^{-\beta}$$

in one dimension. We shall report on his results at a later date.

Another class of quenching problems which is not well studied is that for systems. We have in mind problems of the form

$$u_t = \Delta u + \epsilon_1(\alpha_1 - v)^{-\beta_1}$$
$$v_t = \Delta v + \epsilon_2(\alpha_2 - u)^{-\beta_2}$$

where ϵ_i, $\alpha_i > 0$. Now Chen and Chan, [4], have made a very restricted study of systems in which the first component of the solution is known to quench. (In [4], $\alpha_1, < \alpha_2$). However, we are interested in obtaining deeper information about systems. For example, we consider the simple system

$$u_t = u_{xx} + \epsilon(1 - v)^{-\beta}$$
$$v_t = v_{xx} + \epsilon(1 - u)^{-\beta}$$

with $0 < x < 1$, $u = v = 0$ at $x = 0, 1$. We ask: Can we characterize the set of stationary solutions? What are the stability-instability properties of these solutions? Given initial values, $u(x, 0)$, $v(x, 0)$ (< 1 on $[0, 1]$), do we have infinite time quenching? What about "beyond quenching?" Other interesting questions will doubtless occur to the reader.

Finally, it would be nice to have proofs of all of the numerical results for hyperbolic quenching found by Axtell.

PARABOLIC PROBLEMS

1. A. Ackor, B. Kawohl, "Remarks on Quenching," Nonlinear Analysis, TMA **13** (1989), 53–61.

2. C. Bandle, C. M. Brauner, "Singular Perturbation method in a parabolic problem with free boundary," Proc BAIL IVth Conf. Novosibirsk, eds, S. K. Godunov, J. J. H. Miller, V. A. Novikov, Boole Press, Dublin (1987), 7–14.

3. P. Baras, L. Cohen, "Complete blow up for the solution of a semilinear heat equation," J. Funct. Anal. **71** (1987), 142–174.

4. C. Y. Chan, C. S. Chen, "Critical lengths for global existence of solutions for coupled semilinear singular parabolic problems." (preprint)

5. C. Y. Chan, H. G. Kaper, "Quenching for semilinear singular parabolic problems," SIAM J. Math. Anal. **20** (1989), 558–566.

6. C. Y. Chan, C. S. Chen, "A numerical method for semilinear singular parabolic quenching problems," Quart. Appl. Math. **47** (1989), 45–57.

7. C. Y. Chan, M. K. Kwong, "Existence results of steady states of semilinear reaction- diffusion equations and their applications," JDE **77** (1989), 304–321.

8. C. Y. Chan, M. K. Kwong, "Quenching phenomena for singular nonlinear parabolic equations," Nonlinear Analysis **12** (1988), 1377–1383.

9. K. Deng, H. A. Levine, "On the blow up of u_t at quenching," Proc. A.M.S. **106** (1989) 1049–1056.

10. M. Fila, J. Hulshof, "A note on the quenching rate." (preprint).

11. M. Fila, B. Kawohl, "Is quenching in infinite time possible?" Quail. Appl. Math., (in press).

12. M. S. Floater, "Blow up at the boundary for degenerate semilinear parabolic equations," (in press).

13. A. Friedman, B. McLeod, "Blow-up of positive solutions of semilinear heat equations," Indiana University Math. J. **34** (1985), 425–447.

14. J. S. Guo, "On the semilinear elliptic equation $\Delta w - \frac{1}{2}y \cdot \nabla w + \lambda w - w^{-\beta} = 0$ in R^N." (manuscript).

15. J. S. Guo, "On the quenching behavior of the solution of a semilinear parabolic equation," J. Math. Anal. Appl., (in press).

16. H. Kawarada, "On solutions of the initial-boundary value problem for $u_t = u_{xx} + 1/(1-u)$", RIMS Kyoto U. **10** (1975), 729–736.

17. H. A. Levine, "Quenching, nonquenching, and beyond quenching for solutions of some parabolic equations," Annali di Mat. Pura et Applic. (in press).

18. H. A. Levine, "The phenomenon of quenching: A survey," in *Proc. VIth Int. Conf. on Trends in the Theory and Practice of Nonlinear*

Analysis, North Holland, N.Y., (1985).

19. S. R. Park, "The phenomenon of quenching in the presence of convection," Ph.D. Thesis, Iowa State University, (1989).

20. D. Phillips, "Existence of solutions to a quenching problem," Applicable Analysis **24** (1987), 253–264.

HYPERBOLIC PROBLEMS

21. J. Axtell, "A numerical study of the derivatives of solutions of the wave equation with a singular forcing term at quenching," Num. Meth. for PDE **1** (1989), 53–76.

22. P. H. Chang, H.A. Levine, "The quenching of solutions of semilinear hyperbolic equations," SIAM J. Math Anal. **12** (1982), 893–903.

23. M. A. Rammaha, "On the quenching of solutions of the wave equation with a nonlinear boundary condition," J. Reine Ang. Mat. (in print).

24. R. A. Smith, "On a Hyperbolic Quenching problem in several dimensions," SIAM J. Math Anal. **20** (1989), 1081–1094.

25. N. Sternberg, "Blowup near higher modes of nonlinear wave equations," TAMS, **296** (1986), 315–325.

OTHER REFERENCES

26. Dziuk and Kawohl, "On radially symmetric mean curvature flow," (to appear).

27. Y. Giga, R. V. Kohn, "Nondegeneracy of blowup for semilinear heat equations," Hokkaido U. Preprint Series #46, October 1988.

28. H. A. Levine, "The quenching of solutions of linear parabolic and hyperbolic equations with nonlinear boundary conditions," SIAM J. Math. Anal. **14** (1983) 1139–1153.

29. H. A. Levine, "Some nonexistence and instability theorems for solutions of formally parabolic equations of the form $Pu_t = -Au + F(u)$, Arch. Rat. Mech. Anal. **51** (1973) 371–386.

30. Jong-Sheng Guo, "On the quenching rate estimate," Quarterly of Applied Mathematics (submitted).

Howard A. Levine
Department of Mathematics
Iowa State University
Ames, IA 50010

On Some Almost Everywhere Symmetry Theorems

JOHN L. LEWIS and ANDREW VOGEL

1. Introduction

In this note we consider problems of the following type: Let D be a bounded domain in Euclidean n space (R^n) and suppose u is a positive solution to $Lu = 0$ in D with $u(x) \to 0$, $|\nabla u(x)| \to$ constant, as $x \to \partial D$, in an appropriate sense. Show that D is a ball and u is radially symmetric about the center of D. This problem was solved very elegantly by Serrin [25] under the assumption that ∂D is of class C^2, $u \in C^2(\bar{D})$, and

$$Lu = a(u, |\nabla u|)\Delta u + b(u, |\nabla u|)u_{x_i}u_{x_j}u_{x_i x_j} - c(u, |\nabla u|) = 0, \qquad (1.1)$$

where a, b, c, are continuously differentiable in each variable. Also L is elliptic and repeated indices denote summation from 1 to n. Immediately following Serrin's article, Weinberger [30] gave another proof of Serrin's theorem when $Lu = \Delta u + 1 = 0$. From Weinberger's proof it is clear that the assumption, $\partial D \in C^2$ is unnecessary in this special case, provided the boundary conditions are interpreted as,

(A) Given $\epsilon > 0$ there exists a neighborhood N of ∂D such that
$u(x) < \epsilon$, $||\nabla u(x)| - a| < \epsilon$, when $x \in N$.

In (A), a denotes a positive constant.

Garafalo and the first author [17] extended Weinberger's proof to weak solutions u in D of

$$\text{div}(|\nabla u|^{-1}f'(|\nabla u|)\nabla u) = -1, \qquad (1.2)$$

where $f \in C^2(0, \infty)$ and for some p, $1 < p < \infty$, $c_1, c_2 > 0$,

$$c_1(t^p - 1) \leq tf'(t) \leq c_2(t^p + 1), \ t > 0, \qquad (1.3)$$

$$c_1 \leq tf''(t)/f'(t) \leq c_2, \ t > 0, \qquad (1.4)$$

$$\int_D |\nabla u|^p dx < +\infty. \qquad (1.5)$$

Thus for u satisfying (1.2), (1.5), with boundary values as in (A) it is still true that D is a ball and u is radially symmetric about the center of D.

Supported in part by the NSF and the Commonwealth of Kentucky through the Kentucky EPSCoR program

The second author [29] has recently shown that the same conclusion is valid when u satisfies (1.1) with boundary values as in (A). His results are also valid for equations of the form

$$\operatorname{div}(|\nabla u|^{-1} f'(|\nabla u|)\nabla u) = d(u, |\nabla u|), \tag{1.6}$$

where f, u, are as in (1.3)-(1.5), provided f has continuous third derivatives and d is continuously differentiable in each of its arguments. Details of this proof will appear elsewhere, however we give a brief outline here.

Most of the effort in the proof is devoted to formalizing the statement: If near the boundary of a general region, u is C^2 and satisfies an elliptic equation, as well as (A), then this is enough to prove some boundary regularity. More specifically,

Theorem A. *Let u satisfy (1.1) in D with boundary values as in (A). Then given $x_0 \in \partial D$ there exists $r > 0$ so that $B(x_0, r) \cap D$ consists of at most two components D', D'', satisfying*
(i) $x_0 \in \partial D' \cap \partial D''$
(ii) $\partial D' \cap B(x_0, r)$, $\partial D'' \cap B(x_0, r)$ *are $C^{2,\alpha}$ surfaces for some α, $0 < \alpha < 1$*
(iii) $\partial D \cap B(x_0, r) = (\partial D' \cap B(x_0, r)) \cup (\partial D'' \cap B(x_0, r))$.

In Theorem A, $B(x_0, r) = \{x : |x - x_0| < r\}$. Note that (i)-(iii) does not imply ∂D is $C^{2,\alpha}$ in the usual sense, but intuitively, ∂D is $C^{2,\alpha}$ from "each side".

To begin the proof of Theorem A, the second author first studies curves in D along which ∇u is tangent (the trajectories of u). If $t > 0$ is small enough it follows from (A) and positivity of u that for each x_0 in $\{x : u(x) = t\}$ there exists exactly one trajectory beginning at x_0 and ending at $\psi(x_0)$ in ∂D, along which u decreases. Moreover, it is easily seen that ψ is a continuous function from $\{x : u(x) = t\}$ onto ∂D. Note from (A) and the implicit function theorem that $\{x : u(x) = t\}$ has a finite number of components. Hence ∂D has a finite number of components, since ψ is continuous. Using these trajectories and estimates on u, it can also be shown for given $x_0 \in \partial D$ and $\sigma > 0$ small, that there exists $r > 0$, a direction ν, and a region D' with
(a) $x_0 \in \partial D'$
(b) $D' \cap B(x_0, r) \subseteq D \cap B(x_0, r) \cap H(\sigma r)$
(c) $D' \cap H(-\sigma r) \cap B(x, r) = H(-\sigma r) \cap B(x_0, r)$
where $H(t) = \{x : (x - x_0) \cdot \nu < t\}$. (a)-(c) are conditions similar to the flatness condition introduced by Alt and Cafferelli in [3] (see also [4]). The method used in these papers can be suitably modified to show that preliminary flatness in the above sense can be improved in a smaller ball about x_0 of radius $c_0 r$ $(0 < c_0 < 1)$.

Iterating this result it follows that ∂D is $C^{1,\alpha}$ from each side at x. To improve the regularity a hodograph type transformation of Kinderlehrer and Nirenberg [21] is then used so that regularity results of Agmon-Douglis-Nirenberg [1] can be applied to obtain ∂D is $C^{2,\alpha}$ from each side. Finally, Serrin's argument can be adapted, to obtain that D is a ball.

2. Almost Everywhere Symmetry Theorems

Next we consider some symmetry problems where boundary condition (A) can be weakened at the expense of assuming more about ∂D. To this end we say that a bounded domain D is Lipschitz provided for each $y \in \partial D$ there is $r = r(y) > 0$ and a truncated right circular cylinder Z, with radius r, y on the axis at the center of the cylinder, and the property: If the axis of the cylinder is chosen parallel to the x_n axis, then there exists a Lipschitz function ψ on R^{n-1} $n \geq 2$, such that $\partial D \cap Z = Z \cap \{(x', \psi(x')) : x' \in R^{n-1}\}$ and $D \cap Z = Z \cap \{(x', x_n) : x_n > \psi(x')\}$. Also the bases of Z are at some positive distance from ∂D. From compactness we see that ∂D is contained in the union of a finite number of such cylinders. We say that u defined on $D \cap Z$ has a radial limit at $y = (y', \psi(y'))$ in $\partial D \cap Z$ provided $\lim_{t \to 0} u(y', \psi(y') + |t|)$ exists finitely. If $0 < r_0 < r$ and $(x', \psi(x') + 2r_0)$ is in D whenever $(x', \psi(x')) \in \partial D \cap Z$ put

$$u^*(y) = u^*(y', \psi(y')) = \sup_{|t| < r_0} |u(y', \psi(y') + |t|)|.$$

u^* is called the radial maximal function of u relative to r_0. Let H^k denote k dimensional Hausdorff measure. We note that on $\partial D \cap Z$ we have for $y \in \partial D$,

$$dH^{n-1}y = \sqrt{1 + |\nabla \psi|^2(y')}dy'.$$

For D a bounded Lipschitz domain replace boundary condition (A) by

(A*) $\lim_{x \to y} |\nabla u(x)| = a$, radially, for H^{n-1} a.e. $y \in \partial D$, while $u(x) \to 0$
continuously as $x \to \partial D$ (as in (A)).

We remark that the assumption $u \to 0$ continuously, is true in a Lipschitz domain for a solution u to any of the above partial differential equations provided u can be approximated arbitrarily close in the norm of a certain Sobolev space (depending on the P.D.E.) by infinitely differentiable functions with compact support in D. In case u is Green's function for D we can prove almost everywhere symmetry theorems rather easily. More specifically,

Theorem 1. *Let $D \subseteq R^n$ be a bounded Lipschitz domain and suppose that u is Green's function for D with pole at $0 \in D$. If u satisfies boundary condition (A*), then D is a ball with center at 0 and u is radially symmetric about 0.*

Our proof of Theorem 1 will involve, among other things, a simple barrier argument.

Proof. Let $\omega_2 = (2\pi)^{-1}$, and $\omega_n = \{(n-2)H^{n-1}[\partial B(0,1)]\}^{-1}$ if $n > 2$. Then by definition, $u(x) - \omega_n|x|^{2-n}$ is harmonic and bounded in D for $n > 2$ (if $n = 2$, replace $|x|^{2-n}$ by $\log(|x|^{-1})$) while $u(x) \to 0$ as $x \to \partial D$ in the sense of Perron-Wiener-Brelot, so continuously for a Lipschitz domain (see [18, section 1.6.3]). The first step is to show that

$$aH^{n-1}(\partial D) = 1. \tag{2.1}$$

To prove (2.1) let Ω be a smooth domain with $\bar{\Omega}$ (the closure of Ω) contained in D. Then from the divergence theorem and the usual limiting argument it follows that

$$-\int_{\partial\Omega} \nabla u \cdot \nu dH^{n-1} = 1 \tag{2.2}$$

provided $0 \in \Omega$. Here ν is the outer unit normal to Ω. The idea now is to use (A*) and choose a sequence of smooth domains $(\Omega_j)_1^\infty$ such that $\bar{\Omega}_j \subseteq \Omega_{j+1}$, $\bigcup_{j=1}^\infty \Omega_j = D$, and with the property that (2.2) with $\Omega = \Omega_j$ approaches (2.1) as $j \to \infty$. To justify the limits we shall need to know that:

(+)   The radial maximal function of ∇u for some $r_0 > 0$, taken componentwise and denoted $(\nabla u)^*$, is square integrable with respect to H^{n-1} measure on $\partial D((\nabla u)^* \in L_2(\partial D))$,

(++) $\nabla u(x) \to -an(y)$ radially for H^{n-1} almost every $y \in \partial D$ where $n(y) = \frac{(\nabla\psi(y'),-1)}{\sqrt{1+|\nabla\psi(y')|^2}}$, $y \in \partial D$, is the outer unit normal to D.

(+) and (++) were proved by Dahlberg in [9, Thm. 3], [10, Thm. 2]. The $L_2(\partial D)$ norm of $(\nabla u)^*$ depends only on the Lipschitz norm of the functions ψ defining ∂D. Another way to prove (+) and (++) is to note that $u(x) - \omega_n|x|^{2-n}$ has tangential derivatives on ∂D which are in $L_2(\partial D)$. It then follows from a theorem of Verchota (see [28, Cor. 3.5]) that u can be represented as a single layer potential for which (+) and (++) hold. To get $(\Omega_j)_1^\infty$, we first smoothly approximate ψ locally from above in such a way that the sequence of approximants has uniformly bounded Lipschitz norm. Second piece together the resulting graphs to obtain $(\Omega_j)_1^\infty$ (see [28, Thm. 1.12] for more details). From (+), (++), and Lebesgue dominated convergence we deduce that (2.1) is the limit of (2.2) as $j \to \infty$.

Next given $y \in \partial D$ we claim that

$$\limsup_{x \to y} |\nabla u(x)| \le a \tag{2.3}$$

To prove this claim, we observe that $|\nabla u|$ is subharmonic in $D \cap B(y,r)$ for $r > 0$ small enough. Let $(\delta_j)_1^\infty$ be a sequence of small positive numbers with $\lim_{j\to\infty} \delta_j = 0$, and put

$$D + (0, \delta_j) = \{x + (0, \delta_j) = (x', x_n + \delta_j) : x \in D\}.$$

Choose a sequence of smooth domains $(\Omega_j)_1^\infty$ whose boundaries are locally the graphs of Lipschitz functions with uniformly bounded Lipschitz norm and for which

$$(D + (0, \delta_j)) \cap B(y, 2r) \subseteq \Omega_j \subseteq D \cap B(y, 8r), \quad j = 1, 2, \ldots$$

Let h_j be the least harmonic majorant of

$$(|\nabla u| - a)^\wedge = \max[|\nabla u| - a, 0]$$

in Ω_j and let $g_j(\cdot, w)$, $j = 1, 2, \ldots$, be Green's function for Ω_j with pole at $w \in \Omega_j$. Then from the classical Poisson integral formula for smooth domains (see [18, section 1.5]) we have

$$(|\nabla u| - a)^\wedge(x) \le h_j(x) = \int_{\partial \Omega_j} |\nabla g_j|(z, x)(|\nabla u| - a)^\wedge(z) dH^{n-1} z \quad (2.4)$$

whenever $x \in \Omega_j$. From (+) we see for given $x \in B(y, r) \cap D$ and $j \ge j_0$ large, that

$$\int_{\partial \Omega_j} |\nabla g_j|^2(z, x) dH^{n-1} z \le k,$$

where k may depend on x, r, and D but is independent of $j \ge j_0$. Using this inequality, (+), (++), (A*), and letting $j \to \infty$ in (2.4) we see from Hölder's inequality and Lebesgue dominated convergence that for properly chosen $(\Omega_j)_1^\infty$, we have

$$(|\nabla u| - a)^\wedge(x) \le h(x), \quad x \in D \cap B(y, r),$$

where h is harmonic in $D \cap B(y, r)$ with boundary value 0 on $\partial D \cap B(y, r)$ in the sense of Perron-Wiener-Brelot. Because a Lipschitz domain is regular for the Dirichlet problem, it follows first that h has a continuous extension to $(D \cup \partial D) \cap B(y, r)$ with $h \equiv 0$ on $\partial D \cap B(y, r)$ and there upon that (2.3) is true.

Finally we are in a position to prove Theorem 1. Let d be the distance from 0 to ∂D and let G be Green's function for $B(0, d)$ with pole at 0. Then clearly from the minimum principle for harmonic functions either $u - G$ is a positive harmonic function in $B(0, d)$ or $u \equiv G$ (so $D = B(0, d)$). Let $x_0 \in \partial B(0, d) \cap \partial D$ and observe from the mean value theorem of calculus that

$$-\frac{\partial}{\partial t} u(t x_0) \ge -\frac{\partial}{\partial t} G(t x_0),$$

for some $t < 1$ and arbitrarily near 1. Using this inequality, (2.1), and (2.3) we get

$$\frac{1}{H^{n-1}(\partial D)} = a \ge -\limsup_{t \to 1} \frac{\partial}{\partial t} u(t x_0) \ge -\lim_{t \to 1} \frac{\partial G}{\partial t}(t x_0) = a^*. \quad (2.5)$$

Since G satisfies boundary condition (A*) we can repeat the argument leading to (2.1) to obtain

$$\frac{1}{H^{n-1}(\partial B(0,d))} = a^*.$$

This equality and (2.5) yield

$$H^{n-1}(\partial D) \le H^{n-1}(\partial B(0,d)). \tag{2.6}$$

From the classical isoperimetric inequality:

$$H^n(D)^{1-1/n} \le (\nu_n)^{-1/n} n^{(1/n-1)} H^{n-1}(\partial D),$$

where $\nu_n = H^{n-1}(\partial B(0,1))$, we see that (2.6) can hold only if $D = B(0,d)$. Another way to see that (2.6) implies $D = B(0,d)$ is to project ∂D radially onto $\partial B(0,d)$ and use the fact that surface area decreases under this projection, unless $\partial D = \partial B(0,d)$. Q.E.D.

We remark that a proof of Theorem 1 for smooth domains can be given using Serrin's original argument or as in [23].

Theorem 1 generalizes to certain domains in Hyperbolic and Spherical n space (denoted H_n, S_n, respectively). In the usual way we identify H_n with $B(0,1)$ under the Riemannian metric,

$$g_{ij}(x) = 4\delta_{ij}(1 - |x|^2)^{-2}, \quad x \in B(0,1),$$

and S_n with R^n under the Riemannian metric,

$$g_{ij}(x) = 4\delta_{ij}(1 + |x|^2)^{-2}, \quad x \in R^n, \quad 1 \le i, \ j \le n.$$

Here, δ_{ij} denotes the Kronecker delta and we also use $(g^{ij}) = (g_{ij})^{-1}$, $g = \det(g_{ij})$. The definition of a bounded Lipschitz domain D is unchanged, provided bounded is interpreted with respect to the usual distance function for H_n, S_n. By definition, if u denotes Green's function for $D \subseteq H_n$ or $D \subseteq S_n$ with pole at $0 \in D$, then $0 = \tilde{\Delta} u(x)$ for $x \in D - \{0\}$, where

$$\tilde{\Delta} u = g^{-1/2} \frac{\partial}{\partial x_i}(g^{1/2} g^{ij} u_{x_j}) = \frac{1}{4}(1 \pm |x|^2)^n \frac{\partial}{\partial x_i}\left[(1 \pm |x|^2)^{2-n} \frac{\partial u}{\partial x_i}\right]. \tag{2.7}$$

Here the $+$ sign is taken if $D \subseteq S_n$ and the $-$ sign if $D \subseteq H_n$. Moreover, if θ has compact support in D, then

$$\int_D (\nabla\theta \cdot \nabla u)(1 \pm |x|^2)^{2-n} dH^n x = \theta(0). \tag{2.8}$$

Again the $+$ sign is taken if $D \subseteq S_n$ and the $-$ sign if $D \subseteq H_n$. Computing the Hyperbolic and Spherical gradients with respect to each (y_{ij}) we find that (A^*) should be replaced with

(A^{**}) $\displaystyle\lim_{x \to y} |\nabla u(x)| = a(1 \pm |y|^2)^{-1}$, for H^{n-1} a.e. $y \in \partial D$, while $u(x) \to 0$
continuously as $x \to \partial D$.

The sign convention is the same as above. With this notation we prove

Theorem 2. *Let u be Green's function for a bounded Lipschitz domain D with pole at $0 \in D$. If $D \subseteq H_n$, then D is a Hyperbolic ball while if $D \subseteq S_n$ and $H^n(D) \leq \frac{1}{2} H^n(S_n)$ (considered as sets on the unit sphere in R^{n+1}), then D is a Spherical ball.*

Proof. We argue as in Theorem 1. In place of (2.1) we show

$$a \int_{\partial D} (1 \pm |x|^2)^{1-n} dH^{n-1} x = 1, \tag{2.9}$$

where again the $+$ sign is taken if $D \subseteq S_n$ and the $-$ sign if $D \subseteq H_n$. The proof of (2.9) is essentially the same as the proof of (2.1) once we show $(+)$ and $(++)$ (with $\nabla u(x)$ replaced by $(1 \pm |x|^2)\nabla u(x)$) are valid in this situation. One way to prove $(+)$ and $(++)$ is to use the mapping, $(x', \psi(x') + \lambda) \to (x', \lambda)$, $x' \in R^{n-1}$, $\lambda > 0$, to map, $B(y, r) \cap D$, $y \in \partial D$, onto a portion of a half space. If $q(x', \lambda) = u(x', \psi(x') + \lambda)$, then q satisfies a divergence type equation for which the results of Fabes, Jerison, and Kenig [15] can be applied (see also [16, Thm. 3.5]). Doing this, we get $(+)$, $(++)$. Another proof of $(+)$, $(++)$, can be given, by using Rellich-Necas-Pohožaev type inequalities (see [17, 19, 24,28]) to show that $|\nabla u|$ has a certain weak limit in $L_2(\partial D)$. The rest of Dahlberg's proof can then essentially be repeated to get $(+)$, $(++)$. Using $(+)$, $(++)$, we obtain (2.1) in the same way as previously. To prove

$$\limsup_{x \to y} |\nabla u(x)| \leq a(1 \pm |y|^2)^{-1}, \tag{2.10}$$

for all $y \in \partial D$, let $v(x) = (\epsilon + |\nabla u(x)|^2)^{1/2}$, $x \in D$, for given $\epsilon > 0$. We differentiate (2.7) with respect to x_k, $1 \leq k \leq n$. From the resulting equalities and (2.7) we deduce for r, $\epsilon > 0$ small enough,

$$\Delta v(x) \geq -c(|\nabla u(x)| + 1), \quad x \in D \cap B(y, 4r). \tag{2.11}$$

Using (2.11), the Riesz representation formula for subharmonic functions (see [18, Thm. 3.14]) and arguing as in the proof of (2.3) we find that

$$v(x) \leq p(x) + b(x), \quad x \in B(y, r) \cap D,$$

where b is harmonic in $B(y, r) \cap D \subseteq R^n$ with

$$\lim_{x \to z} b(x) = [a^2(1 \pm |z|^2)^{-2} + \epsilon]^{1/2}, \quad z \in \partial D \cap B(y, r), \tag{2.12}$$

and

$$p(x) = c \int_{B(y,2r)\cap D} g(y,x)(|\nabla u|(y) + 1)dy, \quad x \in D \cap B(y,2r).$$

Here, $g(\cdot,x)$ is Green's function for $D \subseteq R^n$ with pole at x. Now from (2.8) and (A**) it is easily seen for $r > 0$ small enough that $|\nabla u|$ is square integrable with respect to H^n measure on $D \cap B(y,2r)$. Using this fact and Sobolev's Theorem (see [27, Ch. 5]) we obtain p is integrable to the m-th power in $B(y,2r)$ where $m = 2n/(n-2)$ if $n > 2$ and $n < \infty$ if $m = 2$. Since b has continuous boundary values on $B(y,r) \cap \partial D$, we conclude that v is integrable to the m-th power in $B(y,r/2)$. Using Sobolev's Theorem again and repeating the above argument a finite number of times we see that $|\nabla u| \leq k < \infty$ in $B(y,r/2)$. From this inequality, (A**), and Harnack's inequality (see (4.25)) it follows easily that $\lim_{x\to z} p(x) = 0$, $z \in \partial D \cap B(y,r)$. Using this equality, (2.12), and letting $\epsilon \to 0$, we find (2.10) is true.

To prove Theorem 2, let G be Green's function for $B(0,d)$ with pole at 0, where d is the distance from 0 to ∂D in each geometry. Then G satisfies (2.7), (2.8), with u replaced by G and D by $B(0,d)$. From uniqueness of G, we see that G is radially symmetric. From the maximum principle for elliptic P.D.E.'s we also have $G \leq u$ in $B(0,d)$. Hence if $x_0 \in \partial B(0,d) \cap \partial D$, then from (2.10) we deduce as in (2.5)

$$a(1 \pm |x_0|^2)^{-1} \geq -\limsup_{t\to 1} \frac{\partial}{\partial t}u(tx_0) \geq -\lim_{t\to 1} \frac{\partial G}{\partial t}(tx_0) = a^*(1 \pm |x_0|^2)^{-1}, \tag{2.13}$$

so $a \geq a^*$. We note that (2.9) also holds with a replaced by a^* and D by $B(0,d)$ because G satisfies the same hypotheses as u. Thus

$$1 = a^* \int_{\partial B(0,d)} (1 \pm |x|^2)^{1-n}dH^{n-1}x = a \int_{\partial D} (1 \pm |x|^2)^{1-n}dH^{n-1}x. \tag{2.14}$$

Recall that the $+$ sign is taken if $D \subseteq S_n$ and the $-$ sign if $D \subseteq H_n$. Projecting ∂D onto $\partial B(0,d)$ and using (2.13) we see for $D \subseteq H_n$ that (2.14) can only hold when $D = B(0,d)$. If $D \subseteq S_n$, we identify D with its spherical image by way of stereographic projection. Then from the classical spherical isoperimetric inequality we have

$$H^{n-1}(\partial D) \geq H^{n-1}(\partial P), \tag{2.15}$$

where P is a spherical ball (cap) with the same H^n measure as D. Also from (2.13), (2.14), we see that

$$H^{n-1}(\partial D) \leq H^{n-1}(\partial B(0,d)), \tag{2.16}$$

for D, $B(0,d)$, contained in the unit sphere of R^{n+1}. Finally observe that if $P_1 \subseteq P_2 \subseteq Q$, where P_1, P_2 are spherical balls (caps), and Q is a hemisphere, then $H^{n-1}(\partial P_1) \leq H^{n-1}(\partial P_2)$. In view of this fact, (2.15), and

(2.16) we conclude $D = B(0, d)$, whenever $H^n(D) \leq \frac{1}{2} H^n(S_n)$.   Q.E.D.

3. Parabolic Symmetry Theorems

Let $D \subseteq R^n$ be a bounded Lipschitz domain, as in section 2. Let u be a function defined on $D \times (0, T)$, $0 < T < \infty$. If $(y, t) \in \partial D \times (0, T)$, define the radial limit of u as in section 2 relative to $D \times \{t\}$. Replace (A*) by

(A$^+$) $\lim\limits_{x \to y} |\nabla u|(x, t) \quad = \quad a(t)$, radially for H^n almost every

$(y, t) \in \partial D \times (0, T)$, $0 < t < T$, while $u(x, t) \to 0$ continuously as $(x, t) \to \partial D \times [0, T)$.

We prove

Theorem 3. *Let D be a Lipschitz domain, $0 \in D$, and suppose that u is Green's function for the heat equation in $D \times [0, T)$ with pole at $(0, 0)$. Then D is a ball with center at $(0, 0)$ and for fixed t, $0 < t < T$, $u(\cdot, t)$ is radially symmetric about the center of D.*

We remark that by definition

$$p(x, t) = u(x, t) - (4\pi t)^{-n/2} e^{-(|x|^2/4t)}, \quad (x, t) \in D \times [0, T),$$

is a bounded solution to the heat equation in $D \times [0, T) (\Delta_x p = p_t)$ and u has boundary value zero in the Perron-Wiener-Brelot sense. Here Δ_x and ∇_x denote the Laplacian and gradient with respect to $x \in R^n$, only. Since every point in $\partial D \times [0, T)$ is regular for the heat equation, it follows that the assumption $u(x) \to 0$ continuously as $x \to \partial D \times [0, T)$ is automatic in this case. As motivation for Theorem 3, we also remark that Alessandrini and Garofalo generalized Serrin's theorem to smooth cylinders in [2], so Theorem 1 should have a similar generalization.

Proof. The proof of Theorem 3 is similar to the proof of Theorem 1. In place of (2.1) we show

$$H^{n-1}(\partial D) \left(\int_0^{T_1} a(t) dt \right) + \int_D u(x, T_1) dH^n x = 1, \qquad (3.1)$$

whenever $0 < T_1 < T$. For this purpose let Ω be a smooth domain with $\bar{\Omega} \subset D$. Applying the divergence theorem in $[\Omega \times (0, T)] - E$, to $\nabla_x u$ where

$$E = \{(x, t) : |t|^{1/3}, \ |x| \leq \epsilon\},$$

and letting $\epsilon \to 0$ we get

$$-\int_0^{T_1} \int_{\partial \Omega} \nabla_x u \cdot \nu dH^{n-1} x dt + \int_\Omega u(x, T_1) dH^n x = 1, \qquad (3.2)$$

where ν is the outer unit normal to Ω. As in section 2 the idea now is to approximate D by a sequence of smooth domains $(\Omega_j)_1^\infty$ in such a way that (3.2) with $\Omega = \Omega_j$ approaches (3.1) as $j \to \infty$. To do this we need to show as in section 2 that

$\quad(\cdot)\ (\nabla_x u)^* \in L_2[\partial D \times (0,T)]$
$\quad(\cdot\cdot)\ \nabla_x u(x,t) \to -a(t)n(y)$ for H^n a.e. $(y,t) \in \partial D \times (0,T)$, where
$\qquad n(y)$ is as in $(++)$.

$(\cdot)$ and $(\cdot\cdot)$ follow from the work of Fabes and Salsa [14, Thms. 1.3, 1.4, and 3.2]. They generalized Dahlberg's Theorem to the heat equation in cylinders of the form $D \times (0,T)$. $(\cdot)$ and $(\cdot\cdot)$ also are a consequence of a theorem of Brown (see [5, Thm. 6.1], [6]) who among other results obtained analogues of Verchota's work for the heat equation in cylinders. From $(\cdot)$ and $(\cdot\cdot)$ we conclude that (3.1) is the limit of (3.2) with $\Omega = \Omega_j$ as $j \to \infty$. Next we claim for H^1 almost every $t \in (0,T)$ that

$$\limsup_{(x,t)\to(y,t)} |\nabla u(x,t)| \le a(t), \quad \text{radially, for all } y \in \partial D. \qquad (3.3)$$

The proof of (3.3) is essentially the same as the proof of (2.3) if we assume for example that $a(t)$ is continuous on $(0,T)$. Otherwise, we must use slightly deeper results of the above authors: Let h be the unique solution to the heat equation in $D \times (0,T)$, with $h^* \in L_2[\partial D \times (0,T)]$, $h(x,0) = 0$ continuously, $x \in D$, and

$$\lim_{(x,t)\to(y,t)} h(x,t) = a(t) \quad \text{radially, for } H^n \text{ a.e. } (y,t) \text{ in } \partial D \times (0,T).$$

The existence of h follows from $(\cdot)$ and either [14, Thm. 3.2] or [5, Thm. 8.1]. Since $|\nabla u|$ is a subsolution to the heat equation it follows as in section 2 that for given $\epsilon > 0$ and $(y,t) \in \partial D \times (0,T)$, there exists $r_1 > 0$ with

$$|\nabla u|(x,s)| \le h(x,s) + \epsilon, \quad (x,s) \in [B(y,r_1) \cap D] \times (0,T).$$

From this inequality we see it suffices to prove

$$\lim_{x\to y} h(x,t) = a(t), \quad \text{radially, for a.e. } t \in (0,T), \text{ and all } y \in \partial D, \qquad (3.4)$$

in order to get (3.3). From $(\cdot),(\cdot\cdot)$, we deduce that parabolic measure (see [31, section 2]) with respect to $(0,\mathrm{T})$ is equal to

$$a(T - t)dH^{n-1}ydt \quad \text{for } H^n \text{ a.e. } (y,t) \in \partial D \times (0,T).$$

From this deduction and a theorem of Kemper ([20, Thm. 2.6]) it follows that (3.4) actually holds whenever

$$\lim_{\delta\to 0} \left[\int_0^\delta |a(s) - a(t)|a(T - s)ds \Big/ \left(\int_0^\delta a(T - s)ds \right) \right] = 0, \qquad (3.5)$$

$0 < t < T$. Finally, (3.5) follows from the usual Lebesgue differentiation Theorem, ($\cdot$), and ($\cdot\cdot$), once it is shown $u(t) \not\equiv 0$ for a.c. $t \subset (0,T)$. This last inequality, again by the above deduction, is equivalent to the assertion that H^n measure on $\partial D \times (0,T)$ is absolutely continuous with respect to parabolic measure at (0,T) which likewise is true by [14, Thm. 3.1]. We conclude from (3.5), (3.4), that (3.3) is true. To complete the proof of Theorem 3, let G be Green's function with pole at $(0,0)$ for the heat equation in $B(0,d) \times (0,T)$. Again d is the distance from $(0,0)$ to ∂D. Then clearly, $G \leq u$, so from (3.3) we see as in section 2 that

$$a^*(t) = \lim_{x \to y} |\nabla G(x,t)| \leq a(t), \quad \text{radially for a.e. } t \in (0,T), \tag{3.6}$$

whenever $y \in \partial B(0,d)$. Thus from (3.1),

$$H^{n-1}(\partial B(0,d)) \left(\int_0^{T_1} a^*(t)dt \right) + \int_D G(x,T_1)dH^n x \tag{3.7}$$

$$\leq H^{n-1}(\partial D) \left(\int_0^{T_1} a(t)dt \right) + \int_D u(x,T_1)dH^n x = 1,$$

with equality only if $D = B(0,d)$. Now equality must hold in (3.7) because (3.1) is also true with u replaced by G, D by $B(0,d)$, and $a(t)$ by $a^*(t)$. Hence $D = B(0,d)$. Since the boundary values of u are invariant under rotations in x, we conclude from uniqueness of u, that $u(\cdot,t)$ is radial, $t \in (0,T)$. $\hfill$ Q.E.D.

Next we note that if

$$k(x,t) = \begin{cases} (4\pi t)^{-n/2} \exp\left[-\frac{|x|^2}{4t}\right], & \text{if } x \in R^n, \quad t > 0; \\ 0, & \text{if } x \in R^n, \quad t \leq 0. \end{cases}$$

denotes the Green's function for the heat equation in $R^n \times R$, then for given $\lambda > 0$,

$$|\nabla_y k|(y,t) = \lambda|y|/(2t), \quad \text{on } \{(x,t) : k(x,t) = \lambda\}. \tag{3.8}$$

Using (3.8) we shall obtain a different generalization of Theorem 1 to domains whose boundaries can be rough in the time variable. To this end suppose now $D \subseteq R^n \times R$ is bounded and for given $(y,t) \in \partial D \cap [R^n \times (0,T)]$ there exists $r > 0$ such that after a possible rotation in the x variable:

$$Z \cap \partial D = \{(x', x_n, s) : x_n = \psi(x',s), x' \in R^{n-1}, s \in R\} \cap Z$$

$$Z \cap D = \{(x', x_n, s) : x_n > \psi(x',s), x' \in R^{n-1}, s \in R\} \cap Z$$

where $Z \subset R^n \times R$ is a truncated circular cylinder of radius r with $(y, t) = (y', \psi(y', t), t)$ at the center of the cylinder and axis parallel to the x_n axis. Also the bases of Z have a positive distance to ∂D. In case $n = 1$ delete x' from the above equation. Here ψ is a function on $R^{n-1} \times R$ with compact support and the following properties: For each fixed t, $\psi(\cdot, t)$ is Lipschitz on R^{n-1} with

$$||\psi(\cdot, t)||^* \leq a_1 < \infty, \tag{3.9}$$

while for each fixed $x' \in R^{n-1}$,

$$\psi(x', t) = \int_R |s - t|^{-1/2} b(x', s) ds, \quad t \in R,$$

where $b(x', \cdot)$ is of bounded mean oscillation on R with

$$||b(x', \cdot)||^\wedge \leq a_2 < \infty \tag{3.10}$$

Again if $n = 2$, remove equation (3.9) and delete x' from (3.10). Also, $||\ ||^*$, $||\ ||^\wedge$, denote the Hölder and BMO norms, respectively. Let

$$D^* = D \cap [R^n \times (0, T)] \text{ and } \partial' D^* = \partial D \cap [R^n \times (0, T)]$$

and put

$$d\sigma(y', t) = \sqrt{1 + |\nabla_{y'} \psi|^2(y', t)} \, dy' dt, \quad (y, t) \in \partial' D^* \cap Z.$$

It is easily seen that σ is well defined on $\partial' D^*$ independently of y. In fact if $D(t) = D \cap (R^n \times \{t\})$ and f is integrable with respect to σ, then $\int_{\partial' D^*} f d\sigma = \int_0^T \left(\int_{\partial D(t)} f dH^{n-1} \right) dt$. The radial limit of a function u at $(y, t) \in \partial' D^*$ is defined to be $\lim_{\alpha \to 0} u(y', \psi(y', t) + |\alpha|, t)$ provided this limit exists, and if $(x', \psi(x', s) + 2r_0, s) \subseteq D^* \cap Z$, whenever $(x', \psi(x', s), s)$ is in $\partial' D^* \cap Z$, then the nontangential maximal function, u^*, relative to $r_0 > 0$ at (y, t) is,

$$u^*(y, t) = \sup_{|\alpha| \leq r_0} |u(y', \psi(y', t) + |\alpha|, t)|.$$

Replace (A*) by

(A$^\sim$) $\lim_{(x,s) \to (y,t)} |\nabla u|(x, s) = a|y|/t$, radially, for σ almost every (y, t) in $\partial' D^*$ while u has continuous boundary values 0 on $\partial' D^*$

Since the only case of interest in the theorem to follow is when $(0, 0) \in \partial D$, we must extend the definition of the Green's function with pole at $(0,0)$. For this purpose suppose $(0, 0) \in \bar{D}$ and for some $\tau > 0$ that

$$\{(x, t) : k(x, t) > \tau\} \cap [R^n \times (0, T)] \subseteq D^*. \tag{3.11}$$

Let $u > 0$ be the positive solution to the heat equation in D for which

$$u(x,t) = k(x,t) + q(x,t), \quad (x,t) \in D,$$

where q is the bounded solution to the heat equation in D with boundary values: $q = -k$ on $\partial D - \{0\}$, in the sense of Perron-Wiener-Brelot. Existence of q, $-\tau \leq q \leq 0$, follows from the usual Perron family argument, thanks to (3.11). Note that u is just the Green's function for D with pole at $(0,0)$ when $(0,0) \in D$. Finally we point out that for D satisfying the above conditions every point in $\partial' D^*$ is regular (see [31, section 1]). Hence the assumption, $u(x) \to 0$ as $x \to \partial' D^*$, continuously, is unnecessary for the Green's function of D with pole at $(0,0)$. With this notation we prove

Theorem 4. *Let D be as above and let u be Green's function for the heat equation in D with pole at $(0,0) \in \bar{D}$. There exists $a_0 > 0$ such that if $a_1, a_2 \leq a_0$ in (3.9), (3.10), and u has boundary values as in ($A^{\sim}$), then for some $\lambda > 0$, $\partial' D^* \subseteq \{(x,t) : k(x,t) = \lambda\}$ and $u \equiv k - \lambda$ in D^*.*

Note that Theorem 1 could be restated, as above.

Proof. The proof of Theorem 4 is similar to the proof of Theorem 3. In place of (3.1) we want to show for arbitrary $T_1, T_2, \ 0 < T_1 < T_2 < T$,

$$a \int_{T_1}^{T_2} \left(\int_{\partial D(t)} |x| dH^{n-1} x \right) \frac{dt}{t} + \int_{D(T_2)} u(x, T_2) dH^n x = \int_{D(T_1)} u(x, T_1) dH^n x$$

$$\tag{3.12}$$

Let Ω be a smooth domain with $\bar{\Omega} \subseteq D$. Then from the divergence theorem we deduce for $\Omega^* = \Omega \cap (R^n \times [T_1, T_2]), \ 0 < T_1 < T_2$, and $\Omega(t) = \Omega \cap (R^n \times \{t\}), \ 0 < t < T$,

$$-\int_{\partial \Omega^*} \nabla_x u \cdot \nu_x dH^n - \int_{\partial \Omega^*} u \nu_t dH^n = 0 = \int_{\Omega(T_2)} u(x, T_2) dH^n x - \int_{\Omega(T_1)} u(x, T_1) dH^n x$$

$$\tag{3.13}$$

$$- \int_{T_1}^{T_2} \left(\int_{\partial \Omega(t)} \nabla_x u \cdot \nu_x' dH^{n-1} x \right) dt - \int_{T_1}^{T_2} \left(\int_{\partial \Omega(t)} u \nu_t' dH^{n-1} x \right) dt,$$

where $\nu = (\nu_x, \nu_t)$ is the outer unit normal to Ω, $\nu_x' = |\nu_x|^{-1}(\nu_x, 0)$, $\nu_t' = (0, \nu_t)|\nu_x|^{-1}$. Again we shall obtain (3.12) as the limit of (3.13) when $\Omega \in (\Omega_j)_1^\infty$. To justify the limit we need a stronger version of $(\cdot)$, $(\cdot\cdot)$, namely for a_0 small enough,

    $(-)$ $\nabla_x u$ taken componentwise is locally square integrable with respect to σ on $\partial' D^*$

    $(--)$ $\nabla_x u \to -a(|y|/t) n(y, t)$, radially, for σ almost every $(y, t) \in \partial' D^*$

        where $n(y, t) = \dfrac{(\nabla_{y'} \psi(y', t), -1, 0)}{\sqrt{1 + |\nabla_{y'} \psi(y', t)|^2}}$

$(-), (--)$ follow from the work of Murray and the first author in [22]. The sequence $(\Omega_j)_1^\infty$ can be obtained by piecing together smooth approximates to ψ from above locally; since by compactness, $\partial' D^* \cap (R^n \times [T_1, T_2])$ is contained in a finite union of cylinders. It should be noted that convolution of ψ with an approximant identity in both the x and t variables separately, gives a smooth function with Lipschitz and BMO norms still bounded by a_1, a_2, respectively. Using this fact, $(-), (--)$, dominated convergence, and taking a limit in (3.13) with $\Omega = \Omega_j$ as $j \to \infty$, we get (3.12).

Next, we let $T_1 \to 0$ in (3.12) and use the fact that

$$\int_{D(T_1)} u(x, T_1) dH^n x \leq \int_{R^n \times \{T_1\}} k(x, T_1) dH^n x = 1,$$

to deduce

$$a \int_0^{T_2} \left(\int_{\partial D(t)} |x| dH^{n-1} x \right) t^{-1} dt + \int_{D(T_2)} u(x, T_2) dH^n x \leq 1. \tag{3.14}$$

Clearly, (3.14) implies that $0 \in \partial D$. We shall also need

$$\limsup_{(x,s) \to (y,t)} |\nabla u|(x, s) \leq a|y|/t, \quad \text{whenever } (y, t) \in \partial' D^*. \tag{3.15}$$

This inequality follows from the work in [22] in a way similar to the proof of (2.3). We omit the details. To continue the proof of Theorem 4, let $\lambda(\epsilon)$, $0 < \epsilon < \frac{1}{4}$, be the smallest of the numbers γ such that

$$q(x, t) \geq -\gamma + \epsilon \ln|x|, \quad \text{for all } (x, t) \in D^* \cap [R^n \times (0, T_2)],$$

where q is as in the definition of u. Using (3.11), and the fact that $u < k$ in D^* we see there exists $\epsilon_0 > 0$ (small) and $0 < b_1 < b_2 < \infty$, such that

$$b_1 < \lambda(\epsilon) < b_2, \quad 0 \leq \epsilon \leq \epsilon_0. \tag{3.16}$$

Moreover, since $q(x, t) + \lambda(\epsilon) - \epsilon \ln|x|$, is a supersolution to the heat equation in D^*, and q is continuous and bounded on $D - \{0\}$, we see from the minimum principle for supersolutions to the heat equation that

$$q(x, t) + \lambda(\epsilon) - \epsilon \ln|x| = 0,$$

for some $x = x(\epsilon)$, $t = t(\epsilon)$, with $(x, t) \neq (0, 0)$ and $(x, t) \in \partial D \cap \{R^n \times [0, T_2]\}$. Moreover, since $q \equiv 0$ on $(R^n \times \{0\}) \cap [\partial D - \{(0, 0)\}]$ we see from (3.16) that for $\epsilon_0 > 0$ small enough we have $t(\epsilon) > 0$, $0 < \epsilon \leq \epsilon_0$. Using this fact, (3.15), and radial symmetry of $k(\cdot, t)$ as in (2.5) we obtain at $(x, t) = (x(\epsilon), t(\epsilon))$,

$$a\frac{|x|}{t} \geq \frac{|x|}{2t} k(x, t) - \frac{\epsilon}{|x|} \tag{3.17}$$

We now let $\epsilon \to 0$ and consider two cases: either (a) $\lim_{\epsilon \to 0} t(\epsilon) = 0$, which by the above reasoning implies $\lim_{\epsilon \to 0} x(\epsilon) = 0$ or (b) $\lim_{\epsilon \to 0} t(\epsilon) = t_0 > 0$, $\lim_{\epsilon \to 0} x(\epsilon) = x_0 \neq 0$, and $\lim_{\epsilon \to 0} \lambda(\epsilon) = \lambda_0$, for $\epsilon \in (\epsilon_i)_1^\infty$. In case (b) we see that $\lambda_0 = \lambda(0)$ and $k(x_0, t_0) = \lambda_0$. From (3.17) we conclude that in case (b),

$$\{(y, s) : k(y, s) > \lambda_0\} \cap [R^n \times (0, T_2)] \subseteq D^* \tag{3.18}$$

and

$$a \geq \frac{\lambda_0}{2}. \tag{3.19}$$

If case (a) occurs observe from (3.11) and (3.16) that for ϵ_1 small enough, $0 < \epsilon \leq \epsilon_1 < \epsilon_0$, $x = x(\epsilon)$, $t = t(\epsilon)$, we have

$$-\tau \leq -k(x, t) = q(x, t) = -\lambda(\epsilon) + \epsilon \ln|x| \leq -\frac{1}{2}b_1.$$

From this inequality we see for ϵ_1 small enough that there exists b_3, $0 < b_3 < \infty$, with

$$||x|^2 + 2nt \ln t| \leq b_3 t, \quad 0 < \epsilon \leq \epsilon_1.$$

Hence, $\lim_{\epsilon \to 0} (t/|x|^2) = 0$. Using this inequality in (3.17) and the fact that $k(x, t) \geq \lambda(\epsilon)$, ϵ small, we obtain for $\lambda_0 = \limsup_{\epsilon \to 0} \lambda(\epsilon)$ that (3.19) is still true. Also (3.18) remains valid, as is easily seen. Let

$$W = \{(y, s) : k(y, s) > \lambda_0\},$$

$$W(t) = W \cap (R^n \times \{t\}).$$

From (3.18) and the maximum principle for the heat equation we observe first that $u \geq k - \lambda_0$ in $D \cap [R^n \times (0, T_2)]$ and second that

$$\min_{x \in \partial D(t)} |x| \geq \max_{x \in \partial W(t)} |x|, \quad 0 < t < T_2.$$

Since $W(t)$ is a ball in $R^n \times \{t\}$ it follows from (3.19), the above observations, the isoperimetric inequality, and (3.14) that

$$\frac{\lambda_0}{2} \int_0^{T_2} \left(\int_{\partial W(t)} |x| dH^{n-1}x \right) t^{-1} dt + \int_{W(T_2)} (k(x, T_2) - \lambda_0) dH^n x$$

$$\leq a \int_0^{T_2} \left(\int_{\partial D(t)} |x| dH^{n-1}x \right) t^{-1} dt + \int_{D(T_2)} u(x, T_2) dH^n x \leq 1,$$

with equality only if $u \equiv k - \lambda_0$ in $D \cap [R^n \times (0, T_2)]$. Moreover equality must hold in this inequality, as it follows from (3.18) and the same argument used in proving (3.14) that

$$\frac{\lambda_0}{2} \int_0^{T_2} \left(\int_{\partial W(t)} |x| dH^{n-1}x \right) t^{-1} dt + \int_{W(T_2)} (k(x, T_2) - \lambda_0) dH^n x = 1.$$

Thus, $u \equiv k - \lambda_0$ in $D \cap [R^n \times (0, T_2)]$ and because T_2 is arbitrary, $0 < T_2 < T$, the proof is complete.   Q.E.D.

4. Sets of Finite Perimeter

In this section we suppose that D is a bounded domain of finite perimeter: By definition D is of finite perimeter if whenever ϕ is a smooth vector field defined on R^n, and $|\phi(x)| \leq 1$, $x \in R^n$, then

$$\int_D \nabla \cdot \phi \, dH^n \leq M < \infty,$$

If D is of finite perimeter it follows (see [13, section 5.8]) that

$$\int_D \nabla \cdot \phi \, dx = \int_{\partial^* D} \phi \cdot n(x) \, dH^{n-1} x, \tag{4.1}$$

where $\partial^* D$ (the reduced boundary of D), is the set of points where a certain measure has a derivative with respect to another. For our purposes it is enough to know that $\partial^* D$ is H^{n-1} a.e. equivalent to

$$\partial_* D = \{x \in \partial D : \limsup_{r \to 0} r^{-n} \min[H^n(B(x,r) \cap D), \ H^n(B(x,r) - D)] > 0\},$$

the so called measure theoretic boundary (see [13, section 5.8]) of D. We note that $n(y)$, $y \in \partial_* D$, is a measure theoretic outer normal in the sense that for H^{n-1} a.e. $y \in \partial_* D$,

$$\lim_{r \to 0} \{r^{-n} H^n[B(y,r) \cap D \cap K^+(y)]\} = 0 \tag{4.2}$$

$$\lim_{r \to 0} \{r^{-n} H^n[(B(y,r) - D) \cap K^-(y)]\} = 0,$$

where

$$K^+(y) = \{x \in R^n : n(y) \cdot (x - y) > 0\},$$

$$K^-(y) = \{x \in R^n : n(y) \cdot (x - y) < 0\}.$$

Approximating $n(y)$, $y \in \partial_* D$ by smooth functions on R^n we see from (4.1) that, $H^{n-1}(\partial_* D) < +\infty$. This inequality is also sufficient for D to be of finite perimeter [13, section 5.8]. We shall discuss possible extensions of Theorem 1 in domains D of finite perimeter with

$$H^{n-1}(\partial D - \partial_* D) = 0. \tag{4.3}$$

Observe that a Lipschitz domain clearly satisfies these conditions. One way to state boundary condition (A) which avoids the definition of radial limits

and is equivalent for Lipschitz domains due to Dahlberg's theorem; is to
require that for each Borel subset $E \subsetneq \partial D$,

$(\mathrm{A}^\wedge)$ $\mu(E) = aH^{n-1}(E) = aH^{n-1}(E \cap \partial_* D)$, where μ is harmonic measure of ∂D relative to 0, while $u(x) \to 0$ as $x \to \partial D$, continuously.

More specifically, let f be a continuous function on ∂D and suppose H_f
is the harmonic solution to the Dirichlet problem obtained by way of the
usual Perron family argument. Then

$$|H_f(x)| \leq \max_{x \in \partial D} |f(x)|, \quad x \in \partial D.$$

From the Riesz representation theorem it follows that there exists a regular Borel measure μ on ∂D so that the functional, $f \to H_f(0)$, can be
represented as

$$H_f(0) = \int f d\mu. \tag{4.4}$$

We would like to be able to extend Theorem 1 with (A) replaced by $(\mathrm{A}^\wedge)$
to domains of the above type. This it turns out is impossible. In fact
there exists simply connected domains $D \subset R^2$ (other than disks), which
are bounded by a rectifiable Jordan curve (so D is of finite perimeter and
(4.3) holds) for which harmonic measure with respect to 0 is a constant
multiple of H^1 on ∂D (condition $(\mathrm{A}^\wedge)$). For construction of these domains,
classically called non Smirnov domains see [11, section 10.4] for references.
An examination of the proof of Theorem 1 reveals that (2.3) must fail since
from (4.4) and $(\mathrm{A}^\wedge)$ we deduce

$$1 = \mu(\partial D) = aH^{n-1}(\partial D),$$

which is (2.1). Indeed, for a non Smirnov domain it is true that $|\nabla u(x)| \to$
$+\infty$ as $x \to y$ for some $y \in \partial D$. Therefore we assume for some $\lambda > 0$ that
$B(0, 2\lambda) \subseteq D$ and

$$|\nabla u(x)| \leq k < \infty, \quad x \in D - B(0, \lambda). \tag{4.5}$$

We prove

Theorem 5. *Let $D \subseteq R^n$ be a bounded domain of finite perimeter for
which (4.3) is valid. Let u be Green's function for D with pole at $0 \in D$
and suppose that u satisfies $(\mathrm{A}^\wedge)$, (4.5). Then D is a ball with center at
zero and u is radially symmetric about 0.*

As mentioned above the same argument as in Theorem 1 can be used
to prove Theorem 5, once we show (2.3) holds in this situation. Thus we
only prove (2.3).

Proof. The proof is essentially due to Alt and Caffarelli [3, Thm. 6.3].
Let $y \in \partial D, r < |y|/2$, and put

$$\sigma(r, u) = \sigma(r, u, y) = [H^{n-1}(\partial B(y, r))]^{-1} \left(\int_{\partial B(y, r)} u dH^{n-1} \right). \tag{4.6}$$

Observe from (4.6) and (4.5) that

$$\sigma(r, u) \le ckr. \tag{4.7}$$

From the Riesz representation formula for subharmonic functions (see [18, (3.9.1), (3.9.4)]) we have for $\nu_n = H^n(\partial B(0,1))$ as in section 1,

$$0 = u(y) = \nu_n \sigma(r, u) - \int_0^r \mu(B(y,t) \cap \partial D) t^{1-n} dt. \tag{4.8}$$

Using $(A^\wedge)$, (4.6), (4.7), and (4.8) it follows that

$$H^{n-1}(B(y, r/2) \cap \partial D) r^{1-n} \le cr^{-1} \int_{r/2}^r H^{n-1}(B(y,t) \cap \partial D) t^{1-n} dt \tag{4.9}$$

$$\le a^{-1} c \sigma(r, u) r^{-1} \le c.$$

In (4.7), (4.9), as in the rest of this section, c denotes a positive constant depending only on n, k, a, not necessarily the same at each occurrence. We note from (4.7)-(4.9) and $(A^\wedge)$ that for $\delta > 0$ small and $y \in \partial D$ that

$$cr H^{n-1}[B(y, r) \cap \partial D](\delta r)^{1-n} \ge \int_{\delta r}^r t^{1-n} \mu[B(y,t) \cap \partial D] dt$$

$$= \sigma(r, u) - \sigma(\delta r, u) \ge \sigma(r, u) - c\delta r.$$

From this inequality we see that if $k_1 \sigma(r, u) \ge r$, then there exists $\delta = \delta(k_1) > 0$ such that

$$H^{n-1}[B(y, r) \cap \partial D] \ge c(k_1) r^{n-1}, \tag{4.10}$$

where $c(k_1)$ is a positive constant depending on k_1, n, k, a.

Next let $(r_m)_1^\infty$ be a decreasing sequence of positive numbers with $\lim_{m \to \infty} r_m = 0$. Extend u to a continuous subharmonic function on $R^n - \{0\}$ by defining $u \equiv 0$ on $R^n - D$. For fixed $y \in \partial D$ let

$$v_m(x) = r_m^{-1} u[y + r_m x], \quad x \in R^n - \{-y/r_m\}, \ m = 1, 2, ...$$

Then from (4.5) we see that $(v_m)_1^\infty$ is a sequence of uniformly bounded Lipschitz functions in $R^n - \{-y/r_m : m = 1, 2, ...\}$. Thus a subsequence converges uniformly on compact subsets to a Lipschitz function v on R^n. In fact we claim for H^{n-1} a.e. $y \in \partial D$ that $v(x) = 0$ when $n(y) \cdot x > 0$, and

$$v(x) = -a(n(y) \cdot x), \quad \text{when} \ n(y) \cdot x \le 0. \tag{4.11}$$

To prove this claim we need the fact that for H^{n-1} a.e. $y \in \partial_* D$ ([13, section 5.7, Cor. 1]),

$$\lim_{r \to 0} [r^{1-n} H^{n-1}(\partial_* D \cap B(y, r))] = \alpha_n \tag{4.12}$$

where α_n is the volume of the unit ball in $R^{n-1}(\alpha_2 = 2)$. In view of (4.3) we can replace $\partial_* D$ by ∂D in this inequality. Now suppose that $y \in \partial D$ is a point where (4.2) and (4.12) hold. From (4.2) we see that $v \geq 0$ is subharmonic on R^n with $v \equiv 0$ on $\{x : n(y) \cdot x \geq 0\}$. Also from (4.8), (4.12), and ($A^\wedge$), we find

$$\sigma(\rho, v) = \sigma(\rho, v, 0) = \alpha_n a \rho \nu_n^{-1}, \quad 0 < \rho < \infty, \tag{4.13}$$

while from (4.5) it follows that for H^n a.e. x,

$$|v(x)| \leq c|x|, \quad x \in R^n. \tag{4.14}$$

From (4.14) and the Riesz representation formula for subharmonic functions in a halfspace we see that

$$v(x) = -\beta(n(y) \cdot x) - q(x), \quad n(y) \cdot x < 0, \tag{4.15}$$

where q is a Green's potential and $\beta \geq 0$. Now it follows from essentially the Phragmén-Lindelöf theorem (see [12]) that $\lim_{\rho \to \infty} \rho^{-1}\sigma(\rho, -q) = 0$. This equality and (4.13) imply $q \equiv 0$. Putting (4.15) with $q \equiv 0$ into (4.13) and using the divergence theorem we get $\beta = a$, so (4.11) is true.

Since each subsequence of $(v_m)_1^\infty$ converges to v and $(r_m)_1^\infty$ is arbitrary we conclude from (4.5) that

$$|u(x + y) + a(n(y) \cdot x)| \, |x|^{-1} \to 0, \tag{4.16}$$

uniformly as $|x| \to 0$. We note that if h is harmonic in $B(x_0, s)$ then from the Poisson integral formula it is easily shown that

$$|\nabla h(x)| \leq c s^{-(n+1)} \left(\int_{B(x_0,s)} |h| dH^n \right), \quad x \in B(x_0, s/2). \tag{4.17}$$

Let $d(x, \partial D)$ denote the distance from x to ∂D. Using (4.17) in (4.16) we conclude that if $k_2 d(x + y, \partial D) > |x|$, k_2 large, and $\eta > 0$ is given, then there exists $r_0 = r_0(\eta, k_2, y) > 0$ such that

$$||\nabla u(x + y)| - a| \leq \eta, \quad |x| \leq r_0. \tag{4.18}$$

Since (4.18) holds for H^{n-1} a.e. $y \in \partial D$, we see for fixed η, k_2, and

$$E(\epsilon) = \{y \in \partial D : r_0(\eta, k_2, y) \geq \epsilon\},$$

that

$$\lim_{\epsilon \to 0} H^{n-1}[\partial D - E(\epsilon)] = 0. \tag{4.19}$$

Next put $w(x) = \max[|\nabla u(x)| - a, 0]$, $x \in D - \{0\}$ and observe that w is subharmonic in $D - \{0\}$. Let $g(\cdot, y)$ be Green's function for D with pole at $y \in D$ and recall that $u = g(\cdot, 0)$. For fixed $x_0 \neq 0$ in $D - B(0, 3/2\lambda)$ let $r > 0$ be such that

$$B(0, \lambda) \cap \{x : g(x, x_0) \leq r\} = \{\phi\}.$$

If $D_1 = \{x : g(x, x_0) > r\} - B(0, \lambda)$, we also choose r so that $|\nabla g(\cdot, x_0)| \neq 0$ on ∂D_1, and $x_0 \in D_1 - B(0, \frac{3}{2}\lambda)$. Then from Green's second identity and subharmonicity of w we deduce

$$w(x_0) \leq -\int_{\partial D_1} w(y)\frac{\partial g}{\partial \nu}(y, x_0)dH^{n-1}y + \int_{\partial D_1} (g(y, x_0) - r)|\nabla w(y)|dH^{n-1}y$$

$$\tag{4.20}$$

$$\leq -\int_{\{y:g(y,x_0)=r\}} w(y)\frac{\partial g}{\partial \nu}(y, x_0)dH^{n-1}y + \int_{\partial B(0,\lambda)} (w(y)|\nabla g(y, x_0)|$$

$$+ g(y, x_0)|\nabla w(y)|)dH^{n-1}y = I_1(x_0) + I_2(x_0),$$

where ν is the outer unit normal to D_1. From Harnack's inequality we have $g(y, x_0) \leq cu(x_0)$ for $y \in B(0, \lambda)$ and $x_0 \in D - B(0, \frac{3}{2}\lambda)$. From this inequality, (4.17), and $(A^{\wedge})$ we get

$$I_2(x_0) \to 0 \quad \text{continuously, as } x_0 \to \partial D. \tag{4.21}$$

From (4.20) and (4.21) we see that in order to prove (2.3) it suffices to show for fixed $x_0 \in D - B(0, 3/2\lambda)$ that

$$I_1(x_0) \to 0 \quad \text{as } r \to 0. \tag{4.22}$$

As for (4.22) suppose $g(y, x_0) = r$, $d(y, \partial D) \geq k_3 r$. Then for r sufficiently small, say $0 < r \leq r_1$, we see from Harnack's inequality that there exists $c_1 = c_1(x_0, r_1, n) > 0$ such that

$$(c_1)^{-1}r = (c_1)^{-1}g(y, x_0) \leq u(y) \leq c_1 g(y, x_0) = c_1 r, \tag{4.23}$$

so from (4.17) we have for $r_1 > 0$ small, $0 \leq r \leq r_1$,

$$|\nabla u(y)| \leq \frac{c}{k_3 r}\sigma\left(\frac{1}{2}k_3 r, u, y\right) \leq \frac{c}{k_3 r}u(y) \leq \frac{c\, c_1}{k_3}.$$

Thus if $k_3 = k_3(a, x_0, r_1, n)$ is large enough, then $w(y) = 0$ and it follows that

$$I_1(x_0) = -\int_F w(y)\frac{\partial g}{\partial \nu}(y, x_0)dH^{n-1}y \tag{4.24}$$

where

$$F = \{y : g(y, x_0) = r, \ d(y, \partial D) < k_3 r\}.$$

Fix $k_3 > 0$ to be the smallest number such that (4.24) holds. If $y \in F$ observe from (4.23) and (4.5) that

$$kd(y, \partial D) \geq u(y) \geq (c_1)^{-1} g(y, x_0) = (c_1)^{-1} r. \qquad (4.25)$$

Given $\epsilon > 0$, let $k_2 = 8kk_3c_1$, $r < \epsilon/k_2$, and let F_1 be the set of all $y \in F$ such that there exists $z \in E(\epsilon)$ with $y \in B(z, 8k_3r)$. Then from (4.25), (4.18), we find that $w(y) \leq \eta$. Hence

$$- \int_{F_1} w(y) \frac{\partial g}{\partial \nu}(y, x_0) dH^{n-1}y \leq -\eta \int_{\{y: g(y, x_0) = r\}} \frac{\partial g}{\partial \nu}(y, x_0) dH^{n-1}y = c\eta. \qquad (4.26)$$

To handle the integral over $F - F_1$ we use a well known covering lemma (see [13, 1.5.2]) to get (z_m), $z_m \in F - F_1$, such that $F - F_1 \subseteq \cup B(z_m, 4k_3r)$ and each point in the union is contained in at most $N = N(n)$ balls. Now from (4.17), (4.23), and (4.5) we deduce for $y \in D$ and $s = d(y, \partial D) \leq 8k_3r$,

$$|\nabla g(y, x_0)| \leq cs^{-1} \sigma \left(\frac{s}{2}, g(\cdot, x_0), y \right) \leq c\, c_1 s^{-1} \sigma \left(\frac{s}{2}, u, y \right) \leq c_2, \qquad (4.27)$$

for some $c_2 = c_2(x_0, r_1, n, k) > 0$. Let $L_m = B(z_m, 4k_3r) \cap F$. Then from (4.27), (4.5), and the divergence theorem we obtain

$$- \int_{L_m} w(y) \frac{\partial g}{\partial \nu}(y, x_0) dH^{n-1}y \leq -k \int_{L_m} \frac{\partial g}{\partial \nu}(y, x_0) dH^{n-1}y \qquad (4.28)$$

$$\leq k \int_{\partial B(z_m, 4k_3r)} |\nabla g(y, x_0)| dH^{n-1}y \leq kc\, c_2(k_3r)^{n-1} = c_3 r^{n-1},$$

where c_3 depends on a, x_0, r_1, n, k, and k_3. Again by well known estimates for subharmonic functions and (4.23) we see there exists z_m^* in ∂D with $|z_m^* - z_m| < k_3r$ and

$$\sigma(2k_3r, u, z_m^*) \geq cu(z_m) \geq c(c_1)^{-1} g(z_m, x_0) = c_4 r.$$

Hence if r is replaced by $2k_3r$ in (4.10) and $k_1 = 2k_3/c_4$, then from (4.10) we obtain

$$r^{n-1} \leq c_5 H^{n-1}[B(z_m^*, 2k_3r) \cap \partial D] \leq c_5 H^{n-1}[B(z_m, 4k_3r) \cap \partial D],$$

where $c_5 = c_5(a, x_0, r_1, n, k, k_1, k_3) > 0$. Using this inequality in (4.28) we conclude

$$- \int_{L_m} w(y) \frac{\partial g}{\partial \nu}(y, x_0) dH^{n-1}y \leq c_6 H^{n-1}[B(z_m, 4k_3r) \cap \partial D],$$

where c_6 has the same dependence as c_5. Summing this inequality it follows that

$$-\int_{F-F_1} w(y)\frac{\partial g}{\partial \nu}(y,x_0)dH^{n-1}y \leq -\sum \int_{L_m} w(y)\frac{\partial g}{\partial \nu}(y,x_0)dH^{n-1}y \tag{4.29}$$

$$\leq c_6\left(\sum_m H^{n-1}[B(z_m,4k_3r)\cap \partial D]\right)$$

$$\leq c\, c_6 H^{n-1}\left\{[\bigcup_m B(z_m,4k_3r)]\cap \partial D\right\}$$

$$\leq c\, c_6 H^{n-1}(\partial D - E(\epsilon)),$$

because

$$\partial D \cap \bigcup_m B(z_m,4k_3r) \subseteq \partial D - E(\epsilon).$$

Combining (4.29), (4.26), and (4.24), we conclude

$$I_1(x_0) \leq c\eta + c\, c_6 H^{n-1}(\partial D - E(\epsilon)).$$

Since the right-hand side is independent of r, $0 < r < r_1$, we have

$$\xi = \limsup_{r\to 0} I_1(x_0) \leq c[\eta + c_6 H^{n-1}(\partial D - E(\epsilon))].$$

Next we let $\epsilon \to 0$ and use (4.19) to obtain $\xi \leq c\,\eta$. Finally letting $\eta \to 0$ we get (4.22). Thus (2.3) is valid and Theorem 5 follows from our earlier work. $\hspace{2cm}$ Q.E.D.

5. Remarks and Problems

(1) As mentioned in section 1, the second author in his thesis, requires f to have continuous third partials and d to have continuous first partials in each variable; in order to conclude that a solution u to (1.6) under boundary condition (A) is radially symmetric. Can the same conclusion be made under weaker regularity assumptions on f, d? If for example, d is only bounded while f is C^∞ and uniformly convex, then classical Schauder type estimates give $u \in C^{1,\alpha}(D)$ for $0 < \alpha < 1$ and it can be shown as outlined in section 1 that ∂D is $C^{1,\alpha}$ from each side. However to use Serrin's argument we need ∂D to be C^2. Also if $f'(0) = 0$ or ∞, and f is only C^2, then it is not known for some functions d whether a Hopf boundary maximum principle holds for solutions to (1.6). This maximum principle is needed in Serrin's argument.

(2) When can the Green's function in Theorems 1,2,5, be replaced by a solution u to either (1.1) or (1.6)? For a general L as in (1.1) this question could be difficult, since an answer appears linked with determining the sets of L elliptic measure zero. If

$$Lu = \Delta u + f(u, |\nabla u|) = 0,$$

where $f > 0$ is C^1 in both u and $|\nabla u|$, then in R^2, it can be shown (using the fact that a certain function of $|\nabla u|$ is a super solution to a uniformly elliptic P.D.E.) that boundary condition (A*) forces ∂D to be C^2 when D is Lipschitz. Serrin's method can then be applied. In R^n, $n > 2$, super solution estimates are no longer available and the procedure for showing ∂D smooth is much more involved. However it appears likely that a new method of Caffarelli [7,8] can be used in the Lipschitz case (Theorem 1) to show that boundary condition (A*) forces ∂D to be C^2. Serrin's argument can then be applied to get D is a ball. If Caffarelli's method works, then parabolic analogues of Theorem 3 in Lipschitz cylinders for

$$u_t = \Delta u + f(u, \nabla u)$$

should also hold. Still, though, a more direct approach to these problems which requires only subsolution estimates, would be preferable.

Also, for more general domains D, in Theorem 5, it is probably not possible to first show that a boundary condition similar to (A$^\wedge$) forces ∂D to be smooth. In fact we do not know how to show this, even in R^2. One essential difference between this case and the Lipschitz case is that $|\nabla u|$, a priori, need not be bounded away from zero in a neighborhood of a boundary point, so super solution estimates appear difficult. Moreover in R^3, Alt and Caffarelli [3, section 2.7] point out that there exists a positive Lipschitz harmonic function u in the exterior of a cone K with $u = 0$, $|\nabla u| = a$, continuously on K, except at the vertex of the cone, and

$$k^{-1} \le |\nabla u| \le k \text{ for some } k, \ 0 < k < +\infty,$$

in a neighborhood of the vertex. Clearly K is not smooth in any neighborhood of its vertex. The above authors also show for a similar problem that ∂D is locally smooth for H^{n-1} a.e. $y \in \partial D$. Thus can Serrin's argument be extended to domains that are locally smooth outside of a small exceptional set. We have had no luck in trying this approach. If

$$Lu = \Delta u + 1 = 0, \tag{5.1}$$

analogues of Theorems 1,4, can be obtained using Weinberger's original method, and arguments similar to those for the Green's function. We briefly sketch the proof of Theorem 1 for a solution u to (5.1) satisfying (A*).

In place of (2.2) it can be shown that

$$\int_D (n|\nabla u|^2 + 2u)dH^n = na^2 H^n(D).\qquad(5.2)$$

For (5.2) we use the Rellich-Necas-Pohoẑaev formula [17,19,24,28]

$$-\int_{\partial\Omega}[(x\cdot\nu)|\nabla u|^2 - 2(\nabla u\cdot\nu)(x\cdot\nabla u) - 2(x\cdot\nu)u + (2n-2)u(\nabla u\cdot\nu)]dH^{n-1}$$

$$= \int_\Omega (n|\nabla u|^2 + 2u)dH^n,\qquad(5.3)$$

where ν is the outer unit normal to the smooth domain Ω with $\Omega \subseteq D$. Choosing a sequence of smooth domains $(\Omega_j)_1^\infty$ as in section 1 and using the radial limit theorems mentioned there, we obtain (5.2) as the limit of (5.3) with $\Omega = \Omega_j$ as $j \to \infty$. (2.3) remains true in this case and its proof is essentially unchanged, since $|\nabla u|$ is subharmonic and its radial maximal function is in $L_2(\partial D)$. Now

$$\Delta(n|\nabla u|^2 + 2u) = 2n\sum_{i,j}(u_{x_i x_j})^2 - 2 \geq 2(\Delta u)^2 - 2 \geq 0,\qquad(5.4)$$

where we have used Schwarz's inequality. Thus $n|\nabla u|^2 + 2u$ is subharmonic in D and so from (A*), (2.3), and the maximum principle for subharmonic functions we have

$$n|\nabla u|^2 + 2u \leq na^2\qquad(5.5)$$

in D with equality at any point of D only if $n|\nabla u|^2 + 2u \equiv na^2$. In view of (5.2), (5.5), it follows that $n|\nabla u|^2 + 2u \equiv na^2$. From the case of equality in Schwarz's inequality, we conclude from (5.4) that $\left(u + \frac{1}{2n}|x|^2\right)_{x_i x_j} \equiv 0$ in D, $1 \leq i,j \leq n$, which clearly implies D is a ball and u is radially symmetric about the center of D.

To obtain a version of Theorem 3 for solutions u to $u_t - \Delta u - 1 = 0$, where u satisfies boundary condition (A*) and $u(x,0) \equiv 0$, $x \in D$, continuously, we argue as in Garafalo and Alessandrini [2] to get

$$u(x,t) = \psi(x) + \sum_{m=1}^\infty b_m e^{-\lambda_m t}\phi_m(x), \quad x \in D, \ 0 < t < T,\qquad(5.6)$$

where $\Delta\psi = -1$ in D with $\psi = 0$ continuously on ∂D. Here $(\phi_m)_1^\infty$ is a complete set of orthonormal eigenfunctions for the Laplacian with Dirichlet boundary conditions and $\lambda_m \leq \lambda_{m+1}$, $m = 1, 2, ...$, are the eigenvalues for the Laplacian. Also

$$b_m = -\int_D \phi_m\psi dH^n, \quad m = 1, 2, ...$$

For fixed m we write

$$\psi_m = h + p$$

where h is harmonic in D and p is a solution to $\Delta p = -\lambda_m \phi_m$ in some ball B with $\bar{D} \subset B$ and $p = 0$ continuously on ∂B (Define $\phi_m \equiv 0$ outside D). The $L_2(\partial D)$ norm of the tangential derivatives of h can be estimated in terms of those of p, which in turn follow from well known estimates on λ_m, ϕ_m. Doing this and using Verchota's theorem again (see section 2) we get

(i) $(\nabla \phi_m)^* \in L_2(\partial D)$ with norm $\leq cm^l$, for some $l = l(n) > 0$

(ii) $\nabla \phi_m(y) = \lim_{x \to y} \nabla \phi_m(x) = \pm |\nabla \phi_m(y)| n(y)$ radially for H^{n-1} a.e. y in ∂D, $m = 1, 2, \ldots$

Similar statements are true for ψ. Using (i), (ii), (A*), and (5.6) we deduce

$$-a(t) = \nabla \psi(y) \cdot n(y) + \sum_{m=1}^{\infty} b_m e^{-\lambda_m t} \nabla \phi_m(y) \cdot n(y), \qquad (5.7)$$

for H^{n-1} a.e. $y \in \partial D$. If y_1, y_2, satisfy this equality, then

$$\nabla \psi(y_1) \cdot n(y_1) - \nabla \psi(y_2) \cdot n(y_2)$$

$$= \sum_{m=1}^{\infty} b_m e^{-\lambda_m t} [\nabla \phi_m(y_1) \cdot n(y_1) - \nabla \phi_m(y_2) \cdot n(y_1)] \qquad (5.8)$$

(5.8) holds for $t > 0$ since the right hand side is real analytic in t for $t > 0$. Letting $t \to \infty$ we get, $|\nabla \psi(y)| = a$ for H^{n-1} a.e. $y \in \partial D$. From our previous proof of Theorem 1 for ψ we now conclude that D is a ball and by uniqueness of u that $u(\cdot, t)$ for $0 < t < T$ is symmetric about the center of D.

(3) Is the assumption, $H^n(D) \leq \frac{1}{2} H^n(S_n)$, necessary in Theorem 2? Although we know of no counterexamples, it should be pointed out here that Serrin's Theorem is false when D is not contained in a hemisphere. The authors would like to thank Robert Molzon for pointing out this fact to us, by way of the following example, which is apparently due to Carlos Berenstein. Let

$$x_1 = \rho \cos \theta, \quad x_2 = \rho \sin \theta \, \sin \phi, \quad x_3 = \rho \sin \theta \, \cos \theta,$$

$$0 < \theta \leq \pi, \quad 0 \leq \phi < 2\pi, \quad \text{and}$$

$$\rho = (x_1^2 + x_2^2 + x_3^2)^{1/2}$$

be spherical coordinates in three space and define α, $0 < \alpha < \pi/2$, by $\cos \alpha = \frac{1}{\sqrt{3}}$. Let

$$u(\theta, \phi) = \frac{1}{2} - \frac{3}{2} \cos^2 \theta, \quad \alpha < \theta < \pi - \alpha, \quad 0 < \phi \leq 2\pi.$$

If $D = \{(\theta, \phi) : \alpha < \theta < \pi - \alpha, \quad 0 \leq \phi < 2\pi\}$, then $u > 0$ in D and

$$\tilde{\Delta} u = -6u \ \text{ in } D \subseteq S_3,$$

where $\tilde{\Delta}$ denotes the spherical Laplacian, while $u = 0$, $|\nabla u| = 3\sin\alpha\cos\alpha$ on ∂D. Clearly D is not a spherical ball.

(4) Can a_0 be replaced in Theorem 4 by ∞?

(5) Does Theorem 5 remain valid if we do not assume (4.3) but still (a) D is of finite perimeter, (b) μ is a constant multiple of H^{n-1} measure on ∂D, and (c) (4.5) holds? Shapiro asks in [26] whether there exists "a pseudosphere in 3 space, that is, a surface homeomorphic (but not congruent) to a sphere with respect to which the average of each harmonic function equals the value of the function at some fixed point." In a future paper we shall give an affirmative answer to Shapiro's question.

The reader is invited to state and prove parabolic analogues of Theorem 5 for the Green's function and solutions u to $u_t = \Delta_x u + 1$. For this latter equation it appears difficult to use the argument of Alessandrini and Garofalo, since it is hard to see how (4.5) (where ∇ is replaced by ∇_x)can be used to estimate the eigenfunctions of the Laplacian in (5.7).

REFERENCES

[1] S. Agmon, A. Douglis, and L. Nirenberg, *Estimates near the boundary for solutions of elliptic partial differential equations satisfying general boundary conditions*, Comm Pure Appl. Math. **12**, 623-727 (1959).

[2] G. Alessandrini and N. Garafalo, *Symmetry for degenerate parabolic equations*, to appear.

[3] H. Alt and L. Caffarelli, *Existence and regularity for a minimum problem with free boundary*, J. Reine Angew. Math. **325**, 105-144 (1981).

[4] H. Alt, L. Caffarelli, and A. Friedman, *A free boundary problem for quasi-linear elliptic equations*, Ann. Scoula Norm. Sup. Pisa (4) **11**, 1-44 (1984).

[5] R. Brown, *The method of layer potentials for the heat equation in Lipschitz cylinders*, Amer. J. Math. **111**, 339-379 (1989).

[6] R. Brown, *The initial-Neumann problem for the heat equation in Lipschitz cylinders*, to appear.

[7] L. Caffarelli, *A Harnack inequality approach to the regularity of free boundaries. PartI: Lipschitz free boundaries are $C^{1,\alpha}$*, Revista Mathematica Iberoamericana **3**, 139-162 (1987).

[8] L. Caffarelli, *A Harnack inequality approach to the regularity of free boundaries. Part II: Flat free boundaries are Lipschitz*, Comm. on Pure and Appl. Math. **42**, 55-78 (1989).

[9] B. Dahlberg, *Estimates on harmonic measure*, Arch. Rational Mech. Anal. **65**, 275-288 (1977).

[10] B. Dahlberg, *On the Poisson integral for Lipschitz and C^1 domains*, Studia Math. **66**, 7-24 (1979).

[11] P. Duren, *Theory of H^p spaces*, Academic Press, 1970.

[12] M. Essén and J. Lewis, *The generalized Ahlfors-Heins theorem in certain d-dimensional cones*, Math. Scand. **33**, 113-129 (1973).

[13] L. Evans and R. Gariepy, *Lecture notes on measure theory and fine properties of functions*, EPSCoR preprint series, University of Kentucky.

[14] E. Fabes and S. Salsa, *Estimates of caloric measure and the initial-Dirichlet problem for the heat equation in Lipschitz cylinders*, Trans. Amer. Math. Soc. **279**, 635-650 (1983).

[15] E. Fabes, D. Jerison, and C. Kenig, *Necessary and sufficient conditions for absolute continuity of elliptic harmonic measure*, Annals of Math. **119**, 121-141 (1984).

[16] R. Fefferman, C. Kenig, and J. Pipher, *The theory of weights and the Dirichlet problem for elliptic equations*, to appear.

[17] N. Garafalo and J. Lewis, *A symmetry result related to some overdetermined boundary value problems*, American Journal Math. **111**, 9-33 (1989).

[18] Hayman and Kennedy, *Subharmonic functions 1*, Academic Press, (1976).

[19] D. Jerison and C. Kenig, *The Neumann problem on Lipschitz domains*, Bull. Amer. Math. Soc. **4**, 203-207 (1981).

[20] J. Kemper, *Temperatures in several variables: Kernel functions, representations and parabolic boundary values*, Trans. Amer. Math. Soc. **167**, 243-262 (1972).

[21] D. Kinderlehrer and L. Nirenberg, *Regularity in free boundary problems*, Ann. Scoula Norm. Sup. Pisa (4) **4**, 372-391 (1977).

[22] J. Lewis and M. Murray, *The Dirichlet and Neumann problems for the heat equation in domains with time dependent boundaries*, in preparation.

[23] Payne and Schaeffer, *Duality theorems in some overdetermined boundary value problems*, to appear in Math. Methods Appl. Sci.

[24] P. Pucci and J. Serrin, *A general variational identity*, Indiana Univ. Math. J. **35**, (1986), no. 3.

[25] J. Serrin, *A symmetry problem in potential theory*, Arch. Rational Mech. Anal. **43**, 304-318 (1971).

[26] H. Shapiro, *Remarks concerning domains of Smirnov type*, Michigan Math. J. **13**, 341-348 (1966).

[27] E. Stein, *Singular integrals and differentiability properties of functions*, Princeton University Press, 1970.

[28] G. Verchota, *Layer potentials and regularity for the Dirichlet problem for Laplace's equation in Lipschitz domains*, Jour. of Functional Analysis **59**, 572-611 (1984).

[29] A. Vogel, *Regularity and symmetry for general regions having a solution to certain overdetermined boundary values problems*, thesis, University of Kentucky, 1989.

[30] H. Weinberger, *Remark on the preceeding paper of Serrin*, Arch. Rat. Mech. Anal. **43**, 319-320 (1971).

[31] J. Wu, *On parabolic measures and subparabolic functions*, Trans. Amer. Math. Soc. **251**, 171-185 (1979).

Symmetry Properties of Finite
Total Mass Solutions
of Matukuma Equation

YI LI

Introduction

Let us consider the following problem in $\mathbb{R}^n$ $(n \geq 3)$

(1)
$$\begin{cases} \Delta u + f(|x|, u) = 0 & \text{in } \mathbb{R}^n \\ u > 0 & \text{everywhere} \\ u \to 0 & \text{at } \infty \end{cases}$$

and because of its invariance under orthonormal transformation, a natural question about the properties of solutions of (1) arises:

(Q) Is a solution of (1) radially symmetric?

In 1981, by using the "Moving Plane" method, Gidas, Ni and Nirenberg ([GNN2]) proved a series of remarkable theorems concerning (Q). Here is one of their results.

Theorem A. (Gidas, Ni and Nirenberg). *Let $u > 0$ be a C^2 solution of*

(2)
$$\Delta u + K(|x|)u^p = 0 \quad \text{in } \mathbb{R}^n$$

If $K(r)$ is either a positive constant or a strictly decreasing positive function of r and if $u(x) = O(|x|^{-\alpha})$ at ∞ such that $p\alpha > n + 1$, then u is radially symmetric and if r is the radial variable, then $\frac{\partial u}{\partial r} < 0$ for $r > 0$.

Furthermore

(3)
$$\lim_{x \to \infty} |x|^{n-2} u(|x|) = \text{some positive constant.}$$

The following technical lemma was crucial to the proof of Theorem A.

Supported in part by the National Science Foundation.

Lemma B. (Gidas, Ni and Nirenberg). *Let $u \in C^2$ be given by*

$$u(x) = \int_{\mathbb{R}^n} \frac{f(y)}{|x-y|^{n-2}} \, dy$$

with $f(y) = O(|y|^{-q})$ at ∞ for some $q > n$. Then

$$(4) \qquad \lim_{x \to \infty} |x|^{n-2} u(x) = \int f(y) \, dy.$$

Furthermore, if $q > n+1$, then

$$(5) \qquad \lim_{x_1 \to \infty} \frac{|x|^n}{x_1} \frac{\partial u}{\partial x_1}(x) = -(n-2) \int f(y) \, dy$$

and if $\{\lambda^i\} \in \mathbb{R} \to \lambda \in \mathbb{R}$ and $\{x^i\}$ is a sequence in $\mathbb{R}^n$ going to infinity with $x_1^i < \lambda^i$, then

$$(6) \qquad \frac{|x^i|^n}{\lambda^i - x_1^i} (u(x^i) - u(x^{i\lambda^i})) \to 2(n-2) \int f(y)(\lambda - y_1) \, dy$$

as $i \to \infty$, and where x^λ is the reflection point of x about the hyperplane $\{x_1 = \lambda\}$, i.e., $x^\lambda = (2\lambda - x_1, x_2, \ldots, x_n)$.

Later on, many applications of this kind of result as well as extensions have been made (see, e.g., [FL], [L1], [CS], [BN1,2], [CGS], [LN2,3,4] [Li] and a survey paper [N] on a general aspect of (1).)

In particular, by making use of decay of $K(|x|)$ at infinity in (2), Li and Ni ([L1] and [LN2]) made the following improvement.

Theorem C. (Li and Ni). *Let u and K be as in Theorem A. Assume that $K(|x|) = O(|x|^{-\beta})$ for some $\beta \geq 0$ at infinity. Then the conclusions of Theorem A still hold provided $\beta + p\alpha > n+1$.*

This paper is intended to give a report on a new development of symmetric properties of positive solutions of (2), which is a joint work with Wei-Ming Ni ([LN3,4]). To make this report concise we will confine ourselves to the *Matukuma* equation.

In 1915, A. S. Eddington introduced the following equation to describe the dynamics of globular clusters of stars [E].

$$(E) \qquad \Delta u + \frac{e^{2u}}{1 + |x|^2} = 0 \quad \text{in } \mathbb{R}^3$$

where $u > 0$ is the gravitational potential, $\rho = -\frac{1}{4\pi}\Delta u = \frac{1}{4\pi(1+|x|^2)}$ is the density and $\int_{\mathbb{R}^3} \rho(x)dx$ is the total mass. Fifteen years later, in 1930, T. Matukuma proposed the following equation to improve Eddington's model ([M]).

$$\text{(M)} \qquad \Delta u + \frac{1}{1+|x|^2}u^p = 0 \quad \text{in } \mathbb{R}^3$$

where $p > 1$ and $\int_{\mathbb{R}^3} \frac{1}{4\pi(1+|x|^2)}u^p dx$ represents the total mass. Since the physical models are derived under the assumption that u is radial, (E) and (M) reduce to the following

$$\text{(E}_\alpha\text{)} \qquad \begin{cases} u_{rr} + \dfrac{2}{r}u_r + \dfrac{1}{1+r^2}e^{2u} = 0 \text{ in } [0,\infty) \\ u(0) = \alpha > 0, \ u_r(0) = 0 \end{cases}$$

and

$$\text{(M}_\alpha\text{)} \qquad \begin{cases} u_{rr} + \dfrac{2}{r}u_r + \dfrac{1}{1+r^2}u^p = 0 \quad \text{in } [0,\infty) \\ u(0) = \alpha > 0, \ u_r(0) = 0 \end{cases}$$

For $\alpha > 0$, we use $u(r,\alpha)$ to denote the solution of (E$_\alpha$) or (M$_\alpha$) with initial value α. T. Matukuma conjectured for (M$_\alpha$) that

(i) if $p < 3$, then $u(r,\alpha)$ has a finite zero for every $\alpha > 0$,
(ii) if $p = 3$, then $u(r,\alpha)$ is a positive entire solution with finite total mass for every $\alpha > 0$,
(iii) if $p > 3$, then $u(r,\alpha)$ is a postive entire solution with infinite total mass for every $\alpha > 0$.

Later, in 1938, he discovered an exact solution of (E)

$$u(r;\sqrt{3}) = \left(\frac{3}{1+r^2}\right)^{1/2}$$

with $p = 3$ which confirms part of his conjecture.

Recently, Ni and Yotsutani ([NY1,2]) studied (E$_\alpha$) and (M$_\alpha$) systematically—the first time since 1938. And they found that the Eddington equation (E$_\alpha$) does not have any positive radial entire solution (this may show that (M) is perhaps a better model). And for (M) they proved

Theorem D. (Ni and Yotsutani) *Let u be the solution of (M$_\alpha$).*
(i) *If $1 < p < 5$, then $u(r;\alpha)$ has a finite zero for every sufficiently large $\alpha > 0$.*

(ii) *If $1 < p < 5$, then $u(r; \alpha)$ is a positive entire solution with infinite total mass for every sufficiently small $\alpha > 0$.*

(iii) *if $p \geq 5$, then $u(r; \alpha)$ is a positive entire solution with infinite total mass for every $\alpha > 0$.*

But the existence of positive radial solutions with *finite total mass* was later treated in [LN2] using a variational argument.

Theorem E. *For every $1 < p < 5$, there is an $\alpha^* > 0$ such that the solution $u(r; \alpha^*)$ of (M_α) is positive in $[0, \infty)$ and has finite total mass, i.e.,*

$$\int_{\mathbb{R}^3} \frac{1}{4\pi(1 + |x|^2)} u^p(|x|; \alpha^*) \, dx < \infty.$$

Remark 1. It is shown that a positive entire solution of (M_α) must satisfy either

$$c \leq ru(r) \leq c^{-1} \quad \text{in } [1, \infty) \tag{7}$$

or

$$c \leq (\log r)^{\frac{1}{p-1}} u(r) \leq c^{-1} \quad \text{in } [e, \infty) \tag{8}$$

for some positive constant c.

Later in [L2] it was established that such a solution must satisfy

$$\lim_{r \to \infty} (\log r)^{\frac{1}{p-1}} u(r) \equiv u_\infty = \begin{cases} \left(\frac{n-2}{p-1}\right)^{\frac{1}{p-1}} & \text{or} \\ 0 \end{cases} \tag{9}$$

and if $u_\infty = 0$, then

$$\lim_{r \to \infty} ru(r) \text{ exists and is positive.}$$

Now we have (see [NY2]) that

(i) If u is an entire solution of (M_α) with infinite total mass, then

$$\lim_{r \to \infty} (\log r)^{1/(p-1)} u(r) = \left(\frac{n-2}{p-1}\right)^{1/(p-1)}.$$

(ii) If u is an entire solution of (M_α) with finite total mass, then

$$\lim_{r \to \infty} ru(r) = \int_{\mathbb{R}^3} \frac{1}{4\pi(1 + |x|^2)} u^p(x)\, dx,$$

i.e., the limit of $|x|u(x)$ at ∞ is the total mass.

Remark 2. For various models related to (E) and (M), see [BFH].

However, it is mathematically interesting to understand whether or not (M) has *only* the radial solutions. As a consequence of Theorem B (Theorem A.1 in [LN2]), Li and Ni proved the following results concerning the finite total mass solutions based on an analysis of such solutions at infinity (see section 1).

Theorem F. (Li and Ni) (i) *If $2 < p < 5$, then every bounded positive entire solution u of equation (M) with finite total mass is radially symmetric about the origin and $u_r < 0$ in $r > 0$. Furthermore*

$$lim_{r \to \infty} ru(r) = k > 0.$$

(ii) *If $p \geq 5$, then every bounded positive entire solution of (M) has infinite total mass.*

Comparing Theorems E and F, we find that there is still a gap between the existence of finite total mass solutons and the symmetric properties of such solutions. For $1 < p \leq 2$, the total decay of nonlinearity is at most $2 + p \leq 4$ $(= n + 1$; and see (5) and (6) in Lemma B) so that $(\lambda - y_1)\frac{1}{1 + |y|^2} u^p(y) \notin L^1(\mathbb{R}^n)$. Thus new ideas and techniques are developed in [LN3,4] to handle these cases. By looking further into the asymptotic behavior of solutions near infinity and using some more precise estimates, we are able to show

Theorem 1. *Let u be a bounded positive entire solution of (M) with finite total mass. Then u must be radially symmetric about the origin and $\frac{\partial u}{\partial r} < 0$ for $r > 0$. Furthermore,*

$$(10) \quad r^{n-2} u(r) = c_0 + c_1 r^{-(p-1)(n-2)} + \cdots + c_m r^{-m(p-1)(n-2)} + r^{-1} R(\frac{1}{r})$$

where c_0 is the total mass, $c_1, \ldots, c_m$ are constants, $R(\cdot)$ is a C^1 function in a neighborhood of the origin such that $R(0) = 0$ and m is the first integer satisfying $m(p - 1)(p - 2) \geq 2$.

This paper is organized as follows. In section 1 an asymptotic expansion of finite total mass solutions is obtained which is essential in the proof of Theorem 1. In section 2, a sketch of the symmetry proof is given. We would like to mention here that our methods apply to a more general class of equations than (M) and the interested reader may find a more complete and detailed description in [LN3,4].

Finally, (Q) can also be raised for the Dirichlet and Neumann boundary value problem over a finite region. In such cases, many nice results have been obtained, see e.g., [S], [W], [GNN1], [GL], [BP] and [LKK].

1. Asymptotic expansion

From now on, we will assume that u is a bounded positive entire solution of (M) with finite total mass in $\mathbb{R}^n$, i.e.,

$$
(1.1) \qquad
\begin{cases}
\Delta u + \frac{1}{1+|x|^2} u^p = 0 & \text{in } \mathbb{R}^n, 1 < p < \frac{n+2}{n-2} \\
\int_{\mathbb{R}^n} \frac{1}{1+|x|^2} u^p(x)\, dx < \infty & \text{and } u > 0, \text{ in } \mathbb{R}^n.
\end{cases}
$$

First, we will recall some results from [LN2,3].

Lemma 1.1. *Let u be a solution of (1.1). Then*

$$
(1.2) \qquad u(x) \le c|x|^{2-n} \quad \text{at } \infty \text{ for some } c > 0
$$

Proof. See Lemma 2.3, [LN2].

Now, let $f(x) = \frac{1}{n(n-2)\omega_2} \frac{1}{1+|x|^2} u^p(x)$ where ω_n is the area of the unit sphere in $\mathbb{R}^n$. Then we have (see [LN1], for example)

$$
(1.3) \qquad u(x) = \int_{\mathbb{R}^n} \frac{f(y)}{|x-y|^{n-2}}\, dy
$$

and because of (1.2), we have

$$
(1.4) \qquad f(y) = O(|y|^{-2-p(n-2)}) \quad \text{where } 2 + p(n-2) > n.
$$

Therefore (4) in Lemma B implies

$$
(1.5) \qquad \lim_{x \to \infty} |x|^{n-2} u(x) = \int f(y)\, dy \equiv c_0 - \text{ total mass.}
$$

Remark 1.1. Because the case when $2 + p(n - 2) > n + 1$ is already covered by Theorem F in [LN2], we consider only the exponent p for which $2 + p(n - 2) \leq n + 1$, i.e.,

$$p \leq \frac{n-1}{n-2}$$

Lemma 1.2. *Let u be a solution of (1.1). Then*

$$(1.6) \quad u(x) - \frac{c_0}{|x|^{n-2}} = \begin{cases} O(|x|^{-(n-1)} \log |x|) & \text{at } \infty \text{ if } p = \frac{n-1}{n-2} \\ O(|x|^{-p(n-2)}) & \text{at } \infty \text{ if } 1 < p < \frac{n-1}{n-2}. \end{cases}$$

Proof. See [LN3].

Now we are ready to prove the following theorem on the asymptotic behavior of solutions of (1.1).

Theorem 1.3. *Let u be a solution of (1.1). then*

$$(1.7) \quad \begin{cases} u(x) = \dfrac{c_0}{|x|^{n-2}} + \dfrac{c_1}{|x|^{p(n-2)}} + \cdots + \dfrac{c_{2k+2}}{|x|^{[1+2(k+1)(p-1)](n-2)}} \\[2mm] \qquad + \dfrac{\vec{a}x}{|x|^n} \left(1 + \dfrac{d_1}{|x|^{(p-1)(n-2)}} + \cdots + \dfrac{d_{k+1}}{|x|^{(k+1)(p-1)(n-2)}} \right) \\[2mm] \qquad \dfrac{1}{|x|^{n-1}} R(\dfrac{x}{|x|^2}) \quad \text{at } \infty \end{cases}$$

where k is an integer such that $k(p-1)(n-2) \leq 1 < (k+1)(p-1)(n-2)$, c_0 is defined by (1.5), $c_1, \ldots, c_{2k+2}$, $d_1, \ldots, d_{k+1}$ are constants, $\vec{a}$ is a constant vector and R is Lipschitz near 0 with $R(0) = 0$.

Proof. Let $v(x) = |x|^{2-n} u(\frac{x}{|x|^2})$ be the Kelvin-inversion of u. Then it is standard to verify that

$$(1.8) \qquad \Delta v(x) + \frac{|x|^{p(n-2)-n}}{1 + |x|^2} v^p(x) = 0 \quad \text{in } B_1(0) \setminus \{0\}$$

while (1.6) implies that

$$(1.9) \qquad v(x) = c_0 + v_1(x)$$

with $v_1(x) = O(|x|^{(p-1)(n-2)-\varepsilon})$ at 0 for all $\varepsilon > 0$.

Claim.

$$v(x) = c_0 + c_1|x|^{(p-1)(n-2)} + \cdots + c_{2k+2}|x|^{2(k+1)(p-1)(n-2)}$$
$$+ \vec{a}x(1 + d_1|x|^{(p-1)(n-2)} + \cdots + d_{k+1}|x|^{(k+1)(p-1)(n-2)}) + |x|R(x)$$

where the c_i, d_i are constants, $a \in \mathbb{R}^n$ and $R \in C^{0,1}(B_1(0))$ with $R(0) = 0$.

Estimate (1.9) implies that

$$\Delta v(x) + \frac{|x|^{p(n-2)-n}}{1 + |x|^2}(c_0 + v_1(x))^p = 0$$

or

$$\Delta v(x) + c_0^p|x|^{p(n-2)-n} + R_1(x) = 0 \quad \text{in } B_1(0) \setminus \{0\}$$

with $R_1(x) = O(|x|^{(2p-1)(n-2)-n-\varepsilon})$ at 0 for all $\varepsilon > 0$.

Then the uniqueness theorem on continuous harmonic functions (see [GT]) implies that

$$(1.10) \qquad v(x) = c_0 - \frac{c_0^p}{(p-1)p(n-2)^2}|x|^{(p-1)(n-2)} + v_2(x)$$

where

$$(1.11) \qquad v_2(x) = \begin{cases} O(|x|^{2(p-1)(n-2)-\varepsilon}), \ \forall\varepsilon & \text{if } 2(p-1)(n-2) \leq 1 \\ O(|x|) & \text{if } 2(p-1)(n-2) > 1 \end{cases}$$

Now we can prove the claim inductively. Suppose that for $m \leq k - 1$, we have

$$(1.12) \ v(x) = c_0 + c_1|x|^{(p-1)(n-2)} + \cdots + c_{m-1}|x|^{(m-1)(p-1)(n-2)} + v_m(x)$$

with $v_m(x) = O(|x|^{m(p-1)(n-2)-\varepsilon})$ at 0 for all $\varepsilon > 0$. Then by putting (1.12) back into (1.8), we obtain

$$\Delta v(x) + \frac{|x|^{p(n-2)-n}}{1 + |x|^2}(c_0 + \cdots + c_{m-1}|x|^{(m-1)(p-1)(n-2)} + v_m(x))^p = 0$$

or
(1.13)
$$\begin{cases} \Delta v(x) + |x|^{p(n-2)-n}(c_0^p + c_1'|x|^{(p-1)(n-2)} + \cdots + c_{m-1}'|x|^{(m-1)(p-1)(n-2)}) \\ \qquad + R_m(x) = 0 \quad \text{in } B_1(0) \setminus \{0\} \end{cases}$$

by a binomial expansion, where

$$(1.14) \qquad R_m(x) = O(|x|^{[m(p-1)+p](n-2)-n-\varepsilon}) \quad \text{at } 0 \ \forall\varepsilon > 0.$$

As above, we thus have

$$(1.15) \quad \begin{cases} v(x) = c_0 + c_1|x|^{(p-1)(n-2)} + \cdots + c_{m-1}|x|^{(m-1)(p+1)(n-2)} \\ \qquad + c_m|x|^{m(p-1)(n-2)} + v_{m+1}(x) \end{cases}$$

with $v_{m+1}(x) = O(|x|^{(m+1)(p-1)(n-2)-\varepsilon})$ at 0, for all $\varepsilon > 0$.

Therefore, this gives us by induction from (1.10) (1.12) and (1.15) that

$$(1.16) \quad v(x) = c_0 + c_1|x|^{(p-1)(n-2)} + \cdots + c_k|x|^{k(p-1)(n-2)} + v_{k+1}(x)$$

with $v_{k+1}(x) = O(|x|)$ at 0, because $(k+1)(p-1)(n-2) > 1$. (See (1.11))

As in (1.13), we have now that

$$(1.17)$$
$$\Delta v(x) + |x|^{p(n-2)-n}(c_0^p + \cdots + c_k'|x|^{k(p-1)(n-2)}) + R_{k+1}(x) = 0 \text{ in } B_1(0)\setminus\{0\}$$

where $R_{k+1}(x) = O(|x|^{p(n-2)+1-n})$ at 0.

Since $p(n-2) + 1 - n > -1$, we have from (1.17) tht

$$(1.18) \quad v(x) = c_0 + \cdots + c_{k+1}|x|^{(k+1)(p-1)(n-2)} + v_{k+2}(x)$$

where $v_{k+1} \in C^{1,\alpha}(B_1(0))$ for all $\alpha \in (0, (p-1)(n-2))$ and $v_{k+2}(0) = 0$. Therefore, $v_{k+2}(x) = a \cdot x + v_{k+3}(x)$ where $a = \nabla v_{k+2}(0)$ and $v_{k+3}(x) = O(|x|^{1+(p-1)(n-2)-\varepsilon})$ at 0 for all $\varepsilon > 0$.

As before, putting (1.18) back into (1.8), we obtain that

$$(1.19) \quad \begin{cases} \Delta v(x) + |x|^{p(n-2)-n}(c_0^p + \cdots + c_{k+1}'|x|^{(k+1)(p-1)(n-2)} + d_1'(ax)) \\ \qquad + R_{k+3}(x) = 0 \quad \text{in } B_1(0) \setminus \{0\} \end{cases}$$

where $R_{k+3}(x) = O(|x|^{(2p-1)(n-2)+1-n-\varepsilon})$ at 0 for all $\varepsilon > 0$. This implies, once again, that

$$(1.20)$$
$$v(x) = c_0 + \cdots + c_{k+1}|x|^{(k+2)(p-1)(n-2)} + ax(1 + d_1|x|^{(p-1)(n-2)}) + v_{k+4}(x)$$

where $v_{k+4}(x) = O(|x|^{2(p-1)(n-2)+1-\varepsilon})$ at 0 for all $\varepsilon > 0$.

Now by iteration as before, we have

$$(1.21)$$
$$\begin{aligned} v(x) = c_0 &+ \cdots + c_{2k+2}|x|^{2(k+1)(p-1)(n-2)} \\ &+ ax(1 + d_1|x|^{(p-1)(n-2)} + \cdots + d_{k+1}|x|^{(k+1)(p-1)(n-2)}) \\ &+ |x|R(x) \end{aligned}$$

where $R(x)$ is a Lipschitz function in $B_1(0)$ with $R(0) = 0$ because $2(k+1)(p-1)(n-2) > 2$.

If we let

$$I_m \equiv \frac{1}{|x|^{[1+m(p-1)](n-2)}} - \frac{1}{|x^\lambda|^{[1+m(p-1)](n-2)}}$$

and

$$I'_m \equiv \frac{\vec{a} \cdot x}{|x|^{n+m(p-1)(n-2)}} - \frac{\vec{a} \cdot x^\lambda}{|x^\lambda|^{n+m(p-2)(n-2)}}$$

then from (2.4) and (2.5) we have for $\lambda > 0$

(2.7)

$$\begin{cases} \dfrac{4\lambda(\lambda - x_1)}{|x|^{[1+m(p-1)](n-2)}|x^\lambda|(|x| + |x^\lambda|)} \leq I_m \leq \dfrac{4\lambda[1 + m(p-1)](n-2)(\lambda - x_1)}{|x|^{[1+m(p-1)](n-2)}|x^\lambda|(|x| + |x^\lambda|)} \\[4mm] |I'_m| \leq \dfrac{4[n + m(p-1)(n-2)]|\vec{a}|\lambda(\lambda - x_1)}{|x|^{n+m(p-1)(n-2)}|x^\lambda|(|x| + |x^\lambda|)} + \dfrac{2|\vec{a}|(\lambda - x_1)}{|x^\lambda|^{n+m(p-1)(n-2)}} \\[4mm] \qquad \text{for } |x| \geq 1, \ x_1 < \lambda \end{cases}$$

and
(2.8)

$$\begin{cases} \left| \dfrac{1}{|x|^{n-1}} R\left(\dfrac{x}{|x|^2}\right) - \dfrac{1}{|x^\lambda|^{n-1}} R\left(\dfrac{x^\lambda}{|x^\lambda|^2}\right) \right| \\[4mm] \leq 2(LipR)\left[\dfrac{2(n-1)\lambda}{|x|^n|x^\lambda|(|x\| + |x^\lambda|)} + \dfrac{4\lambda}{|x||x^\lambda|^n(|x| + |x^\lambda|)} + \dfrac{1}{|x^\lambda|^{n+1}} \right](\lambda - x_1) \end{cases}$$

Therefore, combining (2.6), (2.7), and (2.8) we have

(2.9)
$$u(x) - u(x^\lambda)$$
$$\geq \frac{4c_0\lambda(\lambda - x_1)}{|x|^{n-2}|x^\lambda|(|x| + |x^\lambda|)}[1 - \alpha(n, C'_{is}, d'_{is}, |\vec{a}|, (LipR))(\frac{1}{|x|^{(p-1)(n-2)}} + \frac{1}{\lambda})].$$

Therefore, if $\lambda \geq 4\alpha \equiv \lambda_0$ and $|x^\lambda| > |x| \geq r_0 \equiv \max\{1, (4\alpha)^{\frac{1}{(p-1)(n-2)}}\}$ we have
(2.10)

$$u(x) - u(x^\lambda) \geq \frac{2c_0\lambda(\lambda - x_1)}{|x|^{n-2}|x^\lambda|(|x| + |x^\lambda|)} \quad \text{if } |x| \geq r_0, x_1 < \lambda \text{ with } \lambda \geq \lambda_0.$$

On the other hand, we have

(2.11) $$u(x) - u(x^\lambda) > 0 \quad \text{if } |x| \leq r_0, x_1 < \lambda \text{ with } \lambda \text{ large .}$$

Therefore Claim 1 is proven.

Claim 2 Λ *is open in* $\mathbb{R}^+$.

This step is essentially the same as the one in [GNN2] and [L1], except here Lemma B(6) is not valid. However, by using the knowledge (1.7) of u at ∞, we have (2.2) instead of (6), which will permit us to proceed. Thus the proof is omitted.

Claim 3. Λ *is closed in* $\mathbb{R}^+$ *and hence* $\Lambda = \mathbb{R}^+$.

To show that Λ is closed in $\mathbb{R}^+$, let $\{\lambda'\} \in \Lambda$ and $\lim_{i\to\infty} \lambda^i = \lambda \in \mathbb{R}^+$. Then the continuity of u gives

$$(2.12) \qquad u(x) \geq u(x^\lambda), \; x \in \mathbb{R}^n \text{ and } x_1 \leq \lambda$$

Let $v(x) = u(x) - u(x^\lambda)$. In $\Omega_\lambda \equiv \{x_1 < \lambda\}$ we have

$$\Delta v(x) + \frac{1}{1+|x|^2}u^p(x) - \frac{1}{1+|x^\lambda|^2}u^p(x^\lambda) = 0 \text{ in } \Omega_\lambda$$

and hence (2.4) and (2.12) imply

$$(2.13) \qquad \begin{cases} \Delta v(x) < 0 & \text{in } \Omega_\lambda \\ v(x) = 0 & \text{on } \partial\Omega_\lambda. \end{cases}$$

Therefore, the maximum principle and Hopf lemma imply

$$(2.14) \qquad v(x) > 0 \quad \text{in } \Omega_\lambda$$

i.e., u satisfies (I_λ) and

$$(2.15) \qquad \left.\frac{\partial v}{\partial x_1}\right|_{x_1=\lambda} = 2\left.\frac{\partial u}{\partial x_1}\right|_{x_1=\lambda} < 0.$$

Therefore, we have shown that $\Lambda = \mathbb{R}^+$ and (2.15) for $\lambda > 0$. But the continuity of u gives again

$$(2.16) \qquad u(x_1, x') \geq u(-x_1, x') \quad \forall x \in \mathbb{R}^n, x_1 \leq 0$$

by letting λ go to zero in (2.12).

Since equation (1.1) is invariant under orthonormal transformation, we could take any direction as x_1-axis and therefore conclude that

$$(2.17) \qquad u(x_1, x') = u(-x_1, x').$$

This implies u is radially symmetric about the origin and (2.15) implies $\frac{\partial u}{\partial r} < 0$ for all $r > 0$. This completes the proof.

NOTE. While preparing this paper the author learned that Eiji Yanagida has given a proof of the uniqueness of finite total mass solutions of (M_α), namely there is only one $\alpha^* > 0$ such that $u(r; \alpha^*)$ has finite total mass.

REFERENCES

[BN1] H. Berestycki, L. Nirenberg, *Monotonicity, symmetry and antisymmetry of solutions of semilinear elliptic equations*, JGP **5** (1988), 237–275.

[BN2] ______, *Some qualitative properties of soluitons of semilinear elliptic equations in cylindrical domains*, Analysis, et cetera (1990), 115–164.

[BP] H. Berestycki, F. Pacella, *Symmetry properties for positive solutions of elliptic equations with mixed boundary*, Comm. Pure. Appl. Math. **42**, 271–297.

[CGS] A.Caffarelli, B. Gidas, J. Spruck, *Asymptotic symmetry and local behavior of similinear elliptic equations with critical sobolev growth*, preprint.

[CS] W. Craig, P. Sternberg, *Symmetry of solitary waves*, Commun. P.D.E., vol 13, no. 5 (1988), 603–633.

[FL B. Franchi and E. Lanconelli] *Radial symmetry of the ground states for a class of quasilinear elliptic equations*, Nonlinear Diffusion Equations and their Equilibrium States (W.-M. Ni, L.A. Peletier and J. Serrin, eds.) **1** (1988), 287–292.

[GL] N. Garofalo and J. L. Lewis, *A symmetry result related to some overdetermined boundary value problems*, Amer. J. Math. **111** (1989), 9–33.

[GNN1] B. Gidas, W.-M. Ni, L. Nirenberg, *Symmetry and related properties via the maximum principle*, Comm. Math. Phys. **68** (1979), 209–243.

[GNN2] ______, *Symmetry of positive solutins of nonlinear elliptic equations in $\mathbb{R}^n$*, Math. Anal. and Applications, Part A, Advances in Math. Suppl. Studies 7A, (Ed. L. Nachbin), Academic Pr., (1981), pp 369–402.

[Li C. Li] *Some qualitative properties of fully nonlinear elliptic and parabolic equations*, Ph.D. Thesis, New York University, 1989.

[L1] Y. Li, *On the semilinear elliptic equations in $\mathbb{R}^n$*, Ph.D. Thesis, University of Minnesota, 1988.

[L2] ______, *Asymptotic behavior of positive solutions of equation $\Delta u + K(x)u^p = 0$ in $\mathbb{R}^n$*, to appear in J. Diff. Eqns..

[LKK] Y. Li, M.-K. Kwong and H.G. Kaper,, *On the positive solutions of the free-boundary problem for Emden-Fowler type equations*, preprint.

[LN1] Y. Li and W. -M. Ni, *On conformal scalar curvature equations in $\mathbb{R}^n$*, Duke Math. J. **57** (1988), 895–924.

[LN2] ______ , *On the existence and symmetry properties finite total mass solutions of Matukuma equation, Eddington oquation and their generalizations*, Arch. Rational Mech. Anal **108** (1989), 175–194.

[LN3] ______ , *On the asymptotic behavior and radial symmetry of positive solutions of semilinear elliptic equations in $\mathbb{R}^n$, part I. Asymptotic behavior*, preprint.

[LN4] ______ paper On the asymptotic behavior and radial symmetry of positive solutions of semilinear elliptic equations in $\mathbb{R}^n$, part II. Radial symmetry, preprint.

[N] W.-M. Ni, *Some aspects of semilinear elliptic equations on $\mathbb{R}^n$*, (W.-M. Ni, L. A. Peletier and J. Serrin ed.), Springer-Verlag, Nonlinear Diffusion Equations and their Equilibrium States.

[NY1] W.-M. Ni and S. Yotsutani, *On Matukuma's equation and related topics*, Proc.Japan Acad. (Series A) **62** (1986), 260–263.

[NY2] ______ , *Semilinear elliptic equations of Matukuma-type and related topics*, Japan J. Appl. Math. **5** (1988), 1–32.

[PW] M. Protter and H. Weinberger, *Maximum principles in differential equations*, Prentice-Hall, 1967.

[S] J. Serrin, *A symmetry problem in potential theory*, Arch. Rational Mech. Anal. **43** (1971), 304–318.

[W] H. Weinberger, *Remark on the preceding paper of Serrin*, Arch. Rat. Mech. Anal. **43** (1971), 319–320.

Yi Li
Department of Mathematics
University of Chicago
Chicago, IL 60637

An Exact Reduction
of Maxwell's Equations

J.B. McLEOD, C.A. STUART and W.C. TROY

1. Introduction

This paper is concerned with the existence and properties of solutions of the following problem,

$$u''(r) + \frac{u'(r)}{r} - \frac{u(r)}{r^2} + u^3(r) - u(r) = 0 \quad \text{for } r > 0, \tag{1.1}$$

$$\lim_{r \to 0} u(r) = \lim_{r \to \infty} u(r) = 0. \tag{1.2}$$

The main results are Theorems 1 and 2 in section 2, establishing the existence of solutions having any prescribed number of zeros and with the further property that the zeros of u and u' interlace. The solutions of (1.1)-(1.2) lead to a description of beams of light which, due to the nonlinearity of the medium in which they propagate, remain concentrated (self-trapped) near the axis of propagation. In any plane transverse to the axis of propagation the intensity of illumination is radially symmetric with respect to the axis and the zeros and turning points of u correspond to circles of zero and maximal intensity. Our analysis and conclusions are similar to those in [1] for another model for self-trapped light.

More precisely, solutions of (1.1)-(1.2) can be used to construct exact solutions of Maxwell's equations for a nonlinear dielectric medium, which have the characteristics associated with self-trapped light beams. In a charge-free, dielectric medium, Maxwell's equations in CGS units are [2],

$$\nabla \wedge E = -\frac{1}{c}\frac{\partial B}{\partial t}, \qquad \nabla . D = 0,$$

$$\nabla \wedge H = -\frac{1}{c}\frac{\partial D}{\partial t}, \qquad \nabla . B = 0,$$

where c is the speed of light in a vacuum. The fields are considered to be functions of Cartesian coordinates $(x, y, z, t) \in \mathbb{R}^4$. For a non-magnetic medium, $B = H$, and the remaining constitutive assumption is a relationship between D and E. In the discussion of nonlinear materials exhibiting self-trapped beams, it is usually assumed that

$$D(x, y, z, t) = \{1 + 4x\chi_o + 4\pi\chi_3 < E^2 > (x, y, z)\}E(x, y, z, t), \tag{1.3}$$

where χ_o and χ_3 are the linear and third-order dielectric susceptibilities and $< E^2 > (x, y, z)$ denotes the time-average of the intensity, $|E(x, y, z, t)|^2$, of the electric field. This is sometimes called the Kerr nonlinearity and for the material to exhibit self-trapping behaviour we should treat the case where

$$1 + 4\pi\chi_o > 0 \quad \text{and} \quad \chi_3 > 0. \tag{1.4}$$

It follows from (1.3) that the medium is homogeneous and isotropic. See [3,4,5] for a discussion of (1.3) and (1.4).

In this kind of medium, we seek a solution of Maxwell's equations in the form of a TE (transverse electric) mode propagating along the z-axis with high frequency and short wavelength. Hence, although the medium really forms a cylindrical waveguide, we suppose that it occupies all of $\mathbb{R}^3$. To reduce Maxwell's equations to a (nonlinear) Helmholtz equation we must find a situation in which $\nabla.D$ and $\nabla.E$ both vanish identically. In view of (1.3), this can be achieved by seeking a solution in the form

$$E(x, y, t) = v(r)\cos(kz - \omega t)\, i_\theta, \tag{1.5}$$

where $r = \sqrt{x^2 + y^2}$ and $i_\theta = (-y/r, x/r, 0)$. Then $< E^2 > (x, y, z) = \frac{1}{2}v^2(r)$ and, by (1.3),

$$D(x, y, z, t) = \{1 + 4\pi\chi_o + 2\pi\chi_3 v^2(r)\}v(r)\cos(kz - \omega t)\, i_\theta. \tag{1.6}$$

From (1.5),(1.6), we obtain a solution of Maxwell's equations, in the region $r > 0$, if

$$v''(r) + \frac{v'(r)}{r} - \frac{v(r)}{r^2} + \frac{4\pi\omega^2}{c^2}\{\chi_o + \frac{1}{2}\chi_3 v^2(r)\}v(r) + \left(\frac{\omega^2}{c^2} - k^2\right)v(r) = 0 \tag{1.7}$$

for $r > 0$,
and $B(x, y, z, t) = H(x, y, z, t)$

$$= -\frac{kc}{\omega}v(r)\cos(kz - \omega t)\, i_r + \frac{c}{\omega}\{v'(r) + \frac{v(r)}{r}\}\sin(kz - \omega t)\, i_z, \tag{1.8}$$

where $i_r = (x/r, y/r, 0)$ and $i_z = (0, 0, 1)$. The fields defined by (1.5), (1.6), (1.8) can be extended smoothly on to the z-axis provided that

$$\lim_{r \to 0} v(r) = 0 \qquad \lim_{r \to 0} v'(r) \quad \text{exists}, \tag{1.9}$$

and $\lim_{r \to 0}[v'(r) + \frac{v(r)}{r}]' = 0$. However, $[v'(r) + \frac{v(r)}{r}]' = v''(r) + \frac{v'(r)}{r} - \frac{v(r)}{r^2}$ for $r > 0$, so that when v satisfies (1.7) it is enough to verify (1.9). The idea that the fields (1.5), (1.6), (1.8) constitute a self-trapped beam amounts to requiring that

(i) they all decay to zero far from the axis of propagation,

(ii) the total electromagnetic energy in each plane perpendicular to the z-axis is finite. The requirements (i) and (ii) are sometimes referred to as guidance conditions and for the fields (1.5), (1.6), (1.8) they reduce to

$$(i)\quad \lim_{r\to\infty} v\left(r\right) = \lim_{r\to\infty} v'\left(r\right) = 0, \tag{1.10}$$

$$(ii)\quad \int_0^\infty r[v'^2\left(r\right) + v^2\left(r\right)]dr < \infty, \tag{1.11}$$

provided that (1.9) is satisfied.

Thus establishing the existence of self-trapped beams of the form (1.5), (1.6), (1.8) reduces to finding non-trivial solutions of (1.7) that satisfy the conditions (1.9), (1.10), (1.11). The intensity of illumination of such a beam at (x, y, z) on the transverse plane is given by the time-average of the component of Poynting's vector, $\frac{1}{c}\left(E \wedge H\right)$, in the direction i_z. This reduces to $\frac{k}{2\omega}v^2\left(r\right)$ and hence zeros and turning points of v correspond to circles of zero and maximal intensity on a transverse plane.

As the subsequent analysis shows, the formulation (1.7), (1.9), (1.10), (1.11) contains a certain amount of redundancy. For (1.7) to have a solution that decays to zero as $r \longrightarrow \infty$, we must have $\gamma = k^2 - \frac{\omega^2}{c^2}\{1 + 4\pi\chi_0\} > 0$. Replacing r by $\gamma^{-1/2}r$ and then setting $u\left(r\right) = \frac{\omega}{c}\sqrt{\frac{2\pi\chi_3}{\gamma}}v\left(r\right)$ reduces equation (1.7) to (1.1) provided that (1.4) is satisfied. The requirements (1.9), (1.10), (1.11) translate to analogous conditions on u. However our analysis shows that any solution of (1.1), (1.2) satisfies all of these conditions. In particular, u and u' decay exponentially to zero as $r \longrightarrow \infty$.

Our equation (1.1) appears as equation (5.2) in [3] and was first put forward by Pohl [6]. However, except for [7], there appears to be no mathematical discussion of its solutions. In fact, most of the analysis of self-trapping for light beams is based on the problem, [3, 4, 5, 8, 9],

$$w''\left(r\right) + \frac{w'\left(r\right)}{r} + w^3\left(r\right) - w\left(r\right) = 0 \qquad \text{for } r > 0, \tag{1.12}$$

$$\lim_{r\longrightarrow 0} w'\left(r\right) = \lim_{r\longrightarrow 0} w\left(r\right) = 0, \tag{1.13}$$

but, as is remarked in [3], its solutions do not lead to exact solutions of Maxwell's equations. For (1.12), (1.13), results analogous to ours are well-known [1, 10, 11]. Furthermore, it is known [12, 13] that (1.12), (1.13) has exactly one positive solution. It would be interesting to obtain a similar uniqueness result for (1.1), (1.2) but due to the non-monotonicity of positive solutions of (1.1), (1.2) the techniques developed recently to establish uniqueness for generalisations of (1.12), (1.13) do not seem to apply [13,

14, 15]. In the absence of such a uniqueness result, and perhaps as a step towards proving one, one might try to show that any positive solution of (1.1), (1.2) has exactly one maximum, but even this seems difficult. For both (1.1), (1.2) and (1.12), (1.13), there is strong numerical evidence that there is a unique (up to sign) solution with any given positive number of zeros in $(0, \infty)$, but so far this has not been proved.

Finally we make some additional remarks about (1.3) and generalisations of (1.1). The optical behaviour of a medium is sometimes specified in terms of its refractive index, n, which is related to the dielectric susceptibility, χ, by $n = \sqrt{1 + 4\pi\chi}$. For nonlinear materials, the coefficients in the development of n as a function of $< E^2 >$,

$$n = n_0 + n_2 < E^2 > + \cdots \tag{1.14}$$

are tabulated [4, 8, 16]. The values of χ_0 and χ_3 are then obtained by $4\pi\chi_0 = n_0^2 - 1$ and $4\pi\chi_3 = 2n_0 n_2$. However, using a truncation such as (1.3) leads to a susceptibility that becomes infinite as $< E^2 >$ becomes infinite, whereas the phenomena causing the nonlinear response of the material lead to a susceptibility that should approach some finite value as $< E^2 >$ becomes infinite [3, 4]. Furthermore, the guidance properties of an optical fibre are often enhanced by using concentric layers of material with different composition. The generalisation of (1.3) that allows for both saturation at large intensity and nonhomogeneous composition is

$$D(x, y, z, t) = \{1 + 4\pi\chi (r, < E^2 > (x, y, z))\} E(x, y, z, t), \tag{1.15}$$

where $\lim\limits_{s \to \infty} \chi(t, s)$ is finite. With this kind of constitutive assumption the existence of self-trapped solutions of the form (1.5), (1.6), (1.8) is studied in [7] using a variational method. The equation (1.7) becomes

$$v''(r) + \frac{v'(r)}{r} - \frac{v(r)}{r^2} + \frac{4\pi\omega^2}{c^2} \chi\left(r, \frac{1}{2}v^2(r)\right) v(r) + \left(\frac{\omega^2}{c^2} - k^2\right) v(r) = 0, \tag{1.16}$$

with the same conditions (1.9), (1.10), (1.11) on v for guidance. For this more general equation, the parameters cannot be removed by rescaling the variables and a new aspect of the discussion involves determining the behaviour of the total intensity of the light beam

$$\frac{k}{2\omega} \int_0^\infty r v^2(r) \, dr$$

as the wavelength $2\pi/k$ is varied [7]. Variational methods can also be used to establish the existence of an infinite number of solutions of (1.16) with different numbers of zeros [17]. It would also be interesting to know under what conditions on χ the solutions of (1.16) have the same qualitative behaviour as is established here for (1.1).

2. Statement of Results

The problem is to solve

$$u'' + \frac{u'}{r} - \frac{u}{r^2} + u^3 - u = 0 \tag{2.1}$$

along with the boundary conditions

$$u(0) = 0, \tag{2.2}$$

$$u(\infty) = 0. \tag{2.3}$$

In this connection we prove the following theorems.

Theorem 1. *There exists at least one solution of* (2.1) - (2.3) *which satisfies the additional conditions that, in* $(0, \infty)$, $u > 0$ *and* u' *has precisely one zero.*

Theorem 2. *Let* $J \geqslant 1$ *be an integer. Then there exists at least one solution of* (2.1) - (2.3) *which has precisely* J *zeros in* $(0, \infty)$ *and whose derivative has precisely* $J + 1$ *zeros.*

Although uniqueness of solutions is hard, we would at least like to have proved that every solution of (2.1)-(2.3) has the property that between each pair of zeros where is just one zero of the derivative, but this we have been unable to do. However, some of the qualitative behaviour of solutions of (2.1) (without necessarily the boundary conditions (2.2) and (2.3)) is given in the following list of properties.

Property 1. *If a solution of (2.1) has a positive maximum A followed by a postive minimum B, then there must follow a further positive maximum C and $C < A$.*

Property 2. *If a solution of (2.1) has a positive maximum A followed by a zero and then a negative minimum $-C$, we have $C < A$.*

Property 3. *For any solution $u(r)$ of (2.1), the maxima of $|u|$ decrease as r increases. (This is an immediate consequence of Properties 1 and 2 and requires no separate proof.)*

Property 4. *If a maximum of $|u|$ does not exceed $\sqrt{2}$, then u has no further zeros and $u \longrightarrow \pm 1$ as $r \longrightarrow \infty$.*

Property 5 *If all maxima of $|u|$ exceed $\sqrt{2}$ (and only a finite number can do so), then $u \longrightarrow 0$ as $r \longrightarrow \infty$.*

Property 6. *If a solution $u(r)$ of (2.1) has $u' = u'' = 0$ at $r = r_0$, then for $r < r_0$ we cannot have $ru' \pm u = 0$. In particular, no such solution can vanish for $r < r_0$.*

We first give the proof of Properties 1,2,4-6 in §§3-7, and then the proofs of Theorem 1 and 2 is the final sections.

3. Proof of Property 1

Setting $t = \log r$, and $\delta u = du/dt$, we transform (2.1) into the form

$$\delta^2 u - u = e^{2t}\left(u - u^3\right). \tag{3.1}$$

We suppose that the maximum A and the minimum B correspond to t_1 and t_2. It is immediate that $A > 1$ (since we must have $\delta^2 u \leq 0$), although we may have either $B > 1$ or $B < 1$. It is also clear that there must be a further maximum following the minimum at t_2. For otherwise we have $\delta u \geq 0$ and u increasing for $t > t_2$. Clearly then either $u \longrightarrow 1$ or $u \longrightarrow \infty$. The second alternative is impossible because $\delta^2 u$ becomes large and negative, contradicting $\delta u \geq 0$, and the first alternative is impossible because $u \longrightarrow 1$(and $u \leq 1$) implies $\delta^2 u \geq u$, which is in consistent with $u \longrightarrow 1$. Let this further maximum have value C at $t = t_3$. Certainly, $C > 1$.

Now suppose $B < 1$, and let t_4, t_5 be defined by

$$u(t_4) = u(t_5) = 1, \quad t_1 < t_4 < t_2, \quad t_2 < t_5 < t_3.$$

We consider along with (3.1) the equations

$$\delta^2 u - u = e^{t_4}\left(u - u^3\right), \tag{3.2}$$

$$\delta^2 u - u = e^{t_5}\left(u - u^3\right). \tag{3.3}$$

In all equations we take u as the independent variable, and, with $p(u) = \delta u$, the equations (3.1)-(3.3) become

$$p\frac{dp}{du} - u = e^{2t}\left(u - u^3\right), \tag{3.4}$$

$$p\frac{dp}{du} - u = e^{2t_4}\left(u - u^3\right), \tag{3.5}$$

$$p\frac{dp}{du} - u = e^{2t_5}\left(u - u^3\right). \tag{3.6}$$

Let p_1, p_2, p_3 be the solutions of (3.4)-(3.6) satisfying $p(B) = 0$. Integrating with respect to u, we have

$$\frac{1}{2}p_1^2 - \frac{1}{2}p_2^2 = \int_B^u \left(e^{2t(v)} - e^{2t_4}\right)\left(v - v^3\right)dv, \tag{3.7}$$

where $t(u)$ is defined as the inverse function to the solution $u(t)$ of (3.1) in (t_1, t_2). Thus $t(v) = t_4$ at $v = 1$, and the integrand on the right of (3.7) is positive both for $v < 1$ and for $v > 1$, and so we cannot have p_1 vanishing before p_2. Thus $A > A_2$, where A_2 is the maximum attained by the solution of (3.2) which has a minimum B. Of course, both (3.2) and (3.3) are autonomous equations, and it is an easy integration of (3.2) to obtain

$$A_2^2 + B^2 = 2\left(1 + e^{-t_4}\right).$$

Similarly,

$$\frac{1}{2}p_1^2 - \frac{1}{2}p_3^2 < 0,$$

and so $C < A_3$, where

$$A_3^2 + B^2 = 2\left(1 + e^{-t_5}\right).$$

Since $t_4 < t_5$, we have $A_3 < A_2$, and so

$$C < A_3 < A_2 < A.$$

This completes the proof of Property 1, provided that $B < 1$. If $B \geq 1$, the argument is the same except that we define

$$t_4 = t_5 = t_2.$$

It should be added that it is easy to adapt the proof of Property 1 to give a proof of the first part of Property 4. For if we write the equation in the form (3.4) and integrate with respect to u from a zero of the solution to the previous maximum U, we have

$$-\frac{1}{2}U^2 - \frac{1}{2}p^2(0) = \int_0^U e^{2t}\left(u - u^3\right) du. \tag{3.8}$$

If we suppose for contradiction that $U \leq \sqrt{2}$, then

$$\int_0^U \left(u - u^3\right) du \geq 0. \tag{3.9}$$

Further, in the integrand in (3.8), the values of t corresponding to positive values of the integrand ($u < 1$) are greater than the values corresponding to negative values of the integrand. Hence (3.9) implies that the integral on the right of (3.8) is positive, which is a contradiction. In particular this shows that after a maximum not exceeding $\sqrt{2}$, the solution cannot decrease to 0 at infinity.

4. Proof of Property 2

To prove Property 2, we repeat the proof of Property 1, but now taking $B = 0, u(t_1) = A, u(t_2) = 0, u(t_3) = -C, u(t_4) = 1, u(t_5) = -1$. Let our solution have $\delta u(t_2) = -k$, and let p_1, p_2, p_3 be the solution of (3.4)-(3.6) satisfying $p^2(0) = k^2$. Then as in §3, we see that p_1 cannot vanish before p_3, and so $A > C$.

5. Proof of Property 4

We have already, in §3, given one proof of the first part of Property 4. An alternative is to introduce the energy

$$Q = \frac{1}{2}u'^2 + \frac{1}{4}u^4 - \frac{1}{2}u^2\left(1 + \frac{1}{r^2}\right). \tag{5.1}$$

It is trivial to verify that

$$Q' = -\frac{u'^2}{r} + \frac{u^2}{r^3}, \tag{5.2}$$

$$\delta\left(e^{2t}Q\right) = e^{2t}\left(\frac{1}{2}u^4 - u^2\right), \tag{5.3}$$

$$\delta\{e^{2t}\left(Q + \frac{1}{4}\right)\} = \frac{1}{2}e^{2t}\left(u^2 - 1\right)^2. \tag{5.4}$$

At a maximum of $|u|$ not exceeding $\sqrt{2}$, we have, from (2.1),

$$1 + \frac{1}{r^2} \leq u^2 \leq 2,$$

so that $r \geq 1$ and $Q \leq 0$. Since Properties 1 and 2 imply that $u^2 < 2$ from now on, we have from (5.3) that $e^{2t}Q$ is negative decreasing, and this precludes a further zero of u, where necessarily $Q > 0$. It precludes also the possibility that u decreases to 0 at infinity since $e^{2t}Q$ tends to a strictly negative limit, and linearisation about $u = 0$ at infinity easily shows that $u(r)$ vanishes exponentially as $r \longrightarrow \infty$, and so $e^{2t}Q = r^2Q \longrightarrow 0$.

Further, from (5.2), Q' consists of a negative term and a term integrable as $r \longrightarrow \infty$, and so Q tends to a limit, at first possibly $-\infty$, although (5.1) then makes it clear that Q is bounded below. So $Q \longrightarrow Q_0, Q_0$ finite. Writing (5.1) in the form

$$Q + \frac{1}{4} = \frac{1}{2}u'^2 + \frac{1}{4}\left(u^2 - 1\right)^2 - \frac{1}{2}\frac{u^2}{r^2},$$

we see that

$$\frac{1}{2}u'^2 + \frac{1}{4}\left(u^2 - 1\right)^2 \longrightarrow Q_0 + \frac{1}{4}.$$

Clearly, we must have $Q_0 \geq -\frac{1}{4}$, and from what has been said above, $Q_0 \leq 0$. If $Q_0 = -\frac{1}{4}$, clearly $u' \longrightarrow 0$ and $u^2 \longrightarrow 1$, as required.

If u is ultimately monotonic, then (2.1) makes it clear that, since $u(\infty) = 0$ is already excluded, we must have $u(\infty) = \pm 1$, as required. If u is not ultimately monotonic, then the decreasing maxima of u (assuming without loss of generality that u is ultimately positive) tend to a limit u_0 and, since $u \geq 1$ for a maximum, we have $| \leq u_0 < \sqrt{2}$, and

$$\frac{1}{4}\left(u_0^2 - 1\right)^2 = Q_0 + \frac{1}{4}.$$

If $u_0 = 1$, we are back to the case $Q_0 = -\frac{1}{4}$. If $1 < u_0 < \sqrt{2}$, then in the limit u oscillates between u_0 and u_1, where $u_1 \, (< u_0)$ is the other positive root of

$$\frac{1}{4}\left(u^2 - 1\right)^2 = Q_0 + \frac{1}{4}.$$

It is easy to see that, for such a solution,

$$\int^{\infty} \frac{u'^2}{r} dr = \infty, \tag{5.5}$$

which through (5.2) contradicts Q bounded. This contradiction completes the proof of the lemma.

6. Proof of Property 5

We can prove as in the last section that Q tends to a finite limit Q_0. Suppose for contradiction that there are an infinite sequence of decreasing maxima exceeding $\sqrt{2}$. Then these maxima of $|u|$ must tend to a limit, say u_0, and $u_0 \geq \sqrt{2}$. Further

$$\frac{1}{4}\left(u_0^2 - 1\right)^2 = Q_0 + \frac{1}{4}.$$

If $u_0 > \sqrt{2}$, then in the limit u oscillations between u_0 and $-u_0$, we have (5.5) as before, and again a contradiction.

If $u_0 = \sqrt{2}$, then $Q_0 = 0$ and

$$\frac{1}{2}u'^2 + \frac{1}{4}u^4 - \frac{1}{2}u^2 \longrightarrow 0.$$

Thus, for large r, the solution behaves in any bounded interval of r like a solution of

$$\frac{1}{2}u'^2 + \frac{1}{4}u^4 - \frac{1}{2}u^2 = 0, \tag{6.1}$$

and since the solution of (6.1) which satisfies, at any r_0,

$$u(r_0) = \sqrt{2}, \quad u'(r_0) = 0$$

has

$$u(\pm\infty) = 0,$$

we see that, after a maximum close to $\sqrt{2}$ at large r, the solution of (2.1) takes a long distance to reach 0, and then a further long distance to reach to next maximum, and during most of this distance $|u| < \sqrt{2}$ and so, from (5.3), r^2Q is decreasing. Thus r^2Q is less at each succeeding maximum of $|u|$, and ultimately r^2Q is large and negative. But at a maximum of $|u|$, always exceeding $\sqrt{2}$,

$$r^2Q \geq -\frac{1}{2}u^2,$$

and so cannot become large and negative. This contradiction proves that there are only a finite number of maxima of $|u|$ exceeding $\sqrt{2}$.

Thus our solution u is ultimately monotonic and so tends to a limit as $r \longrightarrow \infty$. From (2.1) it is clear that this limit must be 0 or ± 1, and ± 1 can be ruled out because linearisation about ± 1 shows that the solution must be oscillatory. This completes the proof of Property 5.

7. Proof of Property 6

Since $u' = u'' = 0$ at $r = r_0$, we have from (2.1) that

$$u^2 = 1 + \frac{1}{r_0^2},$$

$$Q = -\frac{1}{4}\left(1 + \frac{1}{r_0^2}\right)^2.$$

In particular, $Q(r_0) < -\frac{1}{4}$. Also, from (5.2), $Q'(r_0) > 0$. So Q decreases as we decrease r. Thus, if there exists a previous point where $ru' \pm u = 0$, we must have there $Q < -\frac{1}{4}$, contradicting the fact that at such a point $Q = \frac{1}{4}u^4 - \frac{1}{2}u^2 \geq -\frac{1}{4}$.

If we suppose without loss of generality that $u(r_0) > 0$, then $ru' - u < 0$ at $r = r_0$ and clearly $ru' - u \geq 0$ at the previous zero of u (if such exists). Hence there exists a zero previous to r_0 of $ru' - u$, and the first part of the proof shows that no such previous zero can exist.

8. Proof of Theorems 1 and 2

These depend on three lemmas which we now state. We will then show how Theorems 1 and 2 follow from the lemmas and give the proofs of the lemmas themselves in later sections.

Lemma 1. *If $(u(0), u'(0)) = (0, \alpha)$, and $\alpha > 0$ is sufficiently small, then u has no zeros.*

Lemma 2. *If $(u(0), u'(0)) = (0, \alpha)$ and $J \geq 0$ is any given integer, then we can choose $\alpha > 0$ sufficiently large that u has at least $J + 1$ zeros,* *at $r_1 > 0, r_2, \cdots, r_{J+1}$, say, and u' has precisely one zero in each of the intervals $(0, r_1), (r_1, r_2), \cdots, (r_J, r_{J+1})$.*

To prove Theorem 2 (or Theorem 1) we start with a value of α sufficiently large that, by Lemma 2, u has at least $J + 1$ zeros. As we decrease α continuously to 0, these zeros must disappear. A zero cannot disappear at a finite value of r, since that would imply $u = u' = 0$ and so $u \equiv 0$. Thus the zeros must disappear at ∞, and when the $(J + 1) - th$ zero first disappears, we have the required solution. (We note that, by Property 6, u' cannot collect any additional zeros during the continuation process, and so the solution has the properties demanded in Theorem 2.) We must merely check that when a zero disappears in this way, then only one zero disappears. This follows immediately from

Lemma 3. *If $(u(R), u'(R)) = (0, \alpha)$, where R is large and $\alpha > 0$ small, then u has no zeros in (R, ∞) and $u(\infty) \neq 0$.*

9. Proof of Lemma 1

If $(u(0), u'(0)) = (0, \alpha)$ and α is small, then u remains small until r is large. Further, (2.1) written in the form

$$(ru')' = r\{u\left(1 + \frac{1}{r^2}\right) - u^3\}$$

makes it clear that u' certainly does not vanish until $u > 1$, and so there exists a large value of r, say r_1, at which $u = 1$. We can further give the approximate value of $u'(r_1)$. For if we consider the situation starting at r_1, and consider it for decreasing r, then for a long range of r the equation is approximately

$$u'' + u^3 - u = 0,$$

and the solution has to be positive and monotonic. This shows that $u'(r_1)$ is close to the value a where a is such that

$$u'' + u^3 - u = 0$$

with

$$u(-\infty) = 0, \quad u(0) = 1$$

has

$$u'(0) = a.$$

It is trivial to check that $a = 1/\sqrt{2}$. Thus

$$u(r_1) = 1, \quad u'(r_1) \sim 1/\sqrt{2}. \tag{9.1}$$

For any bounded distance beyond r_1, the solution approximates the solution of

$$u'' + u^3 - u = 0, \tag{9.2}$$

$$u(r_1) = 1, \quad u'(r_1) = 1/\sqrt{2}. \tag{9.3}$$

Using the energy integral

$$\frac{1}{2}u'^2 + \frac{1}{4}u^4 - \frac{1}{2}u^2 = 0,$$

we see that the solution of (9.2)-(9.3) rises to $\sqrt{2}$ in a finite interval, and then returns to 1 at $r = r_2$, say, and the same is approximately true for the solution of (2.1) and (9.1). If it rises above $\sqrt{2}$, it does so only briefly, and so from (5.3) we conclude that

$$r^2 Q < 0 \quad \text{for } r \leq r_2.$$

As u now decreases for $r > r_2$, we see that it cannot reach $u = 0$, since $r^2 Q$ continues to decrease and $Q > 0$ when $u = 0$. The solution thus turns up again before reaching $u = 0$, but, if it rises as high as $\sqrt{2}$, then it must do so with $u' = O(1/r)$, since we must have $Q < 0$. If therefore it rises above $\sqrt{2}$, it does so only briefly, we continue to have $Q < 0$, and again the solution cannot go down as far as $u = 0$. Repetition of this argument proves the lemma.

10. Proof of Lemma 2

We rescale (2.1) by melting

$$\tau = \alpha^{1/2} r, \quad u(r) = \alpha^{1/2} q(\tau),$$

so that

$$q'' + \frac{q'}{\tau} - \frac{q}{\tau^2} + q^3 - \frac{q}{\alpha} = 0, \tag{10.1}$$

with

$$q(0) = 0 \quad \text{and} \quad q'(0) = 1. \tag{10.2}$$

This shows that for any bounded range of τ we can approximate our original solution by the solution of

$$q'' + \frac{q'}{\tau} - \frac{q}{\tau^2} + q^3 = 0, \tag{10.3}$$

$$q(0) = 0, \quad q'(0) = 1, \tag{10.4}$$

and the lemma is proved if we can show that the solution of (10.3)-(10.4) has as many zeros as we please if we take the range of τ large enough, and that between these zeros there is precisely one zero of q'. Setting

$$\delta q = \frac{dq}{d(\log \tau)}$$

(since $\log \tau$ and $\log r$ differ only by a constant, this is the same δ as in §3), we see that

$$\left(\delta^2 - 1\right) q + \tau^2 q^3 = 0,$$

and if we further set

$$q = v/\tau,$$

then

$$\delta \left(\delta - 2\right) v + v^3 = 0. \tag{10.5}$$

This equation is now autonomous and so can be represented in the phase plane. It is easily verified that the phase portrait is a spiral, and since there is the energy integral

$$\frac{1}{2} \left(\delta v\right)^2 + \frac{1}{4} v^4 = 2 \int \left(\delta v\right)^2,$$

it is an expanding spiral. The fact that it is a spiral proves the lemma, except for showing that between the zeros of q there is precisely one zero of q'.

To do this, we proceed as in the proof of Property 6. We define

$$Q^* = \frac{1}{2} q'^2 + \frac{1}{4} q^4 - \frac{1}{2} \frac{q^2}{\tau^2},$$

with

$$Q^{*\prime} = -\frac{q'^2}{\tau} + \frac{q^2}{\tau^3}.$$

If $q' = q'' = 0$ at $\tau = \tau_0$, then $q^2 = 1/\tau_0^2$ and

$$Q^* = -\frac{1}{4}/\tau_0^4, \quad Q^{*\prime} > 0.$$

As we decrease τ from τ_0, we cannot reach a point where $\tau q' \pm q = 0$, since Q^* would have decreased from a negative value to a positive one. In particular, q cannot vanish, or, equivalently, if $q\left(\tau_1\right) = 0$, then we cannot have $q'\left(\tau\right) = q''\left(\tau\right) = 0$ for $\tau > \tau_1$.

If therefore we consider the solution such that $q\left(\tau_1\right) = 0, q'\left(\tau_1\right) = \beta > 0$, and vary β, then if, for some β_0, the solution has precisely one zero of q' before $q = 0$ again (and we must have a subsequent point at which $q = 0$ by the discussion of v above), then, since q' cannot collect extra zeros, we must have precisely one zero of q' before the next zero of q, for all β. Furthermore, because of the autonomous nature of the equation, the value of τ_1 is irrelevant. We will therefore have proved what is required (that between any two zeros of q there is precisely one of q') if we can prove the following three results:

(i) given τ_1, we can choose β so that the next zero is as large as we please;

(ii) given τ_1, we can choose β so that the next zero is as close to τ_1 as we please;

(iii) there is some β for which q' has precisely one zero before the next zero of q.

Result (i) is proved by taking β small. Then the solution is, for a long range of τ, like the corresponding solution of

$$q'' + \frac{q'}{\tau} - \frac{q}{\tau^2} = 0,$$

and it is easy to see that this does not vanish.

Result (ii) is proved by taking β large. It is in fact best to do this in the context of the equation (10.5) for v. If we set $\log \tau = z$, then this equation is

$$\frac{d^2v}{dz^2} - 2\frac{dv}{dz} + v^3 = 0. \tag{10.6}$$

Consider $v(z_1) = 0, v'(z_1) = \gamma$, where γ is large. Set $v = \gamma^{1/2}w, z = \gamma^{-1/2}s$, and (10.6) becomes

$$\frac{d^2w}{ds^2} - 2\gamma^{-1/2}\frac{dw}{ds} + w^3 = 0, \quad w(s_1) = 0, \quad w'(s_1) = 1.$$

For large γ, this can be compared with

$$\frac{d^2w}{ds^2} + w^3 = 0, \quad w(s_1) = 0, \quad w'(s_1) = 1, \tag{10.7}$$

the solution of which vanishes again a finite distance from s_1 and so (in terms of v and z) close to z_1. This proves result (ii), and also, in fact, result (iii). For, for the solution of (2.8), $\frac{d^2w}{ds^2} < 0$ when $w > 0$, and so the solution has only one zero of w' before w vanishes again. Thus the same is true for (10.6) when γ is sufficiently large, as required. This completes the proof of the lemma.

11. Proof of Lemma 3

This is a repeat of the proof of Lemma 1, rendered easier by the fact that we are dealing only with $u(r)$ where r is large.

REFERENCES

[1] Troy, W.C., *Multiple solutions of a nonlinear boundary value problem*, Proc. Roy. Soc. Edin., **113A** (1989), 191–209.

[2] Born, M. and Wolf, E., *Principles of Optics* Pergamon Press, fifth edition, Oxford, 1975.

[3] Akhmanov, S.A., Khokhlov, R.V. and Sukhorukov, A.P., "Self-focusing, self-defocusing and self modulation of laser beams" in *Laser Handbook*, Vol. 2, F.T. Arecchi and E.O. Schulz-DuBois, eds. North Holland, Amsterdam, 1972.

[4] Svelto, O., "Self-focusing, self-trapping, and self-phase modulation of laser beams" in *Progress in Optics*, Vol. 12, E. Wolf, ed. North-Holland, Amsterdam, 1974 .

[5] Agrawal, G.P., *Nonlinear Fiber Optics*, Academic Press, Boston, 1989 .

[6] Pohl, D., *Vectorial theory of self-trapped light beams* Optics Comm., **2** (1970), 305–308 .

[7] Stuart, C.A., *Self-trapping of an electro-magnetic field and bifurcation from the essential spectrum*, Arch. Rational Mech. Anal., to appear .

[8] Chiao, R.Y., Garmire, E. and Townes, C.H., *Self-trapping of optical beams* Phys. Rev. Lett., **13**(1964), 479–482 .

[9] Strauss, W.A., "The nonlinear Schrodinger equation" in *Contemporary Developments in Continuum Mechanics and P.D.E.*, G.M. de La Penha and L.A. Medeiros, eds. North-Holland, Amsterdam, 1978 .

[10] Jones, C. and Küpper, T., *On the infinitely many solutions of a semi-linear elliptic equation* SIAM J. Math. Anal., **17**(1986) 803–835 .

[11] McLeod, K., Troy, W.C. and Weissler, F.B., *Radial solutions of $\Delta u + f(u) = 0$ with prescribed numbers of zeros* J. Diff. Equat., to appear .

[12] Coffman, C.V., *Uniqueness of the ground state solution for $\Delta u - u + u^3 = 0$ and a variational characterisation of other solutions* Arch. Rational Mech. Anal., **46**(1972) 81–95 .

[13] McLeod, K. and Serrin, J., *Uniqueness of positive radial solutions of $\Delta u + f(u) = 0$ in $\mathbb{R}^N$* Arch. Rational Mech. Anal., **99**(1987) 115–145 .

[14] Kwong, M.K., *Uniqueness of positive solutions of $\Delta u - u + u^p = 0$ in $\mathbb{R}^N$* Arch. Rational Mech. Anal., **105**(1989) 243–266 .

[15] McLeod, K., *Uniqueness of positive radial solutions of $\Delta u + f(u) = 0$ in $\mathbb{R}^N$*, II, preprint .

[16] Smith, W.L. "Nonlinear refractive index in *Handbook of Laser Science and Technology*, Vol. 3, M.J. Weber, ed. CRC Press, Boca Raton, 1986.

[17] Vuille, R., thesis, EPFL, in preparation .

J.B. McLeod
Dept. Math
Univ. Pittsburgh
Pittsburgh, PA 15260

C.A. Stuart
Dépt. de Math
École Polytechnique Fédérale
de Lausanne
CH-1015 Lausanne-Ecublens
Switzerland

W.C. Troy
Dept. Math
Univ. Pittsburgh
Pittsburgh, PA 15260

A General I-Theorem
for Semilinear Elliptic Equations

KEVIN McLEOD

1. Introduction and Statement of Results

In these notes, we will give a slight generalisation of a uniqueness result [6] for the semilinear initial-boundary value problem

$$u'' + \frac{n-1}{r}u' + f(u) = 0 \text{ for } r > 0, \tag{1}$$

$$u'(0) = 0, \ u(r) \to 0 \text{ as } r \to \infty, \tag{2}$$

$$u(r) > 0 \text{ for } r \geq 0. \tag{3}$$

Here, $f : [0, \infty) \to \mathbb{R}$ satisfies

(i). $f \in C^1([0, \infty))$, $f(0) = 0$, $f'(0) = -m < 0$, and
(ii). for some $\alpha > 0$, $f(u) < 0$ for $0 < u < \alpha$, $f(u) > 0$ for $u > \alpha$ and $f'(\alpha) > 0$,

and $n > 1$ is a real parameter. The problem (1)–(3) will be referred to as (GS), since a solution of (1)–(3) can be considered (at least when $n \geq 2$ is an integer) as a positive, radially symmetric solution $u(X) = u(\|X\|)$ of $\Delta u + f(u) = 0$ in $\mathbb{R}^n$, and in many physical situations such a solution will represent the state of lowest energy, or ground state, of the system.

Due to its large number of applications, (GS) has been extensively studied in recent years, and the question of existence is now well understood. We will not review the results here (see [1], [2], [4], [8]), but we will recall that for the important model case $f(u) = -u + u^p$ $(p > 1)$ existence holds if and only if $1 < p < \frac{n+2}{n-2}$. The uniqueness problem was first attacked by Coffman [3], who considered the model case with $n = p = 3$. His methods were later generalised by McLeod and Serrin [7] to other values of p and n and to more general functions, but the full solution in the model case is due to Kwong [5]. Subsequently McLeod [6] was able to simplify Kwong's proof and apply it to the class of functions considered in [7]. The main Theorem of [6] does not strictly contain that of [7] however, but it is easy to add an extra parameter to remedy this, and it is this modified Theorem whose proof we wish to sketch here. The details of the proof are similar to those of [6] and so we will frequently refer the reader to that paper.

We now state our main result. We introduce a version of the I-function from [7]: for $\lambda > 0$, $c \in \mathbb{R}$ we define

$$I(u, \lambda, c) = \lambda(u - c)f'(u) - (2 + \lambda)f(u). \tag{4}$$

Theorem 1 (I-Theorem). *Let f satisfy conditions* (i) *and* (ii)*, and suppose that for each $U > \alpha$ there are numbers λ, $\overline{\lambda} > 0$ and c, $\overline{c} \in \mathbb{R}$, continuously depending on U, such that*

$$I(u, \lambda, c) \geq 0 \ for \ 0 < u < U \ and \tag{5}$$

$$I(u, \overline{\lambda}, \overline{c}) \leq 0 \ for \ u > U. \tag{6}$$

Then (GS) *has at most one solution.*

Remark. Under assumptions (i) and (ii), the possible values of c are in fact quite restricted. For example, letting $u \to \alpha$ in (5) gives $\lambda(\alpha-c)f'(\alpha) \geq 0$, and since $\lambda > 0$ and $f'(\alpha) > 0$, we deduce that $c \leq \alpha$. Similarly, letting $u \to 0$ in (5) shows that $c \geq 0$.

The main hypothesis of the I-Theorem can be hard to check directly, and so we will state a corollary Theorem which can be easily applied to a wide class of functions. (See [6] for the proof.)

Theorem 2. *Suppose f satisfies* (i) *and* (ii) *and that there is some $\tau \geq 1$ such that*

$$u^{-\tau}f(u) \ is \ increasing \ for \ u > 0, \ and \tag{7}$$

$$u(u^{-\tau}f(u))' \ is \ decreasing \ for \ u > \alpha. \tag{8}$$

(In case $\tau = 1$, we require $u(u^{-\tau}f(u))'$ to be strictly decreasing for $u > \alpha$.) Then (GS) *has at most one solution.*

As an example, let f be a general "polynomial",

$$f(u) = \sum_{k=1}^{\nu} a_k u^{p_k},$$

where $1 = p_1 < p_2 < \cdots < p_\nu = p$. It can be easily checked by differentiation that f will satisfy the conditions of Theorem 2 with the choice $\tau = p$, provided $a_1 < 0$, $a_k \leq 0$ for $2 \leq k \leq \nu - 1$ and $a_\nu > 0$. Existence for (GS) is known to hold under these conditions if and only if $p < \frac{n+2}{n-2}$, and we obtain uniqueness for this range of p.

It is worth noting that the proof of Theorem 1 (and hence also of Theorem 2) does not depend on any growth conditions on f, and so can be applied to solutions of an "exterior Neumann problem", see [6], e existence holds in the model cases for all $p \in (1, \infty)$. The proof also gives uniqueness results for the Dirichlet problem

$$\Delta u + f(u) = 0 \text{ in } B, \ u = 0 \text{ on } \partial B,$$

where B is a ball in $\mathbb{R}^n$.

2. Elementary Results

In this section, we collect some well-known results concerning the solutions of (GS). The results of this section do not require the main hypothesis of the I-Theorem, but we continue to assume that conditions (i) and (ii) hold.

It has become standard to study the uniqueness problem for (GS) by considering the initial value problem

$$u'' + \frac{n-1}{r} u' + f(u) = 0 \text{ for } r > 0, \tag{9}$$

$$u(0) = a > 0, \ u'(0) = 0, \tag{10}$$

$$u \text{ extends maximally to the right with } u \geq 0. \tag{11}$$

The solution of this problem (which is unique [2]) will be denoted either by $u(r)$ or $u(r, a)$. We define

$$S^+ = \{a > 0 : u(r, a) \text{ remains bounded away from } 0\}$$
$$S^0 = \{a > 0 : u(r, a) \text{ solves (GS)}\}$$
$$S^- = \{a > 0 : u(R, a) = 0 \text{ for some (first) } R = R(a) > 0\}.$$

In case $a \in S^0$, we will also set $R(a) = \infty$. The variation $\delta(r) = \delta(r, a) = \frac{\partial u}{\partial a}$ satisfies

$$\delta'' + \frac{n-1}{r} \delta' + f'(u)\delta = 0 \tag{12}$$

$$\delta(0) = 1, \ \delta'(0) = 0 \tag{13}$$

The following two Lemmas describe the behaviour, including asymptotic behaviourf solutions of (9) and (12). For proofs of these results, the reader may consult [2] or [9]. **Lemma 1.** (a) *The sets S^+, S^- and S^0 are disjoint and cover the interval $(0, \infty)$, with $(0, \alpha] \in S^+$. In particular, if u is a solution of (GS), then $u(0) > \alpha$. Also, S^+ and S^- are both open in $(0, \infty)$.*

(b) *Any solution u with $u(0) \in S^0 \cup S^-$ is monotone decreasing. In particular, any solution of (GS) is monotone decreasing.*

(c) *If u is a solution of* (GS), *then for any $\epsilon \in (0, m)$,*

$$\limsup_{r \to \infty} u(r)e^{r\sqrt{m-\epsilon}} < \infty, \quad \limsup_{r \to \infty} |u'(r)|e^{r\sqrt{m-\epsilon}} < \infty, \quad and$$

$$\frac{|u'(r)|}{u(r)} \to \sqrt{-m} \ as \ r \to \infty.$$

(d) *If u is a solution of* (9) *with $u(0) \in S^+$ then u has an infinite number of local maxima and minima. Furthermore, if $u(r_0)$ is a local minimum of u then $u(r) > u(r_0)$ for $r > r_0$, while if $u(r_0)$ is a local maximum then $u(r) < u(r_0)$ for $r > r_0$.*

Lemma 2. (a) *If $u(0) \in S^0 \cup S^-$, then δ has only a finite number of critical points in $(0, R)$.*

(b) *If $u(0) \in S^0$, then as $r \to \infty$ either*

$$\delta(r) \to \pm\infty, \ \delta'(r) \to \pm\infty \ with \ \delta(r)\delta'(r) > 0 \ for \ large \ r, \ or$$

$$\delta(r) \to 0, \ \delta'(r) \to 0 \ with \ \delta(r)\delta'(r) < 0 \ for \ large \ r.$$

In the second case, for any $\epsilon \in (0, m)$,

$$\limsup_{r \to \infty} |\delta(r)|e^{r\sqrt{m-\epsilon}} < \infty, \quad \limsup_{r \to \infty} |\delta'(r)|e^{r\sqrt{m-\epsilon}} < \infty.$$

Our next Lemma gives a first relation between the sign of $\delta(R)$ and the local behaviour of solutions of (9)–(11).

Lemma 3. (a) *Let $a_0 \in S^-$ and suppose that $\delta(R(a_0)) < 0$. Then $R(a)$ is a decreasing function of a near a_0.*

(b) *Let $a_0 \in S^0$ and suppose that $\delta(r) \to -\infty$ as $r \to \infty$. Then for some $\epsilon > 0$ the interval $(a_0, a_0 + \epsilon)$ is contained in S^-, while $(a_0 - \epsilon, a_0) \subset S^+$. Similarly, if $\delta(r) \to +\infty$, some right neighbourhood of a_0 is contained in S^+, while a left neighbourhood is contained in S^-.*

Proof. (a) From the implicit function theorem applied to the equation $u(R(a), a) = 0$, we see that $R(a)$ is a differentiable function of a for $a \in S^-$, and that

$$u'(R(a), a)R'(a) + \delta(R(a), a) = 0.$$

Since $u'(R(a), a) < 0$, the assumption $\delta(R(a_0), a_0) < 0$ gives $R'(a_0) < 0$.

(b) Suppose $f'(u) < 0$ for $u \in [0, \gamma)$, and let R_1 be fixed so large that $u(r, a_0) \leq \frac{1}{2}\gamma$ when $r \geq R_1$. If $\delta(r) \to -\infty$ as $r \to \infty$, we see from Lemma 2(b) that for some $R_2 > R_1$ we have $\delta(R_2) < 0$ and $\delta'(R_2) < 0$. Thus, for $a > a_0$ but close to a_0, we have

$$u(R_2, a) < u(R_2, a_0), \ u'(R_2, a) < u'(R_2, a_0). \tag{14}$$

If $a \in S^+$, there would be a subsequent point $R_3 > R_2$ at which $u(R_3, a) = u(R_3, a_0)$, while if $a \subset S^0$, both $u(r, a)$ and $u(r, a_0)$ approach 0 as $r \to \infty$. In either case, by (14), the function $w(r) = u(r, a) - u(r, a_0)$ must have a negative minimum after R_2. However, w satisfies

$$w'' + \frac{n-1}{r} w' + f'(\theta(r))w = 0,$$

where $\theta(r)$ is between $u(r, a)$ and $u(r, a_0)$, so that $f'(\theta(r)) < 0$ in (R_2, R_3), and any negative critical point of w in this interval could only be a maximum. Thus, $a \notin S^+ \cup S^0$ and so $a \in S^-$.

The other cases are all similar. In the two cases in which $u(R_2, a) > u(R_2, a_0)$, a must be taken sufficiently close to a_0 so that $u(R_2, a) < \gamma$.

Lemma 3 shows that information about $u(r, a)$ can be obtained from an analysis of $\delta(r, a)$. In particular, information on the number of zeros of δ plays a crucial role in the proof of Theorem 1. The following terminology is due to Kwong [5].

Definition. Let $a \in S^0 \cup S^-$. a is said to be *admissable* if $\delta(r, a)$ has exactly one zero in $[0, R)$. a is said to be *strictly admissable* if a is admissable and in addition $\delta(R, a) < 0$ (or $\delta(r) \to -\infty$ as $r \to \infty$, in case $a \in S^0$).

In the final section of the paper we will show that, under the hypotheses of Theorem 1, every $a \in S^0 \cup S^-$ is strictly admissable. Theorem 1 then follows easily. The heart of the argument is Lemma 9, in which we show that admissability implies strict admissability. Not surprisingly, the proof of this result requires the full hypotheses of the I-Theorem, but we can finish the present section with two results related to admissibility. The first of these is a Lemma which implies Lemma 9 in case $\delta(r) = 0$ at some r where $u(r) \leq \alpha$, while the second, due to Zhang [10], states that the smallest a in $S^0 \cup S^-$ is admissable. (Of course, if no such a exists, then $S^+ = (0, \infty)$, $S^0 = \emptyset$ and Theorem 1 is trivially true.)

Lemma 4. *Suppose that $a \in S^0 \cup S^-$ and that $\delta(r) \to 0$ as $r \to R$. Then if $r_1 \in (0, R)$ is such that $\delta(r_1) = 0$, we have $u(r_1) > \alpha$.*

Proof. Assume for contradiction that $\delta(r_1) = 0$ and $u(r_1) \leq \alpha$. Without loss of generality, we may assume that r_1 is the last zero of δ in $(0, R)$, and that $\delta < 0$ in (r_1, R). From the equations satisfied by u and δ, we obtain

$$[(r^{n-1}u')(r^{n-1}\delta')]' = -r^{2n-2}[f(u)\delta' + f'(u)u'\delta]$$
$$= -r^{2n-2}[f(u)\delta]'.$$

Integrating from r_1 to some $r_2 \in (r_1, R)$ and then integrating by parts, we obtain

$$r_2^{2n-2}u'(r_2)\delta'(r_2) - r_1^{2n-2}u'(r_1)\delta'(r_1) = -[s^{2n-2}f(u(s))\delta(s)]_{r_1}^{r_2}$$
$$+(2n-2)\int_{r_1}^{r_2} s^{2n-3}f(u(s))\delta(s)\,ds.$$

Now let $r_2 \to R$. If $R < \infty$ we note that $\delta'(R) > 0$ (since $\delta < 0$ in (r_1, R) and $\delta(R) = 0$), while if $R = \infty$ we apply Lemmas 1(c) and 2(b). Noting also that $f(u(r)) < 0$ in (r_1, R), we obtain

$$0 > R^{2n-2}u'(R)\delta'(R) - r_1^{2n-2}u'(r_1)\delta'(r_1) = (2n-2)\int_{r_1}^{R} s^{2n-3}f(u(s))\delta(s)\,ds > 0.$$

(If $R = \infty$, the term $R^{2n-2}u'(R)\delta'(R)$ is to be interpreted as 0.) This is a contradiction, and the Lemma is proved.

The proof of Zhang's Lemma relies on the Sturm Oscillation Theorem. Since we will need this result repeatedly in Section 3, we state here a simple form which is sufficient for our purposes, together with a simple disf may be found in [6].

Lemma 5. *Let Y and Z be non-trivial solutions of*

$$Y'' + \frac{n-1}{r}Y' + g(r)Y = 0, \tag{15}$$

$$Z'' + \frac{n-1}{r}Z' + G(r)Z = 0 \tag{16}$$

respectively on some interval $(\mu, \nu) \subset (0, \infty)$, where g and G are continuous on (μ, ν), $G \geq g$ on (μ, ν) and $G \not\equiv g$. If either

(a) $\mu > 0$ *and* $Y(\mu) = Y(\nu) = 0$, *or*

(b) $\mu = 0$, Y *and* Z *are continuous at* μ *and*
$Y'(\mu) = Z'(\mu) = Y(\nu)$,

then Z has at least one zero in (μ, ν). The same conclusions hold if $G \equiv g$ on (μ, ν), provided Y and Z are linearly independant.

Remark. In the situation of Lemma 5, we will say that Z oscillates faster than Y (or that Y oscillates slower than Z) on (μ, ν).

Definition. Suppose that (15) has at least one solution which does not vanish in some neighbourhood of ∞. Define

$$\rho = \inf\{r \in (0, \infty) : \text{there is a solution of (15) with no zeroes in } (r, \infty)\}.$$

The interval (ρ, ∞) will be called the *disconjugacy interval* of (15).

Clearly, no solution of (15) can have two zeroes in (ρ, ∞), for if a solution vanishes at r_1 and r_2 in (ρ, ∞) then any linearly independant solution would vanish in (r_1, r_2) and the disconjugacy interval could not be any larger than (r_1, ∞). On the other hand, if $\rho > 0$ any solution of (15) with a zero in $(0, \rho)$ must have a subsequent zero, or the disconjugacy interval would be larger than (ρ, ∞). Thus the last zero of any solution of (15) must lie in $[\rho, \infty)$. In the next Lemma, we distinguish those solutions whose last zero is precisely at ρ.

Lemma 6. *Let $g(r)$ be continuous on $(0, \infty)$, and suppose that $g(r) < 0$ for large r. Let the disconjugacy interval of (15) be (ρ, ∞) with $\rho > 0$, and suppose that (15) has a solution which goes to 0 as $r \to \infty$. If Y is a non-trivial solution of (15) such that $Y(\rho) = 0$, then Y has no subsequent zeroes and $Y(r) \to 0$ as $r \to \infty$. Conversely, if Y is a non-trivial solution of (15) with a zero in (ρ, ∞), then Y does not approach zero as $r \to \infty$.*

Remark. If the disconjugacy interval of (15) is (ρ, ∞) and $g \leq G < 0$ on this interval, then the disconjugacy interval of (16) cannot be any larger than (ρ, ∞). Specifically, if Y and Z are solutions of (15) and (16) with $Y(\rho) = Z(\rho) = 0$, then $Y \to 0$ as $r \to \infty$, by Lemma 6, and an analysis of the Wronskian of Y and Z shows that if $Z > 0$ on (ρ, ∞) then $Z \to 0$ as $r \to \infty$. In case Y and Z both decay exponentially fast at ∞ (which is the only case we will use) the proof is similar to the standard proof of Lemma 5; if either Y or Z decays less rapidly, the analysis is slightly more delicate.

Lemma 7. (Zhang) *The value $a_0 = \inf(S^0 \cup S^-)$ is admissable.*

Proof. Note that in fact $a_0 \in S^0$ (since S^- is open) and assume that $\delta(r, a_0)$ has two or more zeroes in $(0, \infty)$; let the first two zeroes of $\delta(r, a_0)$ be at r_1 and r_2. Then $\delta(r, a_0) > 0$ in $(0, r_1)$, $\delta(r, a_0) < 0$ in (r_1, r_2) and $\delta(r, a_0) > 0$ in some interval to the right of r_2. For $a < a_0$ but close enough to a_0, then, the solution $u(r, a)$ will intersect $u(r, a_0)$ at least twice. Let the first two intersections be at $y_1(a)$ and $y_2(a)$. Note that $y_2(a) < r_0(a)$, where $r_0(a)$ is the first minimum of $u(r, a)$, for if $u(r_0, a) > u(r_0, a_0)$ then by Lemma 1(b)(d) there can be no further intersection after r_0.

Now decrease a continuously to α. The intersection points y_1 and y_2 will vary continuously with a and cannot coalesce (for otherwise $u(r, a)$ and $u(r, a_0)$ would become tangent, contradicting uniqueness). Note that the function $w(r) = u(r) - \alpha$ satisfies

$$w'' + \frac{n-1}{r}w' + f'(\theta(r))w = 0, \tag{17}$$

where $\theta(r)$ is between α and $u(r)$. For a close to α, the solutions of (17) behave very much like the solutions of

$$\phi'' + \frac{n-1}{r}\phi' + f'(\alpha)\phi = 0.$$

In particular, solutions of (17) will oscillate more quickly than solutions of

$$\psi'' + \frac{n-1}{r}\psi' + \frac{1}{2}f'(\alpha)\psi = 0$$

(recall that $f'(\alpha) > 0$). But this means that as $a \to \alpha$ the second intersection of $u(r, a)$ with the horizontal line $u = \alpha$ remains bounded, and hence so do $r_0(a)$ and $y_2(a)$. Since y_1 and y_2 neither coalesce nor become unbounded, we see that for all $a \in (\alpha, a_0)$, $u(r, a)$ intersects $u(r, a_0)$ at least twice.

However, for a close to α, $u(r, a)$ remains close to $u(r, \alpha) \equiv \alpha$ in any bounded interval, and its derivative remains close to $u'(r, \alpha) \equiv 0$. Since $u'(r, a_0)$ is bounded away from 0 in any compact interval not containing the origin, it follows that for a close to α, $u(r, a)$ can intersect $u(r, a_0)$ only once, and the Lemma is proved.

3. Proof of Theorem 1

Recall that our aim is to show that every $a \in S^0 \cup S^-$ is strictly admissable. As a first step in this direction, we will show that for every such a the variation $\delta(r, a)$ has at least one zero in $(0, R)$. In this proof, we first use an auxiliary function which will prove to be very useful in the remainder of the paper. For $\lambda > 0$ and $c \in \mathbb{R}$, we define

$$v(r) = v_{\lambda,c}(r) = v_{\lambda,c}(r, a) = ru'(r) + \lambda(u(r) - c). \tag{18}$$

A calculation using (9) shows that $v_{\lambda,c}$ is a solution of

$$v'' + \frac{n-1}{r}v' + f'(u)v = I(u, \lambda, c), \tag{19}$$

which can be written as

$$v'' + \frac{n-1}{r}v' + [f'(u) - \frac{I(u, \lambda, c)}{v}]v = 0 \tag{20}$$

in any interval in which $v \neq 0$.

Lemma 8. *Under the assumptions of Theorem 1, for any $a \in S^0 \cup S^-$ the variation $\delta(r, a)$ has at least one zero in $(0, R)$.*

Proof. Suppose δ never vanishes in $(0, R)$. Then $\delta > 0$ in $(0, R)$, since $\delta(0) = 1$. Let $\lambda = \lambda(a)$ and $c = c(a)$ be the values corresponding to $U = a$ in the hypothesis of Theorem 1. Note that $a > \alpha$ by Lemma 1(a). Then $I(u, \lambda, c) \geq 0$ for $0 < u < a = u(0)$, so if $I(u, \lambda, c) \not\equiv 0$, comparison of (12) and (20) shows that oscillates more slowly than δ as long as $v_{\lambda,c} > 0$. But $v_{\lambda,c}(0) > 0$ and $v_\lambda(R) < 0$ (recall that $c \in [0, \alpha eR = \infty)$, so $v_{\lambda,c}$ must have a zero in $(0, R)$, which is a contradiction. (In case $I(u, \lambda, c) \equiv 0$, δ must be a positive multiple of $v_{\lambda,c}$ and so again $\delta(R) < 0$.)

Lemma 9. *Suppose f satisfies the hypotheses of Theorem 1 and let $a \in S^0 \cup S^-$ be admissable. Then a is strictly admissable.*

Proof. Suppose that a is admissable but not strictly admissable. Then δ has exactly one zero in $(0, R)$ and $\delta(r) \to 0$ as $r \to R$. Let r_1 be the zero of δ, and note by Lemma 4 that $u(r_1) > \alpha$. Let $\lambda_1 = \overline{\lambda}(u(r_1))$, $c_1 = \overline{c}(u(r_1))$. Tr $u > u(r_1)$, so comparison of (12) and (20) shows that if $v_{\lambda_1,c_1} > 0$ in $(0, r_1]$ it would oscillate faster than δ in this interval. But this contradicts Lemma 5(b), and we conclude that the first zero of v_{λ_1,c_1} occurs in $(0, r_1]$.

Now let $\lambda_2 = \lambda(u(0))$ and $c_2 = c(u(0))$. As in the proof of Le, v_{λ_2,c_2} oscillates more slowly than δ in $(0, r_1)$ (as long as $v_{\lambda_2,c_2} > 0$) and so the first zero of v_{λ_2}, occurs at or after r_1. It follows that there are values of λ between λ_1 and λ_2 and c between c_1 and c_2 such that the first zero of $v_{\lambda,c}$ occurs exactly at r_1. Since $\lambda(U)$ and $c(U)$ depend continuously on U, these values of λ and c can be chosen as $\lambda(u(r))$, $c(u(r))$ for some $r \in [0, r_1]$, and then $I(u, \lambda, c$ is non-negative at r_1 and remains non-negative in $[r_1, \infty)$. In some interval to the right of r_1, then, we have $I(u, \lambda, c) \geq 0$ and $v_{\lambda,c} < 0$, and as long as these inequalities persist, $v_{\lambda,c}$ oscillates more quickly than δ.

Suppose first that $v_{\lambda,c}$ has no zero beyond r_1. Then neither has δ, which oscillates more slowly than $v_{\lambda,c}$. If $R < \infty$, this shows that a is strictly admissable. If $R = \infty$ then $v_{\lambda,c} \to -\infty$ as $r \to \infty$ by Lemma 1(c) so, by Lemma 6, r_1 is an interior point of the disconjugacy interval for (20). The disconjugacy interval of the less oscillatory equation (12) cannot be any shorter than that of (20) so, by Lemma 6 again, $\delta(r) \to -\infty$ as $r \to \infty$, and again a is strictly admissable.

If on the other hand $v_{\lambda,c}(r_2) = 0$ for some $r_2 > r_1$, then $v_{\lambda,c}$ oscillates more slowly than δ between r_2 and any subsequent zero of $v_{\lambda,c}$. Since δ has no subsequent zeroes neither does $v_{\lambda,c}$, so $v_{\lambda,c}(r) > 0$ for r close to R. It is clear from (18), however, that $v_{\lambda,c} < 0$ for r close to R, so this case cannot occur and the Lemma is proved.

Lemma 10. *The set of (strictly) admissable $a \in S^0 \cup S^-$ is both open and closed in $S^0 \cup S^-$.*

Proof. By Lemma 8, the set of non-admissable a's consists of those a for which $\delta(r, a)$ has at least two zeroes in $(0, R)$. This set is open, by the continuity of δ on a, so the set of admissable a's is closed. However the set of strictly admissable a's is open. This is clear from continuity if $a \in S^-$, while if $a \in S^0$ we can find R_0 so large that $\delta(r, a) < 0$, $\delta'(r, a) < 0$ and $f'(u(r)) < 0$ for $r \geq R_0$. By continuity, if we perturb a slightly we will still have $\delta'(R_0) < 0$, and since δ can then have no subsequent critical point the perturbed a is still strictly admissable. The result now follows from Lemma 9.

The proof of Theorem 1 is now straightforward. Since $a_0 = \inf(S^0 \cup S^-)$ is admissable, it is strictly admissable by Lemma 8. By Lemma 3(b), for some $\epsilon > 0$ the interval $(a_0, a_0 + \epsilon)$ is contained in S^-. Using Lemma 10, we see that every $a \in (a_0, a_0 + \epsilon)$ is strictly admissable, so $R(a)$ is a strictly decreasing function of a on this interval, by Lemma 3(a). As we continue to raise a, Lemmas 10 and 3(b) show that a continues to be strictly admissable and $R(a)$ continues to decrease. Thus $(a_0, \infty) \subset S^-$, and every $a \in S^0 \cup S^-$ is strictly admissable. The proof of Theorem 1 is complete.

REFERENCES

[1] H. Berestycki and P.-L. Lions. *Non-linear scalar field equations I, existence of a ground state; II, existence of infinitely many solutions.* Arch. Rational Mech. Analysis, 82:313–375, 1983.

[2] H. Berestycki, P.-L. Lions, and L. A. Peletier. *An ODE approach to the existence of positive solutions for semilinear problems in* $\mathbb{R}^n$. Indiana University Math. J., 30:141–167, 1981.

[3] C. V. Coffman. *Uniqueness of the ground state for* $\Delta u - u + u^3 = 0$ *and a variational characterization of other solutions.* Arch. Rational Mech. Analysis, 46:12–95, 1972.

[4] C. K. R. T. Jones and H. Küpper. *On the infinitely many solutions of a semilinear elliptic equation.* SIAM J. Math. Anal., 17:803–835, 1986.

[5] M. K. Kwong. *Uniqueness of positive radial solutions of* $\Delta u - u + u^p = 0$ *in* $\mathbb{R}^n$. Arch. Rational Mech. Analysis, 105:243–266, 1989.

[6] K. McLeod. *Uniqueness of positive radial solutions of* $\Delta u + f(u) = 0$ *in* $\mathbb{R}^n$, *II.* Submitted to Trans. A.M.S.

[7] K. McLeod and J. Serrin. *Uniqueness of positive radial solutions of* $\Delta u + f(u) = 0$ *in* $\mathbb{R}^n$. Arch. Rational Mech. Analysis, 99:115–145, 1987.

[8] K. McLeod, W. C. Troy, and F. B. Weissler. *Radial solutions of* $\Delta u + f(u) = 0$ *with prescribed numbers of zeroes.* J. Differential Equations, 83:368–378, 1990.

[9] L. A. Peletier and J. Serrin. *Uniqueness of positive solutions of semilinear equations in* $\mathbb{R}^n$. Arch. Rational Mech. Analysis, 81:181–197, 1983.

[10] L. Zhang. *Uniqueness of positive solutions of* $\Delta u + f(u) = 0$ *in* $\mathbb{R}^n$. preprint.

On Supercritical Phenomena

F. MERLE and L.A. PELETIER

1. Introduction

Let us consider the problem

$$(\text{I}) \quad \begin{cases} -\Delta u = g(u) & \text{in} \quad \Omega \\ u > 0 & \text{in} \quad \Omega \\ u = 0 & \text{on} \quad \partial\Omega, \end{cases}$$

where Ω is a bounded domain in $\mathbf{R}^N$ ($N > 2$) with a smooth boundary $\partial\Omega$ and g a function from $\mathbf{R}$ to $\mathbf{R}$ such that $g(0) = 0$. We consider the problem of existence of a solution of (I) when

$$g(s) \sim \lambda s^q \quad \text{as} \quad s \to 0 \tag{1.1a}$$

and

$$g(s) \sim \mu s^p \quad \text{as} \quad s \to \infty, \tag{1.1b}$$

where $\lambda, \mu \in \mathbf{R}$ and $p > q > 1$. In particular we consider the function

$$g(s) = \lambda s^q + s^p, \tag{1.2}$$

although most of the results for (1.2) continue to hold when g merely satisfies (1.1).

When $g(s)$ has subcritical growth, that is when $p < (N + 2)/(N - 2)$, the problem is classical and has been studied by many authors (see for instance Ambrosetti & Rabinowitz [AR], Brezis & Nirenberg [BN] and Rabinowitz [R]). When the growth of $g(s)$ at infinity is stronger, then Problem (I) may not have a solution. Indeed, when $g(s) = s^p$ and $p \geq (N + 2)/(N - 2)$, and Ω is a star-shaped domain, then (I) has no solution (see Pohozaev [P]). If on the other hand we add to s^p some correction or perturbation term, then in some cases (I) will have a solution. We consider two types of function g:

(A) $$g(s) = s^p + \lambda s^q \quad \text{with} \quad q < \frac{N + 2}{N - 2} < p,$$

(B) $$g(s) = s^p - \lambda s^q \quad \text{with} \quad \frac{N + 2}{N - 2} < p < q.$$

From Pohozaev's identity we know that when $\lambda \leq 0$ and Ω is star shaped, Problem (I) has no solutions when g is given by either (A) or (B). In addition, other types of arguments show that there do exist values of λ such that (I) has a solution when g is given by (A) or (B). In this note we shall describe some recent work on the question as to for what values of λ (I) has a solution u_λ, either for (A) or for (B), and how u_λ changes as $|u_\lambda|_{L^\infty} \to \infty$.

For $p = p_N = (N+2)/(N-2)$ some related problems have been studied. The first example is

$$g(s) = \lambda s + s^{p_N}.$$

It was shown in the famous article of Brezis and Nirenberg [BN] that for this function Problem (I) has a variational solution if and only if $\lambda \in (\lambda^*, \mu_1)$, where μ_1 is the principal eigenvalue of the Laplacian and $0 \leq \lambda^* < \mu_1$, with $\lambda^* > 0$ if $N = 3$ and $\lambda^* = 0$ if $N \geq 4$.

For a solution u_ε of (I) in which

$$g(s) = (\lambda^* + \varepsilon)s + s^{p_N}, \quad \varepsilon > 0$$

the asymptotic behaviour as $\varepsilon \to 0$ was studied by Rey [Re1] when $N \geq 5$ and by Brezis & Peletier [BP] in a ball when $N = 3$. Similarly the asymptotics of the solution u_ε of (I) in which

$$g(s) = s^{p_N - \varepsilon}, \quad \varepsilon > 0$$

as $\varepsilon \to 0$ was studied in [BP] and [Re2], and by Han [H] for star shaped domains. For spherical domains detailed results were obtained by Atkinson & Peletier [AP1,2,3].

Returning to general star-shaped domains, we assume that the family of solutions u_ε has the property

$$\frac{\int_\Omega |\nabla u_\varepsilon|^2}{\left(\int_\Omega u_\varepsilon^{p_N+1-\varepsilon}\right)^{2/(p_N+1-\varepsilon)}} \to S_N \quad \text{as} \quad \varepsilon \to 0,$$

where S_N is the best Sobolev constant for the embedding $L^{p_N+1} \subset H^1$. It was shown that $u_\varepsilon(x)$ concentrates at a single point x_0 as $\varepsilon \to 0$ and that x_0 is a critical point of the function $\varphi(y) = g(y, y)$, where $g(x, y)$ is the regular part of the Green function $G(x, y)$ which solves

$$\begin{cases} -\Delta G = \delta_y & \text{in} \quad \Omega \\ \quad\;\; G = 0 & \text{in} \quad \partial\Omega \end{cases}$$

and is given by the relation

$$G(x, y) = \frac{1}{(N-2)\sigma_N |x-y|^{N-2}} + g(x, y), \tag{1.3}$$

in which σ_N denotes the surface area of the unit ball in $\mathbf{R}^N$.

In the case when Ω is not star-shaped and $g(s) = s^{p_N}$ it is possible for Problem (I) to have solutions. In fact, if Ω has non-trivial topology, then (I) does have a solution [BC]. For related asymptotics we refer to Bandle & Peletier [BaP] and Lewandowski [L].

2. Subcritical perturbation

In this section we consider the problem

$$\text{(II)} \begin{cases} -\Delta u = u^p + \lambda u^q & \text{in} \quad B_1 \\ \quad u > 0 & \text{in} \quad B_1 \\ \quad u = 0 & \text{on} \quad \partial B_1, \end{cases}$$

where

$$q < \frac{N+2}{N-2} < p$$

and B_1 denotes the unit ball in $\mathbf{R}^N$. Thus, by [GNN], any solution of (II) has radial symmetry.

To begin with we set $q = 1$ (similar results will hold for $1 < q < N/(N-2)$). Then it is well known that if $\lambda \notin (0, \mu_1)$, Problem (II) has no solution, and that $(\mu_1, 0)$ is a bifurcation point from which emanates a continuous and unbounded branch $\mathcal{C}$ of solutions (λ, u).

We first need to introduce the notion of a radial singular solution of (II). By this we mean a function w defined on $B_1 \setminus \{0\}$ which satisfies (II), except at the origin, and behaves near the origin as

$$|x|^{2/(p-1)} w(x) \to A(p, N) \quad \text{when} \quad x \to 0, \tag{2.1}$$

where

$$A = \left\{ \frac{2}{p-1} \left(N - 2 - \frac{2}{p-1} \right) \right\}^{1/(p-1)}. \tag{2.2}$$

It was shown in [MP2], and under additional conditions also in [BuN] and [Bu], that there exists a unique pair (λ^*, w^*) which solves (II) such that w^* has the behaviour (2.1) near the origin.

Question 1. Is w the only singular solution of (II)? Or, phrased differently, are solutions of (II) which satisfy (2.1), necessarily radially symmetric?

Let $\{(\lambda_n, u_n)\}$ be a family of solutions on $\mathcal{C}$ such that $|u_n|_\infty \to \infty$ (Here $|\cdot|_\infty$ denotes the norm on $L^\infty(B_1)$). Then we have the following asymptotic estimates.

Theorem 1. *There exists a number* $\lambda^* > 0$ *such that*

(a) $\lambda_n \to \lambda^*$ as $n \to \infty$,

(b) $u_n \to w^*$ as $n \to \infty$

in $H^1(B_1) \cap L^{p+1}(B_1)$ as well as uniformly on compact subsets in $\overline{B}_1 \setminus \{0\}$.

Remark. Similar results can be proved for radial solutions with k zeros, where $k = 1, 2, \ldots$.

Theorem 1 offers an interesting contrast to the situation when p is critical $(p = p_N)$. In that case $u_n(x) \to 0$ as $n \to \infty$ at any point $x \in B_1$, except the origin. Thus, the solution *concentrates* at the origin, whereas by Theorem 1, no such thing happens when $p > p_N$.

Question 2. What is the limiting behaviour of the solutions u_n as $|u_n|_\infty \to \infty$ when the domain Ω is not a ball? Do they converge to a singular solution, and what is the characterization of a singular solution?

As a corollary we observe that there exists a number $\lambda^{**}(p, N) > 0$ such that (II) has a solution if and only if $\lambda \in (\lambda^{**}, \mu_1)$, where in some cases there may exist a solution if $\lambda = \lambda^{**}$ as well.

3. Supercritical perturbation

In this section we consider the problem

$$(\text{III}) \begin{cases} -\Delta u = u^p - \lambda u^q & \text{in} \quad \Omega \\ \quad u > 0 & \text{in} \quad \Omega \\ \quad u = 0 & \text{on} \quad \partial\Omega, \end{cases}$$

where

$$\frac{N+2}{N-2} < p < q. \tag{3.1}$$

We establish results for Problem (III) using variational arguments. This allows us to prove them for general bounded domains Ω. For that purpose we use the variational structure of (III) and show that it is well-posed in some suitable function space. On the other hand, in case of radial symmetry, simpler proofs of the results below can be given by means of techniques from ordinary differential equations.

Problem (III) has a well-posed variational character due to the fact that the norm of the space L^{p+1} can be interpolated between the norms of L^{q+1} and $L^{2N/(N-2)}$ when p and q satisfy (3.1).

We first study existence and uniqueness of solutions of (III) in the whole space $\mathbf{R}^N$. Let us define for $\alpha > 0$

$$K_{N,\alpha} = \inf\{K(u) : |u|_{p+1} = \alpha \text{ and } u \in L^{q+1}, \nabla u \in L^2\},$$

where

$$K(u) = \frac{1}{2} \frac{\int |\nabla u|^2}{\int |u|^{p+1}} + \frac{1}{q+1} \frac{\int |u|^{q+1}}{(\int |u|^{p+1})^\nu}$$

in which

$$\nu = \frac{2(q+1) - (p-1)N}{2(p+1) - (p-1)N}.$$

Theorem 2. *For every $\alpha > 0$, the infimum $K_{N,\alpha}$ is achieved. The minimizer u_α is unique up to translation and has the following properties:*

(a) $\quad u_\alpha > 0$ *in* $\mathbf{R}^N$ *and* $\quad u_\alpha(x) = O(|x|^{-(N-2)})$ *as* $\quad |x| \to \infty.$

(b) *There exists a constant $c_\alpha > 0$ such that*

$$-\Delta u_\alpha = u_\alpha^p - c_\alpha u_\alpha^q.$$

(c) $\quad\quad\quad u_\alpha(x) = u_\alpha(|x|)$ *and* $u_\alpha'(r) < 0$ *for* $r > 0.$

The uniqueness of u_α has been proved by Kwong, McLeod, Peletier & Troy [KMPT]. See [MP2,3] for the proofs of the other statements of Theorem 2.

Remark. Using a scaling argument it is possible to find a unique $\alpha = \alpha(p, q, N) > 0$ such that $u_\alpha(0) = 1$.

We denote u_α and c_α by respectively U^* and c^*.

Remark. If we consider the solution u_a of the following problem

$$\begin{cases} -u'' - \dfrac{N-1}{r} u' = u^p - c^* u^q \\ u(0) = a, \quad u_a'(0) = 0, \end{cases}$$

we find for every $a \in (0, (c^*)^{-1/(q-p)})$ that $u_a > 0$ and $u_a \to 0$ as $r \to \infty$. Thus, the condition on the growth at infinity in Theorem 2 is crucial.

For Problem (III) we have the following existence theorem.

Theorem 3. (i) *The radial case: $\Omega = B_1$.*
For λ positive and small. Problem (III) has at least two solutions u_λ and v_λ such that

$$\lim_{\lambda \to 0} \lambda |u_\lambda|_\infty^{q-p} = c^*(p, q, N) \quad and \quad \lim_{\lambda \to 0} \lambda |v_\lambda|_\infty^{q-p} = 1.$$

(ii) *The general case: Ω is an arbitrary bounded star shaped domain. There exists a sequence $\lambda_n \to 0$ such that for $\lambda = \lambda_n$, Problem (III) has a variational solution u_{λ_n} and*

$$K(u_{\lambda_n}) \to 0, \quad |u_{\lambda_n}|_{p+1} \to 0 \quad as \quad n \to \infty.$$

Remark. In the radial case, u_λ is the variational solution. For v_λ we can show that $v_\lambda/|v_\lambda| \to 1$ uniformly on compact sets.

Using variational techniques and scaling arguments we obtain the following asymptotic behaviour for variational solutions as $\lambda \to 0$ [MP3].

Theorem 4. *Let Ω be a bounded star shaped domain, and let u_λ be a family of solutions of Problem (III) such that*

$$K(u_\lambda) \to 0 \quad and \quad |u_\lambda|_{p+1} \to 0 \quad as \quad \lambda \to 0.$$

We then have up to a subsequence

(i) $$\lambda^{-\theta} u_\lambda(x) \to MG(x, x_o) \quad as \quad \lambda \to 0.$$

Here $\theta = [(N-2)p - N]/[2(q-p)]$, M is a positive constant which only depends on p, q and N and x_o a critical point of the function $\phi(x) = g(x, x)$, where $g(x, y)$ is the regular part of the Green function $G(x, y)$ given by (1.3).

(ii) $$\lambda |u_\lambda|_\infty^{q-p} \to c^*(p, q, N) \quad as \quad \lambda \to 0.$$

(iii) $$\gamma^{-1} u_\lambda(\gamma^{-(p-1)/2}(x - y)) \to U^*(x - x_o) \quad as \quad \lambda \to 0,$$

where $\gamma = |u_\lambda|_\infty$.

Remark. In the radial case, similar results can be proved for nodal solutions with any given number of zeros.

Question 3. Is it possible to make an analysis for Problem (III) comparable to the one made by Rey [Re1] and Glangetas [G] for the corresponding problem involving functions of type (A) with p critical and $\lambda \to 0$? For instance, is it true that for each nondegenerate critical point y of the regular part of the Green function there exists a unique solution which concentrates at y?

References

[AR] Ambrosetti, A. & P.H. Rabinowitz, *Dual variational methods in critical point theory and applications*, J. Funct. Anal. **14** (1973) 349-381.

[AP1] Atkinson, F.V. & L.A. Peletier, *Emden-Fowler equations involving critical exponents*, Nonlinear Anal. TMA. **10** (1986) 755-776.

[AP2] Atkinson, F.V. & L.A. Peletier, *Large solutions of elliptic equations involving critical exponents*, Asymptotic Anal. **1** (1988) 139-160.

[AP3] Atkinson, F.V. & L.A. Peletier, *Elliptic equations with nearly critical growth*, J. Diff. Equ. **70** (1987) 349-365.

[BC] Bahri, A. & J.M. Coron, *On a nonlinear elliptic equation involving the critical Sobolev exponent: the effect of the topology of the domain*, Comm. Pure Appl. Math. **41** (1988) 253-294.

[BaP] Bandle, C. & L.A. Peletier, *Nonlinear elliptic problems with critical exponent in shrinking annuli*, Math. Ann. **280** (1988) 1-19.

[BN] Brezis, H. & L. Nirenberg, *Positive solutions of nonlinear elliptic equations involving critical Sobolev exponents*, Comm. Pure Appl. Math. **36** (1983) 437-477.

[BP] Brezis, H. & L.A. Peletier, *Asymptotics for elliptic equations involving critical Sobolev exponents*, In *Partial differential equations and the calculus of variations*, (Eds. F. Colombini, A. Marino, L. Modica & S. Spagnolo), pp. 149-192, Birkhäuser, 1989.

[B] Budd, C., *Applications of Shilnikov's theory to semilinear elliptic equations*, SIAM J. Math. Anal. **20** (1989) 1069-1080.

[BuN] Budd, C. & J. Norbury, *Semilinear elliptic equations and supercritical growth*, J. Diff. Equ. **68** (1987) 169-197.

[G] Glangetas, L., C.R.A.S.S. **312** (1991) 807–810.

[H] Han, Zheng Chao, Thesis, Courant Institute, 1990.

[KMPT] Kwong, M.K., J. B. McLeod, L.A. Peletier & W.C. Troy, *On ground state solutions of* $-\Delta u = u^p - u^q$. To appear in J. Diff. Equ.

[L] Lewandowski, R., Thesis, Paris, 1990.

[MP1] Merle, F. & L.A. Peletier, *Positive solutions of elliptic equations involving supercritical growth*. Proc. Royal Soc. Edinburgh. **119A**(1991) 49–62.

[MP2] Merle, F. & L.A. Peletier, *Asymptotic behaviour of positive solutions of elliptic equations with critical and supercritical growth I. The radial case.*. Arch. Rational Mech. Anal. **112** (1990) 1–19.

[MP3] Merle, F. & L.A. Peletier, *Asymptotic behaviour of positive solutions of elliptic equations with critical and supercritical growth II. The general case.*. To appear.

[P] Pohozaev, S.I., *Eigenfiuctions of the equation* $\Delta u + \lambda f(u) = 0$, Dokl. Akad. Nauk. SSSR **165** 36-39 and Sov. Math. **6** (1965) 1408-1411.

[R] Rabinowitz, P.H., *Variational methods for nonlinear eigenvalue problems*, Indiana Math. J. **23** (1974) 729-754.

[Re1] Rey, O., *The rôle of the Green function in a nonlinear elliptic equation involving the critical Sobolev exponent*. J. Funct. Anal. **89** (1990), 1–52.

[Re2] Rey, O., *Proof of two conjectures of H. Brezis and L.A. Peletier*, Manuscripta Math. **65** (1989) 19-37.

Ecole Normale Supérieure   Mathematical Institute
Paris, France   Leiden, The Netherlands

On the Existence and Shape of Solutions to a Semilinear Neumann Problem

WEI-MING NI and IZUMI TAKAGI

1. Introduction

In this article we shall review some recent progress in the study of the Neumann problem for a semilinear elliptic equation. Let Ω be a bounded domain in $\mathbf{R}^N$, $N \geq 2$, with smooth boundary $\partial\Omega$ and let ν denote the unit outer normal to $\partial\Omega$. We consider the Neumann problem

$$(1.1) \qquad d\Delta u - u + f(u) = 0 \quad and \quad u > 0 \quad in \ \Omega,$$

$$(1.2) \qquad \frac{\partial u}{\partial \nu} = 0 \quad on \ \partial\Omega,$$

in which $\Delta = \sum_{j=1}^{N} \frac{\partial^2}{\partial x_j{}^2}$ is the Laplace operator, d is a positive constant and throughout the article we assume that

$$(1.3) \qquad f(t) = t^p \quad with \quad p > 1$$

unless it is explicitly stated otherwise. (Most of the results below do generalize to a certain class of functions f including t^p , and the reader is referred to the original papers cited.)

Background of the problem. Until quite recently the Neumann problem (1.1)-(1.2) seems to have attracted much less attention than its Dirichlet counterpart

$$(1.4) \qquad d\Delta v - v + f(v) = 0 \quad and \quad v > 0 \quad in \ \Omega,$$

$$(1.5) \qquad v = 0 \quad on \quad \partial\Omega.$$

Research supported in part by NSF grant DMS 88-1587.

As will be shown at the end of this section, any solution to (1.1)-(1.2) is *unstable* when viewed as a stationary solution to the corresponding parabolic equation. Thus the Neumann problem might have been regarded as not important or not interesting. However, (1.1)-(1.2) arises naturally when we consider a certain type of *systems* of equations in biological pattern formation theory.

For example, let us consider the chemotactic aggregation model due to E. Keller and L. Segel [KS]. Cellular slime molds (amoebae) release a certain chemical (c-AMP) and move toward places of its higher concentration, eventually forming aggregates. If $v = v(x,t)$ and $w = w(x,t)$ denote the population of amoebae and the concentration of c-AMP, respectively, then the simplified Keller- Segel system is as follows:

$$(1.6) \qquad v_t = D_1 \Delta v - \chi \nabla \cdot (v \nabla log w) \quad in \quad \Omega \times (0, \infty),$$

$$(1.7) \qquad w_t = D_2 \Delta w - aw + bv \quad in \quad \Omega \times (0, \infty),$$

$$(1.8) \qquad \frac{\partial v}{\partial \nu} = \frac{\partial w}{\partial \nu} = 0 \quad on \quad \partial\Omega \times (0, \infty),$$

where D_1, D_2, χ, a and b are positive constants. Since the integral $\int_\Omega v dx$ is conserved by virtue of (1.6) and the boundary condition, the stationary problem may be posed as follows: Given a constant $A > 0$, find a pair of positive funcitons $(v, w) = (v(x), w(x))$ satisfying (1.6)-(1.8) and $\int_\Omega v dx = A$. Observe that the right-hand side of (1.6) is $D_1 \nabla \cdot [v \nabla (log\, v - \chi D_1^{-1} log w)]$, so that $v = \lambda w^{\chi/D_1}$ for some positive constant λ because of the boundary condition. Hence any stationary solution to (1.6)-(1.8) is expressed as

$$(1.9) \qquad v(x) = A(\int_\Omega u^p dx)^{-1} u(x)^p,$$

$$(1.10) \qquad w(x) = \frac{b}{a} A(\int_\Omega u dx)^{-1} u(x),$$

where $p := \chi/D_1$ and u is a solution to (1.1)-(1.2).

Another important example is the activator-inhibitor system in developmental biology proposed by Gierer and Meinhardt [GM]. In the study of this reaction-diffusion system (1.1)-(1.2) also plays a crucial role. For details, see [T] and [LNT].

Various numerical studies have been done on the system (1.6)-(1.8) as well as the activator-inhibitor systems. One of the most interesting features expected from those numerical simulations is that the solution seems to exhibit "point-condensation" phenomena, i.e. it tends to zero as d approaches zero except at a finite number of points. Therefore it seems important to know not only the existence of solutions to (1.1)-(1.2) but also the shape of solutions. In §2 we review recent developments in the study of the existence of solutions. Significant progress has been made in [NT1,2] with regard to the shape of a certain family of solutions and §3 will deal with this result. Finally in §4 we conclude our survey with some remarks and open problems.

Before closing this section we would like to make a simple remark on the instability of solutions to (1.1)-(1.2).

Remark 1.1. Suppose that $f \in C^1$, $f(0) = 0$ and $f(t)/t$ is strictly increasing in the interval $(0, +\infty)$. Then there is a unique $\bar{u} > 0$ such that $f(\bar{u}) = \bar{u}$. Suppose further that $f'(\bar{u}) \neq 1$. (Hence $\bar{u}f'(\bar{u}) - f(\bar{u}) > 0$.) Let u be an arbitrary solution to (1.1)-(1.2), and let $\lambda_0(u) < \lambda_1(u) \leq \lambda_2(u) \leq \ldots \leq \lambda_j(u) \leq \ldots$ be the eigenvalues of the linearized operator $d\Delta - 1 + f'(u)$:

$$(1.11) \qquad d\Delta\phi_j - \phi_j + f'(u)\phi_j + \lambda_j(u)\phi_j = 0 \quad in \ \Omega,$$

$$(1.12) \qquad \partial\phi_j/\partial\nu = 0 \quad on \ \partial\Omega.$$

Then $\lambda_0(u) < 0$. (See Lemma 2.10 of [LN, p. 164].)

Therefore, any solution to (1.1)-(1.2) is unstable as a stationary solution to the corresponding parabolic problem.

2. Existence and nonexistence of spatially inhomogeneous solutions

Since there is a unique constant solution $u \equiv 1$ for all $d > 0$, we are interested in the existence of nonconstant solutions.

2.1. Bifurcation analysis.

First of all, we point out that for any $p > 1$ and for any bounded smooth domain Ω , there always exist nonconstant solutions for certain small values of d .

We regard the problem (1.1)-(1.2) as that of finding a pair (d, u) which satisfies (1.1)-(1.2), and hence the set $\{(d,1)|d > 0\}$ is called the branch of constant solutions. Let

$$0 = \ell_o < \ell_1 \leq \ell_2 \leq \ldots \leq \ell_j \leq \ldots \uparrow +\infty$$

be the eigenvalues of $-\Delta$ under homogeneous Neumann boundary conditions and put

$$(2.1) \qquad d^{(j)} := (p-1)/\ell_j \qquad (j=1,2,3,\ldots).$$

By applying the well-known bifurcation theorem for potential operators (see, e.g., Theorems 11.4 and 11.32 of [R2]), one obtains

Proposition 2.1. *Each point $(d^{(j)},1) \in \mathbf{R}_+ \times W^{1,2}(\Omega)$ is a bifurcation point for (1.1)-(1.2), i.e., there exist at least two distinct one parameter families of solutions $(d(\epsilon), u(\epsilon))_{0<\epsilon<\epsilon_0}$ such that $d(\epsilon) \to d^{(j)}$ as $\epsilon \downarrow 0$, $\|u(\epsilon) - 1\|_{W^{1,2}(\Omega)} = \epsilon$.*

In fact, we apply the bifurcation theorem to (1.1)-(1.2) with $f(u)$ replaced by a truncated function $\tilde{f}(u)$ for u sufficiently large. Then by the elliptic regularity theorem, the weak solution $u(\epsilon)$ is indeed a classical solution and it is close to 1 in the C^o-topology, hence resulting in a classical solution to the original equation.

2.2. Variational approach.

We extend $f(t) = t^p$ over $\mathbf{R}$ by putting $f(t) = 0$ for $t < 0$; and we set

$$F(t) := \int_o^t f(s)ds.$$

For $v \in W^{1,2}(\Omega)$, let

$$(2.2) \qquad J_d(v) := \int_\Omega \{\frac{1}{2}(d|\nabla v|^2 + v^2) - F(v)\}dx.$$

Then it is known that J_d is a C^2-functional on $W^{1,2}(\Omega)$ if $1 < p \le (N+2)/(N-2)$, and any critical point u of J_d is a classical solution to (1.1)-(1.2) without the positivity condition. By the maximum principle it turns out that either $u \equiv 0$ or $u > 0$ on $\bar{\Omega}$. Here and in what follows, the condition $1 < p < (N+2)/(N-2)$ stands for $1 < p < +\infty$ if $N = 2$.

Subcritical exponent. If $1 < p < (N+2)/(N-2)$, then the well-known mountain pass lemma due to Ambrosetti and Rabinowitz [AR] applies to J_d and

$$(2.3) \qquad c_d := \inf_{\gamma \in \Gamma} \max_{0 \le t \le 1} J_d(\gamma(t)) > 0$$

is a critical value of J_d . Here, $\Gamma := \{\gamma \in C^o([0,1]; W^{1,2}(\Omega))|\gamma(0) = 0$, $\gamma(1) = e\}$ and $e > 0$ is a element in $W^{1,2}(\Omega)$ such that $J_d(e) = 0$. (It is not difficult to see that for any $v \in W^{1,2}(\Omega)$ with $v \ge 0$ and $v \not\equiv 0$, there exists a unique $t_c > 0$ such that $h(t) := J_d(tv)$ is strictly increasing for

$t \in (0, t_c)$, strictly decreasing for $t > t_c$ and $h(t) \to -\infty$ as $t \to \infty$. Hence $J_d(Tv) = 0$ for some $T > 0$.)

An important observation is the following

Lemma 2.2. *The critical value c_d given by (2.3) does not depend on the choice of the nonzero element e with $J_d(e) = 0$ and c_d is the smallest positive critical value of J_d . It is characterized also as*

$$(2.4) \qquad c_d = inf\{\max_{t \geq 0} J_d(tv) | v \neq 0, 0 \leq v \in W^{1,2}(\Omega)\}$$

For the proof, see [NT, Lemma 2.1]. Thus a critical point $u_d \in J^{-1}(c_d)$ is said to be a *least–energy solution* to (1.1)-(1.2). We summarize some of the basic properties of least-energy solutions, the proof of which may be found in [LNT] and [LN].

Theorem 2.3. *Suppose $1 < p < (N+2)/(N-2)$ and let u_d denote a least-energy solution to (1.1)-(1.2). Then*

(i) $u_d \equiv 1$ *if d is sufficiently large;*

(ii) $u_d \not\equiv 1$ *if $0 < d < d^{(1)}$, where $d^{(1)}$ is defined by (2.1);*

(iii) *There exists a positive constant C_o independent of d such that*

$$1 \leq \|u_d\|_{L^\infty(\Omega)} \leq C_o$$

for all $d > 0$;

(iv) for each $r \in [0, \infty)$ there exist positive constant $C_j(r)$ $(j = 1, 2)$ independent of d, such that

$$C_1(r)d^{\frac{N}{2}} \leq \int_\Omega u_d^r dx \leq C_2(r)d^{\frac{N}{2}}$$

for $d \in (0, d^{(1)})$;

(v) $\lambda_o(u_d) < 0 \leq \lambda_1(u_d)$ for all $d > 0$, where $\lambda_j(u_d)$ is the $(j+1)$-st eigenvalue of the linearized problem (1.11)-(1.12).

From (iii) and (iv) one can see that for any $\eta > 0$ the set

$$(2.5) \qquad \Omega_{\eta,d} := \{x \in \Omega | u_d(x) > \eta\}$$

may be covered by at most m balls of radius $\sqrt{d}$, where m is independent of $d > 0$. This means that u_d localizes arond a finite number of points, a property which we shall call a "point-condensation phenomenon". In §3 we shall analyze the asymptotic behavior of u_d as $d \downarrow 0$ in detail.

Critical exponent. Very recently the following result has been obtained by X.-J. Wang:

Theorem 2.4 ([W, Theorem 3.1]). *Let* $p = \frac{N+2}{N-2}$. *Then (1.1)-(1.2) has a nonconstant solution for each d sufficiently small.*

His proof is also variational and makes use of the approach by Brezis and Nirenberg [BN] to get around the lack of compactness.

The existence of *nonconstant* solutions is verified in both cases $1 < p < (N+2)/(N-2)$ and $p = (N+2)/(N-2)$ by showing that $\max_{t \geq 0} J_d(t\phi) < J_d(1)$ for an appropriate nonnegative function $\phi \in W^{1,2}(\Omega)$ when d is sufficiently small.

2.3. Nonexistence of spatially inhomogeneous solutions.

As a matter of fact, assertion (i) of Theorem 2.3 is a consequence of the following observation.

Theorem 2.5. ([LNT, Theorem 3]).

(i) *Let $f \in C^1$ and suppose that there exists a positive constant C independent of $d \geq d_1$ such that*

$$(2.6) \qquad\qquad \|u\|_{L^\infty(\Omega)} \leq C$$

for any solution to (1.1)-(1.2) with $d \geq d_1$. Then there exists a $d_2 \geq d_1$ such that (1.1)-(1.2) does not possess any nonconstant solution if $d > d_2$.

(ii) *If $f = t^p$ and $1 < p < (N+2)/(N-2)$, then all the solution to (1.1)-(1.2) are uniformly bounded, hence the conclusion of* (i) *holds.*

For the critical exponent, (2.6) does not hold in general as has been shown by Adimurthi and Yadava [AY] and Budd, Knaap and Peletier [BKP]. They prove that if Ω is the unit ball in $\mathbf{R}^N$ with $4 \leq N \leq 6$ and $p = (N+2)/(N-2)$, then there exists a one parameter family $(d_\gamma, u_\gamma)_{\gamma > 1}$ of radial, decreasing solutions to (1.1)-(1.2) with $d = d_\gamma$ such that $u_\gamma(0) = \gamma$ and $d_\gamma \to +\infty$ as $\gamma \to +\infty$.

2.4. Spherically symmetric solutions.

Although the least-energy solution u_d for subcritical exponent p is *not* spherically symmetric when Ω is a ball or annulus and d is sufficiently small (see the next section), it is still somewhat interesting to consider radial solutions because it could shed some light on the nature of the problem in the case of $p \geq (N+2)/(N-2)$.

Let $\Omega = B_1$, the unit ball in $\mathbf{R}^N$, and consider radial solutions to (1.1)-(1.2):

$$(2.7) \qquad d\left(u'' + \frac{N-1}{r}u'\right) - u + u^p = 0, \ u > 0 \text{ for } r \in (0,1),$$

$$(2.8) \qquad\qquad u'(0) = u'(1) = 0.$$

First, we remark that Proposition 2.1 may be considerably refined due to the global result by Rabinowitz [R1]. Let $0 = \ell_r^{(0)} < \ell_r^{(1)} < \ldots < \ell_r^{(j)} < \ldots \uparrow +\infty$ by the eigenvalues of $-\Delta$ subject to homogeneous Neumann boundary conditions in B_1 which have radial eigenfunctions. Put

$$(2.9) \qquad\qquad d_r^{(j)} := (p-1)/\ell_r^{(j)}, \quad j = 1, 2, 3, \ldots.$$

Let $\mathcal{S}$ be the closure in $\mathbf{R}_+ \times C^o[0,1]$ of the set of nonconstant solutions. Then the connected component $\mathcal{C}^{(j)}$ of $\mathcal{S}$ containing the bifurcation point $(d_r^{(j)}, 1)$ is noncompact in $\mathbf{R}_+ \times C^o[0,1]$ and $\mathcal{C}^{(j)} \cap \mathcal{C}^{(k)}$ is empty if $j \neq k$. Moreover, if (d, u) is a nonconstant solution in $\mathcal{C}^{(j)}$, then $u'(r)$ has exactly $j - 1$ zeros in the interval $(0,1)$ which are all simple. Note also that $\mathcal{C}^{(j)}$ consists of two parts $\mathcal{C}_+^{(j)}$ and $\mathcal{C}_-^{(j)}$: $u'(r) > 0$ for $r > 0$ near 0 if $(d, u) \in \mathcal{C}_+^{(j)}$, while $u'(r) < 0$ near 0 if $(d, u) \in \mathcal{C}_-^{(j)}$.

Adimurthi and Yadava have obtained the following result on the behavior of $\mathcal{C}^{(1)}$.

Theorem 2.6 ([AY]). *Let $Proj_{\mathbf{R}_+}\mathcal{C}^{(j)}$ be the projection of $\mathcal{C}^{(j)}$ on $\mathbf{R}_+$ and $p = (N+2)/(N-2)$. Then*

(i) *$Proj_{\mathbf{R}_+}\mathcal{C}_-^{(1)} \supset [d_r^{(1)}, +\infty)$ if $4 \leq N \leq 6$;*

(ii) *$Proj_{\mathbf{R}_+}\mathcal{C}_-^{(1)}$ is a bounded interval if $N = 3$.*

It is easy to see that $Proj_{\mathbf{R}_+}\mathcal{C}^{(1)} \supset (0, d_r^{(1)}]$ for any $N \geq 3$. See also [BKP] for the asymptotic behavior of the branch $\mathcal{C}_+^{(1)}$ for the equation

$$d\left(u'' + \frac{N-1}{r}u'\right) - u^q + u^p = 0$$

with $0 < q < p - 1$.

The above result, (i) in particular, is interesting when compared to the noncritical case:

Proposition 2.7.

(i) *If $1 < p < (N+2)/(N-2)$, then for each $j \geq 1$, $Proj_{\mathbf{R}_+}\mathcal{C}^{(j)}$ is a bounded interval containing $(0, d_r^{(j)}]$ and is included in an interval $(0, M)$ independently of j ;*

(ii) *If $p > (N+2)/(N-2)$, then there exists an $M' > 0$ such that $Proj_{\mathbf{R}_+}\mathcal{C}^{(j)} \subset (0, M')$ for any $j \geq 1$.*

Assertion (i) follows from Theorem 2.5, while the proof of assertion (ii) is found in [LN, Theorem 3.2].

3. Shape of least-energy solutions

Assume that $1 < p < (N+2)/(N-2)$; and let c_d be the smallest positive critical value of J_d given by (2.3) and u_d be a least-energy solution to (1.1)-(1.2), that is $u_d \in J_d^{-1}(c_d)$. This solution is of particular interest since it is the "most stable" solution. In fact, Theorem 2.3 (v) implies that u_d has only one dimensional unstable direction as a stationary solution to the corresponding parabolic equation; and, for example, the Keller-Segel system (1.6)-(1.8) has one quantity conserved along the solution, hence there is a good chance to obtain a stable stationary solution by taking $u = u_d$ in (1.9) and (1.10).

We know that $u_d \equiv 1$ if d is sufficiently large (Theorem 2.3 (i)). For d sufficiently small, the following result has been obtained recently:

Theorem 3.1 ([NT1,2]). *Let $1 < p < (N+2)/(N-2)$. Then a least-energy solution has only one local maximum on $\bar{\Omega}$ (hence it is the maximum) and it is achieved exactly at one point P_d on the boundary, provided that $d > 0$ is sufficiently small. Moreover, we have $H(P_d) \to \max_{P \in \partial\Omega} H(P)$ as $d \to 0$ where H is the mean curvature of $\partial\Omega$.*

More can be said about the shape of u_d . It is known that the problem

$$(3.1) \qquad \Delta w - w + f(w) = 0 \ and \ w > 0 \ for \ z \in \mathbf{R}^N,$$

$$(3.2) \qquad w \to 0 \ as \ |z| \to +\infty$$

has a unique solution w ([K], [MS], [C] and [Z]) and that w is radial, $w = w(r), r = |z|$, and is monotonically decreasing, $w'(r) < 0$ if $r > 0$, and decays exponentially at infinity; i.e., there are positive constants C and μ such that

$$|D^\alpha w(z)| \le C e^{-\mu|z|} \ for \ z \in \mathbf{R}^N$$

with $|\alpha| \le 2$ ([GNN]).

Let $P_d \in \partial\Omega$ be the point at which u_d attains its maximum. To see what happens around P_d , we introduce a diffeomorphism which straightens a boundary portion near P_d . Through translation and rotation of the coordinate system, one may assume that P_d is the origin and the inner normal to $\partial\Omega$ at P_d is pointing in the direction of the positive x_N-axis. Then there is a smooth function $\psi_d(x')$, $x' = (x_1, \ldots, x_{N-1})$, defined near $x' = 0$ such that (i) $\psi_d(0) = 0$ and $\nabla\psi_d(o) = 0$; (ii) $\partial\Omega \cap \mathcal{N} = \{(x', x_N)|x_N = \psi_d(x')\}$ and $\Omega \cap \mathcal{N} = \{(x', x_N)|x_N > \psi_d(x')\}$, in which $\mathcal{N}$ is a neighborhood of P_d . For $y \in \mathbf{R}^N$ with $|y|$ small, we define a mapping $x = \Phi_d(y) = (\Phi_{d,1}(y), \ldots, \Phi_{d,N}(y))$ by

$$\Phi_{d,j}(y) := \begin{cases} y_j - y_N \frac{\partial\psi_d}{\partial x_j}(y') & for \ j = 1, \ldots, N-1, \\ y_N + \psi_d(y') & for \ j = N. \end{cases}$$

Since $\nabla\psi_d(0) = 0$, the differential map $D\Phi_d$ is equal to the identity at $y = 0$, and hence Φ_d has the inverse mapping $y = \Phi_d^{-1}(x) =: \Psi_d(x)$.

For a domain U in $\mathbf{R}^N$ and $d > 0$, put

$$\|v\|_{C_d^2(\bar{U})} := \sum_{|\alpha|\leq 2} d^{\frac{|\alpha|}{2}} \|D^\alpha v\|_{C^\circ(\bar{U})}.$$

We are now ready to state

Theorem 3.2 ([NT 1]). *Let* $1 < p < (N+2)/(N-2)$ *and* $P_d \in \partial\Omega$ *be the point where the maximum of the least-energy solution* u_d *is achieved. Then, given* $\epsilon > 0$, *there exists a subdomain* $\Omega_d^{(i)} \subset \Omega$ *and a* $d_o > 0$ *such that the following statements holds true if* $0 < d < d_o$:

(i) $P_d \in \partial\Omega_d^{(i)}$ *and* $\mathrm{diam}(\Omega_d^{(i)}) \leq C\sqrt{d}$;

(ii) $\|u_d(\cdot) - w(\Psi_d(\cdot)/\sqrt{d})\|_{C_d^2(\overline{\Omega_d^{(i)}})} \leq \epsilon$;

(iii) $0 < u_d(x) \leq C_1\epsilon e^{-\mu_1\delta(x)/\sqrt{d}}$ *for* $x \in \Omega_d^{(0)} := \Omega\backslash\Omega_d^{(i)}$,

where $\delta(x) := \min\{dist(x,\partial\Omega_d^{(i)}),\eta_o\}$ *and* C, C_1, μ_1, η_o *are positive constants depending only on* Ω .

Hence u_d is close to $w(\Psi_d(x)/\sqrt{d})$ in the inner region $\Omega_d^{(i)}$ and it can be made arbitrarily small in the outer region $\Omega_d^{(0)}$ as $d \downarrow 0$. The set $\Omega_{\eta,d}$ defined by (2.5) has only one connected component because of Theorem 3.1, and the following corollary might be helpful in considering the level set of u_d near P_d .

Corollary 3.3. *For each* $\eta \in (0, \max w)$, *one has*

$$\frac{1}{\sqrt{d}}\Psi(\Omega_{\eta,d}) \to B_{w^{-1}(\eta)}^+ \text{ as } d \downarrow 0,$$

where $B_{w^{-1}(\eta)}^+ := \{z \in \mathbf{R}^N | |z| < w^{-1}(\eta) \text{ and } z_N > 0\}$.

Note that $\Psi(\Omega_{\eta,d})$ is close to $\Omega_{\eta,d}$ since $D\Psi(0) = I$ and $\Omega_{\eta,d} \to P_d$ as $d \downarrow 0$.

We shall explain briefly the idea of the proof of Theorems 3.1 and 3.2. First, we state a key lemma:

Lemma 3.4 *As* $d \downarrow 0$, $c_d \leq d^{\frac{N}{2}}\{\frac{1}{2}I(w) + o(1)\}$ *where*

$$I(w) = \int_{\mathbf{R}^N} \{\frac{1}{2}(|\Delta w|^2 + w^2) - F(w)\}dz.$$

Sketch of Proof. Take a point $P \in \partial\Omega$ and the diffeomorphism $y = \Psi(x)$ straightening a boundary portion near P as above. Put

$$\varphi(x) := \zeta(\Psi(x))w(\Psi(x)/\sqrt{d})$$

with an appropriate cut-off function ζ . By a series of straightforward computations, one obtains that

$$\sup_{t>0} J_d(t\varphi) = d^{\frac{N}{2}}\{\frac{1}{2}I(w) + o(1)\}$$

as $d \downarrow 0$. Then the conclusion follows by virtue of Lemma 2.2.

Now the proof of Theorems 3.1 and 3.2 breaks down into three steps. In the first step we show that if u_d assumes a local maximum at P_d , then $dist(P_d, \partial\Omega) = O(\sqrt{d})$ as $d \downarrow 0$. Suppose $dist(P_{d_j}, \partial\Omega)/\sqrt{d_j} \to +\infty$ along some sequence $d_j \downarrow 0$. Then one can show that $w_{d_j}(z) := u_{d_j}(P_{d_j} + \sqrt{d_j}z)$ is approximated by $w(z)$ in $C^2(B_R)$ if d_j is sufficiently small. This results in the estimate $c_{d_j} \geq d_j^{\frac{N}{2}}\{I(w) + o(1)\}$ as $d_j \downarrow 0$, contradicting Lemma 2.2.

The second step is to prove that $dist(P_d, \partial\Omega) = O(\sqrt{d})$ implies $P_d \in \partial\Omega$. Suppose $P_d \to P \in \partial\Omega$ as $d \downarrow 0$. Then one can show that $w_d(z) := \tilde{u}_d(\Phi(\sqrt{d_j}z))$ is approximated by $w(z)$ in $C^2(B_R)$ for sufficiently small $d > 0$, where $x = \Phi(y)$ is the diffeomorphism flattening a boundary portion near P and $\tilde{u}_d(\Phi(y))$ is the extension of u_d by reflection with respect to $y_N = 0$. Since w is radial and strictly decreasing, w_d cannot have two local maxima, yielding $P_d \in \partial\Omega$. If u_d has two local maxima at P and Q on the boundary then by computing the respective contributions from the neighborthoods of P and Q separately one obtains $c_d \geq d^{\frac{N}{2}}\{I(w) + o(1)\}$, a contradiction.

The last step is to obtain a more exact estimate for c_d (up to order $o(\sqrt{d})$) which involves the boundary mean curvature at P_d.

4. Concluding remarks

Significant advances have been made during the past five years in the study of the semilinear Neumann problem (1.1)-(1.2). However, we are still far from a complete understanding of the whole picture. In particular, the following questions are left open.

(I) **The subcritical exponent.** $1 < p < (N + 2)/(N - 2)$. Let u_d denote a least-energy solution.

(Q1) If d is sufficiently large, then $u_d \equiv 1$ and hence its maximum is achieved on the boundary, too. Does u_d attain the maximum on the boundary $\partial\Omega$ for all $d > 0$?

(Q2) For d sufficiently small, u_d has only one peak. Is there a solution with two peaks? In general, find multi peak solutions.

(II) **The critical or supercritical exponent** $p \geq (N+2)/(N-2)$.

Very recently it was shown in [NPT] that a least-energy solution for the critical exponent also has only one (global) maximum in $\bar{\Omega}$ and it must also lie on $\partial\Omega$.

(Q3) For supercritical exponents and general domains, find solutions which are far from the constant solution.

As was pointed out in the introduction, it is very important to obtain qualitative behavior of solutions to (1.1)-(1.2), and particular attention should be paid to solutions with point-condensation character (due to their mathematical interest as well as the possible biological significance). In fact, [NT1,2] and a major portion of [LNT] are devoted to this purpose in the subcritical case $1 < p < (N+2)/(N-2)$. In case $p > (N+2)/(N-2)$ or, $p = (N+2)/(N-2)$ with $N \geq 6$, it is not hard to see that *there exists a positive constant α , independent of $d > 0$, such that* $\inf_\Omega u \geq \alpha$ *for all radial solutions u of (1.1)-(1.2) with $u(0) \geq 1$ when Ω is the unit ball.* (The proof of this fact consists of a simple application of Pohozaev identity.) This eliminates the possibility of existence of *radial spiky solutions* (which tend to zero everywhere except at a finite number of prints as d approaches 0) in the supercritical case as well as in the critical case with dimension N large. For the non-radial case with such p's and N's, this remains open.

REFERENCES

[AY] Adimurthi and S.L. Yadava, *Existence and non existence of positive radial solutions for Sobolev critical exponent problem with Neumann boundary condition*, preprint, 1990.

[AR] A. Ambrosetti and P.H. Rabinowitz, *Dual variational methods in critical point theory and applications*, J. Funct. Anal. **14** (1973), 349-381.

[BN] H. Brezis and L. Nirenberg, *Positive solutions of nonlinear elliptic equations involving critical Sobolev exponents*, Comm. Pure Appl. Math. **36** (1983), 437-477.

[BKP] C. Budd, M.C. Knaap and L.A. Peletier, *Asymptotic behaviour of solutions of elliptic equations with critical exponents and Neumann boundary conditions*, preprint, 1990.

[C] C.V. Coffman, *Uniqueness of the ground state solution for* $\Delta u - u + u^3 = 0$ *and a variational characterization of other solutions*, Arch. Rational Mech. Anal. **46** (1972), 81-95.

[GNN] B. Gidas, W.-M. Ni and L. Nirenberg, *Symmetry of positive solutions of nonlinear elliptic equations in* $\mathbf{R}^n$, Advances in Math., Supplementary Studies **7A** (1981), 369-402.

[GM] A. Gierer and H. Meinhardt, *A theory of biological pattern formation*, Kybernetik (Berlin) **12** (1972), 30-39.

[KS] E.F. Keller and L.A. Segel, *Initiation of slime mold aggregation viewed as an instability*, J. Theo. Biol. **26** (1970), 399-415.

[K] M.K. Kwong, *Uniqueness of positive solutions of $\Delta u - u + u^p = 0$ in $\mathbf{R}^n$*, Arch. Rational Mech. Anal. **105** (1989), 143-266.

[LN] C.-S. Lin and W.-M. Ni, *On the diffusion coefficient of a semilinear Neumann problem, in Calculus of Variations and Partial Differential Equations* (S. Hildebrandt, D. Kinderlehrer, M. Miranda, Ed.) 160-174, Lecture Notes in Math. **1340**, Springer-Verlag, 1988.

[LNT] C.-S. Lin, W.-M. Ni and I. Takagi, *Large amplitude stationary solutions to a chemotaxis system*, J. Differential Equations **72** (1988), 1-27.

[MS] K. McLeod and J. Serrin, *Uniqueness of positive radial solutions of $\Delta u + f(u) = 0$ in $\mathbf{R}^n$*, Arch. Rational Mech. Anal. **99** (1987), 115-145.

[N] W.-M. Ni, *Recent progress in semilinear elliptic equations*, RIMS Kokyuroku **679** (1989), Kyoto University, 1-39.

[NPT] W.-M. Ni, X.-B. Pan and I. Takagi, *Singular behavior of least-energy solutions of a semilinear Neumann problem involving critical Sobolev exponents*, Duke Math. J., to appear.

[NT1] W.-M. Ni and I. Takagi, *On the shape of least-energy solutions to a semilinear Neumann problem*, Comm. Pure Appl. Math., **44**(1991), 819-851.

[NT2] W.-M. Ni and I. Takagi, *Locating the peaks of least-energy solutions to a semilinear Neumann problem*, preprint.

[R1] P.H. Rabinowitz, *Some global results for nonlinear eigenvalue problems*, J. Funct. Anal. **7** (1971), 487-513.

[R2] P.H. Rabinowitz, *Minimax Methods in Critical Point Theory with Applications to Differential Equations*, CBMS Regional Conference Series in Mathematics **65**, Amer. Math. Soc. 1986.

[T] I. Takagi, *Point-condensation for a reaction- diffusion system*, J. Diff. Equat. **61** (1986), 208-249.

[W] X.-J. Wang, *Neumann problem of semilinear elliptic equations involving critical Sobolev exponents.* J. Differential Equations, to appear.

[Z] L. Zhang, *Uniqueness of ground state solutions*, Acta Math. Scientia **6** (1988), 449-468.

Wei-Ming Ni
School of Mathematics
University of Minnesota
Minneapolis, MN 55455

Isumi Takagi
Mathematical Institute
Tohoku University
Sendai 980, Japan

Global Asymptotic Stability for Strongly Nonlinear Second Order Systems

PATRIZIA PUCCI and JAMES SERRIN

1. Introduction

The global asymptotic stability of the rest point for *nonlinear* equations has been studied by Levin and Nohel [3], by Artstein and Infante [1]. These studies have been extended to extremals of scalar variational problems in Section 5 of [5], where the Euler - Lagrange equation exhibits even stronger nonlinearities.

Here we consider vector extremals $u = u(r)$ of variational problems, whose trajectories lie in $N + 1$ dimensional space, that is $u : I \to \mathbf{R}^N$, where I is a half open interval of the form $[R, \infty)$. The typical problem we consider is

$$(1.1) \qquad \delta \int_{\mathbf{R}^+} g(r)[G(u') - F(r, u)]\, dr = 0,$$

where $u = (u_1, \ldots, u_N)$, $u' = (u'_1, \ldots, u'_N)$, and where $g : \mathbf{R}^+ \to \mathbf{R}^+$, $G : \mathbf{R}^N \to \mathbf{R}$, $F : \mathbf{R}^+ \times \mathbf{R}^N \to \mathbf{R}$ satisfy the regularity assumptions explicitly stated in the following Section 2. The most important of these conditions is that G is strictly convex in $\mathbf{R}^N$, that $(\nabla_u F(r, u), u) > 0$ for r large and $u \neq 0$, and that $g(r) \geq$ Const. r^β for some $\beta > 0$. Here $(\cdot, \cdot)$ denotes the inner product in $\mathbf{R}^N$.

An extremal u of (1.1) is defined here as a weak solution of the Euler - Lagrange system

$$(1.2) \qquad \big(g(r)\,\nabla G(u')\big)' + g(r)\, f(r, u) = 0,$$

where $f(r, u) = \nabla_u F(r, u)$ and $' = d/dr$. We shall show that the results of [1] [3], [5] carry over to the vector case in a surprisingly close way, enough even to suggest that it is the variational character of the nonlinear system, more than anything else, which produces the desired asymptotic stability.

The approach here depends on the construction of an appropriate Liapunov function for the Euler - Lagrange system (1.2), based on the general theory of variational identities proposed by the authors in [4]. In general the Euler-Lagrange system for vector extremals involves coupling of the second derivatives in each equation of the system. Traditionally this is treated by writing the system in a canonical form, i.e. solved for the second derivatives, from which a final canonical first order system is obtained (see, for instance, Bliss [2, pp. 16 and Appendix]). Such an approach tends to obscure the essential variational nature of the system, making the determination of a Liapunov function far from transparent; in addition it fails completely for variational problems whose integrand is not suitably smooth or presents other singularities. By introducing the Liapunov function directly in terms of a variational identity, these difficulties are avoided.

There are a number of situations that can be represented by the general system (1.2). When G is of class $C^2(\mathbf{R}^N)$ system (1.2) takes the form

$$(1.3) \qquad \sum_{j=1}^{N} \frac{\partial^2 G}{\partial p_i \partial p_j}(u') u_j'' + \delta(r) \frac{\partial G}{\partial p_i}(u') + f_i(r, u) = 0, \quad i = 1, \ldots, N,$$

where $\delta(r) = g'(r)/g(r)$. Another immediate example occurs when $G(p) = |p|^m/m$, $m > 1$. The corresponding system is

$$(1.4) \quad |u'|^{m-4} \sum_{j=1}^{N} \big[|u'|^2 \delta_{ij} + (m-2)u_i'u_j'\big] u_j'' + \delta(r) |u'|^{m-2}u_i' + f_i(r, u) = 0,$$

where δ_{ij} denotes the Kronecker symbol, which when $m \neq 2$ is singular at points where $u' = 0$. A similar example is $G(p) = \sum_{i=1}^{N} |p_i|^m/m$, $m > 1$, the related system then being

$$(1.5) \qquad\qquad |u_i'|^{m-2}\big[(m-1)u_i'' + \delta(r) u_i'\big] + f_i(r, u) = 0;$$

the second derivatives are uncoupled, but again the system is singular if at least one $u_i' = 0$ and $m \neq 2$. If $m = 2$ both (1.4) and (1.5) reduce to the familiar form

$$u'' + \delta(r) u' + f(r, u) = 0.$$

When $G(p) = \sqrt{1 + |p|^2} - 1$, that is, G is the mean curvature operator, the system (1.3) becomes

$$(1.6) \qquad (1 + |u'|^2)^{-3/2}\bigg\{ \sum_{j=1}^{N} \big[(1 + |u'|^2)\delta_{ij} - u_i'u_j'\big] u_j'' \bigg\}$$
$$+ \delta(r)(1 + |u'|^2)^{-1/2}u_i' + f_i(r, u) = 0.$$

The system (1.2) can also be considered as the *radial* version of the partial differential system in $\mathbf{R}^n$

$$\operatorname{div} \nabla G(Du) + f(r, u) = 0, \qquad r = |x|,$$

where Du denotes the Jacobian matrix $(\partial u_i/\partial x_j)$, $i = 1,\ldots,N$, $j = 1,\ldots,n$ and where $C(Du)$ has the special form $\bar{C}(|\nabla u_1|,\ldots,|\nabla u_N|)$. To place this in the context of (1.1) we take $g(r) = r^{n-1}$ and $G(p) = \bar{G}(|p_1|,\ldots,|p_N|)$.

Various examples illustrating the results for the scalar case have been given in [1], [3] and [5]; see also the remarks in Sections 3, 4 and 5 below.

The following section is concerned with preliminary material, including the statement of the important identity (2.11) for extremals of (1.1). In Section 3 we formulate the main result of the paper, stating that all extremals of (1.1) which are bounded at ∞ tend to zero together with their first derivatives as $r \to \infty$. Section 4 is devoted to global stability for a slightly wider class of nonlinearities $F(r,u)$ than those treated in Section 3. Finally in Section 5 we present stronger versions of the previous results when G has further specialized structure, of the type possessed in particular by systems (1.4)–(1.6).

2. Preliminaries

We consider vector extremals $u = (u_1,\ldots,u_N)$ of the functional

$$(2.1) \qquad \int_{\mathbf{R}^+} g(r)\big[G(u') - F(r,u)\big]\,dr.$$

We suppose throughout the paper that:

(a) $g \in C^1(\mathbf{R}^+)$, $\qquad g(r) > 0$ for all $r > 0$;

(b) $G \in C^1(\mathbf{R}^N)$, $\qquad G$ is strictly convex in $\mathbf{R}^N$, and there exists a positive constant γ such that

$$(2.2) \qquad \big(\nabla G(p),p\big) \geq \gamma\,|\nabla G(p)|\cdot|p| \quad \text{for all } p \in \mathbf{R}^N,$$

where $(\cdot,\cdot)$ denotes the Euclidean inner product of $\mathbf{R}^N$ and $|\cdot|$ the corresponding norm (note that necessarily $\gamma \leq 1$);

(c) $F \in C^1(\mathbf{R}^+ \times \mathbf{R}^N)$.

The first parts of assumptions (a), (b), (c) imply that the integrand in (2.1) is of class $C^1(\mathbf{R}^+ \times \mathbf{R}^N \times \mathbf{R}^N)$. We note, in particular, that condition (2.2) is satisfied whenever $G \in C^1(\mathbf{R}^N)$, $G(p) > 0$ for $p \neq 0$, and G is homogeneous of degree $m > 0$. Indeed, by Euler's identity and the compactness of the unit ball there are positive constants c and c' such that

$$\big(\nabla G(p),p\big) = m\,G(p) \geq c\,m\,|p|^m, \qquad |\nabla G(p)| \leq c'\,|p|^{m-1}.$$

The strict convexity of G in assumption (b) forces $m > 1$.

The fact that G is strictly convex is equivalent to the strong Weierstrass condition for the integrand in (2.1). By setting $p = -t\,\nabla G(0)$ in (2.2) and letting $t \to 0+$ we obtain $\nabla G(0) = 0$. Hence $p = 0$ is the unique absolute minimum of G in $\mathbf{R}^N$. The strict convexity and the regularity of G also imply that ∇G is one-to-one from $\mathbf{R}^N$ onto an open, convex set Ω of $\mathbf{R}^N$, with continuous inverse $(\nabla G)^{-1} : \Omega \to \mathbf{R}^N$; see Theorem 26.5 in [7] and note that G is a convex function of Legendre type on $\mathbf{R}^N$ according to Rockafellar's terminology.

Without loss of generality from here on we assume that G and F are normalized by $G(0) = F(r,0) = 0$, $r > 0$.

Let $H = H(p)$ be the Legendre transform of $G = G(p)$, namely the continuous function defined in $\mathbf{R}^N$ by

$$(2.3) \qquad\qquad H(p) = \big(\nabla G(p), p\big) - G(p).$$

Clearly $H(0) = 0$ and $H(p) > 0$ for $p \neq 0$ since $G(0) = 0$ and G is strictly convex. We also define $G^* : \Omega \to \mathbf{R}$ by

$$(2.4) \qquad\qquad G^* = H \circ (\nabla G)^{-1}.$$

Then, again by Theorem 26.5 of [7], we have $G^* \in C^1(\Omega)$ and

$$(2.5) \qquad\qquad \nabla G^* = (\nabla G)^{-1}.$$

Lemma 1. *For every $p \in \mathbf{R}^N$ we have*
(2.6)
$$H(p) \le \big(\nabla G(p), p\big) \quad \text{and} \quad |\nabla G(p)| \le \frac{1}{\gamma}\{G(p/|p|) + H(p)\} \quad \text{for } p \neq 0.$$

Proof. The first inequality is obvious since $G(p) \ge 0$. Let $p \neq 0$ be fixed so that $p = \tau v$ for some unit vector $v \in \mathbf{R}^N$ and $\tau = |p| > 0$. We now define G_v and H_v on $[0, \infty)$ by $G_v(t) = G(tv)$ and $H_v(t) = H(tv)$. It is easy to check that G_v satisfies the analogous regularity and convexity assumption (b) and that H_v is the Legendre transform of G_v. Therefore by (2.2) we have

$$|\nabla G(p)| \le \frac{\big(\nabla G(p), p\big)}{\gamma|p|} = \frac{1}{\gamma}\frac{dG_v}{dt}(\tau) \le \frac{1}{\gamma}\{G_v(1) + H_v(\tau)\},$$

thanks to Lemma 2 of [6]. Hence we finally obtain

$$|\nabla G(p)| \le \frac{1}{\gamma}\{G(v) + H(p)\}.$$

We now turn to the variational system associated with problem (2.1). Define the vector function $f = (f_1, \ldots, f_N)$ by $f(r, u) = \nabla_u F(r, u)$ for every $(r, u) \in \mathbf{R}^+ \times \mathbf{R}^N$. We say that $u = (u_1, \ldots, u_N)$ is a (weak) *extremal* for (2.1) if u is a C^1 vector function defined on some subinterval I of $\mathbf{R}^+$, such that

$$(2.7) \qquad \nabla G\big(u'(r)\big) \in C^1(I)$$

and u satisfies in I the corresponding Euler - Lagrange system

$$(2.8) \qquad \big(g(r)\, \nabla G(u')\big)' + g(r)\, f(r, u) = 0,$$

or equivalently

$$g(r)\, \nabla G(u') + \int_R^r g(s)\, f(s, u(s))ds = \text{Constant},$$

where R is any fixed point of I.

In spite of the fact that neither $u'(r)$ nor $H(p)$ need be separately differentiable, the composite function $H(u'(r))$ *is* differentiable in I. Indeed $H(u'(r)) = G^*\big(\nabla G(u'(r))\big)$ by (2.4) and so from (2.5) and (2.7) we derive at once

$$(2.9) \qquad \big(H(u'(r))\big)' = \big(u'(r), \frac{d}{dr}\nabla G(u'(r))\big), \quad r \in I.$$

From now on we adopt, for simplicity, the common notation where $u = u(r)$ and $u' = u'(r)$ denote the extremal and its derivative. Thus, along an extremal u of (2.1) in I, from (2.9) and (2.8) we immediately obtain the useful identity

$$(2.10) \qquad \big(H(u') + F(r, u)\big)' = F_r(r, u) - \delta(r)\,\big(\nabla G(u'), u'\big),$$

where $\delta(r) = g'(r)/g(r)$.

Since along an extremal u the function $H(u')$ is differentiable with respect to r, the main identity of Proposition 1 of [5] holds even for C^1 extremals of (2.1), yielding in particular the following useful result.

Let u be an extremal of (2.1) in I and α, k be arbitrary numbers. Then the following identity holds in I

$$(2.11)$$

$$\left\{ r^k \left[H(u') + F(r, u) + \frac{\alpha}{r}\,\big(\nabla G(u'), u\big) \right] \right\}'$$

$$= r^{k-1}\bigg[k\, F(r, u) + r\, F_r(r, u) - \alpha\,\big(f(r, u), u\big) - k\, G(u')$$

$$- \big(r\,\delta(r) - \alpha - k\big)\,\big(\nabla G(u'), u'\big) - \frac{\alpha}{r}\,\big(r\,\delta(r) - k + 1\big)\,\big(\nabla G(u'), u\big)\bigg].$$

The identity (2.10) is the special case of (2.11) when $\alpha = k = 0$.

3. The main theorem

The purpose of this section is to study the asymptotic behavior of extremals of (2.1) which are bounded as $r \to \infty$. We shall treat nonlinearities $F = F(r, u)$ which satisfy the following conditions, for some $R > 0$.

(P) *For all u_0, U with $0 < u_0 \leq U$ there is a constant $\kappa > 0$ for which*

$$\big(f(r, u), u\big) \geq \kappa \quad \text{when } r \geq R \text{ and } |u| \in [u_0, U];$$

(P)′ *For every $U > 0$ there is a nonnegative function $\psi \in L^1[R, \infty)$ such that*

$$F_r(r, u) \leq \psi(r) \quad \text{for a.a. } r \in [R, \infty) \text{ and all } |u| \leq U.$$

When $f = \nabla_u F(r, u) = (f_1, \ldots, f_N)$ is of the form

$$f(r, u) = \rho(r)\,\phi(u),$$

with $\rho : \mathbf{R}^+ \to \mathbf{R}^+$ and $\phi : \mathbf{R}^N \to \mathbf{R}^N$, then property (P) is equivalent to the simple condition

$$\rho(r) \geq \text{Const.} > 0 \quad \text{for } r \geq R, \qquad \big(\phi(u), u\big) > 0 \quad \text{for } u \neq 0.$$

Condition (P)′ obviously holds if and only if $(\rho')^+ \in L^1[R, \infty)$. If F does not depend on r property (P)′ is irrelevant.

Condition (P) implies that $\big(f(r, u), u\big) > 0$ for $r \geq R$ and $u \neq 0$. In turn for those values of r and u we also have by (c)

$$F(r, u) = \int_0^1 \frac{d}{d\tau} F(r, \tau u)\, d\tau = \int_0^1 \big(f(r, \tau u), u\big)\, d\tau = \int_0^1 \frac{1}{\tau}\big(f(r, \tau u), \tau u\big)\, d\tau > 0.$$

Lemma 2. *Suppose that F has property (P). Then for every $U > 0$ there is a positive increasing function $\omega_U : (0, U] \to \mathbf{R}^+$ such that*

$$F(r, u) \geq \omega_U(|u|) \quad \text{for } r \geq R \text{ and } 0 < |u| \leq U.$$

Proof. Let $U > 0$ be fixed and let $\kappa(u_0, U)$ be the value of κ in (P) corresponding to the interval $[u_0, U]$. If this function is not already increasing in the variable u_0, we replace it by the new function

$$\bar{\kappa}(u_0, U) = \sup_{0 < t \leq u_0} \kappa(t, U),$$

which is positive and increasing and for which (P) is also satisfied.

Now for any $r \geq R$ and $0 < |u| \leq U$ we have by (c)

$$F(r,u) = \int_0^1 \frac{1}{\tau}\left(f(r,\tau u), \tau u\right) d\tau \geq \int_{1/2}^1 \frac{1}{\tau}\left(f(r,\tau u), \tau u\right) d\tau$$

$$\geq \int_{1/2}^1 \bar{\kappa}\left(\tfrac{1}{2}|u|, U\right) d\tau = \frac{1}{2}\bar{\kappa}\left(\tfrac{1}{2}|u|, U\right).$$

The proof is thus completed by defining $\omega_U(t) = \dfrac{1}{2}\bar{\kappa}\left(\tfrac{1}{2}t, U\right),\ 0 < t \leq U.$

We now state our main theorem for the general functional (2.1), recalling the relation $\delta(r) = g'(r)/g(r),\ r \in \mathbf{R}^+$.

Theorem 1. *Let u be an extremal of functional (2.1), bounded in $[R, \infty)$. Suppose that there are numbers β, k and K such that for every $r \in [R, \infty)$*

$$(3.1) \qquad r\,\delta(r) \geq \beta, \qquad \int_R^r s^{k-1}\,\delta(s)\,ds \leq K\,r^k,$$

and assume also

$$(3.2) \qquad\qquad\qquad 0 < k < \beta.$$

Then

$$u(r) \to 0 \qquad and \qquad u'(r) \to 0 \quad as\ r \to \infty.$$

Proof. Let u be an extremal of (2.1) with $|u(r)| \leq L,\ r \in [R, \infty)$, for some $L > 0$. Let ψ denote the function given in (P)$'$ corresponding to $U = L$. Then thanks to (2.10), (3.1)$_1$, (b) and property (P)$'$ we have

$$\{H(u') + F(r,u)\}' = F_r(r,u) - \delta(r)\left(\nabla G(u'), u'\right) \leq F_r(r,u) \leq \psi(r)$$

a.e. in $[R, \infty)$. Hence

$$H(u') + F(r,u) + \int_r^\infty \psi(s)\,ds$$

is decreasing in $[R, \infty)$ as a function of r. Since H and F are nonnegative by assumption (b) and property (P), there is a number $\ell \geq 0$ such that

$$(3.3) \qquad\qquad H(u') + F(r,u) \to \ell \quad as\ r \to \infty.$$

Let us first assume that $\ell = 0$. Since $H(0) = 0$ and $H_v(t) = H(tv)$ is strictly increasing in $[0, \infty)$ for any unit vector $v \in \mathbf{R}^N$, it follows easily that $u'(r) \to 0$ as $r \to \infty$. Similarly by Lemma 2 there is a positive increasing function ω_L such that $F(r, u) \geq \omega_L(|u|)$ for any $r \geq R$. Hence also $u(r) \to 0$ as $r \to \infty$, completing the proof when $\ell = 0$.

To complete the proof it is necessary to show that the condition $\ell > 0$ leads to a contradiction. This is in fact the principal effort of the demonstration, and where the identity (2.11) plays a crucial role by providing a key Liapunov function. We refer the reader to [6] for this proof, and indeed for a more general version of Theorem 1.

Remarks.   1. Condition $(3.1)_1$ provides the crucial damping for the Euler-Lagrange system (2.8) to give the asymptotic stability of the rest state. Without some condition of this sort the theorem fails; for example $(3.1)_1$ cannot be weakened even to $\delta \in L^1[R, \infty)$ and still preserve asymptotic stability, as was shown by Levin and Nohel ([3], Remark 2).

2. Condition $(3.1)_2$ is also essentially best possible. For example it fails whenever $\delta(r) \geq r^\epsilon$ for some $\epsilon > 0$, while there exist systems with $\delta(r) \geq r^\epsilon$ for which the rest state is *not* globally asymptotically stable (see Section 5 below).

3. We note that $(3.1)_1$ implies $g(r) \geq \mathrm{Const.}\, r^\beta$. Thus in the functional (2.1) the weight factor $g(r)$ becomes algebraically infinite as $r \to \infty$ under the conditions of Theorem 1.

4. Further results

In this section we consider the functional

$$(4.1) \qquad \int_{\mathbf{R}^+} g(r)\big[G(u') - F(r, u) - E(r, u)\big]\, dr,$$

where as before F satisfies (c), (P) and (P)$'$, while E and $e(r, u) = \nabla_u E(r, u)$ satisfy the further conditions stated below.

Theorem 2. *Let u be an extremal of (4.1) when $E(r, u) = -\big(e(r), u\big)$ and $e = (e_1, \ldots, e_N)$ is such that*

$$(4.2) \qquad e(r) \to 0 \quad as\ r \to \infty \quad and \quad e' \in C^0[R, \infty) \cap L^1[R, \infty).$$

Suppose also that (3.1) and (3.2) hold. If u is bounded in $[R, \infty)$, then

$$u(r) \to 0 \qquad and \qquad u'(r) \to 0 \quad as\ r \to \infty.$$

Proof. For all $r \geq R$ and $u \in \mathbf{R}^N$ we set

$$\hat{F}(r, u) = F(r, u) - \big(e(r), u\big), \qquad \hat{f}(r, u) = \nabla_u \hat{F}(r, u)$$

so that

$$\ddot{F}_r(r,u) = F_r(r,u) - (e'(r),u), \qquad \hat{f}(r,u) = f(r,u) - e(r).$$

Then, as in the first part of the proof of Theorem 1, the function

$$H(u') + \hat{F}(r,u) + \int_r^\infty \left(\psi(s) + L\,|e'(s)| \right) ds$$

is decreasing in $[R,\infty)$. Hence (3.3) holds in view of (4.2). Thus we are done if $\ell = 0$.

The remaining proof is carried out, as for Theorem 1, by using the identity (2.11). We refer the reader to reference [6] for details.

Theorem 3. *Let u be an extremal of the functional (4.1). We assume that (3.1) and (3.2) hold, there exists a nonnegative function $\tilde{\psi} \in L^1[R,\infty)$ such that*

$$(4.3) \qquad |e(r,u)| \leq \tilde{\psi}(r) \quad a.e.\ in\ [R,\infty)\ and\ for\ all\ u \in \mathbf{R}^N,$$

and that there is a positive constant ϑ such that

$$(4.4) \qquad\qquad H(p) \geq \vartheta\,|p| \qquad when\ |p| \geq p_1$$

for some number $p_1 \geq 0$.
If u is bounded in $[R,\infty)$, or if

$$(4.5) \qquad\qquad F(r,u) \to \infty \qquad as\ |u| \to \infty\ uniformly\ in\ r,$$

and (P') holds for all small $u \in \mathbf{R}^n$, then

$$u(r) \to 0 \qquad and \qquad u'(r) \to 0 \quad as\ r \to \infty.$$

Proof. As in Section 2 we see that

$$(4.6) \qquad \begin{aligned} \mathcal{L}'(r) &= \{H(u') + F(r,u)\}' \leq F_r(r,u) + (e(r,u),u') \\ &\leq \psi(r) + \tilde{\psi}(r)\,|u'(r)| \qquad a.e.\ in\ [R,\infty). \end{aligned}$$

Then by (4.4) we obtain

$$\mathcal{L}'(r) \leq \psi(r) + p_1\,\tilde{\psi}(r) + \frac{1}{\vartheta}\,\tilde{\psi}(r)\,\mathcal{L}(r).$$

The Gronwall inequality now gives

$$(4.7) \qquad\qquad \mathcal{L}(r) \leq \text{Constant} \qquad in\ [R,\infty).$$

Hence $H(u')$ is bounded and u' must be bounded thanks to (4.4), say $|u'(r)| \leq L'$ for $r \in [R, \infty)$.

By (4.7) also $F(r, u)$ is bounded along the extremal u. Hence by (4.5), or by assumption, u is bounded, say $|u(r)| \leq L$ in $[R, \infty)$.

Applying (4.6) we see that

$$H(u') + F(r, u) + \int_r^\infty \left(\psi(s) + L'\,\tilde{\psi}(s) \right) ds$$

is decreasing in $[R, \infty)$. Hence

$$H(u') + F(r, u) \to \ell \qquad \text{as } r \to \infty$$

for some $\ell \geq 0$. When $\ell = 0$ the conclusion of the theorem follows immediately. We refer to [6] for the remaining details.

Remarks. Theorems 2 and 3 apply in particular to systems of the form

$$\frac{1}{g(r)} \left(g(r)\, \nabla G(u') \right)' + f(r, u) = e(r)$$

which have nonhomogeneous forcing terms $e(r)$. In particular, when (4.4) holds it is enough that $e \in L^1[R, \infty)$. A condition of this kind, but with the additional assumption that e is bounded, already occurs in work of Levin and Nohel ([3], Theorem 1).

One can also use our results to discuss the behavior of solutions of the more general type of system

$$(4.8) \qquad \left(\nabla G(u') \right)' + h(r, u, u')\, \nabla G(u') + f(r, u) = \ell(r, u, u').$$

Indeed, along any solution $u = u(r)$ of (4.8) in $I = [R, \infty)$ we can consider $h(r, u, u')$ and $\ell(r, u, u')$ as functions solely of r, namely $\delta(r) = h(r, u(r), u'(r))$, $e(r) = \ell(r, u(r), u'(r))$, so that $u(r)$ now becomes a solution of

$$\left(\nabla G(u') \right)' + \delta(r)\, \nabla G(u') + f(r, u) = e(r).$$

This is the Euler - Lagrange system corresponding to extremals of (4.1) with

$$g(r) = \exp\left(\int_R^r \delta(s)\, ds \right), \qquad E(r, u) = \left(e(r), u \right).$$

In the context of Theorem 3, it is now enough to assume that the functions $h(r, u, u')$ and $\ell(r, u, u')$ obey the conditions

$$(4.9) \qquad \beta \leq r\, h(r, u, u') \leq r\, \tilde{\delta}(r), \qquad |\ell(r, u, u')| \leq \tilde{\psi}(r),$$

for all $r \in I$, $u \in \mathbf{R}^N$ and $u' \in \mathbf{R}^N$, where $\beta > 0$, where $\tilde{\delta}(r)$ satisfies

$$\int_R^r s^{k-1}\,\tilde{\delta}(s)\,ds \leq \text{Const. } r^k$$

with $0 < k < \beta$, and where $\tilde{\psi} \in L^1(I)$. Then if (4.4) holds, the result of Theorem 3 shows that

$$u(r) \to 0 \qquad \text{and} \qquad u'(r) \to 0 \quad \text{as } r \to \infty.$$

We note that a damping term of the form $h(r, u, u')$ already appears in the work of Levin and Nohel [3].

5. Special operators

In this section we consider the case when the function G in (2.1) possesses the property that there are numbers $m > 1$ and $\Theta > 0$ for which

$$(5.1) \qquad\qquad |\nabla G(p)| \leq \Theta\,|p|^{m-1} \quad \text{for small } p \in \mathbf{R}^N.$$

In this case the main result established in Theorem 1 can be strengthened as follows

Theorem 4. *Let u be a bounded extremal of (2.1) in $[R, \infty)$. Suppose (5.1) is satisfied and that there are numbers β, k and K such that for every $r \in [R, \infty)$*

$$(5.2) \qquad r\,\delta(r) \geq \beta, \qquad\qquad \int_R^r s^{k-m}\,\delta(s)\,ds \leq K\,r^k,$$

and where

$$(5.3) \qquad\qquad\qquad 0 < k < \beta.$$

Then

$$u(r) \to 0 \qquad and \qquad u'(r) \to 0 \quad as \ r \to \infty.$$

If we take $\beta > m$ in $(5.2)_1$ and $k = m$ in (5.3), then $(5.2)_2$ becomes

$$\int_R^r \delta(s)\,ds \leq K\,r^m,$$

discovered by Artstein and Infante [1] for the case when $N = 1$, $G(p) = |p|^2/2$, $m = 2$, $\delta(r) \geq$ Constant > 0 and when F is independent of r.

When G is homogeneous of degree $m > 1$ condition (5.1) is automatically satisfied, as observed in Section 2. The functions $G(p) = |p|^m/m$ and

$G(p) = \frac{1}{m} \sum_{i=1}^{N} |p_i|^m$, $m > 1$, discussed in the introduction, are special examples of this convex homogeneous behavior. On the other hand the mean curvature integrand $G(p) = \sqrt{1 + |p|^2} - 1$, while not homogeneous, nevertheless satisfies (5.1) with $\Theta = 1$ and $m = \overline{m} = 2$.

It is clear that the results stated in Theorems 2 and 3 continue to hold when G is of the type (5.1) in the functional (4.1) and condition $(3.1)_2$ is weakened to $(5.2)_2$.

Simple examples show that the rest state is not globally asymptotically stable when either one of the hypotheses $(5.2)_1$ or $(5.2)_2$ fails. In particular Levin and Nohel have shown that if $(5.2)_1$ is weakened to $\delta \in L^1[R, \infty)$, then the equation

$$u'' + \delta(r)\, u' + u = 0$$

possesses oscillatory solutions whose amplitudes do not tend to zero as $r \to \infty$; see [3], Remark 2.1.

On the other hand, if $(5.2)_2$ is weakened to

$$(5.4) \qquad\qquad \int_R^r s^{k-m}\, \delta(s)\, ds \leq K\, r^{k+\epsilon}$$

for some $\epsilon > 0$, then rest states need not be stable. Indeed the equation (see (1.4) for $N = 1$)

$$\left(|u'|^{m-2} u'\right)' + \delta(r)\, |u'|^{m-2} u' + u|u|^{m-2} = 0, \qquad m > 1,$$

with

$$\delta(r) = r^{m-1+\epsilon} \left(1 + \frac{m-1+\epsilon}{r^{m+\epsilon}} + \frac{m-1}{r^{m+\epsilon\, m/(m-1)}} \right)$$

satisfies each condition of Theorem 5, with the exception that $(5.2)_2$ is replaced by (5.4). It has, however, the solution

$$u(r) = \exp\left(\frac{m-1}{\epsilon\, r^{\epsilon/(m-1)}} \right)$$

which is a bounded in $[1, \infty)$ but tends to 1 as $r \to \infty$.

REFERENCES

[1] Z. Artstein and E.F. Infante, *On the asymptotic stability oscillators with unbounded damping*, Quart. Applied Math. **34** (1976), 195–199.

[2] G.A. Bliss, *Lectures on the calculus of variations*, Univ. Chicago Press, 1946.

[3] J.J. Levin and J.A. Nohel, *Global asymptotic stability for nonlinear systems of differential equations and applications to reactor dynamics*, Archive Rational Mech. Anal. **5** (1960), 194–211.

[4] P. Pucci and J. Serrin, *A general variational identity*, Indiana Univ. Math. J. **35** (1986), 681–703.

[5] P. Pucci and J. Serrin, *Continuation and limit properties for solutions of strongly nonlinear second order differential equations*, to appear.

[6] P. Pucci and J. Serrin, *Precise damping conditions for global asymptotic stability for nonlinear second order systems*, to appear.

[7] R.T. Rockafellar, *Convex Analysis*, Princeton Univ. Press, 1970.

Patrizia Pucci
Dipartimento di Matematica
Università di Perugia
Italy

James Serrin
Department of Mathematics
University of Minnesota
Minneapolis, MN

The Existence and Asymptotic Behaviour of Similarity Solutions to a Quasilinear Parabolic Equation

YUAN-WEI QI

1. Introduction

In this paper we study the existence and properties of similarity solutions which blow-up in finite time, of the nonlinear parabolic equation

$$u_t = u^\alpha(\triangle u + u^p), \tag{1}$$

where $0 < \alpha < 1$, $p > 1$. This is a transformation of the well-known porous media equation which can be regarded as to describe the propagation of thermal perturbations in a medium with a nonlinear heat conduction coefficient and a heat source term, depending on the temperature. Indeed, let v be a solution of porous media equation

$$v_t = \nabla(v^\sigma \nabla v) + v^q \tag{2}$$

with exponent σ and q. Set

$$u(Cx,t) = v^{\sigma+1}(x,t), \tag{3}$$

where $C = (\sigma + 1)^{1/2}$. Then $u(x,t)$ satisfies

$$u_t = (\sigma + 1)u^\alpha(\triangle u + u^p), \tag{4}$$

where $\alpha = \sigma/(\sigma + 1)$ and $p = q/(\sigma + 1)$. We shall assume throughout this paper that the exponents p and α are in the range $1 < p < p_c \equiv (n + 2)/(n - 2)$, $0 < \alpha < 1$ unless otherwise stated the contrary. We recall that the similarity solutions of (1) are those which have the scaling invariance property and therefore embodied with simple space-time structure. It is easy to verify that the similarity solutions take the form

$$u(x,t) = (T - t)^{\frac{1}{p+\alpha-1}} w(y), \qquad y = x/(T - t)^{\frac{p-1}{2(p+\alpha-1)}}, \tag{5}$$

where $w(y)$ satisfies the following elliptic equation

$$w^\alpha(\triangle w + w^p) - \frac{w}{p + \alpha - 1} - \frac{p - 1}{2(p + \alpha - 1)}y \cdot \nabla w = 0, \qquad y \in R^n. \tag{6}$$

For the special case of $\alpha = 0$, the equation takes the form

$$\triangle w + w^p - \frac{w}{p-1} - \frac{y}{2} \cdot \nabla w = 0, \qquad y \in R^n, \tag{7}$$

which corresponds to the following semilinear heat equation:

$$u_t = \triangle u + u^p. \tag{8}$$

The semilinear equation (7) was studied extensively and many results have been derived. The first result concerning existence was obtained by Giga and Kohn [GK1], where the nonexistence of similarity solutions was proved for $1 < p \leq (n+2)/(n-2)$. Later, Friedman et al [FFM] and Bebernes & Eberly [BE1] gave different proofs on the non-existence of radially symmetric solutions of (7) when $1 < p \leq n/(n-2)$. Giga [Gi] studied the following eigenvalue problem, which includes (7) as a special case with $\lambda = 1/(p-1)$,

$$w'' + \frac{n-1}{r}w' + w^p - \frac{r}{2}w' - \lambda w = 0, \tag{9}$$
$$w'(0) = 0, \qquad w(0) = \eta > 0,$$

where λ is a real parameter. He showed that $\lambda = 1/(p-1)$ is a critical value; for $\lambda > 1/(p-1)$ (9) has positive decreasing solutions, but if $\lambda < 1/(p-1)$, there exists no such solutions. But he could not exclude the existence of positive solutions for $\lambda < 1/(p-1)$, i.e., the ones which are not strictly decreasing. We will show later, as a direct consequence of our approach, that there are indeed no positive solutions for (9) if $\lambda < 1/(p-1)$ and the initial value $\eta > \lambda^{1/(p-1)}$. For results on the existence of positive solutions of (7) for $p > p_c$, see [Tr1], and [BQ1].

The case $0 < \alpha < 1$ is quite different from that of $\alpha = 0$. In particular, Ad'yutov et al [AKM1] proved that if $n = 1$ and $1 < p < \infty$, then there exist positive solutions of (6) which are bounded and decay to zero as $|y| \to \infty$.

In this paper, we shall study the existence of positive solutions to (6) in all dimensions. Our approach is different from that of [AKM1] and our result is stronger. The main result is the following theorem:-

Theorem 1 *Let $0 < \alpha < 1$, $1 < p < \infty$ if $n = 1$, 2 or $1 < p < p_c$ if $n \geq 3$. Then there is a non-increasing bounded positive radially symmetric solution of (6).*

We also study the asymptotic behaviour of bounded positive solutions of (6). More precisely we prove the following result.

Theorem 2 *Every bounded positive radially symmetric solution of (6) must tend to zero as $r \equiv |y| \to \infty$ and there exists a positive constant c such that*

$$w(r)r^{(p-1)/2} \to c \qquad as\ r \to \infty. \tag{10}$$

The structure of this paper is as follows: In section 2 we first derive the initial value problem (12) which is satisfied by the radially symmetric solutions of (6) and then give the proof of the existence of bounded, positive solutions of (12). In section 3 we study the asymptotic behaviour of solutions of (12) as $y \to \infty$.

2. Existence

In this section we study the existence of positive solutions to (6). We shall consider only the radially symmetric solutions. Let $w(y) = w(|y|)$ be a radially symmetric solution of (6). Then the equation for $w(r)$, $r = |y|$, is

$$w^\alpha(w'' + \frac{n-1}{r}w' + w^p) - \frac{p-1}{p+\alpha-1}(\frac{r}{2}w' + \frac{w}{p-1}) = 0. \qquad (11)$$

We observe that for the boundedness of w'' at origin we should take $w'(0) = 0$. So, the problem on which we will concentrate from this point is the following initial value problem on $(0, \infty)$:

$$w^\alpha(w'' + \frac{n-1}{r}w' + w^p) - \frac{p-1}{p+\alpha-1}(\frac{r}{2}w' + \frac{w}{p-1}) = 0, \qquad (12)$$

$$w'(0) = 0, \qquad w(0) = \eta > 0.$$

The purpose of this section is to find the initial values η such that the corresponding solutions $w(r, \eta)$ of (12) are positive and decreasing for all $r > 0$. The major difficulty comes not only from the term $(p-1)rw'/2(p+\alpha-1)$ which makes the " energy " grow rapidly as $r \to \infty$ (see (76)), but also from the quasi-linearity of equation (12). Indeed, if we drop the term $(p-1)rw'/2(p+\alpha-1)$, the equation in (12) becomes

$$w'' + \frac{n-1}{r}w' + w^p - \frac{w^{1-\alpha}}{p+\alpha-1} = 0. \qquad (13)$$

Owing to the sublinearity of $w^{1-\alpha}$, we could not apply the standard variational method developed by Strauss [Str] and Beresticki & Lions [BL] for $\alpha = 0$ to prove the existence of positive solutions of (13). This is because the function term $w^p - w^{1-\alpha}/(p+\alpha-1)$ is no longer locally Lipschitz continuous at $w = 0$. On the other hand, due to the presence of the term $(p-1)rw'/2(p+\alpha-1)$, there is lack of compactness for all $\alpha \geq 0$.

To prove the existence for $\alpha > 0$ we are forced to use the shooting argument which is related to that of [BLP] and [Gi]. But, we use scaling (instead of using a variational approach) and combine it with the linearization about the constant solution $w_0 \equiv \beta^\beta$, $\beta = 1/(p+\alpha-1)$ to distinguish the different oscillation properties of solutions which have initial values close to w_0 from those have large initial values. Then the transitional values will give the desired initial data which correspond to positive solutions.

We begin our study of (6) by investigating the qualitative behaviour of solutions $w(r, \eta)$ which have initial values close to w_0. This qualitative behaviour is (as we will show below) quite different from that of solutions which have large initial values. In fact, the qualitative behaviour of solutions with small initial values is characterized by the linear differential equation problem (17). On the other hand, the qualitative behaviour of the solutions with large initial values can be characterized by studying the related Emden-Fowler equation (41). In that case their asymptotic behaviour is independent of the lower order term $(p-1)(rw'/2+w/(p-1))/(p+\alpha-1)$. This will be shown in Lemma 3.

Let $w = w_0 + \epsilon \tilde{v}$, where ϵ is so chosen that $\tilde{v}(0) = 1$. Then in terms of ϵ we can write the equation (12) as

$$\epsilon(\tilde{v}'' + \frac{n-1}{r}\tilde{v}' + pw_0^{p-1}\tilde{v} - \frac{1-\alpha}{p+\alpha-1}w_0^{-\alpha}\tilde{v} - \tag{14}$$

$$-w_0^{-\alpha}\frac{p-1}{2(p+\alpha-1)}r\tilde{v}') + \epsilon^2 B(\epsilon, v) = 0,$$

where $B(\epsilon, v)$ is some function of (ϵ, v). Dividing (14) by ϵ and letting $\epsilon \to 0$, we obtain, by using the identity $w_0^p = w_0^{-\alpha}/(p+\alpha-1)$, the infinitesimal equation

$$\tilde{v}'' + \frac{n-1}{r}\tilde{v}' + w_0^{-\alpha}\tilde{v} - w_0^{-\alpha}\frac{p-1}{2(p+\alpha-1)}r\tilde{v}' = 0. \tag{15}$$

Evidently, $\tilde{v}$ has initial data

$$\tilde{v}'(0) = 0, \qquad \tilde{v}(0) = 1. \tag{16}$$

After the transformation $v(r) = \tilde{v}(w_0^{\alpha/2}r)$, the equation (16) has the standard formulation

$$v'' + \frac{n-1}{r}v' - \frac{r}{2}v' + \frac{p+\alpha-1}{p-1}v = 0, \tag{17}$$

$$v'(0) = 0, \qquad v(0) = 1.$$

In the following, we will study the eigenvalue problem with a parameter $\lambda > 0$ in place of $(p+\alpha-1)/(p-1)$ in (17):

$$v'' + \frac{n-1}{r}v' - \frac{r}{2}v' + \lambda v = 0, \tag{18}$$

$$v'(0) = 0, \qquad v(0) = 1.$$

Lemma 1 *The solution of linear eigenvalue problem (18) has $[\lambda]$ zeros in $(0, \infty)$, where $[\lambda]$ is defined in what follows. Suppose $m \subset \mathcal{Z}, \lambda > 0$, then*

$$[\lambda] = m + 1 \qquad if \ \ m < \lambda \leq m + 1. \tag{19}$$

Proof. Suppose

$$v_\lambda(r) = 1 + \sum_{k=1}^{\infty} a_k r^{2k} \tag{20}$$

is the solution of (18). Then, substituting (20) into (18), we get, after some calculation, the following recurrence relations

$$
\begin{aligned}
a_1 &= \frac{-\lambda}{2n}, \tag{21} \\
a_{k+1} &= \frac{(k - \lambda)a_k}{(2k + 2)(2k + n)}, \qquad k = 1, 2, \ldots \ .
\end{aligned}
$$

Thus, the series on the right hand side of (20) has an infinite radius of convergence and hence it is the solution of (18). Let us first consider the case of $0 < \lambda < 1$. It can be deduced by using (21) that in this case $a_k < 0$ for all $k \geq 1$. This yields that $v_\lambda(r) \to -\infty$ as $r \to \infty$. But, $v_\lambda(0) = 1$. So, $v_\lambda(r)$ must have a zero on $(0, \infty)$. On the other hand, $v_1(r) = 1 - r^2/(2n)$, which has only one zero on $(0, \infty)$. Therefore the Sturmian Comparison Theorem implies that $v_\lambda(r)$ has only one zero on $(0, \infty)$ for $0 < \lambda < 1$. If $1 < \lambda < 2$, it is apparent from (21) that $a_1 < 0$ and $a_k > 0$ for all $k \geq 2$. This implies that $v_\lambda(r) \to \infty$ as $r \to \infty$. Then by applying Sturmian Comparison Theorem and using the result that $v_\lambda(r)$ has only one zero on $(0, \infty)$ for $0 < \lambda < 1$ we obtain that the solution $v_\lambda(r)$ has at least two zeros. But, as $v_2(r) = 1 - r^2/n + r^4/(8 + 4n)$ has at most two zeros on $(0, \infty)$, therefore it can be shown by a further application of Sturmian Comparison Theorem that $v_\lambda(r)$ has exactly two zeros for $1 < \lambda \leq 2$. In general, if $m < \lambda < m + 1$, where $m \geq 1$ is an integer, we have

$$v_\lambda(r) = 1 + \sum_{k=1}^{m}(-1)^k A_k r^{2k} + (-1)^{m+1} \sum_{k=m+1}^{\infty} A_k r^{2k}, \tag{22}$$

where $A_k, k \geq 1$ are positive constants depending on λ and n. In addition, from (21) we can easily deduce that v_m is a polynomial of degree 2m.

$$v_m(r) = 1 + \sum_{k=1}^{m}(-1)^k a_k r^{2k} \tag{23}$$

with $a_k = (-1)^k |a_k|$. Hence, $v_\lambda(r)$ has different sign from that of v_m for all r large. Then the Sturmian Comparison Theorem implies that $v_\lambda(r)$ has at least one more zero than v_m for all $\lambda > m$. But, since v_{m+1} has at most m+1 zeros on $(0, \infty)$, v_λ must have m+1 zeros for $m < \lambda \leq m + 1$. Q.E.D.

Corollary 1 *Let $\tilde{v}$ be a solution of (15). Then $\tilde{v}$ has $[(p+\alpha-1)/(p-1)]$ zeros.*

Proof. Since $v(r) = \tilde{v}(w_0^{\alpha/2}r)$ is a solution of (17), where the parameter $\lambda = (p+\alpha-1)/(p-1)$, the conclusion follows directly from our Lemma 1.

$$\text{Q.E.D.}$$

Our next result shows that the oscillation number of $\tilde{v}$ around zero is indeed the oscillation number of solutions of (12) which have initial values sufficiently close to w_0.

Lemma 2 *Let $\tilde{v}$ be a solution of (15). Then for any solution w of (12) with initial value $w(0) = \eta$ which is sufficiently close to w_0, $w - w_0$ has at least $[(p+\alpha-1)/(p-1)]$ zeros.*

Proof. To avoid unnecessary complexity of notion, we will only consider the case $w_0 < \eta$. However, our proof can be carried out to cover the case $w_0 > \eta$ without any substantial change. Let $R > 0$ be the last point where $\tilde{v}$ equal to zero, $M = \max_{0 \leq r \leq R+1} |v(r)|$. Let S be the set of extremum points (maximum or minimum points) of $\tilde{v}$ on $[0, R]$ and $\delta \equiv \min_{r \in S} |v(r)|$. If we write $\eta = w_0 + \epsilon$ with $0 < \epsilon \ll 1$, then, by the continuous dependence of solutions to (12) on the initial values, it can be shown that for $\eta > 0$ there exists $\epsilon_0 > 0$ such that for all $0 < \epsilon < \epsilon_0$, $|w - w_0| < \delta$ on $[0,\ R+1]$. The positive constant δ will be fixed later. Recasting (12) and (15) into the equivalent Volterra integral equations we get, after some calculation, that

$$w - w_0 = \frac{p-1}{2(p+\alpha-1)(1-\alpha)} \int_0^r w^{1-\alpha} s\, ds + \tag{24}$$
$$+ \int_0^r G(r,s)\{\frac{1}{p+\alpha-1}[1 - \frac{(p-1)n}{2(1-\alpha)}]w^{1-\alpha} - w^p\}ds,$$

$$\tilde{v} - 1 = \frac{p-1}{2(p+\alpha-1)(1-\alpha)} \int_0^r w_0^{-\alpha}\tilde{v} s\, ds + \tag{25}$$
$$+ \int_0^r G(r,s)\{\frac{1}{p+\alpha-1}[1 - \frac{(p-1)n}{2(1-\alpha)}]w_0^{-\alpha}\tilde{v} - pw_0^{p-1}\tilde{v}\}ds,$$

where

$$G(r,s) = \begin{cases} (r^{2-n} - s^{2-n})s^{n-1} & \text{if } n \neq 2, \\ s\log(r/s) & \text{if } n = 2. \end{cases}$$

It is easy to prove that

$$G(r,s) \leq r, \qquad 0 < s \leq r. \tag{26}$$

Since w_0 is a solution of (12), it satisfies (24) too;

$$\int_0^r G(r,s)\{\frac{1}{p+\alpha-1}(1-\frac{(p-1)n}{2(1-\alpha)})w_0^{1-\alpha}-w_0^p\}ds \ + \tag{27}$$

$$+\frac{p-1}{2(p+\alpha-1)(1-\alpha)}\int_0^r w_0^{1-\alpha}sds \ = \ 0.$$

Multiplying (25) by $\epsilon = w(0) - w_0$ and subtracting from (24), (26), we obtain,

$$|w - w_0 - \epsilon\tilde{v}| \leq I_1 + I_2 + I_3, \tag{28}$$

where

$$I_1 \ = \ \frac{p-1}{2(p+\alpha-1)(1-\alpha)}\int_0^r |w^{1-\alpha} - w_0^{1-\alpha} - \tag{29}$$
$$- \ (1-\alpha)w_0^{-\alpha}\epsilon\tilde{v}|sds,$$

$$I_2 \ = \ \frac{1}{p+\alpha-1}[1-\frac{(p-1)n}{2(1-\alpha)}]\int_0^r G(r,s)\{\frac{1}{p+\alpha-1} \times \tag{30}$$
$$\times \ [1-\frac{(p-1)n}{2(1-\alpha)}]|w^{1-\alpha} - w_0^{1-\alpha} - (1-\alpha)w_0^{-\alpha}\epsilon\tilde{v}|\}sds,$$

$$I_3 = \int_0^r G(r,s)|w^p - w_0^p - pw_0^{p-1}\epsilon\tilde{v}|sds.$$

If we expand w^p and $w^{1-\alpha}$ about $w = w_0$ on $(w_0/2, 3w_0/2)$, we obtain,

$$w^{1-\alpha} = w_0^{1-\alpha} + (1-\alpha)w_0^{-\alpha}(w-w_0) + O(|w-w_0|^2), \tag{31}$$
$$w^p = w_0^p + pw_0^{p-1}(w-w_0) + O(|w-w_0|^2).$$

So,

$$I_1 \leq C\int_0^r (w-w_0)^2 + |w-w_0-\epsilon\tilde{v}|]sds,$$

but as $(w-w_0)^2 = (w-w_0-\epsilon\tilde{v})^2 - \epsilon^2\tilde{v}^2 - 2\epsilon\tilde{v}(w-w_0)$,

$$I_1 \ \leq C\int_0^r [(w-w_0-\epsilon\tilde{v})^2 + |w-w_0-\epsilon\tilde{v}|]sds \tag{32}$$
$$+C(R+1)^2(\epsilon^2 M^2 + 2\epsilon\delta M)$$
$$\leq C\int_0^r |w-w_0-\epsilon\tilde{v}|sds + C(R+1)^2(\epsilon^2 M^2 + 2\epsilon\delta M).$$

Similarly,

$$I_2 \leq C \int_0^r |w - w_0 - \epsilon\tilde{v}|s\,ds + C(R+1)^2(\epsilon^2 M^2 + 2\epsilon\delta M). \qquad (33)$$

$$I_3 \leq C \int_0^r |w - w_0 - \epsilon\tilde{v}|s\,ds + C(R+1)^2(\epsilon^2 M^2 + 2\epsilon\delta M).$$

Therefore

$$|w - w_0 - \epsilon\tilde{v}| \;\leq\; C \int_0^r |w - w_0 - \epsilon\tilde{v}|s\,ds + \qquad (34)$$
$$+ C(R+1)^2(\epsilon^2 M^2 + 2\epsilon\delta M).$$

By applying *Gronwall's* inequality we get

$$|w - w_0 - \epsilon\tilde{v}| \leq \quad C + C(R+1)^2 \times \qquad (35)$$
$$\times(\epsilon^2 M^2 + 2\epsilon\delta M)e^{Cr^2/2}, \qquad 0 \leq r \leq R+1.$$

If we choose δ so small that the right hand side of the above expression is less or equal to $\epsilon\delta/2$. Then,

$$|w - w_0 - \epsilon\tilde{v}| \leq \epsilon\delta/2, \qquad 0 \leq r \leq R+1. \qquad (36)$$

Therefore for all $r \in S$, $w(r) - w_0$ must have the same sign as $\tilde{v}(r)$. Thus, w intersects w_0 at least $[(p + \alpha - 1)/(p - 1)]$ times.

Q.E.D.

Note We have used C in the above to represent various constants which depend on α, p and n only for simplicity of notion. We will keep on doing so in the following without further notification. The reason for this is their values won't make any difference to our proofs.

In the following we shall discuss the behaviour of solutions $w(r, \eta)$ of (12), where $\eta \gg 1$. Let w be a solution of (12) which has initial value η. If we make the transformation

$$u(r) = \frac{w(\eta^{-\frac{p-1}{2}}r)}{\eta}, \qquad (37)$$

then

$$u'(r) = \eta^{-\frac{p+1}{2}}w(\eta^{-\frac{p-1}{2}}r), \qquad u''(r) = \eta^{-p}w(\eta^{-\frac{p-1}{2}}r). \qquad (38)$$

It is easy to verify by using (12), (37) and (38) that u is a solution of the following initial value problem

$$u'' + \frac{n-1}{r}u' + u^p - \eta^{-(p+\alpha-1)}\frac{p-1}{p+\alpha-1} \times \tag{39}$$

$$\times \left(\frac{r}{2}u' + \frac{u}{p-1}\right) = 0 \qquad r > 0,$$

$$u'(0) = 0, \qquad u(0) = 1. \tag{40}$$

If we let $\eta \to \infty$, then we get the well-known Emden-Fowler equation

$$z'' + \frac{n-1}{r}z' + z^p = 0 \qquad r > 0, \tag{41}$$

$$z'(0) = 0, \qquad z(0) = 1.$$

The above equation has been studied by many authors, see Fowler [Fow], Wong [WONG] and Budd & Norbury [BNo] among others. It is well known that if the exponent p is subcritical then the solution of (41) is monotone decreasing and terminates at zero at a finite value $r_z > 0$.

Lemma 3 *Let w be a solution of (12) with initial value $\eta \gg 1$. Then $w(r, \eta)$ is monotone decreasing and terminates at zero at some finite value $r_\eta > 0$.*

Proof. We will work with the function $u(r) = w(\eta^{-\frac{p-1}{2}}r)/\eta$. Let

$$S_u = \max\{r > 0 \,|\, 0 < u < 1 \text{ on } (0, r)\}, \tag{42}$$

$$r_u = \min(S_u, r_z).$$

Then $0 < r_u \leq r_z$ and $0 < u < 1$ on $(0, r_u)$. By writing (39) and (41) into equivalent integral equations, we get

$$u - 1 = -\int_0^r G(r, s)u^p ds + \eta^{-(p+\alpha-1)}\left\{\frac{p-1}{2(p+\alpha-1)(1-\alpha)} \times \right. \tag{43}$$

$$\left. \times \int_0^r u^{1-\alpha}s ds + \frac{1}{p+\alpha-1}[1 - \frac{(p-1)n}{2(1-\alpha)}]\int_0^r G(r, s)u^{1-\alpha}ds\right\},$$

$$z - 1 = -\int_0^r G(r, s)z^p ds,$$

where G(r,s) is as in above. By using the boundedness of $u(r)$, $z(r)$ and the relation $G(r, s) \leq r$ on $(0, r_u)$, we can derive that

$$|u - z| \quad \leq C\eta^{-(p+\alpha-1)} + \int_0^r G(r,s)|u^p - z^p|ds \tag{44}$$

$$\leq C\eta^{-(p+\alpha-1)} + C\int_0^r G(r,s)|u - z|ds$$

$$\leq C\eta^{-(p+\alpha-1)} + C\int_0^r s|u - z|ds.$$

Then an application of *Gronwall's* lemma will yield

$$|u - z| \quad \leq C\eta^{-(p+\alpha-1)}e^{Cr} \tag{45}$$

$$\leq C\eta^{-(p+\alpha-1)} \qquad \text{on } (0, r_u),$$

where C is some positive constant independent of η and u. Similarly, from

$$u'r^{n-1} \quad = \quad -\int_0^r r^{n-1}u^p ds + \eta^{-(p+\alpha-1)}(\frac{p-1}{2(p+\alpha-1)(1-\alpha)} \times \tag{46}$$

$$\times r^n u^{1-\alpha} + \{\frac{1}{p+\alpha-1}\int_0^r [1 - \frac{(p-1)n}{2(1-\alpha)}]u^{1-\alpha}s^{n-1}ds\}),$$

$$z'r^{n-1} = -\int_0^r r^{n-1}z^p ds, \tag{47}$$

we obtain that

$$|u' - z'| \leq C\eta^{-(p+\alpha-1)} \qquad \text{on } (0, r_u). \tag{48}$$

The estimate (48) implies, by recalling the definition of r_u and s_u, that

$$\underline{\lim}_{\eta\to\infty} s_u \geq r_z. \tag{49}$$

So, for any $\epsilon > 0$, there exists an $\overline{\eta} > 0$, such that for all $\eta > \overline{\eta}$, $r_u > r_z - \epsilon$. Thus, by combining (48), (45) with $r_u > r_z - \epsilon$, we obtain

$$u(r_u) \leq \epsilon, \qquad u'(r_u) \leq -m, \qquad \text{for } \eta \gg 1, \tag{50}$$

where m is some positive constant which depends on z' and ϵ is some small positive number. Without loss of generality, we assume $\epsilon < mr_u/8n$.

The equation (39) implies that as long as $u < 1$, $u' < 0$,

$$u'' + \frac{n-1}{r}u' < \eta^{-(p+\alpha-1)}/\alpha. \tag{51}$$

Integrating the above inequality on $[r_u, r]$, where $u' < 0$, we have

$$u'(r) - u'(r_u) \quad + \quad \frac{n-1}{r}(u(r) - u(r_u)) + \int_{r_u}^r \frac{n-1}{s^2} \tag{52}$$

$$< \quad (r - r_u)\eta^{-(p+\alpha-1)}/\alpha.$$

This, in turn, implies that

$$u'(r) \; < \; u'(r_u) + \frac{3(n-1)\epsilon}{r_u} + (r - r_u)\eta^{-(p+\alpha-1)}/\alpha \tag{53}$$

$$< \; -m/2,$$

for $r \le r_1 \equiv r_u + \alpha m \eta^{(p+\alpha-1)}/8$. Therefore on the interval $[r_u, r_1]$,

$$u'(r) < -m/2. \tag{54}$$

We claim, then, u must go to zero at some $r < r_1$ if η is sufficiently large; for otherwise,

$$u(r_1) = u(r_u) + \int_{r_u}^{r_1} u' ds < \epsilon - \frac{\alpha m^2}{16}\eta^{(p+\alpha-1)} < 0. \tag{55}$$

This completes the proof of Lemma 3.

Q.E.D.

Lemma 4 *Let w be a solution of (12) with $w(0) > w_0$. Then w must have at least one intersection with w_0. Furthermore, if $w'(r_0) = 0$, $r_0 \ge 0$, then, there exists $r_1 > r_0$, such that $w(r_1) = w_0$.*

Proof. It is sufficient to prove the second claim. We assume at first that
$w(r_0) < w_0$, and so, $w''(r_0) > 0$. From (12), we may deduce that for all $r > r_0$,

$$w'' + \frac{n-1}{r}w' + w^p > \frac{r(p-1)}{2(p+\alpha-1)}w^{-\alpha}w', \tag{56}$$

as long as $w < w_0$ on $[r_0, r]$. Multiplying (12) by r^{n-1} and integrating over $[r_0, r]$, we obtain that

$$w' r^{n-1} \; = \; \int_{r_0}^{r} \frac{s^n(p-1)}{2(p+\alpha-1)}w^{-\alpha}w' ds + \tag{57}$$

$$+ \; \int_{r_0}^{r} (\frac{w^{1-\alpha}}{p+\alpha-1} - w^p)s^{n-1} ds > 0$$

for all $r > r_0$ so long as $w < w_0$ on $[r_0, r]$. Suppose the contrary, that $w(r) < w_0$ for all $r > r_0$. In this case, $w' > 0$ on (r_0, ∞) and (56) would imply that

$$w'' > crw' \tag{58}$$

for all r sufficiently large, where c is a positive constant. Integrating (58) then yields that

$$w'(R) > c e^{c(R^2 - r^2)/2} w'(r), \qquad R > r > r_0. \tag{59}$$

Therefore $w'(R) \to \infty$ as $R \to \infty$, which contradicts the assumption that $w < w_0$ for all $r > r_0$. So, there must be an $r_1 > r_0$ such that $w(r_1) = w_0$. The case of $w(r_0) > w_0$ can be treated similarly.

Q.E.D.

Lemma 5 *Let p and α be as in Theorem 1. Then there exists a monotone decreasing positive solution of (12).*

Proof. Let

$$S_1 = \{\eta > w_0 | w(r, \eta) \text{ has at least two intersections} \tag{60}$$
$$\text{with } w_0 \text{ on } (0, r_\eta) \},$$

where r_η is the first point at which $w = 0$ (If $w > 0$ on $(0, \infty)$ we define $r_\eta = \infty$). We know from Lemma 2 and the continuous dependence of solutions to (12) on the initial data that S_1 is a non-empty open set. Furthermore, Lemma 3 implies that S_1 is bounded from above. Therefore if we define

$$\eta_1 = \sup_{a \in S_1} \eta, \tag{61}$$

then, $\eta_1 < \infty$ and $\eta_1 \overline{\in} S_1$ as S_1 is open. Hence, $w(r, \eta_1)$ must be monotone decreasing by Lemma 4. If $w(r, \eta_1) = 0$ at some $r = r_0$, then the solutions that have initial values close to η_1 would be all monotone decreasing solutions and decay to zero at some finite value of r. This is clearly a contradiction of the definition of η_1. Thus, $w(r, \eta_1)$ is a monotone decreasing positive solution; $w(r, \eta_1) > 0, w'(r, \eta_1) < 0$ for all $r > 0$.

Q.E.D.

Remark. Many results we proved for the case $0 < \alpha < 1$ so far, such as Lemma 5 above, may not hold for $\alpha \geq 1$. I think it is due to the fact that when $0 < \alpha < 1$, a solution w of (12) which has its first zero at $r = r_0$ has the property that w' is continuous on $[0, r_0]$, but when $\alpha \geq 1$, it is very unlikely that this kind of C^1 continuity will still hold. Therefore there could be a big difference between $0 < \alpha < 1$ and $\alpha \geq 1$.

Remark. We want to point out at this stage that although Lemma 3 in this section has been stated for the case of $\alpha > 0$, it holds for the case of $\alpha = 0$, too.

We conclude this section by giving the following non-existence result for the eigenvalue problem (9) which is related to the similarity problem of the semilinear heat equation (7).

Proposition 1 *Let w be a solution of (9). Let $\lambda < 1/(p-1)$ and $1 < p < p_c$. If the initial value $w(0) = \eta > w_\lambda = \lambda^{1/(n-1)}$, then there is an $r > 0$ such that $w(r) = 0$.*

Proof. Let us suppose the contrary that w is a positive solution of (9) on $(0, \infty)$. It was established in [Gi] by Giga that there exists no monotone decreasing solution of (9). Therefore w must be a non-monotone and so has at least two intersection with the constant solution w_λ by our lemma 4. Thus, the following set is non-empty.

$$S \;=\; \{\eta > w_\lambda | w(r, \eta) \text{ has at least two intersections with} \qquad (62)$$
$$w_\lambda \text{ before } w = 0\}.$$

Furthermore, Lemma 3 implies that S is bounded from above. In that case, the same reasoning as our Lemma 5 implies that there is a $\eta_1 > 0$ for which the corresponding solution $w(r, \eta_1)$ is a monotone decreasing solution of (9). But, this is impossible by the result of [Gi]. This completes the proof of the proposition.

Q.E.D.

3. Asymptotic Behaviour

In this section, we analyse the asymptotic behaviour of positive solutions of (12) as $r \to \infty$. We prove Theorem 2 through a series of lemmas.

Lemma 6 *If w is a bounded positive solution of (12). Then*

$$w'(r) \to 0 \qquad as \; r \to \infty. \qquad (63)$$

Proof. Suppose w is a solution of (12). We define the the following function:

$$f(w, r) \equiv \exp\{-\frac{p-1}{2(p+\alpha-1)} \int_0^r sw^{-\alpha} ds\}. \qquad (64)$$

Multiplying the equation (12) by $r^{n-1} f(w, r)$ and integrating over $[0, r]$, we obtain

$$w'(r) r^{n-1} f(w, r) = \int_0^r s^{n-1} (\frac{w^{1-\alpha}}{p+\alpha-1} - w^p) f(w, s) ds. \qquad (65)$$

Since w is bounded, we assume that $w \leq M$, for an $M > 0$. Then

$$\int_0^r sw^{-\alpha}/2 ds > M^{-\alpha} r^2/4 \qquad r > 0, \qquad (66)$$

which implies that the integral on the right hand side of (65) is convergent as $r \to \infty$ and the integrand decays to zero exponentially fast (faster than any inverse power of r) as $r \to \infty$. Furthermore, the boundedness of w implies that there is a sequence $\{R_l\}_1^\infty$ such that $R_l \to \infty$ as $l \to \infty$ and $|w'(R_l)| \le 1$. So, the integral of the right hand side of (65) tends to zero as $r \to \infty$ and

$$\int_0^r s^{n-1}\left(\frac{w^{1-\alpha}}{p+\alpha-1} - w^p\right)f(w,s)ds = \tag{67}$$
$$-\int_r^\infty s^{n-1}\left(\frac{w^{1-\alpha}}{p+\alpha-1} - w^p\right)f(w,s)ds,$$

which implies that

$$w'(r)r^{n-1}f(w,r) = -\int_r^\infty s^{n-1}\left(\frac{w^{1-\alpha}}{p+\alpha-1} - w^p\right)f(w,s)ds. \tag{68}$$

Therefore the boundedness of w yields that

$$|w'(r)| \le C\frac{\int_r^\infty s^{n-1}f(w,s)ds}{r^{n-1}f(w,r)}. \tag{69}$$

Then an application of $L'H\hat{o}pital's$ rule yields that the right hand side of (69) tends to zero as $r \to \infty$. Thus $w'(r) \to 0$ as $r \to \infty$. This completes the proof of
Lemma 6.

Q.E.D.

Lemma 7 *If w is a bounded positive solution of (12), then $w'(r) < 0$, $w < w_0$ for all r large.*

Proof. Let w be a bounded positive solution of (12). First we prove that $w < w_0$ for all r large. Suppose the contrary, then either $w > w_0$ for all $r \gg 1$ or w intersects w_0 an infinite number of times on $(0,\infty)$.

If $w > w_0$ for all $r \gg 1$, then we may deduce by using Lemma 4 that $w'(r) > 0$ for all $r \gg 1$ and so, $w \to w_1 > w_0$ as as $r \to \infty$. Therefore there exists a $\epsilon > 0$ such that

$$w'' + \frac{n-1}{r}w' < \frac{r(p-1)}{2(p+\alpha-1)}w^{-\alpha}w' - \epsilon, \qquad r \gg 1. \tag{70}$$

Multiplying the equation (70) by $r^{n-1}f(w,r)$ and integrating on $[r,\infty]$, we obtain

$$-w'r^{n-1}f(w,r) < -\epsilon\int_r^\infty s^{n-1}f(w,s)ds \tag{71}$$

or equivalently,

$$w'r > \epsilon \frac{\int_r^\infty s^{n-1} f(w,s)ds}{r^{n-2} f(w,r)}. \tag{72}$$

It can be shown, by using the $L'H\hat{o}pital's$ rule (as we did in Lemma 6) to calculate the asymptotics of the integral on the right hand side of (72), that

$$\frac{\int_r^\infty s^{n-1} f(w,s)ds}{r^{n-2} f(w,r)} \to w_1^\alpha \qquad \text{as } r \to \infty. \tag{73}$$

Therefore

$$w'r > \epsilon w_0^\alpha \qquad \text{for all } r \gg 1, \tag{74}$$

which upon an integration would yield that

$$w' \to \infty \qquad \text{as } r \to \infty. \tag{75}$$

But this is clearly a contradiction of our boundedness assumption, so w cannot be greater than w_0 for all $r \gg 1$.

Suppose w intersects w_0 an infinite number of times on $(0, \infty)$. We define the following function

$$E(r) \equiv \frac{(w')^2}{2} - \frac{w^{2-\alpha}}{(p+\alpha-1)(2-\alpha)} + \frac{w^{p+1}}{p+1}. \tag{76}$$

We claim that because of the boundedness of w, E is an increasing function for all $r \gg 1$. Indeed,

$$\begin{aligned}
E'(r) &= w''w' - \frac{w^{1-\alpha}}{p+\alpha-1}w' + w^p w' \\
&= \frac{r(p-1)}{2(p+\alpha-1)}w^{-\alpha}(w')^2 - \frac{n-1}{r}(w')^2 \\
&\geq (w')^2\left(\frac{r(p-1)}{2(p+\alpha-1)}M^{-\alpha} - \frac{n-1}{r}\right) > 0, \qquad r \gg 1,
\end{aligned} \tag{77}$$

where M is a upper bound of w. On the other hand, our hypothesis of an infinite number of intersections of w with w_0 implies that there is an increasing sequence $\{r_l\}_1^\infty$, such that $\lim_{l\to\infty} r_l = \infty$ and $w = w_0$ at $r = r_l$, $l = 1, 2, \ldots$. Without loss of generality, we assume that $E'(r) > 0$ for all $r > r_1$. This, in turn, implies that

$$|w'(r_{l+1})| > |w'(r_l)| > 0, \qquad l \geq 1. \tag{78}$$

This is impossible. Since we have shown in Lemma 6 that $w'(r) \to 0$ as $r \to \infty$. Thus, $w < w_0$ for all $r \gg 1$ and Lemma 4 implies that $w' < 0$ for all r large.

$$\text{Q.E.D.}$$

Lemma 8 *Suppose $w > 0$ solves (12) in (a, ∞), where $a > 0$, and is monotone decreasing. Then $w \to 0$ as $r \to \infty$ and furthermore, for a given $\theta < 1/(p-1)$,*

$$w(r) < \frac{C}{r^{2\theta}} \qquad r > a, \tag{79}$$

where C is some positive constant independent of r.

Proof. We shall first prove that $w \to 0$ as $r \to \infty$. Suppose the contrary, then Lemma 7 implies that there exists $\mu > 0$, such that $w \to \mu < w_0$ as $r \to \infty$ and $w'(r) < 0$ for all r large. So, there exists $\epsilon > 0$, such that

$$w'' + \frac{n-1}{r}w' > \frac{r(p-1)}{2(p+\alpha-1)}w^{-\alpha}w' + \epsilon, \qquad r \gg 1. \tag{80}$$

As in Lemma 6, we multiply the equation (70) by $r^{n-1}f(w,r)$ and integrate on $[r, \infty]$ to obtain

$$-w'r^{n-1}f(w,r) > \epsilon \int_r^\infty s^{n-1}f(w,s)ds, \tag{81}$$

which is equivalent to

$$w'r < -\epsilon\frac{\int_r^\infty s^{n-1}f(w,s)ds}{r^{n-2}f(w,r)}. \tag{82}$$

Then by using the similar argument as in Lemma 7 it can be shown that

$$w'(r)r < \frac{-\epsilon\mu}{2} < 0, \qquad \text{for } r \gg 1 \tag{83}$$

and so, w would terminate at zero at some finite value of r. This is clearly a contradiction of our hypothesis that $w \to \mu > 0$. Thus, $w \to 0$ as $r \to \infty$.

Since $w \to 0$ and $w' < 0$ for all r large, w'' can not be negative for all r large. In fact, differentiating (12) gives

$$\begin{aligned}
w''' &= (\frac{r}{2}w^{-\alpha} - \frac{n-1}{r})w'' + (\frac{w^{-\alpha}}{2} + \frac{n-1}{r^2})w' - \\
&\quad - \frac{r}{2}w^{-(\alpha+1)}(w')^2 + (\frac{1-\alpha}{p+\alpha-1}w^{-\alpha} - pw^{p-1})w',
\end{aligned} \tag{84}$$

which implies that w'' must be positive for all r large. For otherwise, w'', w''' will be negative for all r large, which in turn implies that w must terminate to zero at some finite point r. But this clearly contradicts our assumption that w is positive on $(0, \infty)$. Thus, w'' is positive for all r large.

Let $\mu < 1/(p-1)$, the conclusion $w \to 0$ and $w' < 0$ yield that for all r large,

$$w'' - \frac{p-1}{p+\alpha-1}(\frac{r}{2}w' + \mu w)w^{-\alpha} > 0 \tag{85}$$

or

$$w^\alpha w'' - \frac{p-1}{p+\alpha-1}(\frac{r}{2}w' + \mu w) > 0. \tag{86}$$

Therefore $w'' > 0$ gives

$$w'' - \frac{p-1}{p+\alpha-1}(\frac{r}{2}w' + \mu w) > 0. \tag{87}$$

For $\theta < \mu$, $W = Kr^{-2\theta}$ solves

$$W'' - \frac{r(p-1)}{2(p+\alpha-1)}W' - (\frac{(p-1)\theta}{p+\alpha-1}W + \frac{2\theta(2\theta+1)}{r^2})W = 0. \tag{88}$$

For r large, W satisfies

$$W'' - \frac{p-1}{p+\alpha-1}(\frac{r}{2}W' + \mu W) < 0. \tag{89}$$

Take K large so that $W(r_1) > w(r_1)$, where r_1 is sufficiently large so that (87) and (89) hold for all $r > r_1$. By Sturmian Comparison Theorem we conclude that $w \leq W$ for $r \geq r_1$, which is the same as (79).

Q.E.D.

Proof of Theorem 2: We use the transformation (following Giga [Gi])

$$z = wr^\beta, \qquad \beta = 2/(p-1). \tag{90}$$

Since

$$w' = (-\frac{2\beta z}{r} + z')r^{-2\beta}, \tag{91}$$

$$w'' = (\frac{2\beta(2\beta+1)}{r^2}z - \frac{4\beta}{r}z' + z''),$$

(12) can be written as

$$z'' - \frac{p-1}{2(p+\alpha-1)}r^{1+2\alpha\beta}z'z^\alpha + \frac{n-1-4\alpha}{r}z' + \tag{92}$$

$$+ \frac{2\alpha(2\alpha+2-n)}{r^2}z + \frac{z^p}{r^2} = 0.$$

468　　　　　　　　　　　　　　YUAN-WEI QI

The estimate (79) yields for $r > 0$,

$$z(r) < Cr^{2\alpha\beta}, \tag{93}$$

where $C = C(\alpha, \beta)$. Since $w' < 0$, (91) yields

$$z'(r) < Cr^{2\alpha\beta - 1} \tag{94}$$

with C independent of r. Applying (93) and (94) to (92) we obtain

$$\left| \frac{z''(r)}{r^{1+2\alpha\beta}} - \frac{p-1}{2(p+\alpha-1)} z' \right| \le \frac{M}{r^2}, \qquad r > 0, \tag{95}$$

where M is independent of r. Integrating by parts gives

$$\int_{r_1}^{r} \left(\frac{z''(s)}{s^{1+2\alpha\beta}} - \frac{p-1}{2(p+\alpha-1)} z' \right) ds = \frac{z'(s)}{s^{1+2\alpha\beta}} \Big|_{r_1}^{r} + \tag{96}$$

$$(1 + 2\alpha\beta) \int_{r_1}^{r} \frac{z'(s)}{s^{2+2\alpha\beta}} ds - \frac{p-1}{2(p+\alpha-1)} (z(r) - z(r_1)).$$

By (94), the first two terms of right hand side converge as $r \to \infty$. This implies that $\lim_{r \to \infty} z(r)$ exists since the left hand side converges as $r \to \infty$ by (95), which means that

$$\lim_{r \to \infty} w(r) r^{\beta} \to C \ge 0. \tag{97}$$

We prove in the following that the limit is indeed positive. Since $w'' > 0$ for all r large,

$$\frac{p-1}{2(p+\alpha-1)} r w' w^{-\alpha} - \frac{n-1}{r} w' > - \frac{w^{1-\alpha}}{(p+\alpha-1)} \tag{98}$$

or

$$\frac{w'}{w} > - \frac{2}{r(p-1)} + \frac{2(p+\alpha-1)(n-1)}{(p-1)r^2} w' w^{\alpha-1}. \tag{99}$$

Integrating the above inequality then yields

$$\log w \Big|_{r_1}^{r} \ge - \frac{2}{(p-1)} \log s \Big|_{r_1}^{r} + \int_{r_1}^{r} \frac{2(p+\alpha-1)(n-1)}{(p-1)s^2} w' w^{\alpha-1} ds, \tag{100}$$

where r, r_1 are sufficiently large. So,

$$w(r) \ge w(r_1) r_1^{\frac{2}{p-1}} r^{-\frac{2}{p-1}} \exp\left(\int_{r_1}^{r} \frac{2(p+\alpha-1)(n-1)}{(p-1)s^2} w' w^{\alpha-1} ds \right). \tag{101}$$

Since $w', w \to 0$ as $r \to \infty$, the integral on the right hand side of (101) is bounded below. Thus,

$$w(r) \ge Cr^{-\frac{2}{(p-1)}}, \tag{102}$$

where $C > 0$. This proves that there is a positive constant C, such that

$$\lim_{r \to \infty} w(r) r^{\beta} \to C > 0. \tag{103}$$

Q.E.D.

REFERENCES

[AKM] M. M. Ad'yutov, Yu. A. Klokov and A. P. Mikhailov, *Self-simulating thermal structures with contracting half-width*, Differential Equations, **19**: 7, (1983), 1107–1114.

[BE1] J. Bebernes and D. Eberly, *A description of self-similar blow-up for dimensions $n \geq 3$*, Ann. Inst. Henri Poincaré, **5** (1988), 1–21.

[BL] H. Berestycki and P.-L Lions, *Nonlinear Scalar field equations, I. Existence of ground states*, Arch. Rational Mech. Anal., **82** (1983), 313–345.

[BLP] H. Berestycki, P.-L. Lions and L. A. Peletier, *An O.D.E. approach to the existence of positive solutions for semilinear problems in R^n*, Indiana Univ. Math. J., **30** (1981), 141–157.

[BNo] C. J. Budd and J. Norbury, *Semilinear elliptic equations with supercritical growth rates*, J. Differential Equations, **68** (1987), 169–197.

[BQ1] C. J. Budd and Y.-W. Qi, *The existence of bounded solutions of a semilinear elliptic equation*, J. Differential Equations, **89** (1989), 207–218.

[Fow] R. H. Fowler, *Further properties of Emden's and similar differential equations*, Quart. J. Math. (Oxford Series), **2** (1931), 259–288.

[FFM] A. Friedman and J. Friedman and J. B. Mcleod, *Concavity of solutions of nonlinear ordinary differential equations*, J. Math. Anal. Appl., **131** (1988), 486–500.

[Gi] Y. Giga, *On elliptic equations related to self-similar solutions for nonlinear heat equations*, Hiroshima Math. J., **16** (1986), 541–554.

[GK1] Y. Giga and R. V. Kohn, *Asymptotically self-similar blowup of semilinear heat equations*, Comm. Pure Appl. Math., **38** (1985), 297–319.

[Str] W. Strauss, *Existence of solitary waves in higher dimensions*, Comm. Math. Phys., **55** (1977), 149–162.

[Tr1] W. Troy, *The existence of bounded solutions for a semilinear heat equation*, SIAM J. Math. Anal., **18** (1987), 332–336.

[WONG] J. S. W. Wong, *On the generalized Emden-Fowler equation*, SIAM Review, **17** (1975), 339-360.

Yuan-wei Qi
Math. Institute
University of Oxford
Oxford, OX1 3LB
U.K.

Maximal Solutions of Singular Diffusion Equations with General Initial Data

ANA RODRÍGUEZ and JUAN LUIS VÁZQUEZ

Abstract

In the range $-1 < m \le 0$ the Cauchy problem

$$u_t = (u^{m-1}u_x)_x, \qquad \text{for } (x,t) \in \mathbf{Q}_T = \mathbf{R} \times (0,T)$$
$$u(x,0) = u_0(x), \qquad \text{if } x \in \mathbf{R}$$

admits infinitely many solutions if for instance u_0 is nonnegative and integrable. We show existence and uniqueness of a maximal solution of the problem for initial data $u_o \in \mathcal{M}_+(\mathbf{R})$, the set of nontrivial, nonnegative and locally bounded Borel measures. These solutions are characterized in terms of a suitable decay rate as $|x| \to \infty$ (good solutions). Since we also show that every good solution defined in a strip $\mathbf{Q}_T$ possesses a uniquely defined trace in $\mathcal{M}_+(\mathbf{R})$ at $t = 0$, a complete theory of maximal solutions for our equation is obtained.

Introduction

We study the questions of existence and uniqueness of positive solutions to the Cauchy problem

$$(0.1) \qquad u_t = (u^{m-1}u_x)_x \qquad \text{for } (x,T) \in \mathbf{Q}_T = \mathbf{R} \times (0,T)$$
$$(0.2) \qquad u(x,0) = \mu(x), \qquad \text{if } x \in \mathbf{R}$$

where $0 < T \le \infty$, $\mu \in \mathcal{M}_+(\mathbf{R})$, the set of nontrivial, nonnegative and locally bounded Borel measures, and the exponent range is precisely $-1 < m \le 0$, as well as the problem of initial traces for solutions of (0.1).

It is known that in this range of exponents the Cauchy problem is *not* well-posed. More specifically, we have recently shown in [RV] that one can uniquely solve (0.1), (0.2) with initial data in $L^1(\mathbf{R})$ if we add *flux* at infinity

$$(0.3a) \qquad -u^{m-1}u_x \to f(t) \qquad \text{as } x \to \infty$$
$$(0.3b) \qquad u^{m-1}u_x \to g(t) \qquad \text{as } x \to \infty$$

Partially supported by DGCICYT Project PB86-0112-C.
Partially supported by EEC Contract SC1-0019-C.

in a strip $\mathbf{Q}_T$ with $T = \sup\{t \geq 0 : \int u_0 > \int_0^t f + g\}$, under the assumptions that $u(x,0) \in L^1(\mathbf{R})$, $u(x,0) \geq 0$ and $f, g \in L_{loc}^\infty(0,\infty)$, $f, g \geq 0$, $u(x,0), f$ and g being otherwise arbitrary. Moreover, if $T < \infty$ then $u(\cdot, t) \to 0$ as $t \to T$, i.e. *extinction* in finite time occurs. Putting $f \equiv g \equiv 0$ we obtain a solution which turns out to be the *maximal* element in the set of solutions of (0.1), (0.2) and exists for $0 < t < \infty$. The existence and properties of maximal solutions with initial data in $L_{loc}^1(\mathbf{R})$ have been investigated in [ERV], where in particular they are characterized by the following growth condition:

$$(0.4) \qquad u^m(x,t) = o(|x|) \qquad \text{as } |x| \to \infty \quad \text{loc. uniformly in } t > 0$$

(For $m = 0$, u^m is replaced by $\log(u)$). The condition is sharp since solutions with nonzero flux data $u^m(x,t)$ behave like $O(|x|)$ as $|x| \to \infty$ as a consequence of (0.3).

The above solutions are C^∞ for $x \in \mathbf{R}$, $t > 0$ and take the initial data in the sense: $u(\cdot, t) \to u_0$ in $L_{loc}^1(\mathbf{R})$ as $t \to 0$.

Our goal in this paper is to provide a complete theory of maximal solutions for equation (0.1) in the above exponent range. We begin by considering an arbitrary positive solution of (0.1) in a strip $\mathbf{Q}_T$ which satisfies the asymptotic condition (0.4). Such a solution will be called a *good* solution. We prove that a good solution admits an initial trace as $t \to 0$, which as expected is a nontrivial, nonnegative and locally bounded Borel measure in $\mathbf{R}$.

We then prove that for every such measure as initial data, the Cauchy problem has one and only one good solution. Finally, we identify the good solution with the maximal element of the set of solutions of the Cauchy Problem. Therefore, maximal solutions of (0.1), (0.2) are always characterized by condition (0.4) for arbitrary initial data.

Technically there appear some differences between the case $-1 < m < 0$ and $m = 0$. They arise from the fact that while (0.1) can be written as

$$u_t = \left(\frac{u^m}{m}\right)_{xx}$$

for $m \neq 0$, the corresponding formula for $m = 0$ is $u_t = (\log u)_{xx}$.

The study of initial traces for $m > 1$ (and equation $u_t = \Delta u^m$ in several space dimensions $N \geq 1$) was done in [AC]. Existence and uniqueness of a solution for general initial conditions follows from [BCP], [P] and [DK1]. We will be using several techniques introduced in those works. Recently, the paper [DK2] treats the same problems for fast diffusion $u_t = \Delta\phi(u)$ with assumptions on ϕ such that in the case of power nonlinearities $\phi(u) = u^m$ they imply that m lies in the *good* range $\min\{(N-2)/N, 0\} < m < 1$. We recall that in all of these cases there is uniqueness of suitably defined weak solutions. Finally, a discussion of the general situation for $m \in \mathbf{R}$ and $N \geq 1$ is contained in [V].

1. Preliminaries

Our investigation will be concerned with C^ω smooth and positive functions u defined in a strip $\mathbf{Q}_T = \mathbf{R} \times (0, T)$, $0 < T \le \infty$, which solve (0.1) and satisfy condition (0.4). We shall for the moment call such a function a *good solution* of (0.1). Our aim is to show that they coincide with the maximal solutions of a general Cauchy problem. Indeed, good solutions are related to maximal solutions in a clear way: for every $\tau > 0$, $(\tau < T)$, $u^\tau(x, t) = u(x, t + \tau)$ is the maximal solution of problem (0.1), (0.2) with initial data $u(x, \tau) \in L^1_{loc}(\mathbf{R})$. This is proved in ERV, Theorem 6.1].

As a consequence, u^τ enjoys all the properties derived in [ERV], in particular pointwise estimates for its derivatives. Letting $\tau \to 0$ we obtain the same estimates for u, namely

Lemma 1. *Good solutions of* (0.1) *satisfy the estimates*

$$(1.1) \qquad -\frac{u}{(1+m)t} \le u_t \le \frac{u}{(1-m)t}$$

and

$$(1.2) \qquad -\frac{1}{(m+1)t} \le v_{xx} \le \frac{1}{(1-m)t}$$

everywhere in $\mathbf{Q}_T$. *Here* $v = u^{m-1}/(m-1)$ *is the usual "pressure" function.*

We show next that good solutions satisfy sharper estimates as $|x| \to \infty$ than just condition (0.4).

Lemma 2. *For every good solution* u *of* (0.1) *and every* $\tau \in (0, T)$ *there exists a positive function* $g(x)$ *such that* $g(x) = O(|x|^2)$ *as* $|x| \to \infty$ *and*

$$(1.3) \qquad u^{m-1} \le g(x)/t \qquad if \quad x \in \mathbf{R}, \ 0 < t < \tau$$

We may choose g *symmetric and increasing for* $x > 0$. *Moreover,* $g(x)/|x|^2 \to \frac{1-m}{2(m+1)}$ *as* $|x| \to \infty$.

Remark. Notice that for $0 \ge m > -1$, (1.3) is a sharper version of (0.4); in fact it implies a decay rate for u as $|x| \to \infty$ of order $|x|^{-2/(1-m)}$ or less.

On the other hand, (1.3) is exact for the explicit family of self-similar solutions

$$(1.4) \qquad w(x, t; C) = \left(\frac{t}{Ct^{\frac{2}{m+1}} + \frac{1-m}{2(m+1)}|x|^2} \right)^{\frac{1}{1-m}},$$

where $C \geq 0$ is arbitrary.

Proof. Let $u(0,\tau) = c > 0$. The right-hand inequality of (1.1), $u_t \leq u/((1-m)t)$ means that $u(x,t)t^{-1/(1-m)}$ is nonincreasing in t for every fixed $x \in \mathbf{R}$. In particular

$$(1.5) \qquad u(0,t) \geq u(0,\tau)(t/\tau)^{\frac{1}{1-m}} = c_1 t^{\frac{1}{1-m}}$$

for every $0 < t \leq \tau$. This is the desired estimate for $x = 0$. To obtain an estimate for $x > 0$ we consider in the half strip $\mathbf{R}^+ \times (0,\tau)$ the function

$$(1.6) \qquad w_0(x,t) = \left\{ \frac{2(m+1)t}{(1-m)(x+d)^2} \right\}^{\frac{1}{1-m}}, \; d > 0$$

obtained from (1.4) by putting $C = 0$ and displacing the x-origin; w_0 is a solution of (0.1) which satisfies (0.4) as $x \to \infty$ and takes on initial value $w_0(x,0) = 0$ for $x > 0$. Moreover, $w_0(0,t) \leq u(0,t)$ for $0 \leq t \leq \tau$ if

$$d^2 \geq \frac{2(m+1)}{(1-m)c_1^{1-m}}.$$

Choosing d in this way we obtain by a version of the Maximum Principle [ERV] that $u(x,t) \geq w_0(x,t)$ for $x > 0$, $0 < t \leq \tau$, which proves (1.3) for $x > 0$ with $g(x) = A(x+d)^2$. A similar argument works for $x \leq 0$. ♯

Thanks to Lemma 2 we can define the bound

$$(1.7) \quad K = K(u; R_0, \tau) \equiv \sup\{u(x,t)^{m-1}t/|x|^2 : 0 < t < \tau, |x| \geq R_0\}$$

Clearly K is finite and tends to $(1-m)/2(m+1)$ as $R_0 \to \infty$ (for fixed u and τ).

Our next result is an integral form of the Harnack inequality which controls the variation in time of the mass contained in finite x-intervals. The present version improves over Lemmas 6.1 and 6.2 of [ERV] by using estimate (1.3).

Lemma 3. *Let* $-1 < m < 0$. *For every good solution* u *and every* $\tau \in (0,T)$ *there exists a constant* $C = C(K,m) > 0$ *such that*
(1.8)

$$\left(\int_{-R}^{R} u(x,t)dx \right)^{1+m} \leq \left(\int_{-2R}^{2R} u(x,s)dx \right)^{1+m} + CR^{-\frac{(1+m)^2}{1-m}}|t-s|^{\frac{1+m}{1-m}}$$

for every $R \geq R_0 > 0$ *and* $0 < s,t \leq \tau$.

Remark. The improvement over the results of [ERV] consists in the fact that C does *not* blow up for small s and t, though it may depend on

u, R_0 and τ through K defined in (1.7). Moreover, (1.8) holds both for $s < t$ and $s > t$, while a similar formula, (6.7) of [ERV], holds only for $t > s = 0$. Formula (1.8) will be essential in the proof of existence of initial traces given in the next section.

Similar Harnack inequalities are true for $0 < m < 1$[HP], but not for $m > 1$ (slow diffusion) where the condition $t < s$ is essential, [AC].

Proof. As in [ERV, Lemma 6.1] we multiply (0.1) by $\phi \in C_0^\infty(\mathbf{R})$, $\phi \geq 0$ and integrate to obtain

$$\frac{d}{dt} \int \phi(x) u(x,t) dx = \int \phi_{xx}(x) \frac{u^m(x,t)}{m} dx.$$

Setting $n = -m \in (0,1)$ and using Hölder's inequality we get

$$\left| \frac{d}{dt} \int \phi u\, dx \right| \leq \frac{1}{n} \left(\int \phi u\, dx \right)^n \left(\int \frac{|\phi_{xx}|^\beta}{\phi^{\beta n}} u^{2m\beta} dx \right)^{1+m},$$

where $\beta = (1+m)^{-1}$. Calling the last integral $I(t)$ and integrating between s and t gives
(1.9)
$$\left(\int \phi(x) u(x,t) dx \right)^{1+m} \leq \left(\int \phi(x) u(x,s) dx \right)^{1+m} + C(m) \left(\int_s^t I(\theta)^{1+m} d\theta \right)$$

In order to estimate $I(t)$ we put $\phi(x) = \phi_0(x/R)^k$ for some $k \geq 2/(1+m)$ and a cutoff function $\phi_0 \in C_0^\infty(\mathbf{R})$ such that $\phi_0(x) = 1$ for $|x| \leq 1$, $\phi_0(x) = 0$ for $|x| \geq 2$. Thanks to estimate (1.3)

$$I(t)^{1+m} \leq C(m) t^{-\frac{2m}{m-1}} R^{m-1} g(2R)^{\frac{2m}{m-1}} \left(\int \frac{|(\phi_0^k)''|^\beta}{\phi_0^{\beta n k}} dx \right)^{1+m}.$$

Since the last integral is finite we obtain from (1.9)

$$\left| \left(\int \phi(x) u(x,t) dx \right)^{1+m} - \left(\int \phi(x) u(x,s) dx \right)^{1+m} \right|$$
$$\leq C(m) h(R) \left| t^{\frac{1+m}{1-m}} - s^{\frac{1+m}{1-m}} \right|$$

with

$$h(R) = R^{m-1} g(2R)^{\frac{2m}{m-1}} = O(R^{-\frac{(1+m)^2}{1-m}})$$

for large R. From this (1.8) follows. $\sharp$

Remarks. 1) Raising (1.8) to the power $1/(1+m) > 1$ we obtain a version that can be easier to manage

$$(1.10) \qquad \int_{-R}^R u(x,t) dx \leq C \left(\int_{-2R}^{2R} u(x,s) dx + R^{-\frac{1+m}{1-m}} |t-s|^{\frac{1}{1-m}} \right)$$

2) As $R \to 0$ we have $g(R) \approx c > 0$ and instead of (1.8) we obtain

$$\left(\int_{-R}^{R} u(x,t)dx\right)^{1+m} \le \left(\int_{-2R}^{2R} u(x,s)dx\right)^{1+m}$$

$$(1.11) \qquad\qquad + CR^{-(1-m)}|t-s|^{\frac{1+m}{1-m}}$$

3) In case we have two good solutions $u_1 \ge u_2 > 0$ in a strip $\mathbf{Q}_T$ we obtain by the same method of the difference $u_1 - u_2$:

$$\left(\int_{-R}^{R} u(x,t) - u_2(x,t))dx\right)^{1+m}$$

$$(1.12) \le \left(\int_{-2R}^{2R} (u_1(x,s) - u_2(x,s))dx\right)^{1+m} + CR^{-\frac{(1+m)^2}{1-m}}|t-s|^{\frac{1+m}{1-m}}$$

valid for $R \ge 1$, $0 < s, t \le \tau < T$ with C depending on m and $K(u_2)$. This estimate improves Lemma 6.1 of [ERV]. We note in passing that formula (6.5) of [ERV] is slightly incorrect, all its terms should be raised to the power $1 + m$ (otherwise a constant is necessary to multiply the second member as in formula (1.10).

The Harnack inequality for $m = 0$ is as follows:

Lemma 4. *Let u be a good solution of $u_t = (\log u)_{xx}$ in $\mathbf{Q}_T$ and let $0 < \tau < T$ and $0 < r < 1/2$. Then there exists a constant $C > 0$ such that*

$$\left(\int_{-R}^{R} u(x,t)dx\right)^{1-r}$$

$$(1.13) \qquad \le \left(\int_{-2R}^{2R} u(x,s)dx\right)^{1-r} + C(R^{-1}|t-s|)^{1-r}$$

for every $R \ge 1$ and every $0 < s, t \le \tau$ with $C = C(K, m, r)$.

Proof. (i) We assume in a first step that u is smooth down to $t = 0$ and prove (1.13) for $R = 1$, $s + 0$ and $t = 1$. As above the proof begins by multiplying the equation by $\phi \in C_0^\infty(\mathbf{R})$, $\phi \ge 0$ and integrating in x to get

$$\frac{d}{dt}\int \phi u \, dx = \int \phi_{xx} \log u \, dx$$

The presence of the *log* prevents using Hölder's inequality on the right-hand member. Therefore, we replace the log by means of the formula $\log(z) \le c(z-1)^r$, valid for $z \ge 1$ and $0 < r < 1$ with $c = c(r)$. In this way the right-hand member above can be estimated as

$$C\int_{u>1} |\phi_{xx}|(u-1)^r dx + C\int_{u<1} |\phi_{xx}|(\frac{1}{u} - 1)^r dx = I_1 + I_2.$$

We now fix $\phi(x) = \phi_0(x)^k$ with $k \geq 2/(1-r)$ and ϕ_0 as in Lemma 3 and take $r < 1/3$. We have

$$I_1 \leq C(\int \phi u\, dx)^r (\int |\phi_{xx}|^q \phi^{-rq} u^{-2rq} dx^{1-r}$$

$$\leq C'(\int \phi u\, dx)^r (\int |\phi_{xx}|^q \phi^{-rq} |x|^2 dx)^{1-r} t^{-2r}$$

$$\leq C'(\int \phi u\, dx)^r t^{-2r},$$

where C' depends not only on m and r but also on the bound $K(u,1,1)$. Inserting these estimates into (1.14) we get

$$(1.15) \qquad \left| (\int \phi u\, dx) \right| \leq (\int \phi u\, dx)^r (C + C' t^{-2r})$$

Integration of (1.15) between 0 and 1 gives

$$(1.16) \qquad \left| (\int \phi(x) u(x,1) dx)^{1-r} - (\int \phi(x) u(x,0) dx)^{1-r} \right| \leq C'$$

Since $\phi = 0$ for $|x| \geq 2$, $\phi = 1$, for $|x| \leq 1$, (1.13) follows in this case.

(ii) For the general case let us assume e.g. that $0 < s_0 < t_0 < T$. We perform the transformation

$$(1.17) \qquad \hat{u}(x,t) = (R^2/t_1) u(xR, t_1 t + s_0)$$

with $R > 1$ and $t_1 = t_0 - s_0$ which shifts the origin of time to s_0, stretches both space and time and produces a new solution $\hat{u}$ of (0.1) defined in a strip $\mathbf{Q}_{\hat{T}}$, $\hat{T} = (T - s_0)/t_1$. The bound K for $t/(u|x|^2)$, $|x| \geq 1$, is not increased by this transformation as it is easily verified. We apply (1.16) to $\hat{u}$ with $t = (t_0 - s_0)/R^2$ and get with $y = xR$

$$\left| (\int \phi(y/R) u(y,t_0) dy)^{1-r} - (\int \phi(y/R) u(y,s_0) dx)^{1-r} \right| \leq C' |t_1/R|^{1-r}. \ \sharp$$

We end this section with an L^∞ estimate

Lemma 5. *For any solution of* (0.1), (0,2) *we have*

$$(1.18) \qquad u(x,t) \leq C \left(t^{-\frac{1}{m+1}} (\int_{x-R}^{x+R} d\mu) + (\frac{t}{R^2})^{\frac{1}{1-m}} \right)$$

where $C = C(m) > 0$.

Proof. The result has been proved in [ERV, Lemma 6.3] for solutions with initial data in $L^1_{loc}(\mathbf{R})$. For the case where the initial trace is a measure we need only apply [ERV]'s result with origin of time $\tau > 0$ and then let $\tau \to 0$. ♯

Remark. Clearly this estimate is true for all solutions of (0.1), not only good solutions. The same happens with the forward L^1_{loc}-estimate established in [ERV, Lemma 6.2].

2. The Initial Trace

We are now in a position to establish that every good solution has an initial trace $t = 0$ which happens to be a measure.

Theorem 1. *Let u be a good solution of (0.1) in a strip $\mathbf{Q}_T$. Then there exists a locally finite, nonnegative Borel measure, μ such that*

$$(2.1) \qquad \lim_{t \to 0} \int u(x,t)\phi(x)dx = \int \phi(x)d\mu(x)$$

for every test function $\phi \in C_0(\mathbf{R})$.

Proof. By virtue of Lemma 3 for $m < 0$ or Lemma 4 for $m = 0$ the mass contained in a finite interval, $\int_{-R}^{R} u(x,t)dx$, remains bounded as $t \to 0$. Therefore, there is a subsequence $t_j \to 0$ such that $u(\cdot, t_j)$ converges in the vague topology $\sigma(\mathcal{M}, C_0)$ to a Borel measure μ, i.e. (2.1) holds for $t = t_j \to 0$.

In order to show that this initial trace is unique we will use a slight variation of the above Lemmas. Let, for instance, $m < 0$ and suppose that along another subsequence $s_j \to 0$ we obtain a second limit $\tilde{\mu}$. We take in (1.9) a cutoff function ϕ such that $\phi(x) = 1$ for $|x| \leq R$ and $\phi(x) = 0$ for $|x| \geq (1 + \epsilon)R$ with $\phi_{xx} \leq 1/\epsilon^2 R^2$ and proceed as in the proof of Lemma 3 to obtain

$$\left(\int_{-R}^{R} u(x,t)dx\right)^{1+m} \leq \left(\int_{-(R+\epsilon)}^{R+\epsilon} u(x,s)dx\right)^{1+m}$$
$$(2.2) \qquad\qquad + C\epsilon^{-2}R^{-\frac{(1+m)^2}{1-m}}|t - s|^{\frac{1+m}{1-m}}$$

Letting now $t = t_i$, $s = s_j$ and t_i, $s_j \to 0$ we obtain

$$\int_{-R}^{R} d\mu \leq \int_{-(R+\epsilon)}^{R+\epsilon} d\tilde{\mu}$$

and as $\epsilon \to 0$ we get

$$\int_{-R}^{R} d\mu \leq \int_{-R}^{R} d\tilde{\mu}$$

for every $R > 0$, and the same identity holds in any finite interval $I = [a, b]$ since the equation is invariant under translations. Therefore $\mu \leq \tilde{\mu}$. Likewise $\tilde{\mu} \leq \mu$, hence $\mu = \tilde{\mu}$ and the trace is uniquely determined.

The case $m = 0$ is similar. ♯

3. Existence

We show the following result:

Theorem 2. *For every $\mu \in \mathcal{M}_+(\mathbf{R})$ there exists a good solution of equation (0.1) defined in $\mathbf{Q} = \mathbf{R} \times (0, \infty)$ whose initial trace is μ.*

Proof. *Step 1.* We begin with the case where μ is a finite Borel measure in $\mathbf{R}$.

Let $\mu_n = \mu * \rho_n$, with $\rho_n = n\rho(nx)$ and $\rho \in C^\infty(\mathbf{R})$ is a typical mollifier, nonnegative, symmetric, supported in the ball of radius 1 and with mass 1. Then $\mu_n \in L^1(\mathbf{R})$, $\int \mu_n dx = \int d\mu$ and $\mu_n \to \mu$ in the topology of $\sigma(\mathcal{M}, C_0(\mathbf{R}))$.

We denote by u_n the maximal solution of (0.1) with initial datum μ_n. According to [ERV] the family $\{u_n\}$ is uniformly bounded in $L^\infty(0, T; L^1(\mathbf{R}))$ (the mass is conserved). It also satisfies

$$(3.1) \qquad \int_{-R}^{R} u(x, t) dx \leq \int_{-2R}^{2R} \mu_n(x) dx + C_m R^{-\alpha/\beta} t^\alpha.$$

so that the u_n's are uniformly bounded in $L^\infty_{loc}(0, \infty; L^1_{loc}(\mathbf{R}))$. By Lemma 5 they are uniformly bounded in $L^\infty(\mathbf{R} \times (\tau, \infty))$ for any $\tau > 0$. Moreover, the estimates (1.1) and (1.2) are satisfied and integration of the former gives for every fixed $t > 0$

$$(3.2)$$

$$\|(u_n^m)_x\|_\infty \leq \int |mu_{n,t}| dx \leq K \int u_n(x, t) dx \leq K \int \mu_n dx = K \int d\mu$$

It follows that (along some subsequence) u_n converges to a function $u \in L^\infty_{loc}(\mathbf{Q})$ which satisfies (3.1) and (3.2) with μ instead of μ_n. In particular, (3.2) and (1.1) imply that u is bounded away from 0 uniformly on compact subdomains of $\mathbf{Q}$ so that, by standard quasilinear theory, u is a C^∞ solution of (0.1). On the other hand, (1.1) and (1.2) imply that it is in fact a good solution.

Let us next check its initial values. Arguing as in Lemma 2, we have from (0.1)

$$(3.3)$$

$$\int [u_n(x, t) - \mu_n(x)]\phi(x) dx \leq \int_0^t |(1/m)u_n^m \phi_{xx}| \, dx \, dt \leq C(m, \phi) t^{\frac{1}{1-m}}$$

where ϕ is a smooth cutoff function. Letting $n \to \infty$ and then $t \to 0$ we obtain

$$(3.4) \qquad \lim_{t \to 0} \int u(x, t)\phi(x) dx = \int \phi(x) d\mu$$

which ends the proof in this case.

Step 2. Given a locally finite Borel measure μ which is not finite in $\mathbf{R}$ we put $\mu_n = \psi_n \mu$ with $\psi_n = \psi(|x|/n)$ and $\psi \in C^\infty(\mathbf{R})$ a standard nonnegative cutoff function supported in the unit interval and equal to 1 for $|x| \leq 1/2$. With this definition the μ_n's form a nondecreasing sequence of finite measures which converges to μ as $n \to \infty$.

Let u_n be the good solution to problem (0.1), (0.2) with initial datum μ_n constructed in Step 1. The u_n's form a nondecreasing sequence of functions in $L^\infty_{loc}(0, \infty; L^1_{loc}(\mathbf{R}))$ and we may pass to the limit $n \to \infty$ to obtain a function u in the same space. It is easily shown that u is a good solution to (0.1) which takes on μ as initial trace and satisfies the inequalities (1.1), (1.2), (3.1) and (3.2). ♯

Uniqueness

We show in this section that a good solution is uniquely determined by its initial trace.

Theorem 3. *Let u and v be two good solutions of* (0.1) *in* $\mathbf{Q}_T$, $T > 0$, *such that*

$$(4.1) \qquad \lim_{t \to 0} \int (u(x,t) - v(x,t))\phi(x)dx = 0$$

for every $\phi \in C_0^\infty(\mathbf{R})\, \phi \geq 0$. Then $u = v$ identically in $\mathbf{Q}_T$.

Our proof follows ideas introduced by Pierre [P] and Dahlberg and Kenig [DK1] for the porous medium equation, combined with the estimates of Section 1. We begin by establishing the following version of the Maximum Principle.

Lemma 6. *Let u and v be two good solutions of* (0.1) *in* $\mathbf{Q}_T$ *and assume that for every $R > 0$*

$$(4.2) \qquad \lim_{t \to 0} \int_{|x| \leq R} |u(x,t) - v(x,t)|\, dx = 0.$$

Then $u = v$ in $\mathbf{Q}_T$.

Proof. Since both u and v are smooth we can apply Kato's inequality and obtain

$$(4.3) \qquad \frac{\partial}{\partial t}|u - v| - \frac{\partial^2}{\partial x^2}\left|\frac{u^m}{m} - \frac{v^m}{m}\right| \leq 0$$

Let $\psi \in C_0^\infty(\mathbf{R})$, $0 \leq \psi \leq 1$. Multiplying (4.3) by ψ and integrating by parts give

$$\frac{d}{dt}\left(\int \psi|u - v|(t)dx\right) \leq |1/m| \int |\psi_{xx}||u^m - v^m|(t)dx$$

$$\leq C\left(\int \psi|u-v|(t)dx\right)^{-m}\left(\int (|\psi_{xx}|u^m v^m \psi^m)^{\frac{1}{m+1}}dx\right)^{m+1}.$$

Integration of this expression with respect to t between 0 and t produces an inequality similar to (1.8), from which we get by the same methods

$$\int_{-R}^{R} |u - v|(t)\,dx \leq C\left(\int_{-2R}^{2R} |u - v|(0)\,dx + R^{\frac{1+m}{1-m}} t^{\frac{1}{1-m}} \right) \leq C R^{\frac{1+m}{1-m}} t^{\frac{1}{1-m}}$$

Let now $R \to \infty$ for fixed $t > 0$ to end the proof of our Lemma. ♯

The next step consists in proving that solutions which are bounded below away from zero are unique.

Lemma 7. *Let u and v be solutions of (0.1) in $\mathbf{Q}_T$ such that $u, v \geq \epsilon$ for some constant $\epsilon > 0$. Then if they have the same initial trace they are identical.*

Proof. Let $u_\epsilon = u - \epsilon$, $v_\epsilon = v - \epsilon$ and $\Phi_\epsilon = \Phi(s + \epsilon) - \Phi(\epsilon)$ where $\Phi(s) = s^m/m$ if $m < 0$, $\Phi(s) = \log(s)$ if $m = 0$. Then u_ϵ and v_ϵ satisfy

$$u_t = (\Phi_\epsilon(u))_{xx}$$

Moreover, $\Phi_\epsilon(u_\epsilon(\cdot, t))\,\Phi_\epsilon(v_\epsilon(\cdot, t)) \in L^1(\mathbf{R})$ and both have the same trace as $t \to 0$. We may in these circumstances apply Pierre's uniqueness result [P] to conclude that $u_\epsilon = v_\epsilon$, hence $u = v$. ♯

Proof of Theorem 3. (i) Since u is a good solution, by Theorem 1 it has an initial trace μ which is a locally finite and nonnegative Borel measure. Following [DK1] we let u_j be the unique maximal solution of (0.1) with initial data

$$(4.4) \qquad\qquad \epsilon + u(x, \frac{1}{j}).$$

By the Maximum Principle, which holds for those solutions (cf. [ERV]), we have

$$(4.5) \qquad u_j(x, t) \geq \epsilon, \qquad u_j(x, t) \geq u(x, t + \frac{1}{j})$$

Due to the L^∞-bound, Lemma 5, the u_j's are uniformly bounded on compact subsets of $\mathbf{Q}$, and since they also satisfy uniform local estimates for their derivatives, after possibly passing to a subsequence they converge uniformly on compact subsets of $\mathbf{Q}$ to a function $w(x, t)$ which satisfies

$$(4.6) \qquad w(x, t) \geq \epsilon \qquad \text{and} \qquad w(x, t) \geq u(x, t)$$

(ii) Next we check that

$$(4.7) \qquad \int w(x, t)\phi(x)\,dx \to_{t \to 0} \int \phi(x)(d\mu + \epsilon dx)$$

for every $\phi \in C_0^\infty(\mathbf{R})$. To that effect let $h \in C_0^\infty(\mathbf{R})$ be such that $h(t) = 1$ if $t \approx 0$, while $h(t) = 0$ if $t > T/2$ and let τ, s be such that $0 \le \tau \le T/2 < s < T$ and $h(\tau) = 1$. Then

$$-\int w(x,\tau)\phi(x)dx = \int \int_\tau^s (\frac{w^m}{m}\psi_{xx} + w\psi_t)dx\,dt$$

where $\psi(x,t) = \phi(x)h(t)$. If we now set

$$(4.8) \qquad A_j(\rho) = \int \int_0^\rho (\frac{u_j^m}{m}\psi_{xx} + u_j\psi_t)dx\,dt$$

we have

$$\int w(x,\tau)\phi(x)dx = \lim_{j\to\infty}(A_j(s) - A_j(\tau)).$$

Since

$$A_j(s) = -\int (u(x,j^{-1}) + \epsilon)\phi(x)dx \to_{j\to\infty} -\int \phi(x)d\mu + \epsilon\phi(x)dx$$

we will arrive at (4.7) if we prove that $A_j(\tau) \to 0$ as $\tau \to 0$ uniformly in j. In order to obtain this convergence we first observe that the last term in (4.8) disappears since $\psi_t = 0$ for $0 < t < \tau$ if τ is small enough.

On the other hand, since u is a good solution and using (4.5) we estimate the first term on the right of (4.8) using (4.5) and Lemma 2 as follows:

$$\int_0^\tau \int_{-R}^R u_j^m(x,t)dx\,dt \le \int_0^\tau \int_{-R}^R (u(x,t+j^{-1}))^m dx\,dt$$
$$\le C(m,R)\int_0^\tau (t+j^{-1})^{-\frac{m}{m-1}}dt \le C(m,R)\tau^{\frac{1}{1-m}}$$

It follows that $A_j(\tau) \to 0$ as $\tau \to 0$ uniformly in j and thus (4.7) holds. We have proved that $w(x,t)$ is a solution of (0.1) which satisfies $w(x,t) \ge \epsilon$ and satisfies the initial condition (4.7), which can be loosely written as $u(x,0)dx = d\mu + \epsilon\,dx$. By Lemma 7, w is uniquely determined by its initial trace, i.e. $w = w(\mu, \epsilon)$.

(iii) To end the proof of the Theorem we will show that $w(\epsilon, \mu)$ decreases to u as $\epsilon \to 0$. Since the solution v has the same trace as u, by step (ii) it generates the same function w, and consequently w will also tend to v, hence $u = v$ and we are done.

Let us take for simplicity $w_k = w(k^{-1}, \mu)$, $k = 1, 2, \ldots$. The sequence $\{w_k\}$ is nonincreasing and by construction $w_k \ge w_{k+1} \ge u$. Hence, there is a limit $w^* = \lim_{k \to \infty} w_k$ and $w^* \geq u$. We need to show that $w^* = u$. For that we take $\psi \in C_0(\mathbf{R})$, $\psi \geq 0$ and compute

$$\int \phi(x)d\mu = \lim_{t \to 0} \int u(x,t)\phi(x)dx$$

$$\leq \lim \int w^*(x,t)\phi(x)dx \leq \lim_{t \to 0} \int w_k(x,t)\phi(x)dx$$

$$= \int \phi(x)(d\mu + \frac{1}{k}dx) = \int \phi(x)d\mu + \frac{C}{k}.$$

This shows that u and w^* have the same initial trace. Since moreover $w^* \geq u$, this is equivalent to saying that

$$\int_{-R}^{R} |u(x,t) - w^*(x,t)|dx \to 0$$

as $R \to 0$ for every fixed $R > 0$. Lemma 6 implies now that $u = w^*$.   ♯

Good Solutions are Maximal Solutions

We end our study with the following result:

Theorem 4. *For every $\mu \in \mathcal{M}_+(\mathbf{R})$ there exists a unique maximal solution of problem* (0.1), (0.2) *which is characterized by condition* (0.4).

Proof. Let u be any (positive and smooth) solution of (0.1) in $\mathbf{Q}_T$ with initial trace μ. For $\tau > 0$ we consider the unique maximal solution $U^\tau(x,t)$, $t \geq \tau$, with initial data $U^\tau(x,\tau) = u(x,\tau)$ which exists by the results of [ERV]. By Lemmas 3 and 4 the family $\{U^\tau\}$ is uniformly locally bounded on compact subsets of $\mathbf{Q}$. Using also the estimates for derivatives and passing if necessary to a subsequence, we know that U^τ converges as $\tau \to 0$ to a function $U = U(x,t)$ which satisfies

$$U(x,t) \geq u(x,t) \qquad \text{for every} \quad t > 0$$

It is also clear that U will satisfy the estimates (1.1), (1.2), so that it is a good solution of (0.1). Finally, arguing as in the uniqueness proof (Theorem 3, part (ii)) we conclude that

$$\lim_{t \to 0} \int U(x,t)\phi(x)dx = \int \phi(x)d\mu$$

for every $\phi \in C_0(\mathbf{R})$. This proves that U, the unique good solution of (0.1) with initial data μ, is the maximal solution of problem (0.1), (0.2).   ♯

REFERENCES

[AC] D.G. Aronson and L.A. Caffarelli, *The initial trace of a solution of the porous medium equation*, Trans. Amer. Math. Soc. **280** (1983), 351–366.

[BCP] Ph. Bénilan, M.G. Crandall and M. Pierre, *Solutions of the PME in $\mathbf{R}^N$ with optimal conditions on the initial data*, Indiana Univ. Math. Jour. **33** (1984), 51–87.

[DK1] D.E.J. Dahlberg and C.E. Kenig, *Nonnegative solution of the porous medium equation*, Comm. Partial Diff. Eqns. **9** (1984), 265–293.

[DK2] D.E.J. Dahlberg and C.E. Kenig, *Nonnegative solutions to fast diffusions*, Revista Mat. Iberoamericana **4**(1988), 11–29.

[ERV] J.R. Esteban, A. Rodríguez and J.L. Vázquez, *A nonlinear heat equation with singular diffusivity*, Comm. Partial Diff. Eqns. **13** (1988), 985–1039.

[HP] M.A. Herrero and M. Pierre, *The Cauchy problem for $u_t = \Delta u^m$ when $0 < m < 1$*, Trans. Amer. Math. Soc. **291** (1985), 145–158.

[P] M. Pierre, *Uniqueness of solutions of $u_t - \Delta\phi(u) = 0$ with initial datum a measure*, Nonlinear Analysis T.M. A. **6** (1982), 175–187.

[RV] A. Rodríguez and J.L. Vázquez, *A well-posed problem in singular Fickian diffusion*, Archive Rat. Mech. Anal., to appear.

[V] J.L. Vázquez, *Nonexistence of finite-mass solutions for fast-diffusion equations*, to appear J. Math. Pures Appl..

A. Rodríguez
E.T.S. Arquitectura
Universidad Politécnica de Madrid
28040 Madrid, Spain

J.L. Vázquez
Departamento de Matemáticas
Universidad Autónoma de Madrid
28049 Madrid, Spain

The Evolution of Harmonic Maps: Existence, Partial Regularity, and Singularities

MICHAEL STRUWE

Abstract

We survey recent results on global existence and partial regularity for the heat flow for harmonic maps. A new proof of short-time blow-up in dimensions $m \geq 3$ is given.

1. Let M and N be compact Riemannian manifolds of dimensions m and ℓ and with metrics γ and g, respectively. We may assume that N is isometrically embedded in some Euclidean $\mathbb{R}^n$.

For C^1-maps $u : M \to N \subset \mathbb{R}^n$ let

$$(1.1) \qquad e(u) = \frac{1}{2}|\nabla u|^2 = \frac{1}{2}\sum_{\alpha,\beta=1}^{m}\sum_{i=1}^{n}\gamma^{\alpha\beta}\frac{\partial}{\partial x_\alpha}u^i\frac{\partial}{\partial x_\beta}u^i$$

be the energy density and - with $dM = \sqrt{\det(\gamma_{\alpha\beta})}dx$ - let

$$E(u) = \int_M e(u)dM$$

be the energy of u. A map u is harmonic if E is stationary at u, which for $u \in C^2$ is equivalent to the condition that the vector $(\Delta_M u)(x)$ at all points $x \in M$ is orthogonal to the tangent space $T_{u(x)}N$ to N at $u(x) \in \mathbb{R}^n$; that is,

$$(1.2) \qquad \Delta_M u \perp T_u N.$$

(Δ_M is the Laplace-Beltrami operator on M.) In local coordinates on N, (1.2) takes the form

$$(1.3) \qquad \Delta_M u = \Gamma(u)(\nabla u, \nabla u)$$

with a bilinear form Γ involving the Christoffel symbols of the metric g on N.

The notion of harmonic map generalizes the concept of closed geodesic to higher dimensions.

In their pioneering work on harmonic maps, Eells and Sampson [8] also introduce the evolution problem

$$(1.4) \qquad \partial_t u - \Delta_M u \perp T_u N \quad \text{on } M \times]0, \infty[$$

with initial condition

$$(1.5) \qquad u = u_0 \quad \text{at } t = 0.$$

Upon multiplying (1.4) by $\partial_t u \in T_u N$ and integrating by parts, we at once obtain the energy inequality

$$(1.6) \qquad E\big(u(T)\big) + \int_0^T \int_M |\partial_t u|^2 dM\, dt \le E(u_0), \quad \forall T > 0$$

for any solution u of (1.4), (1.5). That is, (1.4) is the L^2-gradient flow for E. In local coordinates again, (1.4) takes the form

$$(1.7) \qquad \partial_t u - \Delta_M u = \Gamma(u)(\nabla u, \nabla u).$$

Upon differentiation the latter expression and multiplying by ∇u, we moreover obtain the following Bochner-type diferential inequality for the energy density

$$(1.8) \qquad (\partial_t - \Delta_M)e(u) + C|\nabla^2 u|^2 \le K_N e(u)^2 + C_M e(u),$$

where K_N denotes an upper bound for the sectional curvature of $N, C > 0$, and C_M depends on (the Ricci curvature of) the metric γ. Since standard variational techniques fail, Eells and Sampson used the flow (1.4) to produce examples of harmonic maps:

Theorem 1 (Eells-Sampson [8]). *Suppose $K_N \le 0$, then for any smooth map $u_0 : M \to N$ problem $(1.4), (1.5)$ admits a unique, smooth solution $u(t)$ which, as $t \to \infty$ suitably, converges to a smooth harmonic map u_∞ homotopic to u_0.*

The curvature restriction $K_N \le 0$ is in some sense optimal as Eells and Wood [9] show that Theorem 1 ceases to be true for $M = T^2, N = S^2$ and an initial map u_0 of topological degree 1.

2. In two dimensions ($m = 2$) Lemaire [15] and Sacks-Uhlenbeck [17] independently showed that also the topological condition $\pi_2(N) = 0$ suffices to find harmonic representatives for all homotopy classes of maps $u_0 : M \to N$. A new proof of this result, using the evolution problem (1.4) was given by the author [21], based on the following extension of Theorem

1. (Since by the result of Eells and Wood [9] we must expect singularities, we consider also maps in the Sobolev space

$$H^{1,2}(M;N) = \left\{ u \in H^{1,2}(U;\mathbb{R}); \ u(M) \subset N \right\}$$

of measurable, finite energy maps $u : M \to N$; that is, with distributional derivative in L^2.)

Theorem 2 (Struwe [20; Theorem 4.2]). *Suppose $m = 2$. For any $u_0 \in H^{1,2}(M;N)$ there exists a global weak solution $u : M \times]0, \infty[\to N$ of $(1.4), (1.5)$ which satisfies (1.6) and is $(C^\infty-)$ regular away from finitely many points $(\bar{x}_k, \bar{t}_k)$, $1 \leq k \leq K$. The solution u is unique in this class.*

At a singularity $(\bar{x}, \bar{t})$ a "harmonic sphere" $\bar{u} : S^2 \cong \overline{\mathbb{R}^2} \to N$ separates in the sense that for suitable $x_m \to \bar{x}$, $R_m \searrow 0$, $t_m \nearrow \bar{t}$ we have

$$(2.1) \qquad u_m(x) := u\big(\exp_{x_m}(R_m x), t_m\big) \longrightarrow \tilde{u} \quad \text{in } H^{2,2}_{\text{loc}}(\mathbb{R}^2; N),$$

where $\tilde{u} \not\equiv \text{const.}$ is harmonic, has finite energy and extends to a smooth harmonic map $\bar{u} : S^2 \cong \overline{\mathbb{R}^2} \to N$.

Finally, as $t \to \infty$ suitably, $u(t) \rightharpoonup u_\infty$ weakly in $H^{1,2}(M;N)$ where $u_\infty : M \to N$ is smooth and harmonic. Convergence is strong away from finitely many points $(\bar{x}_\ell, \bar{t}_\ell = \infty)$ where harmonic spheres separate in the sense (2.1).

Theorem 2 was extended to 2-manifolds M with boundary $\partial M \neq \emptyset$ by Chang [1].

To this day it is not known whether in general (in two dimensions) the flow (1.4) will encounter singularities in finite time. (Of course, the result of Eells and Wood provides us with an example where either such singularities exist or the flow fails to converge asymptotically.) However, some recent results of Chang and Ding [2], respectively work by Grayson and Hamilton [11] lend support to the conjecture that in two dimensions the flow (1.4) does not develop singularities in finite time.

3. The situation is quite different in higher dimensions. Progress on the evolution problem (1.4) for $m \geq 3$ and general targets came as a consequence of a peculiar monotonicity formula for (1.4), discovered in [22; Lemma 3.2] for maps $u : \mathbb{R}^m \times [0, T] \to N$ and extended to curved domains by Chen-Struwe [5].

Let

$$(3.1) \qquad G_{z_0}(x, t) = \frac{1}{\sqrt{2\pi(t_0 - t)}} \exp\left(-\frac{|x - x_0|^2}{4(t_0 - t)}\right), \quad \text{if } t < t_0$$

be the fundamental solution to the heat equation on $\mathbb{R}^m \times \mathbb{R}$ with singularity at $z_0 = (x_0, t_0)$, and let $\varphi \in C_0^\infty(\mathbb{R}^m)$ be a smooth cut-off function such that $\varphi \equiv 1$ in a neighbourhood of 0 and such that the support of φ is contained in ball of radius ρ less that the injectivity radius ρ_M of M. For a solution $u : M \times [0, T[\to N$ and $R^2 < t_0 < T$ let

$$\Phi(R; z_0) = \frac{1}{2} R^2 \int_{B_\rho(x_0)} |\nabla u|^2 \varphi^2 G_{z_0} \, dx \big|_{t=t_0 - R^2},$$

in local normal coordinates around x_0 on M.

Theorem 3 (Chen-Struwe [5; Lemma 4.2]). *There exists a constant C depending only on M and N such that for any $T > 0$, any $0 < R^2 < R_0^2 \leq t_0 < T$ and any regular solution $u : M \times]0, T[\to N$ of (1.4) there holds*

$$(3.2) \qquad \Phi(R; z_0) \leq \exp\left(C(R_0 - R)\right) \Phi(R_0; z_0) + C E(u_0)(R_0 - R).$$

A particular consequence of the monotonicity formula (3.2) is the following.

Theorem 4 (Struwe [22; Theorem 5.1]; Chen-Struwe [5; Lemma 4.4]). *There exists a constant $\varepsilon_0 > 0$ depending only on M and N such that for any solution $u : M \times [0, T[\to N$ of (1.4) the following is true:*

If $\Phi(R; z_0) < \varepsilon_0$ for some $z_0 = (x_0, t_0)$, $0 < R^2 \leq t_0 < T, R < \iota_M$, then $|\nabla u(z_0)| \leq C$, with a constant $C = C(M, N, R)$.

Theorem 4 together with a penalization device to obtain approximate solutions for (1.4), due to Chen [3], Keller-Rubinstein-Sternberg [14], and Shatah [19] led to a global existence and partial regularity result for (1.4) in higher dimensions
$(m \geq 3)$.

Theorem 5 (Chen-Struwe [5; Theorem 1.5]). *For any (smooth) map $u_0 : M \to N$ there exists a global weak solution $u : M \times]0, \infty[\to N$ of (1.4), (1.5) satisfying (1.6) and regular off a set of co-dimension ≥ 2 (in the parabolic metric $\delta((x, t), (y, x)) = |x - y| + \sqrt{|s - t|}$). As $t \to \infty$ suitably, $u(t) \to u_\infty$ weakly in $H^{1,2}(M, N)$ where $u_\infty : M \to N$ is harmonic and regular off a set of co-dimension ≥ 2.*

Moreover, u satisfies a variant of the monotonicity formula (3.2). This fact was used by Coron [6] to prove that the solution obtained in Theorem 5 is in general not unique - even among partially regular solutions satisfying (1.6).

The estimate on the co-dimension of the singular set very likely can be improved to 3, as for energy-minimizing (stationary) harmonic maps; see Giaquinta-Giusti [10] and Schoen-Uhlenbeck [18]. However, as was first observed by Coron-Ghidaglia [7], in higher dimensions singularities may appear in finite time.

Subsequently, Chen and Ding [4] gave an argument relating singularities to the fact that in higher dimensions the infimum of the energy in certain non-trivial homotopy classes of maps may be 0, an observation due to B. White [23]. In fact, their reasoning can be considerably simplified by combining Theorem 4 with Moser's [16] weak Harnack inequality for parabolic equations. First note:

Theorem 6. *For any $T > 0$ there exists $\varepsilon_1 > 0$ depending only on T, M and N such that any smooth solution $u : M \times [0,T] \to N$ of $(1.4), (1.5)$ with $E(u_0) < \varepsilon_1$ can be extended to a global, smooth solution $u : M \times [0,\infty[\to N$, converging, as $t \to \infty$ suitably, to a constant harmonic map.*

Proof. Let $R_0^2 = \inf\{\iota_M^2, T\}$. For $0 < R_0^2 \le t_0 \le T$, $x_0 \in M$ we can estimate with constants C depending on M, N, and T only

$$(3.3) \qquad \Phi(R_0; z_0) \le C R_0^{2-m} E\big(u(t_0 - R_0^2)\big) \le C E(u_0) < \varepsilon_0,$$

if $\varepsilon_1 > 0$ is sufficiently small. Hence by Theorem 4 we have

$$|\nabla u(x,t)| \le C \text{ uniformly for } t \ge R_0^2, x \in M,$$

and u can be extended as a smooth solution of (1.4), (1.5) on $M \times [0,\infty[$. By (1.8) and Moser's [16] supremum-estimate for weak sub-solutions of linear parabolic equations, moreover we obtain

$$(3.4) \qquad |\nabla u(x,t)|^2 \le C E\big(u(t - R_0^2)\big) \le C E(u_0) \text{ for } t \ge 2R_0^2, x \in M.$$

From this uniform estimate, asymptotic convergence follows as in Eells-Sampson [8] or Jost [13]. Finally, if $\varepsilon_1 > 0$ is sufficiently small, by (3.4) the image of any map $u(t), t \ge 2R_0^2$, and hence also of the limiting harmonic map u_∞ is contained in a strictly convex geodesic ball on N. It follows that $u_\infty \equiv$ const.; see Jäger-Kaul [12].

$$\text{Q.E.D.}$$

By Theorem 6, of course, for homotopically non-trivial initial data u_0 with $E(u_0) < \varepsilon_1(T)$ the flow (1.4), (1.5) must blow up before time T. In fact, blow-up time approaches 0 as the initial energy decreases to 0.

Finally, we remark that in dimensions $m \ge 3$ singularities - as in the case $m = 2$ - may be related to harmonic spheres or to self-similar solutions $u(x,t) = w\left(\frac{x-x_0}{\sqrt{t_0-t}}\right)$ of (1.4); see Struwe [22; Theorem 8.1]. (The work of Coron-Ghidaglia [7] strongly suggests that solutions of the latter kind in dimensions $m \ge 3$ actually exist.)

REFERENCES

[1] K.-C. Chang, "Heat flow and boundary value problem for harmonic maps", preprint (1988).

[2] K.-C. Chang and W.-Y. Ding, "A result on global existence for heat flows of harmonic maps from D^2 into S^2", preprint (1989).

[3] Y. Chen, "Weak solutions to the evolution problems of harmonic maps", Math. Z.

[4] Y. Chen and W.-Y. Ding, "Blow-up and global existence for heat flows of harmonic maps", preprint.

[5] Y. Chen and M. Struwe, "Existence and partial regularity results for the heat flow for harmonic maps", Math. Z. 201 (1989), 83-103.

[6] J.-M. Coron, "Nonuniqueness for the heat flow of harmonic maps", preprint (1988).

[7] J.-M. Coron and J.-M. Ghidaglia, "Explosion en temps fini pour le flot des applications harmoniques", preprint (1988).

[8] J. Eells and J.H. Sampson, "Harmonic mappings of Riemannian manifolds", Amer. J. Math. 86 (1964), 109-160.

[9] J. Eells and J.C. Wood, "Restrictions on harmonic maps of surfaces", Topology 15 (1976), 263-266.

[10] M. Giaquinta and E. Giusti, "On the regularity of the minima of variational integrals", Acta Math. 148 (1982), 31-46.

[11] M. Grayson and R.S. Hamilton, "The formation of singularities in the harmonic map heat flow", preprint.

[12] W. Jäger and H Kaul, "Uniqueness and stability of harmonic maps, and their Jacobi fields", manusc. math. 28 (1979), 269-291.

[13] J. Jost, "Ein Existenzbeweis für harmonische Abbildungen, die ein Dirichletproblem lösen, mittels der Methode des Wärmeflusses", manusc. math. 34 (1981), 17-25.

[14] J. Keller, J. Rubinstein and P. Sternberg, "Reaction - diffusion processes and evolution to harmonic maps", preprint.

[15] L. Lemaire, "Applications harmoniques de surfaces riemanniennes", J. Diff. Geom. 13 (1978), 51-78.

[16] J. Moser, "A Harnack inequality for parabolic differential equations", Comm. Pure Appl. Math. 17 (1964), 101-134.

[17] J. Sacks and K. Uhlenbeck, "The existence of minimal immersions of 2-spheres", Ann. of Math. 113(1981), 1-24.

[18] R.S. Schoen and K. Uhlenbeck, "A regularity theory for harmonic maps", J. Diff Geom. 17 (1982), 307-335, and 18 (1983), 329.

[19] J. Shatah, "Weak solutions and the development of singularities of the $SU(2)\,\sigma$-model", Comm. Pure Appl. Math. 41 (1988), 459-469.

[20] M. Struwe, "On the evolution of harmonic maps of Riemannian surfaces", Comment. Math. Helv. 60 (1985), 558-581.

[21] M. Struwe, "Heat flow methods for harmonic maps of surfaces and applications to free boundary problems", Lect. Notes Math. 1324, 293-319, Springer (1988)

[22] M. Struwe, "On the evolution of harmonic maps in higher dimensions", J. Diff. Geom. 28 (1988), 485-502.

[23] B. White, "Infima of energy functionals in homotopy classes of mappings", J. Diff. Geom. 23 (1986). 127-142.

Michael Struwe
Mathematik
ETH-Zürich
Zürich, Switzerland

Two Dimensional Emden–Fowler Equation with Exponential Nonlinearity

TAKASHI SUZUKI

1. Introduction

Emden–Fowler equation with the exponential nonlinearity P:

$$-\Delta u = \lambda e^u \qquad (\text{in } \Omega) \tag{1}$$

$$u = 0 \qquad (\text{on } \partial\Omega), \tag{2}$$

arises in the theories of thermonic emission (Gel'fand [10]), isothermal gas sphere (Chandrasekhar [5]), and gas combustion (Mignot–Murat–Puel [17]), where λ is a positive constant and $\Omega \subset R^n$ is a bounded domain with smooth boundary $\partial\Omega$.

Up to now a lot of work has been done for P including higher dimensional cases. We can point out that the structure of the solution set is very senstitive to the domain Ω, its dimension, topology, and geometry. However, in the present paper we shall concentrate on two-dimensions, where the equation has fine structures of complex function theory and differential geometry.

The study may be divided into three parts: local, asymptotic, and global analysis, respectively.

Local analysis is to pick up mild solutions for P, of which character is close to the trivial solution $u = 0$ of $\lambda = 0$. In some sense the method has been established, based on implicit function theory, comparison principle, and so on.

From the work of Keller–Cohen [13], Fujita [9], Laetsch [14], Keener–Keller [12] and Crandall–Rabinowitz [7], we know the following, where $S_\lambda = \{u \in C^2(\Omega) \cap C^0(\overline{\Omega}) | u \text{ solves } P\}$ denotes the solution set for the given $\lambda > 0$:

Fact (i)   *There exists a $\overline{\lambda} \in (0, \infty)$ such that $S_\lambda = \emptyset$ and $S_\lambda \neq \emptyset$ for $\lambda > \overline{\lambda}$ and $0 < \lambda < \overline{\lambda}$, respectively.*

Fact (ii)   *The set S_λ has a minimal element $u = \underline{u}_\lambda$ whenever $S_\lambda \neq \emptyset$.*

Fact (iii)   *There exists no triple $\{u_1, u_2, u_3\} \subset S_\lambda$ satisfying $u_1 \leq u_2 \leq u_3$ and $u_1 \neq u_2 \neq u_3$.*

Fact (iv)   *Minimal solutions $\{(\lambda, \underline{u}_\lambda) | 0 < \lambda < \overline{\lambda}\}$ form a branch $\underline{L}$, i.e., one-dimensional manifold in the $\lambda - u$ plane starting from $(\lambda, u) = (0, 0)$.*

Fact (v)   *$\underline{L}$ continues up to $\lambda = \overline{\lambda}$ and then bends back.*

Fact (vi)   *$\underline{u}_\lambda$ for $\lambda \in (0, \overline{\lambda})$ is linearly stable, i.e., the first eigenvalue $\mu_1 = \mu_1(p, \Omega)$ of the linearized operator $A_p \equiv -\Delta_D(\Omega) - p$ is positive,* *where $-\Delta_D(\Omega)$ denotes the Laplacian $-\Delta$ in Ω with the Dirichlet condition $\cdot|_{\partial\Omega} = 0$ and $p = \lambda e^{\underline{u}\lambda}$. On the other hand $\mu_1(p,\Omega) = 0$ holds for $p = \lambda e^{\underline{u}\lambda}$ with $\lambda = \overline{\lambda}$.*

However, the variational method assures us of the existence of the second solution, where the necessity of other analyses is to be recognized (Crandall–Rabinowitz [7]):

Fact (vii) *For each $\lambda \in (0, \overline{\lambda})$, it holds that $S_\lambda \backslash \{u_\lambda\} \neq \emptyset$.*

The following fact, which will be proved later, suggests that those non-minimal solutions get to bear striking characters as $\lambda \downarrow 0$:

Fact (viii) *For each $\varepsilon > 0$, there exists a constant $C_\varepsilon > 0$ such that $\|u\|_{L^\infty(\Omega)} \leq C_\varepsilon$ for any $u \in S_\lambda$ with $\lambda \geq \varepsilon$.*

Asymptotic analysis is the first step to study nonminimal solutions and is divided into two parts. The first is to classify singular limits, that is, the limiting functions of the classical solutions for P as $\lambda \downarrow 0$. We have already had satisfactory features for these singular limits. In particular, any classical solutions either converge to the trivial solution $u = 0$, make finite point blow-up, or make the entire blow-up as $\lambda \downarrow 0$. We shall describe details in later sections.

The second part is to construct classical solutions close to those singular limits. This procedure is sometimes called the singular perturbation. Up to now, one-point blow-up solutions have been constructed for a generic simply-connected domain Ω by Weston [24] and Moseley [18].

Finally, a global analysis will reveal how those singular limits are connected globally with each other in the $\lambda - u$ plane. In later sections we shall propose a geometric argument to get an answer to the problem.

This paper is composed of four sections. Taking preliminaries in the following section, we shall perform asymptotic and global analysis in Sections 3 and 4 respectively.

2. Preliminaries: complex function theory, differential geomtery, and special functions

2.1. *Liouville integral*

First, we show that the equation

$$-\Delta u = \lambda e^u \qquad (\text{in } \Omega \subset \mathcal{R}^2) \tag{3}$$

has integral essentially discovered by Liouville [16].

To this end we take the complex variable $z = x_1 + \sqrt{-1}x_2$ and $\overline{z} = x_1 - \sqrt{-1}x_2$ for $x = (x_1, x_2) \in \Omega$ and introduce the function

$$s = u_{zz} - \frac{1}{2}u_z^2. \tag{4}$$

From (3) we have $u_{z\bar{z}} = -\frac{\lambda}{4}e^u$ and hence

$$s_{\bar{z}} = u_{zz\bar{z}} - u_z u_{z\bar{z}} = -\frac{\lambda}{4}e^u u_z + \frac{\lambda}{4}e^u u_z = 0.$$

Therefore, $s = s(z)$ is a holomorphic function of $z \in \Omega \subset C$.

Regarding (4) as a Riccati equation we see that $\phi = e^{-u/2}$ satisfies

$$\phi_{zz} + \frac{1}{2}s\phi = 0. \tag{5}$$

Taking a point $x^* = (x_1^*, x_2^*) \in \Omega$, we introduce the fundamental system of solutions $\{\phi_1, \phi_2\}$ of the linear ordinary equation (5) in z such that

$$\phi_1|_{z=z^*} = \frac{\partial}{\partial z}\phi_2|_{z=z^*} = 1 \quad \text{and} \quad \frac{\partial}{\partial z}\phi_1|_{z=z^*} = \phi_2|_{z=z^*} = 0, \tag{6}$$

where $z^* = x_1^* + \sqrt{-1}x_2^*$. Then, $\phi_1 = \phi_1(z)$ and $\phi_2 = \phi_2(z)$ become analytic functions of $z \in \Omega$ and the relation

$$\phi \equiv e^{-u/2} = \overline{f}_1(\bar{z})\phi_1(z) + \overline{f}_2(\bar{z})\phi_2(z) \tag{7}$$

holds for some functions $\overline{f}_1 = \overline{f}_1(\bar{z})$ and $\overline{f}_2 = \overline{f}_2(\bar{z})$ of $\bar{z}$.

We can show that

$$f_1(z) = C_1\phi_1(z) \quad \text{and} \quad f_2(z) = C_2\phi_2(z) \tag{8}$$

with

$$C_1 = e^{-u/2}|_{x=x^*} \quad \text{and} \quad C_2 = \frac{\lambda}{8}e^{u/2}|_{x=x^*} \tag{9}$$

if $x^* = (x_1^*, x_2^*) \in \Omega$ is taken to be a critical point of $u = u(x)$, where $f_1(z) = \overline{\overline{f}_1(\bar{z})}$ and $f_2(z) = \overline{\overline{f}_2(\bar{z})}$.

In fact we have

$$W(\phi_1, \phi_2) \equiv \phi_1\phi_{2z} - \phi_{1z}\phi_2 = 1$$

and hence

$$\overline{f}_1(\bar{z}) = W(\phi, \phi_2) = \phi\phi_{2z} - \phi_z\phi_2$$

and

$$\overline{f}_2(\bar{z}) = W(\phi_1, \phi) = \phi_1\phi_z - \phi_{1z}\phi,$$

which are independent of z. Putting $z = z^*$ we get

$$\overline{f}_1(\bar{z}) = \phi(z^*, \bar{z}) \quad \text{and} \quad \overline{f}_2(\bar{z}) = \phi_z(z^*, \bar{z}).$$

Since $\phi = e^{-u/2}$ is real, it solves also

$$\phi_{\bar{z}\bar{z}} + \frac{1}{2}\bar{s}\phi = 0 \tag{10}$$

for $\bar{s}(\bar{z}) = \overline{s(z)}$ and so do $\bar{f}_1(\bar{z}) = \phi(z^*, \bar{z})$ and $\bar{f}_2(\bar{z}) = \phi_z(z^*, \bar{z})$.

A fundamental system of solutions for (10) is given by $\{\bar{\phi}_1, \bar{\phi}_2\}$ with $\bar{\phi}_1(\bar{z}) = \overline{\phi_1(z)}$ and $\bar{\phi}_2(\bar{z})$ and $\overline{\phi_2(z)}$ satisfying

$$\bar{\phi}_1|_{\bar{z}=\bar{z}^*} = \frac{\partial}{\partial \bar{z}}\bar{\phi}_2|_{\bar{z}=\bar{z}^*} = 1 \quad \text{and} \quad \frac{\partial}{\partial \bar{z}}|_{\bar{z}=\bar{z}^*} = \bar{\phi}_2|_{\bar{z}=\bar{z}^*} = 0.$$

Furthermore we have

$$\bar{f}_1(\overline{z^*}) = \phi(z^*, \overline{z^*}) = e^{-u/2}|_{x=x^*} = C_1,$$

$$\frac{\partial \bar{f}_1}{\partial \bar{z}} = \phi_{\bar{z}}(z^*, \overline{z^*}) = \frac{\partial}{\partial \bar{z}}e^{-u/2}|_{x=x^*} = 0,$$

$$\bar{f}_2(\overline{z^*}) = \phi_z(z^*, \overline{z^*}) = \frac{\partial}{\partial z}e^{-u/2}|_{x=x^*} = 0$$

and

$$\frac{\partial \bar{f}_2}{\partial \bar{z}}(\overline{z^*}) = \phi_{z\bar{z}}(z^*, \overline{z^*}) = \frac{1}{4}\Delta e^{-u/2}|_{x=x^*}$$

$$= -\frac{1}{8}e^{-u/2}(\Delta u)|_{x=x^*} = \frac{\lambda}{8}e^{u/2}|_{x=x^*} = C_2,$$

because of the hypothesis $\nabla u(x^*) = 0$. These relations indicate (8) with (9).

Thus we have proved

$$e^{-u/2} = C_1|\phi_1|^2 + C_2|\phi_2|^2, \tag{11}$$

C_1 and C_2 being defined in (9). Writing $\psi_1 = C_1^{1/2}(\frac{\lambda}{8})^{-1/4}\phi_1$, $\psi_2 = C_1^{-1/2}\left(\frac{\lambda}{8}\right)^{1/4}\phi_2$ and $F = \psi_2/\psi_1$, we obtain from $W(\psi_1, \psi_2) = W(\phi_1, \phi_2) = 1$ that

$$\left(\frac{\lambda}{8}\right)^{1/2}e^{u/2} = \left\{C_1\left(\frac{\lambda}{8}\right)^{-1/2}|\phi_1|^2 + C_1^{-1}\left(\frac{\lambda}{8}\right)^{1/2}|\phi_2|^2\right\}^{-1}$$

$$= \frac{1}{|\psi_1|^2 + |\psi_2|^2} = \frac{|F'|}{1 + |F|^2} \equiv \rho(F). \tag{12}$$

Here, the analytic funtion F of $z \in \Omega \subset C$ is a quotient of two linearly independent solutions for (5) so that satisfies

$$\{F; z\} = -\frac{1}{2}s, \tag{13}$$

where $\{F; z\} = \frac{3}{4}\left(\frac{F''}{F'}\right)^2 - \frac{1}{2}\frac{F'''}{F'}$ denotes the usual Schwarzian derivative.

Conversely, for any analytic function $F = F(z)$ in $z \in \Omega \subset C$ with positively single-valued $\rho(F) \equiv \frac{|F'|}{1+|F|^2}$, the function u defined by (12) with a positive constant λ satisfies (3).

At this point the Emden–Fowler equation P is reduced to finding such an analytic function $F = F(z)$ satisfying

$$\rho(F)|_{\partial\Omega} = \left(\frac{\lambda}{8}\right)^{1/2}. \tag{14}$$

Such an F is not unique for each given solution u of P because $\rho(F)$ has a geometrical meaning invariant under certain group action as we shall see later. However, modulo that action of the group, F is unique ([25]).

2.2 Alexandrov–Bol's inequality

Let $\overline{C}$ be the Riemann sphere with $(0,0,0)$ and $(0,0,1)$ as south and north poles, respectively, and let $d\sigma_1$ be its canonical metric. The meromorphic function F in the previous subsection induces the conformal mapping $\overline{F} = \tau \circ F$ from Ω into $\overline{C}$ through the stereographic mapping $\tau : C \cup \{\infty\} \to \overline{C}$. Under $\overline{F}$, the relation

$$\frac{d\sigma_1}{ds} = \rho(F) \tag{15}$$

is verified, where $ds^2 = dx_1^2 + dx_2^2$ denotes the Euclidean metric on Ω. From this well-known fact, the quantity $\rho(F)$ has been called the spherical derivative of F in the complex function theory. In particular, $\rho(F)$ is invariant under the O(3)-action on $\overline{C}$.

Thus for each subdomain $\omega \subset \Omega$ the quantities

$$\ell_1(\partial\omega) \equiv \int_{\partial\omega} \rho(F)^{1/2} dH^1 \quad \text{and} \quad m_1(\omega) \equiv \int_{\omega} \rho(F) dH^2$$

indicate the length of $\overline{F}(\partial\omega)$ and the area of $\overline{F}(\omega)$ in $\overline{C}$ as immersions so that they enjoy the isoperimetric inequality on the round sphere, i.e.,

$$\ell_1(\partial\omega)^2 \geq 4m_1(\omega)(\pi - m_1(\omega)).$$

This is nothing but a special form of Alexandrov–Bol's inequality for surfaces with Gaussian curvatures dominated by a constant from above. For instance the inequality

$$\ell(\partial\omega)^2 \geq \frac{1}{2}m(\omega)(8\pi - m(\omega)) \tag{16}$$

holds for each subdomain $\omega \subset \Omega$ surrounded by a number of Jordan curves, where

$$\ell(\partial\omega) = \int_{\partial\omega} p^{1/2} dH^1 \quad \text{and} \quad m(\omega) = \int_{\omega} p dH^2, \tag{17}$$

$p = p(x) > 0$ satisfying

$$-\Delta \log p \leq p \qquad (\text{in } \Omega \subset R^2) \tag{18}$$

See Burago–Zalgaller [4] for instance.

This fact has sometimes been taken up by C. Bandle to study the structure of solutions for the Emden–Fowler equation P. For instance, some isoperimetric inequalities for $\overline{\lambda}$, the least upper bound for the existence of solutions described in Section 1, are given in [1] and [2]. Another remarkable estimate is derived from her generalized Schwarz symmetrization, that is, the isoperimetric inequality for the first eigenvalue $\mu_1(p, \Omega)$ of the linearized operator $A_p \equiv -\Delta_D(\Omega) - p$, where $p = \lambda e^u$ with (λ, u) solving P.

In fact, for the quantity

$$\nu_1 = \inf\left\{ \int_\Omega |\nabla v|^2 dH^2 \,|\, v \in H_0^1(\Omega), \int_\Omega v^2 p dH^2 = 1 \right\} \tag{19}$$

the inequality

$$\nu_1(p, \Omega) \geq \nu_1(p^*, \Omega^*) \tag{20}$$

arises, provided that p satisfies (18) and

$$\sum \equiv \int_\Omega p dH^2 < 8\pi, \tag{21}$$

where $\Omega^* = \{|x| < 1\} \subset R^2$ and $p^* = \lambda^* e^{u^*}$ with (λ^*, u^*) satisfying

$$-\Delta u^* = \lambda^* e^{u^*} \qquad (\text{in } \Omega^*) \tag{22}$$

with

$$u^* = 0 \qquad (\text{on } \partial\Omega^*) \tag{23}$$

and

$$\int_{\Omega^*} \lambda^* e^{u^*} dH^2 = \Sigma. \tag{24}$$

As we shall see later, such a pair (λ^*, u^*) exists uniquely for each given $\Sigma \in [0, 8\pi)$.

The derivation of the isoperimetric inequality (20) follows from a natural modification of Faber–Krahn's method. Thus putting $D_t = \{v > t\}$ for a given nonnegative function $v = v(x)$ in Ω and constant $t > 0$, we take the open concentric disc D_t^* of Ω^* through

$$\int_{D_t^*} p^* dH^2 = \int_{D_t} p dH^2 \in [0, 8\pi)$$

to define Bandle's spherically decreasing rearrangement v^* of v in Ω^* by

$$v^*(x) = \sup\{t \,|\, x \in D_t^*\}. \tag{25}$$

Then, it is a kind of equi-measurable rearrangement and the relation

$$\int_\Omega v^2 p\, dH^2 = \int_{\Omega^*} v^{*2} p^*\, dH^2 \tag{26}$$

holds. On the other hand the decrease of Dirichlet integral,

$$\int_\Omega |\nabla v|^2 dH^2 \geq \int_{\Omega^*} |\nabla v^*|^2 dH^2 \tag{27}$$

when v is a nonnegative C^1-function with zero on the boundary $\partial\Omega$, is verified from the co-area formula and the Alexandrov–Bol inequality (16).

The relations (26) and (27) imply (20). We refer to [3] for details.

2.3. *Radial solutions*

In the case where Ω is the disc: $\Omega = B \equiv \{|x| < 1\} \subset R^2$ every solution for the Emden–Fowler equation P is known to be radial (Bandle [3], Gidas–Ni–Nirenberg [11]), while the radial solutions can be computed explicitly. Thus in this case $\Omega = B$, the least upper bound $\overline{\lambda}$ for the existence of classical solutions is 2.

According to $0 < \lambda < 2$ and $\lambda = 2$, the number of solutions is two and one, respectively. The solutions are given as

$$\left(\frac{\lambda}{8}\right)^{1/2} e^{u/2} = \frac{\rho^{1/2}}{|x|^2 + \rho} \quad \text{with} \quad \rho^{1/2} = \rho_\pm^{1/2} = \left(\frac{\lambda}{2}\right)^{-1/2} \left\{1 \mp \sqrt{1 - \frac{\lambda}{2}}\right\}. \tag{28}$$

Therefore, $\lim\limits_{\lambda\downarrow 0} u_-(x) = 0$ and $\lim\limits_{\lambda\downarrow 0} u_+(x) = 4\log\frac{1}{|x|}$ hold for each $x \in B$.

In the $\lambda - u$ plane the solutions form a branch L^* starting from $(\lambda, u) = (0, 0)$, bending at $\lambda = 2$, and growing up to the singular limit $u_0(x) = 4\log\frac{1}{|x|}$ as $\lambda \downarrow 0$.

The Liouville integrals (12) for those radial solution are given as

$$F(z) = C(z) \quad \text{with} \quad C = C_\pm = \left\{\frac{1}{\lambda}\{4 - \lambda \pm 2\sqrt{4 - 2\lambda}\}\right\}^{1/2}. \tag{29}$$

Regarding that the quantities

$$\ell_1(\partial B) = \int_{\partial B} \left(\frac{\lambda}{8} e^u\right)^{1/2} dH^1 = 2\pi \left(\frac{\lambda}{8}\right)^{1/2}$$

and

$$m_1(B) = \int_B \frac{\lambda}{8} e^u\, dH^2 = \frac{1}{8}\Sigma$$

denote the length of the area of $\overline{F}(\partial B)$ and $\overline{F}(B)$, respectively, we can see that Σ grows from 0 to 8π monotonously along the branch L^* with the bending point $\overline{\lambda}$ corresponding to $\Sigma = 4\pi$.

On the contrary, nonradial solutions arise in P when Ω is an annulus: $\Omega = A \equiv \{a < |x| < 1\} \subset R^2$, where $0 < a < 1$ (Lin [15], Nagasaki–Suzuki [19]). In [19] global existence of m-mode solutions has been established for any $m = 0, 1, 2, \ldots$, while symmetry breaking, bifurcation from radial solutions, has been observed by [15]. Those works utilize the fact that radial solutions for P in $\Omega = A$ are classified in terms of the analytic function $F(z) = \beta^{1/2} z^\alpha$ in (12), where $\beta > 0$ and $\alpha > 0$ are parameters.

2.4. Associated Legendre equations

From what has been said in the previous subsection, we are very easy to realize that $\Sigma = \int_B \lambda^* e^{u^*} dH^2 < 4\pi$ in (22) with (23) implies the minimality of the radial solution u^* and hence by Fact (vi) in Section 1, the positivity of the first eigenvalue $\mu_1(P^*, \Omega^*)$ of the linearized operator $A_p^* \equiv -\Delta_D(\Omega^*) - p^*$, where $p^* = \lambda^* e^{u^*}$. Combining this with the isoperimetric inequality (20), we obtain

Proposition 1. *If the positive function $p = p(x)$ satisfies (18) and*

$$\Sigma = \int_\Omega p\,dH^2 < 4\pi, \tag{30}$$

then

$$\mu_1(p, \Omega) > 0 \tag{31}$$

holds.

However, Bandle [3] gave a more analytic proof for

$$\nu_1(p^*, \Omega^*) \equiv \inf\left\{\int_{\Omega^*} |\nabla v|^2 dH^2\, v \in H_0^1(\Omega^*), \int_{\Omega^*} v^2 p^* dH^2 = 1\right\} > 1, \tag{32}$$

which is equivalent to (31), under $\Sigma = \int_{\Omega^*} p^* dH^2 < 4\pi$. Bandle directly studied the eigenvalue problem

$$\Delta\phi + \frac{1}{\Lambda}\zeta_0(x)\phi = 0 \qquad (\text{in } B = \{|x| < 1\} \subset R^2) \tag{33}$$

with

$$\phi = 0 \qquad (\text{on } \partial B), \tag{34}$$

where $\zeta_0(x) = p^* = \lambda^* e^{u^*} = \frac{8\rho}{(r^2+\rho)^2}$ for some $\rho \in (0, \infty)$, where $r = |x|$.

After the separation of variables $\phi = \Phi(r)e^{\sqrt{-1}m\theta}$, Bandle introduced the transformation

$$\xi = (\rho - r^2)/(\rho + r^2) \tag{35}$$

to deduce the associated Legendre equation

$$[(1 - \xi^2)\Phi_\xi]_\xi + [2/\Lambda - m^2/(1 - \xi^2)]\Phi = 0 \, (\xi_p < \xi < 1) \qquad (36)$$

with

$$\Phi(1) : \text{positive}, \qquad \Phi(\xi_\rho) = 0, \qquad (37)$$

where $\xi_p = \frac{\rho - 1}{\rho + 1} \in (-1, 1)$.

Then, $\nu_1(p^*, \Omega^*) > 1$ under $\Sigma = \int_{\Omega^*} p^* dH^2 < 4\pi$ is reduced to $\Phi(0) > 0$ for solutions Φ of (36) with (37), where $\Lambda = 1$ and $m = 0$.

It should be noted that the most technically crucial part of the Weston–Moseley theory described in Section 1 was the asymptotic analysis for this associated Legendre equation derived from Bandle's transformation. Also the work of Lin [5] on symmetry breaking on annului can be read in this context of associated Legendre equation. See [22].

2.5. *Laplace–Beltrami operator*

Noting (29) and (15), we realize that $\nu_1(p^*, \Omega^*)$ is nothing but $\frac{1}{8}$ of the first eigenvalue $\hat\mu_1(\omega)$ of the Laplace–Beltrami operator with the Dirichlet condition in $\omega \subset \overline{C}$, where ω is a disc of area $\frac{1}{8}\Sigma$. This will explain why the associated Legendre equation has arisen in the study of (33). Inequality (32) means that if ω in a chemi-ball of $\overline{C}$, then $\hat\mu_1(\omega) > 8$ holds.

Bandle's spherically decreasing rearrangement can be regarded as the Schwarz symmetrization on a round sphere in this context. Namely, let S be the two-dimensional round sphere with area 8π, and $d\sigma$ be its canonical metric. Given a positive function $p = p(x)$ in a domain $\Omega \subset R^2$ satisfying (18) and $\Sigma = \int_\Omega pdH^2 < 8\pi$, we can take a disc ω in S such that $\int_\omega d\sigma = \Sigma$.

For each nonnegative function $v = v(x)$ in Ω, we can define a function $v^* = v^*(x)$ in ω through

$$v^*(x) = \sup\{t | x \in \omega_t\}, \qquad (38)$$

where ω_t denotes the open concentric disc of ω satisfying

$$\int_{\omega_t} d\sigma^2 = \int_{\{v > t\}} pdH^2. \qquad (39)$$

Then the relations (26) and (27) mean

$$\int_\Omega v^2 pdH^2 = \int_\omega v^* d\sigma^2, \qquad (40)$$

and

$$\int_\Omega |\nabla v|^2 dH^2 \geq \int_\omega |\nabla v^*|^2 d\sigma, \qquad (41)$$

respectively.

In this way Proposition 1 implies the following: Given a domain $\omega \subset S$, the first eigenvalue $\mu_1(\omega)$ of the Laplace–Beltrami operator in ω under the Dirichlet condition is greater than one, provided that $\int_\omega d\sigma < 4\pi$.

3. Asymptotic analysis: classification
of the singular limits

In [20] we have the following:

Theorem 1. *Let $\{(\lambda, u)\}$ be the classical solutions for the Emden–Fowler equation P and set $\Sigma = \int_\Omega \lambda e^u dH^2$. Then $\{\Sigma\}$ accumulates to $8\pi m$ for some $m = 0, 1, 2, \ldots, \infty$ as $\lambda \downarrow 0$. The solutions $\{u\}$ behave as follows:*

(a) In the case $m = 0$, $\|u\|_{L^\infty} \to 0$, i.e., uniform convergence to zero.

(b) In the case $0 < m < \infty$, $u|_S \to \infty$ and $\|u\|_{L^\infty_{loc}(\overline{\Omega}\backslash S)} \in O(1)$ for some set S of m-points, i.e., m-point blow-up.

(c) In the case $m = \infty$, $u(x) \to \infty$ for any $x \in \Omega$, i.e., entire blow-up. In the case (b), the blow-up points $S = \{x_1, \ldots, x_m\} \subset \Omega$ satisfy

$$\frac{1}{2}\nabla R(x_j) + \sum_{\ell \neq j} \nabla_x G(x_\ell, x_j) = 0 \quad (1 \leq j \leq m) \tag{42}$$

and the singular limit $u_0 = u_0(x)$ must be of the form

$$u_0(x) = 8\pi \sum_{j=1}^m G(x, x_j), \tag{43}$$

where $G = G(x, y)$ denotes the Green function for $-\Delta_D(\Omega)$ and $R = R(x)$ denotes the Robin function:

$$R(x) = \left[G(x, y) + \frac{1}{2\pi}\log|x - y|\right]_{x=y}$$

To prove the theorem, we first recall the theory of Gidas–Ni–Nirenberg [11] to have constants $\alpha > 0$ and $t_0 > 0$ independent of u such that $\xi \in R^2$, $|\xi| = 1$ and $\xi \cdot N(x) \geq \alpha$ at $x \in \partial\Omega$ imply

$$\frac{d}{dt}u(x + t\xi) < 0 \qquad (-t_0 < t < 0), \tag{44}$$

where $N = N(x)$ denotes the outer unit normal vector on $\partial\Omega$. This fact implies the following boundary estimate:

If $\|u\|_{L^1_{loc}(\Omega)} \in O(1)$, then for each $k = 0, 1, 2, \ldots$, there exists an $\overline{\Omega}$-neighbourhood ω_k of $\partial\Omega$ such that

$$\|D^\alpha u\|_{L^\infty(\omega_k)} \in O(1) \quad \text{for} \quad |\alpha| \le k. \tag{45}$$

In fact, the case $k = 0$ follows from the argument of deFigueiredo–Lions–Nussbaum [8], which implies the other cases by the bootstrap argument based on the elliptic regularity.

Here we employ Kaplan's argument of taking the first eigenvalue and eigenfunction of $-\Delta_D(\Omega)$, denoted by $\lambda_1 > 0$ and $\phi_1 = \phi_1(x) > 0$, repectively. From the equation P, we obtain

$$\lambda_1 \int_\Omega u\phi_1 dH^2 = J \equiv \int_\Omega \lambda e^u \phi_1 dH^2. \tag{46}$$

Hence $J \in O(1)$ implies $\|u\|_{L^1_{loc}}(\Omega) \in O(1)$. Therefore, the boundary estimate $\|u\|_{L^\infty_{loc}(\omega_0)} \in O(1)$ holds so that $\Sigma \in O(1)$.

Equivalently, $\Sigma \to \infty$ implies $J \to \infty$, which deduces that $u(x) \to \infty$ for any $x \in \Omega$. In fact there exists a constant $\gamma_x > 0$ such that

$$G(x, y,) \ge \gamma_x \phi_1(y) \quad (y \in \overline{\Omega})$$

and hence

$$u(x) = \int_\Omega G(x, y)\lambda e^{u(y)} dH^2(y) \ge \gamma_x J.$$

Therefore, what we have to do is deduce (b) assuming $\Sigma \in O(1)$ and $\|u\|_{L^\infty} \to \infty$.

The idea is very simple. We have only to recall the expression (11) with (9) where $\{\phi_1, \phi_2\}$ denotes the fundamental system of solutions for (5) satisfying (6), with $s = s(z)$ being defined in (4). We can take a maximal point of u as $x^* : u(x^*) = \|u\|_{L^\infty}$.

In the case $\Sigma \in O(1)$ the boundary estimate (45) holds so that $\|s\|_{L^\infty(\omega)} \in 0(1)$ for $\omega = \omega_2$. However, $\{s = s(z)\}$ is a family of holomorphic functions so that its suitable subsequence converges to some holomorphic function $s_0 = s_0(z)$ on any compact subset of Ω by classical Montel's theorem. This convergence reflects to the fundamental system $\{\phi_1, \phi_2\}$. Since $C_1 = e^{-\|u\|_{L^\infty}/2} \to 0$, we have the singular limit that

$$e^{-u_0/2} = C_{20}|\phi_{20}|^2, \tag{47}$$

where $C_{20} = \lim_{\lambda \downarrow 0} C_2 \in [0, \infty]$ and $\phi_{20} = \lim_{\lambda \downarrow 0} \phi_2$ so that the blow-up points of $\{u\}$ correspond to zeros of $C_{20}\phi_{20}$.

The latter must be discrete because $\phi_{20} \ne 0$ is analytic and $C_{20} \ne 0$ from the boundary estimate for $\{u\}$. This proves the finiteness of the blow-up points of $\{u\}$ again the boundary estimate (45).

The singular limit $u_0 = u_0(x)$ is harmonic outside the blow-up points and the measure $-\Delta u_0$ must be a finite sum of δ-functions. Now the form (43) with (42) is verified from the residue analysis, because the function

$$s_0 = u_{0zz} - \frac{1}{2}u_{0z}^2$$

is holomorphic everywhere. See [20] for details.

Fact (vii) described in Section 1 is proved similarly, because ϕ_1 and ϕ_2 cannot have common zero points.

4. Global analysis: a geometric argument

We can show the following:

Theorem 2. *If $\Omega \subset R^2$ is simply connected and $\Sigma = \int_\Omega \lambda e^u dH^2$ approaches 8π from below as $\lambda \downarrow 0$, then the singular limit $u_0(x) = 8\pi G(x, k)$ connects with the trivial solution $u = 0$ in $\lambda - u$ plane through a branch bending just once.*

For the moment we describe Weston–Moseley's branch $\widehat{L}$ of nonminimal solutions. For κ satisfying

$$\nabla R(\kappa) = 0, \tag{48}$$

we take a univalent function $g : B \equiv \{|x| < 1\} \to \Omega$ satisfying $g(0) = \kappa$. Then $g''(0) = 0$ holds. Under some generic assumption other than $|g'''(0)/g'(0)| \neq 2$, Weston–Moseley's branch $\widehat{L}$ can be constructed of which solutions $\{\hat{u}\}$ make one-point blow-up at κ as $\lambda \downarrow 0$.

Then the relation

$$\int_\Omega \lambda e^{\hat{u}} dH^2 = 8\pi + C\lambda + 0(1) \quad \text{as} \quad \lambda \downarrow 0 \tag{49}$$

holds with

$$\frac{C}{\pi} = -|a_1|^2 + \sum_{k=3}^{\infty} \frac{k^2}{k-2}|a_k|^2, \tag{50}$$

where

$$g(z) = \kappa + a_1 z + \sum_{k=3}^{\infty} a_k z^k. \tag{51}$$

Therefore, if $c < 0$ then $\widehat{L}$ connects with the branch of minimal solutions. See [23] for (49) with (50).

The idea for the proof of Theorem 2 can be taken from the radical case described in Section 2.4. We are going to parametrize the solutions $\{(u, \lambda)\}$

in terms of $\Sigma = \int_\Omega \lambda e^u dH^2$. For this purpose we introduce the nonlinear mapping

$$\Psi : \quad \begin{matrix} C_0^{2+\alpha}(\overline{\Omega}) \\ \otimes \\ R \end{matrix} \quad \longrightarrow \quad \begin{matrix} C^\alpha(\overline{\Omega}) \\ \otimes \\ R \end{matrix}$$

through

$$\psi(h, \Sigma) = \begin{pmatrix} \Delta u + \lambda e^u \\ \int_\Omega \lambda e^u dH^2 - \Sigma \end{pmatrix} \quad \text{for} \quad h = \begin{pmatrix} u \\ \lambda \end{pmatrix} \in \begin{matrix} C_0^{2+\alpha}(\overline{\Omega}) \\ \otimes \\ R \end{matrix},$$

where $C^\alpha(\overline{\Omega})$ denotes the usual Schauder space and

$$C_0^{2+\alpha}(\overline{\Omega}) = \{v \in C^{2+\alpha}(\overline{\Omega}) | v = 0 \quad \text{on} \quad \partial\Omega\}$$

for some $\alpha \in (0,1)$. Then each zero of $\Psi(\cdot, \Sigma)$ represents a solution $h = \begin{pmatrix} u \\ \lambda \end{pmatrix}$ of P satisfying

$$\Sigma = \int_\Omega e^u dH^2. \tag{52}$$

We can show the following lemmas concerning the nonlinear mapping Ψ:

Lemma 1. *For each $\delta > 0$ the set $\{h = \begin{pmatrix} u \\ \lambda \end{pmatrix} | \Psi(h, \Sigma) = 0$ for some $\Sigma \in [0, 8\pi - \delta]\}$ is compact.*

Lemma 2. *If $\Psi(h, \Sigma) = 0$ with some $\Sigma \in [0, 8\pi)$, then the linearized operator*

$$d_h \Psi(h, \Sigma) : \quad \begin{matrix} C_0^{2+\alpha}(\overline{\Omega}) \\ \otimes \\ R \end{matrix} \quad \longrightarrow \quad \begin{matrix} C^\alpha(\overline{\Omega}) \\ \otimes \\ R \end{matrix}$$

is an isomorphism.

Only the trivial solution $h = \begin{pmatrix} 0 \\ 0 \end{pmatrix}$ is admitted as zeros of $\Psi(\cdot, 0)$. Therefore, these lemmas imply that in the $\Sigma - h$ plane with $0 < \Sigma < 8\pi$ forms a unique branch C as for zeros of Ψ, starting from $(\Sigma, h) = (0, \begin{pmatrix} 0 \\ 0 \end{pmatrix})$, approaching the hyperplane $\Sigma = 8\pi$, and not bending or bifurcating.

Thus the singular limit $u_0 = 8\pi G(x, k)$ with $\Sigma \uparrow 8\pi$ connects with the trivial solution $u = 0$ in $\lambda - u$ plane through a branch $L = \{(\lambda(\Sigma), u(\Sigma)) | 0 < \Sigma < 8\pi\}$. Namely, $(\lambda(\Sigma), u(\Sigma)) \to (0,0)$ as $\Sigma \downarrow 0$ and $(\lambda(\Sigma), u(\Sigma)) \to (0, 8\pi G(\cdot, \kappa))$ as $\Sigma \uparrow 8\pi$.

Here we can prove the following:

Lemma 3. *If (λ, u) solves P with $\Sigma = \int_\Omega \lambda e^u dH^2 < 8\pi$, then the second eigenvalue $\mu_2(p, \Omega)$ of $A_p \equiv -\Delta_D - p$ is positive, where $p = \lambda e^u$.*

We also have the local theory of Crandall–Rabinowitz [7]. Namely, in case that the first eigenvalue $\mu_1(\bar{p}, \Omega)$ is equal to zero for $\bar{p} = \lambda(\overline{\Sigma})e^{u(\overline{\Sigma})}$ with some $\overline{\Sigma} \in (0, 8\pi)$, the branch L bends at $\lambda = \overline{\lambda} = \lambda(\overline{\Sigma})$, changing the solutions $\{u(\Sigma)\}$ from the minimal to the nonminimal.

Regarding the uniqueness of minimal solutions and Lemma 3, we realize only one possibility of such a degeneracy $\mu_1(\bar{p}, \Omega) = 0$. This proves Theorem 2.

To prove Lemma 1, we recall the existence of the upper bound $\overline{\lambda}$ for λ described in Fact (i), Section 1. From the elliptic regularity the lemma is therefore reduced to the a priori estimate for $\|u\|_{L^\infty}$. We can prove that

$$\|u\|_{L^\infty} \leq -2\log\left(1 - \frac{\Sigma}{8\pi}\right)_+, \tag{53}$$

utilizing Alexandrov–Bol's inequality. See [3] or [23] for the proof.

Lemma 3 is reduced to Proposition 1, Bandle's isoperimetric inequality for $\mu_1(p, \Omega)$. In fact the second eigenfunction ψ_2 of $A_p \equiv -\Delta_D(\Omega) - p$ has just two nodal domains $\Omega_\pm = \{\pm \psi_2 > 0\}$ so that Ω_+ and Ω_- are open connected sets with a number of piecewise C^2 Jordan curves as boundaries by Cheng's argument [6].

Hence, $\mu_2(p, \Omega) = \mu_1(p, \Omega_\pm)$ and also $\Sigma = \int_{\Omega_+} pdH^2 + \int_{\Omega_-} pdH^2 < 8\pi$. Therefore, either $\int_{\Omega_+} pdH^2 < 4\pi$ or $\int_{\Omega_-} pdH^2 < 4\pi$ holds so that $\mu_2(p, \Omega) > 0$.

Lemma 2 is most crucial to prove, but it is obvious when $\Sigma = 0$. If $\Sigma > 0$, $\Psi(h, \Sigma) = 0$ is equivalent to $\Phi(h, \Sigma) = 0$, where

$$\Phi = \Phi(\cdot, \Sigma): \quad \begin{matrix} C_0^{2+\alpha}(\overline{\Omega}) \\ \otimes \\ R \end{matrix} \quad \longrightarrow \quad \begin{matrix} C^\alpha(\overline{\Omega}) \\ \otimes \\ R \end{matrix}$$

with

$$\Phi(h, \Sigma) = \begin{pmatrix} \Delta u + \lambda e^u \\ \int_\Omega e^u dH^2 - \frac{\Sigma}{\lambda} \end{pmatrix} \quad \text{for} \quad h = \begin{pmatrix} u \\ \lambda \end{pmatrix}.$$

Hence the isomorphy of $d_h \Psi(h, \Sigma)$ is reduced to that of $d_h \Phi(h, \Sigma)$.

However, the linearized operator

$$d_h \Phi(h, \Sigma) = \begin{pmatrix} \Delta + \lambda e^u & e^u \\ \int_\Omega e^u \cdot dH^2 & \frac{\Sigma}{\lambda^2} \end{pmatrix} : \quad \begin{matrix} C_0^{2+\alpha}(\overline{\Omega}) \\ \otimes \\ R \end{matrix} \quad \longrightarrow \quad \begin{matrix} C^\alpha(\overline{\Omega}) \\ \otimes \\ R \end{matrix}$$

has a natural self-adjoint extension in $\dfrac{L^2(\Omega)}{R} \otimes$, denoted by $-T$, with the

domain $D(T) = \dfrac{H_0(\Omega)pH^2(\Omega)}{R} \otimes$. By virtue of elliptic regularity the iso-

morphy of $d_h\Phi(h, \Sigma)$ is equivalent to that of T.

The operator T is associated with the bilinear form $A = A(\cdot, \cdot)$ on $\dfrac{H_0^1(\Omega)}{R} \otimes$. That is, for $\xi = \begin{pmatrix} v \\ \kappa \end{pmatrix} \in D(T)$ and $\eta = \begin{pmatrix} w \\ \rho \end{pmatrix} \in \dfrac{H_0^1(\Omega)}{R} \otimes$ we have

$$< T\xi, \eta >= A(\xi, \eta),$$

where $< \xi, \eta >= \int_\Omega vw dH^2 + \kappa\rho$ and

$$A(\xi, \eta) = \int_\Omega \nabla v \cdot \nabla w dH^2 - \int_\Omega \{\lambda e^u vw + e^u \kappa w + e^u v\rho\} dH^2 \qquad (54)$$

$$- \Sigma\kappa\rho/\lambda^2 = \int_\Omega \nabla v \cdot \nabla w dH^2 - \int_\Omega \lambda e^u \left(v + \frac{\kappa}{\lambda}\right)\left(w + \frac{\rho}{\lambda}\right) dH^2$$

by $\Sigma = \int_\Omega \lambda e^u dH^2$.

For each $\lambda > 0$ the mapping

$$\xi = \begin{pmatrix} v \\ \kappa \end{pmatrix} \in \dfrac{H_0^1(\Omega)}{R} \otimes \; \longmapsto v + \frac{\kappa}{\lambda} \in H_c^1(\Omega) \qquad (55)$$

is an isomorphism, where $H_c^1(\Omega) = \{v \in H^1(\Omega)|v =$ constant (on $\partial\Omega$)$\}$. Therefore, the isomorphy of T is reduced to that of $\widehat{A}_p$, the self-adjoint operator in $L^2(\Omega)$ associated with the bilinear form $B|_{H_c^1(\Omega)\times H_c^1(\Omega)}$, where

$$B(v, w) = \int_\Omega \nabla v \cdot \nabla w dH^2 - \int_\Omega pvw dH^2 \qquad (56)$$

for $p = \lambda e^u$. Namely,

$$(\widehat{A}_p v, w)_{L^2} = B(v, w)$$

for $v \in D(\widehat{A}_p) \subset H_c^1(\Omega)$ and $w \in H_c^1(\Omega)$.

The first eigenvalue $\hat{\mu}_1(p, \Omega)$ of $\widehat{A}_p$ is negative because the constant function $\zeta = 1/|\Omega|^{1/2}$ belongs to H_c^1 with $\|\zeta\|_{L^2} = 1$ and hence

$$\hat{\mu}_1(p, \Omega) = \inf\{B(v, v)|v \in H_c^1(\Omega), \|v\|_{L^2} = 1\}$$

$$\leq -\frac{1}{H^2(\Omega)} \int_\Omega p dH^2 < 0.$$

Furthermore, we can prove the following:

Proposition 2. *If a positive function $p = p(x)$ in $\Omega \subset R^2$ satisfies (18) and $\Sigma = \int_\Omega p\,dH^2 < 8\pi$, then*

$$K \equiv \inf\{\int_\Omega |\nabla v|^2 dH^2 \,|\, v \in H_c^1(\Omega), \int_\Omega v^2 p\,dH^2 = 1, \int_\Omega v p\,dH^2 = 0\} > 1. \tag{57}$$

By virtue of Courant's mini-max principle, (57) implies that the second eigenvalue $\hat{\mu}_2(p, \Omega)$ of $\widehat{A}_p$ is positive provided that $\Sigma < 8\pi$. This proves Lemma 2.

In proving Proposition 2, we note that K is nothing but the second eigenvalue of the following *E.P.*

Find $\phi \in H_c^1(\Omega)\backslash\{0\}$ and $K \in \mathcal{R}$ such that

$$\int_\Omega \nabla\phi \cdot \nabla v\,dH^2 = K \int_\Omega \phi v p\,dH^2 \quad \text{for any} \quad v \in H_c^1(\Omega). \tag{58}$$

In fact its first eigenvalue and eigenfunction are 0 and nonzero constants, respectively.

Therefore, the minimizer $\phi \in H_c^1(\Omega)$ of K satisfies

$$-\Delta\phi = Kp\phi \quad (\text{in } \Omega) \tag{59}$$

and

$$\phi = \text{constant (on } \partial\Omega), \int_{\partial\Omega} \frac{\partial\phi}{\partial N} dH^1 = 0, \tag{60}$$

where N denotes the outer unit normal vector on $\partial\Omega$.

Let $\{\Omega_i\}_{i\in I}$ be the nodal domains of ϕ. Then we have

$$\int_{\partial\Omega_i} \frac{\partial\phi_i}{\partial N}\phi_i dH^1 = 0 \quad \text{for each} \quad i \in I, \tag{61}$$

where the simple connectedness of Ω is utilized.

From Å. Pleijel's argument [21] the relation (61) implies $\#I = 2$ for the second eigenfunction ϕ. Let $\Omega_\pm = \{\pm\phi > 0\}$ and $\kappa = \phi|_{\partial\Omega} \in R$.

In the case $\kappa = 0$, $\phi \in H_c^1(\Omega)$ satisfies

$$-\Delta\phi = Kp\phi, \quad \pm\phi > 0 \quad (\text{in } \Omega_\pm) \tag{62}$$

and

$$\phi = 0 \quad (\text{on } \partial\Omega_\pm). \tag{63}$$

On the other hand $\Sigma < 8\pi$ implies either $\Sigma_+ \equiv \int_{\Omega_+} p\,dH^2 < 4\pi$ or $\Sigma_- \equiv \int_{\Omega_-} p\,dH^2 < 4\pi$ and hence $K > 1$ from Proposition 1.

In the case $\kappa \neq 0$ we may suppose that $\pi_1(\Omega_-) = 0$ and $\partial\Omega \subset \partial\Omega_+$ without loss of generality. Either $\Sigma_+ < 4\pi$ or $\Sigma_- < 4\pi$ holds. In the case $\Sigma_- < 4\pi$, $K > 1$ holds again by Proposition 1.

We perform some rearrangement procedure to study the case $\Sigma_+ < 4\pi \leq \Sigma_-$. We set $\Gamma = \partial\Omega$ and $\gamma = \partial\Omega_+ \backslash \Gamma$. Then the function $\phi|_{\Omega_+}$ satisfies

$$-\Delta\phi = Kp\phi, \quad \phi > 0 \quad (\text{in } \Omega_+) \tag{64}$$

with

$$\phi = 0 \quad (\text{on } \gamma), \; \phi = \text{constant (on } \Gamma), \quad \int_\Gamma \frac{\partial\phi}{\partial N} dH^1 = 0 \tag{65}$$

so that

$$K = \inf\left\{ \int_{\Omega_+} |\nabla v|^2 dH^2 \big| v \in H^1(\Omega_+), \, v = 0 \quad (\text{on } \gamma) , \right. \tag{66}$$

$$\left. v = \text{constant (on } \Gamma), \int_{\Omega_+} v^2 p \, dH^2 = 1 \right\}.$$

In fact the minimizer $\psi > 0$ for K in (66) is given by constant $\times \phi|_{\Omega_+}$.

Putting $\tau = \psi|_\Gamma \in R_+$ we divide Ω_+ into $\Omega_1 = \{\psi < \tau\}$ and $\Omega_2 = \{\psi \geq \tau\}$. The latter may be empty but otherwise we adopt Bandle's rearrangement described in Section 2.5 for $\psi_2 \equiv \psi|_{\Omega_2}$ preparing a disc B on the round sphere S of area 8π and metric $d\sigma$ satisfying

$$\int_B d\sigma^2 = \int_{\Omega_2} p \, dH^2.$$

On the other hand the following procedure is taken as for $\psi_1 \equiv \psi|_{\Omega_1}$, which may be called the annular increasing rearrangment. Thus preparing an annulus $A \equiv B_1 \backslash \overline{B_2}$ on S with the concentric discs B_1 and B_2 so that

$$\int_A d\sigma^2 = \int_{\Omega_1} p \, dH^2 \quad \text{and} \quad \int_{B_2} d\sigma^2 = \int_{\Omega_-} p \, dH^2,$$

we put

$$\psi_{1*}(x) = \inf\{t | x \in A_t\},$$

where A_t denotes the closed concentric annulus of $\overline{A}$ in S such that

$$\int_{A_t \cup \overline{B_2}} d\sigma = \int_{\{\psi_1 \leq t\} \cup \overline{\Omega}_-} p \, dH^2.$$

From the assumption that $\Sigma+ < 4\pi \leq \Sigma_-$ and $\Sigma_+ + \Sigma_- < 8\pi$, A and B can be so arranged that concentric, disjoint, and contained in chemi-sphere of S.

We put $\omega_+ \equiv A \cup B$. Furthermore, γ^* and Γ^* denote the inner and the outer boundaries of A, respectively. Then, through these procedures of rearrangement, $K > 1$ can be reduced to the "radial" case. We can prove that

$$K \geq K^* \equiv \inf\left\{ \int_{\omega_-} |\nabla v|^2 d\sigma \,|\, v \in H^1(\omega_+), \, v = 0 \quad (\text{on } \gamma^*), \right.$$

$$\left. v|_{\Gamma^*} = v|_{\partial B} = \text{constant}, \quad \int_{\omega_+} v^2 d\sigma = 1 \right\}. \tag{67}$$

Through Bandle's transformation described in Section 2.4, the proof of $K^* > 1$ under those situations is reduced to the positivity of $\Phi(0)$ for a solution of the Legendre equation. More precisely, according to the cases $\Omega_2 \neq \emptyset$ and $\Omega_2 = \emptyset$, Φ solves for some a, b in $0 < a < b < 1$ that

$$[(1 - \xi^2)\Phi_\xi]_\xi + 2\Phi = 0 \, (0 < \xi < a, \, b < \xi < 1) \tag{68}$$

with

$$\Phi(1) : \text{positive}, \quad \Phi(b) = \Phi(a), \, \Phi'(b) = \Phi'(a), \tag{69}$$

and

$$[(1 - \xi^2)\Phi_\xi]_\xi + 2\Phi = 0 \quad (0 < \xi < a) \tag{70}$$

with

$$\Phi(a) : \text{positive}, \quad \Phi'(a) = 0, \tag{71}$$

respectively.

We can verify the desired inequality $\Phi(\xi) > 0 \, (0 < \xi < a, \, b < \xi < 1)$ utilizing the fundamental system of solutions for (68), that is,

$$\Phi_1 = \xi \quad \text{and} \quad \Phi_2 = -1 + \frac{\xi}{2}\log\frac{1+\xi}{1-\xi}. \tag{72}$$

In this way the proof of Proposition 2 has been completed.

REFERENCES

[1] Bandle C., *Existence theorems, qualitative results and a priori bounds for a class of nonlinear Dirichlet problems*, Arch. Rat. Mech. Anal., **58** (1975), 219–238.

[2] Bandle C., *Isoperimetric inequalities for nonlinear eigenvalue problems*, Proc. Amer. Math. Soc., **56** (1976), 243–246.

[3] Bandle C., *Isoperimetric inequalities and Applications*, Pitman, Boston-London-Melbourne, 1980.

[4] Burago Yu. D. and Zalgaller, V. A., *Geometric Inequalities*, Springer, Berlin-Heidelberg-New York-London-Paris-Tokyo, 1988.

[5] Chandrasekhar S., *An introduction to the study of Stellar Structure*, Chapter 11, Dover, New York, 1957

[6] Cheng S. Y., *Eigenfunctions and nodal sets*, Comment. Math. Helvetici, **51** (1976), 43–55.

[7] Crandall M. G., and Rabinowitz P. H., *Some continuation and variational methods for positive solutions of nonlinear elliptic eigenvalue problems*, Arch. Rat. Mech. Anal., **58** (1975), 207–218.

[8] De Figueiredo D. G., Lions P. L., and Nussbaum R. D., *A priori estimates and existence of positive solutions of semilinear elliptic equations*, J. Math. Pure et Appl., **61** (1982), 41–63.

[9] Fujita H., *On the nolinear equations $\Delta u + e^u = 0$ and $\partial v / \partial t = \Delta t + e^v$*, Bull. Amer. Soc., **75** (1969), 132–135.

[10] Gel'fand I. M., *Some problems in the theory of quasilinear equations*, Amer. Math. Soc. Transl., **29**(2) (1963), 295–381.

[11] Gidas B., Ni W. M., and Nirenberg L., *Symmetry and related properties via the maximum principle*, Comm. Math. Phys., **68** (1979), 209–243.

[12] Kenner J. P. and Keller H. B., *Positive solutions of convex nonlinear eigenvalue problem*, J. Diff. Equat., **16** (1974), 103–125.

[13] Keller H. B. and Cohen D. S., *Some positive problems suggested by nonlinear heat generation*, J. Math. Mech., **16** (1967), 1361–1376.

[14] Laetsch T., *On the number of solutions of boundary value problems with convex nonlinearities*, J. Math. Anal. Appl., **35** (1971), 389–404.

[15] Lin S. S., *On non-radially symmetric bifurcation in the annulus*, J. Diff. Equat., **80** (1989), 251–279.

[16] Liouville J., *Sur l'équation aux différences partielles $\partial^2 log \lambda / \partial u \partial v \pm \lambda / 2a^2 = 0$*, J. Math., **18**(1853), 71–72.

[17] Mignot F., Murat F. and Puel, J. P., *Variation d'un point retourment par rapport au domaine*, Comm. P. D. E., **4**(1979), 1263-1297.

[18] Moseley J. L., *Asymptotic solutions for a Dirichlet problem with an exponential nonlinearity*, SIAM J. Math. Anal., **14** (1983), 719–735.

[19] Nagasaki K. and Suzuki T., *Radial and nonradial solutions for the nonlinear eigenvalue problem $\Delta u + \lambda e^u = 0$ on annuli in R^2*, J. Diff. Equat., **87** (1990), 144–168.

[20] Nagasaki K. and Suzuki T., *Asymptotic analysis for two-dimensional elliptic eigenvalue problems with exponentially-dominated nonlinearities*, Asymptotic Analysis, **3** (1990), 173–188.

[21] Pleijel Å, *Remarks on Courant's nodal line theorem*, Comm. Pure Appl. Math., **9** (1956), 543–550.

[22] Suzuki T., *Radial and nonradial solutions for semilinear elliptic equations*, In: Talenti, G. et. al. (eds.), Geometry of Solutions of Partial Differential Equations, Symposia Mathematics bf 30 (1989), Academic Press, pp. 153–174. .

[23] Suzuki, T. and Nagasaki K., *On the nonlinear eigenvalue problem $\Delta u + \lambda e^u = 0$*, Trans. Amer. Math. Soc., **309** (1988), 591–608.

[24] Weston V. H., *On the asymptotic solution of a partial differential equation with an exponential nonlinearity*, SIAM J. Math. Anal., **9** (1978), 1030–1053.

[25] Yamashita S., *Derivatives and length-preserving maps*, Canad. Math. Bull., **30** (1987), 379–384.

Department of Mathematics,
Tokyo Metropolitan University,
Minamiohsawa 1-1
Hachiôjishi, Tokyo 158,
Japan

Global Bifurcation of Positive Solutions in $\mathbb{R}^n$

ACHILLES TERTIKAS

1. Introduction

We study the global bifurcation diagram of the following semilinear elliptic problem

$$\Delta u + \lambda g(r) f(u) = 0 \quad \text{in } \mathbb{R}^n \tag{1.1}$$

$$(\mathcal{E}) \qquad 0 < u < 1 \quad \text{in } \mathbb{R}^n \tag{1.2}$$

$$\lim_{|x|\to\infty} u(x) = 0 \tag{1.3}$$

mainly in the case $n \geq 3$. Δ is the usual Laplace operator.

This type of problem arises in a two alleles model in population genetics [6], [7], [15]. Condition (1.3) is the most natural for our problem, and corresponds to stable solutions [16].

Various aspects of this problem have been studied by quite a large number of researchers. P.C. Fife and L.A. Peletier [6] considered the $n = 1$ case under different boundary conditions and proved the existence and stability of clines. W.H. Fleming [7] studied the problem in bounded domains under Neumann boundary conditions, where P. Hess and T. Kato [9] studied it under Dirichlet boundary conditions. H. Matano [10] established the existence of L^∞ stable solutions to our problem when the function g stays away from zero at infinity. K.J. Brown, C. Cosner and J. Fleckinger [2] studied the corresponding linear problem under various assumptions on g.

Recently, a variant of our model, has been studied by C. Bandle, M.A. Pozio and A. Tesei [1], L.A. Peletier and A. Tesei [14]. In these cases $f'(0) = \infty$ as a consequence of which multiple solutions having compact support appear for $\lambda > 0$.

The difficulty in studying bifurcation diagrams for problems in unbounded domains is that the linearized problem cannot be formulated as a problem involving compact operators. This is because the linearized operator may have eigenvalues and a nonempty continuous spectrum at the same time. Hence standard bifurcation theory can no longer be applied.

We refer to T. Kupper and C.A. Stuart (in this volume), C.A. Stuart [17] and J.F. Toland [22] for some related problems.

Instead of using standard bifurcation theory, our approach is to use O.D.E.'s methods and study the shooting pattern of the corresponding O.D.E. The shooting pattern can easily be analyzed by using the analysis in [19], [20] or [21], where the uniqueness of radially symmetric solutions has been studied.

Our method is indirect. We follow backwards an easily obtained solution. It turns out that we can continue the branch to the left until it bifurcates from the solution $u \equiv 0$ or bifurcates asymptotically from the solution $u \equiv 1$.

We assume throughout this work that $n \geq 3$ except the last section and that functions f, g satisfy

(F) $f \in C^2(\mathbb{R})$

              $f(0) = f(1) = 0$ and $f(u) > 0$ for $u \in (0, 1)$

              $f''(u) < 0$ for $u \in [0, 1]$,

(G) $g: [0, \infty) \to \mathbb{R}$ is locally Hölder continuous and bounded, and there exist $\alpha, \beta \geq 0$ such that $g(\beta) > 0$, $g(\alpha) = 0$ and $g(r) < 0$ for $r > \alpha$.

We define

$$\Lambda = \{\lambda \mid \lambda > 0 \text{ such that } (\mathcal{E}) \text{ has a solution } u\}$$

and

$$\bar{g} = \int_{\mathbb{R}^n} g \, dx. \tag{1.4}$$

Suppose in addition to our hypotheses, that g stays away from zero at infinity. Then it is shown in [5] that there exists $\lambda_{cr} > 0$ such that $\Lambda = (\lambda_{cr}, \infty)$ and moreover $u \to 0$ in L^∞_{loc} as $\lambda \to \lambda_{cr}^+$. Also in case $n = 1, 2$, $\bar{g} > 0$ and if we have some condition for g at infinity it is shown that $\Lambda = (0, \infty)$ and $u \to 1$ in L^∞_{loc} as $\lambda \to 0^+$.

Our analysis suggests that the global bifurcation diagram does not depend on $\bar{g}$ when $n \geq 3$. Generally, we can conclude that the global bifurcation diagram depends upon two main factors, the space dimension and the sign of $\bar{g}$. More precisely for $n = 1, 2$ and $\bar{g} < 0$ we have bifurcation from the zero solution at a simple positive eigenvalue, but if $\bar{g} > 0$, we have asymptotic bifurcation from $u \equiv 1$ at $\lambda = 0$. Contrary to the previous cases when $n \geq 3$ we always have bifurcation from a simple eigenvalue no matter what the sign of $\bar{g}$ is.

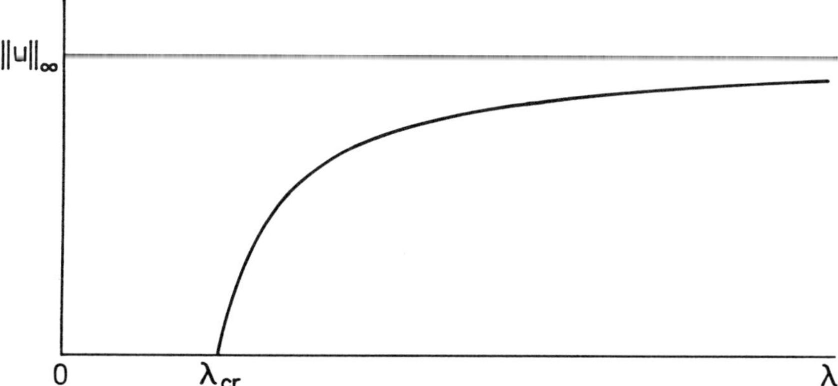

Figure 1. Typical bifurcation diagram of $(\mathcal{E})$ when $\int_{\mathbb{R}^n} g\,dx < 0$ and $n = 1, 2$.

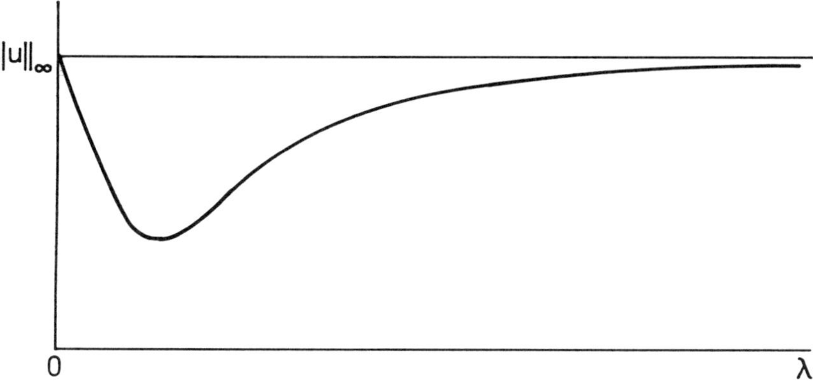

Figure 2. Typical bifurcation diagram of $(\mathcal{E})$ when $\int_{\mathbb{R}^n} g\,dx > 0$ and $n = 1, 2$.

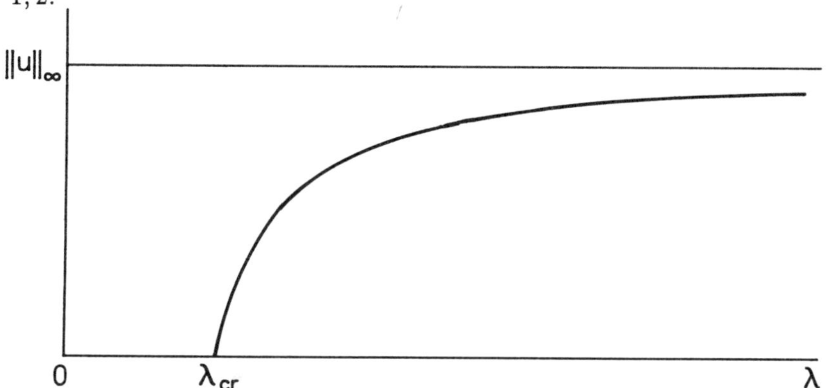

Figure 3. Typical bifurcation diagram of $(\mathcal{E})$ when $n \geq 3$.

We now describe how this work is organized. In Section 2 we include some preliminary results concerning radially symmetric solutions of $(\mathcal{E})$. Unfortunately the function g, in most of the cases that we are interested, is not decreasing. Hence we cannot use the general results of B. Gidas, W.N. Ni and L. Nirenberg [8] to conclude the radial symmetry of solutions. We prove the radial symmetry of solutions using different type of arguments in Section 3. Next we study the existence and nonexistence of H^1 solutions for small $\lambda > 0$. Section 5 contains the continuous dependence of solutions on the parameter λ. Finally we study in Section 6 the bifurcation when $n \geq 3$. In the last section we consider the case $n = 1, 2$.

2. Preliminary Results — The Shooting Pattern

Radially symmetric solutions of our problem satisfy

$$u'' + \frac{n-1}{r}u' + \lambda g(r)f(u) = 0 \quad r > 0 \tag{2.1}$$

$$(\mathcal{E})_\lambda \qquad u'(0) = 0, \quad \text{and} \quad 0 < u(r) < 1, \quad r \geq 0 \tag{2.2}$$

$$\lim_{r \to \infty} u(r) = 0. \tag{2.3}$$

Hence we consider the initial value problem

$$u'' + \frac{n-1}{r}u' + \lambda g(r)f(u) = 0, \quad r > 0 \tag{2.4}$$

$$u'(0) = 0 \tag{2.5}$$

$$u(0) = p \quad \text{for} \quad p \in [0, 1] \tag{2.6}$$

and we denote the solution of this initial value problem by $u(\cdot, p, \lambda)$. From standard theorems on the continuous dependence of solutions on parameters and on initial data, we obtain that

$$(p, \lambda) \mapsto u(\cdot, p, \lambda)$$

is a continuous function in L^∞_{loc}.

We define

$$A(\lambda) = \{p \in (0, 1) \mid \exists R > 0 \text{ such that } 0 < u(r, p, \lambda) < 1$$
$$\text{for } 0 \leq r < R \text{ and } u(R, p, \lambda) = 0\} \tag{2.7}$$

and

$$B(\lambda) = \{p \in (0, 1) \mid \exists R > 0 \text{ such that } 0 < u(r, p, \lambda) < 1$$
$$\text{for } 0 \leq r < R \text{ and } u(R, p, \lambda) = 1\}. \tag{2.8}$$

Then we can easily conclude [18] that $A(\lambda)$, $B(\lambda)$ are open, disjoint and that if $p \in A(\lambda)$ then there exists $\varepsilon > 0$ such that $p \subset \bigcap_{|\mu-\lambda|<\varepsilon} A(\mu)$ (similarly for $B(\lambda)$).

Proposition 2.1. *Problem* $(\mathcal{E})_\lambda$ *has at most one radially symmetric solution. If* $(\mathcal{E})_\lambda$ *has a solution, then* $A(\lambda) = (0,p)$ *for some suitable* p. *Moreover if* $0 < p_1 < p_2 < p$ *with* R_1, R_2 *be the first roots of* $u(\cdot,p_1,\lambda)$, $u(\cdot,p_2,\lambda)$ *respectively, then* $R_1 < R_2$ *and*

$$u(r,p_1,\lambda) < u(r,p_2,\lambda) \ \ for \ \ r \in [0, R_1). \tag{2.9}$$

Proof. The idea of the proof is in the spirit of graph separation technique of Peletier and Serrin [13]. For the sake of completeness we sketch the proof which is contained in [21]. Hence we define

$$W(r) = \frac{r^{n-1}u'(r)}{f(u(r))} - \frac{r^{n-1}v'(r)}{f(v(r))} \tag{2.10}$$

where u,v are two distinct solutions. Then W satisfies the differential equation

$$W'(r) + f'(v(r)) \left\{ \frac{u'(r)}{f(u(r))} + \frac{v'(r)}{f(v(r))} \right\} W(r)$$
$$= r^{n-1} \left\{ \frac{u'(r)}{f(u(r))} \right\}^2 [f'(v(r)) - f'(u(r))] \tag{2.11}$$

Let

$$H(r) = W(r) \exp\left(\int_\xi^r f'(v(t)) \left[\frac{u'(t)}{f(u(t))} + \frac{v'(t)}{f(v(t))} \right] dt \right) \tag{2.12}$$

with ξ large enough to be chosen later on. Suppose $u(0) > v(0)$ and $[0,T)$ is the maximal interval for which

$$u(r) > v(r), \quad r \in [0,T). \tag{2.13}$$

For $r \in [0,T)$ an easy calculation shows that

$$H'(r) > 0. \tag{2.14}$$

But $H(0) = 0$, therefore

$$\lim_{r \to T} H(r) > 0. \tag{2.15}$$

When T is finite, using the maximality of T we conclude

$$u(T) = v(T) \quad \text{and} \quad u'(T) < v'(T). \tag{2.16}$$

Hence

$$H(T) < 0 \tag{2.17}$$

which contradicts (2.15).

Suppose $T = \infty$, (2.1) can be written as

$$\left(r^{n-1}u'(r)\right)' + \lambda r^{n-1}g(r)f(u(r)) = 0 \tag{2.18}$$

and using (2.3) and (G) we conclude that every solution is decreasing in $[\alpha, \infty)$. For r large enough

$$f'(v(r))/f(u(r)) > f'(u(r))/f(u(r)) \tag{2.19}$$

therefore by taking ξ large enough so that (2.19) holds in $[\xi, \infty)$ we obtain

$$\frac{f'(v(r))u'(r)}{f(u(r))} \leq \frac{f'(u(r))u'(r)}{f(u(r))}. \tag{2.20}$$

It is easy then to show that

$$H(r) \leq -\frac{r^{n-1}v'(r)}{f(v(r))}\frac{f(v(r))}{f(v(\xi))}\frac{f(u(r))}{f(u(\xi))}, \quad r \in [\xi, \infty) \tag{2.21}$$

and finally that

$$\lim_{r \to \infty} H(r) = 0 \tag{2.22}$$

which contradicts (2.15).

In a similar way, to prove $A(\lambda) = (0, p)$ we set

$$v(r) = u(r, q, \lambda) \quad \text{for} \quad q \in (0, p) \tag{2.23}$$

and define W by (2.10), to arrive at $q \in A(\lambda)$.   q.e.d.

3. Radial Symmetry of Positive Solutions

To determine the range of applicability of the ODE method, one wishes to know whether all solutions of $(\mathcal{E})$ are radially symmetric or not. For this we initially prove

Lemma 3.1. *Suppose $n \geq 3$. Then for any solution of $(\mathcal{E})$ and R large enough*

$$u(x) \leq \frac{\|u\|_\infty R^{n-2}}{|x|^{n-2}} \quad \text{for} \quad |x| \geq R. \tag{3.1}$$

Proof. The proof of this lemma is by applying the maximum principle in the same lines as in [8], [11]. Since the function g is negative for large $|x|$ we can choose R large enough such that (3.1) holds for $|x| = R$. Then by setting

$$v(x) = u(x) - \frac{\|u\|_\infty R^{n-2}}{|x|^{n-2}} \quad \text{for} \quad |x| \geq R \tag{3.2}$$

we can easily establish that

$$\Delta v = -\lambda g(r) f(u(r)) \geq 0 \quad \text{for} \quad |x| \geq R. \tag{3.3}$$

Hence by the maximum principle, v takes its maximum on the boundary. But $v(x) \leq 0$ for $|x| = R$ and

$$\lim_{|x| \to \infty} v(x) = 0. \tag{3.4}$$

Therefore

$$v(x) \leq 0 \quad \text{for} \quad |x| \geq R. \qquad \text{q.e.d.}$$

Lemma 3.2. *Suppose $(\mathcal{E})$ has a nonradially symmetric solution u. Then $(\mathcal{E})$ also has a radially symmetric solution v, satisfying $v \geq u$.*

Proof. Assume the contrary. When $n \geq 3$, then by using Lemma 3.1 we can construct weak supersolutions of the form

$$\bar{u} = \begin{cases} 1, & |x| \leq R \\ (R/|x|)^{n-2} & |x| > R \end{cases} \tag{3.5}$$

with $\bar{u} > u$. We consider u as a subsolution to conclude by Ni [12] the existence of a radially symmetric solution v satisfying

$$\bar{u} \geq v \geq u. \tag{3.6}$$

Theorem 3.3. *All solutions of $(\mathcal{E})$ are radially symmetric.*

Proof. Assume the contrary and let u be a nonradially symmetric solution. Then Lemma 3.2 implies the existence of a radially symmetric solution v, $v \geq u$. We intend to prove the existence of another radially symmetric solution v^*, which satisfies $v^* \leq u$. This will then violate Proposition 2.1. As a consequence of Proposition 2.1 we obtain $A(\lambda) = (0, p^*)$, for some suitable p^*. By taking p small enough we can produce a radially symmetric weak subsolution $\underline{u}$, $\underline{u} \leq u$ of the form

$$\underline{u} = \begin{cases} u(r, p, \lambda) & \text{for } 0 \leq r \leq R(p) \\ 0 & \text{for } r > R(p) \end{cases} \tag{3.7}$$

where $R(p)$ is the first zero of $u(r, p, \lambda)$. Using once more Ni's results [12] we obtain a radially symmetric solution v^*, satisfying $v^* \leq u$.   q.e.d.

4. Existence and Nonexistence of Solutions for Small $\lambda > 0$

This section contains the existence of solutions for large λ and nonexistence of solutions for small λ.

Proposition 4.1. *Problem $(\mathcal{E})$ has a radially symmetric solution u for large λ. Moreover $\|u\|_\infty \to 1$ as $\lambda \to \infty$.*

Proof. By our assumption on g there exists an annulus or a ball Ω centered at β and a constant $m > 0$ such that

$$g(|x|) \geq m, \quad x \in \Omega. \tag{4.1}$$

Suppose λ_1, φ_1 denote the principal eigenvalue and the corresponding positive radially symmetric normalized eigenfunction of

$$-\Delta\varphi = \lambda\varphi \text{ in } \Omega; \quad \varphi = 0 \text{ on } \partial\Omega \tag{4.2}$$

with

$$\max_\Omega \varphi_1(x) = 1. \tag{4.3}$$

Let $f_1(u) = \lambda m f(u) - \lambda_1 u$, then $f_1(0) = 0$, and $f_1'(0) = \lambda m f'(0) - \lambda_1$. Hence $f_1'(0) > 0$ provided $\lambda > \lambda_1/mf'(0) = A_0$ say. Thus

$$\tau(\lambda) = \sup\{t \in [0, 1]: \lambda m f(u) > \lambda_1 u \text{ for } 0 \leq u \leq t\} > 0 \tag{4.4}$$

for any $\lambda > A_0$. For $\lambda > A_0$ we have

$$\Delta(\tau(\lambda)\varphi_1) + \lambda g(|x|)f(\tau(\lambda)\varphi_1) \geq -\lambda_1\tau(\lambda)\varphi_1 + \lambda m f(\tau(\lambda)\varphi_1) \geq 0 \tag{4.5}$$

when $x \in \Omega$ and so the function

$$\underline{u}(x) = \begin{cases} \tau(\lambda)\varphi_1(x) & \text{for } x \in \Omega \\ 0 & \text{for } x \in \mathbb{R}^n - \Omega \end{cases} \tag{4.6}$$

is a weak subsolution of $(\mathcal{E})$.

We consider weak supersolutions $\bar{u}$ of the form (3.5) to obtain the existence of a radially symmetric solution. The asymptotics follows since

$$\lim_{\lambda \to \infty} \tau(\lambda) = 1. \tag{4.7}$$

When u is a solution of our problem which decays appropriately at infinity, then typically multiplying (1.1) by u and integrating by parts, one obtains that

$$\int_{\mathbb{R}^n} |\nabla u|^2 dx = \lambda \int_{\mathbb{R}^n} g(r)f(u)u\,dx. \tag{4.8}$$

Sometimes unfortunately, solutions of $(\mathcal{E})_\lambda$ do not belong to $H^1(\mathbb{R}^n)$. Nevertheless when $u \in H^1(\mathbb{R}^n)$, then (4.8) suggests that in this case

$$\int_{\mathbb{R}^n} g(r)f(u)u\,dx > 0. \tag{4.9}$$

Initially, we use (4.8) to obtain lower bounds for λ, when u is an $H^1(\mathbb{R}^n)$ solution of $(\mathcal{E})$. Later on we shall see, that we can use (4.8) to obtain lower bounds for λ even in case where solutions do not belong to $H^1(\mathbb{R}^n)$. It has been proved in [4] that

Proposition 4.2. *Suppose $\int_0^\infty r^{n-1}g(r)dr < 0$. Then there exists $\lambda_0 > 0$ such that*

$$\int_{\mathbb{R}^n} |\nabla u|^2 dx \geq \lambda_0 \int_{\mathbb{R}^n} g(r)f(u)u\,dx \tag{4.10}$$

for all $u \in H^1(\mathbb{R}^n)$, with $0 \leq u \leq 1$ a.e. and $\int_{\mathbb{R}^n} gu^2 dx > 0$. Moreover problem $(\mathcal{E})$ has no $H^1(\mathbb{R}^n)$ solutions for small $\lambda > 0$.

Theorem 4.3. *Suppose that $n \geq 3$ and there exists a positive constant k such that $r^2|g(r)| \leq k$ for all $x \in \mathbb{R}^n$. Then there exists $\Lambda_0 > 0$ such that*

$$\int_{\mathbb{R}^n} |\nabla u|^2 dx \geq \Lambda_0 \int_{\mathbb{R}^n} g(r)f(u)u\,dx \tag{4.11}$$

for all $u \in H^1(\mathbb{R}^n)$, with $0 \leq u \leq 1$ a.e. and $\int_{\mathbb{R}^n} g(r)f(u)u\,dx > 0$. Moreover problem $(\mathcal{E})_\lambda$ has no $H^1(\mathbb{R}^n)$ solutions for small $\lambda > 0$.

Proof. The proof will be given by contradiction. As we have already mentioned any $H^1(\mathbb{R}^n)$ solution of $(\mathcal{E})$ satisfies (4.8). Hence it is sufficient to establish only the first statement in the theorem. We therefore suppose that there exist λ_k, u_k such that

$$\int_{\mathbb{R}^n} |\nabla u_k|^2 dx \leq \lambda_k \int_{\mathbb{R}^n} g(r)f(u_k)u_k\, dx, \quad k = 1, 2, \ldots \tag{4.12}$$

where $u_k \in H^1(\mathbb{R}^n)$, $0 \leq u_k \leq 1$ a.e. and

$$\int_{\mathbb{R}^n} g(r)f(u_k)u_k\, dx > 0 \ \ \text{for} \ \ k = 1, 2, \ldots \tag{4.13}$$

and that $\lambda_k \to 0^+$ as $k \to \infty$. Define

$$P_k^2 = \int_{\mathbb{R}^n} |\nabla u_k|^2 dx \tag{4.14}$$

and

$$v_k = \frac{u_k}{P_k} \quad k = 1, 2, \ldots . \tag{4.15}$$

Then

$$\int_{\mathbb{R}^n} |\nabla v_k|^2 dx = 1 \quad k = 1, 2, \ldots . \tag{4.16}$$

We also have

$$\frac{1}{P_k^2} \int_{\mathbb{R}^n} |\nabla u_k|^2 dx \leq \lambda_k \frac{1}{P_k^2} \int_{\mathbb{R}^n} g(r)f(u_k)u_k\, dx \tag{4.17}$$

or

$$1 \leq \lambda_k \int_{\mathbb{R}^n} g(r)\frac{f(u_k)u_k}{P_k^2} dx \quad k = 1, 2, \ldots . \tag{4.18}$$

Since $\lambda_k \to 0^+$, we conclude that

$$\int_{\mathbb{R}^n} g(r)\frac{f(u_k)u_k}{P_k^2} dx \to \infty \ \ \text{as} \ \ k \to \infty. \tag{4.19}$$

But

$$\left| \int_{\mathbb{R}^n} g(r)\frac{f(u_k)u_k}{P_k^2} dx \right| \leq c \int_{\mathbb{R}^n} \frac{v_k^2}{|x|^2} dx \quad k = 1, 2, \ldots \tag{4.20}$$

for some suitable constant c independent of k. We then apply Hardy's inequality

$$\int_{\mathbb{R}^n} \frac{v_k^2}{|x|^2} dx \leq c_1 \int_{\mathbb{R}^n} |\nabla v_k|^2 dx \quad k = 1, 2, \ldots \tag{4.21}$$

to arrive at

$$\int_{\mathbb{R}^n} |\nabla v_k|^2 dx \to \infty \quad \text{as} \quad k \to \infty \tag{4.22}$$

which is a contradiction. $\hspace{3cm}$ q.e.d.

We now use the previous results to obtain lower bounds on $\lambda > 0$.

Theorem 4.4. *Suppose that $\int_0^\infty r^{n-1} g(r) dr < 0$ or $n \geq 3$ and $r^2 |g(r)| \leq k$. Then problem $(\mathcal{E})$ has no solutions $u \in C^2(\mathbb{R}^n)$ when $0 < \lambda < \min(\lambda_0, \Lambda_0)$.*

Proof. Suppose $u \in C^2(\mathbb{R}^n)$ is a solution to $(\mathcal{E})_\lambda$. Then by Proposition 2.1 we have $A(\lambda) = (0, p^*)$, for some suitable p^*. Let $p \in (0, p^*)$ and $u(\cdot, p, \lambda)$ is the solution of the initial value problem (2.4)–(2.6), with R its first root. Thus the Dirichlet problem

$$\Delta u + \lambda g(r) f(u) = 0 \quad \text{in } B_R \tag{4.22}$$
$$u = 0 \quad \text{on } \partial B_R \tag{4.23}$$

has a solution, which we denote by v with $0 < v < 1$ in B_R. We define

$$w = \begin{cases} v & \text{in } B_R \\ 0 & \text{otherwise} . \end{cases} \tag{4.24}$$

Then $w \in H^1(\mathbb{R}^n)$ and moreover

$$\int_{B_R} |\nabla v|^2 dx = \lambda \int_{B_R} g(r) f(v) v \, dx. \tag{4.25}$$

Hence

$$\int_{\mathbb{R}^n} |\nabla w|^2 dx = \lambda \int_{\mathbb{R}^n} g(r) f(w) w \, dx. \tag{4.26}$$

When $n \geq 3$ and $r^2 |g(r)| \leq k$, then by Theorem 4.3 we conclude that $\lambda \geq \Lambda_0$. When $\int_0^\infty r^{n-1} g(r) dr < 0$, we define

$$h = \frac{v^2}{f(v)} \quad \text{in } B_R. \tag{4.27}$$

An easy calculation shows that $h \in H^1(B_R)$ with

$$\nabla h = \frac{v}{f^2(v)} \{2f(v) - v f'(v)\} \nabla v \quad \text{in } B_R. \tag{4.28}$$

Moreover since v satisfies (4.22), (4.23) multiplying by h and integrating, we conclude that

$$-\int_{B_R} \nabla h \nabla v \, dx + \lambda \int_{B_R} g(r)f(v)h \, dx = 0 \qquad (4.29)$$

or

$$\int_{B_R} \frac{v}{f^2(v)}\{2f(v) - vf'(v)\}|\nabla v|^2 dx = \lambda \int_{B_R} g(r)v^2 dx. \qquad (4.30)$$

However $2f(t) - tf'(t) \geq 0$ if $t \in [0,1]$. Therefore

$$\int_{\mathbb{R}^n} g(r)w^2 dx > 0 \qquad (4.31)$$

and so by Proposition 4.2, it follows that $\lambda \geq \lambda_0$.   q.e.d.

5. Continuous Dependence of Solutions

This section contains a discussion of the continuous dependence of solutions of $(\mathcal{E})$ on λ.

Let us suppose u_k satisfies

$$u_k''(r) + \frac{n-1}{r}u_k'(r) + \lambda_k g(r)f(u_k(r)) = 0 \ \text{ for } r > 0 \qquad (5.1)$$

$$u_k'(0) = 0 \qquad (5.2)$$

$$0 < u_k(r) < 1 \ \text{ for } \ r \geq 0, u_k \in C^2([0,\infty)) \qquad (5.3)$$

with $\lambda_k \to \lambda \in [0,\infty)$. The first result reads

Lemma 5.1. *There exists a subsequence $\{u_{n_k}\}_{k\in\mathbb{N}}$ such that*

$$u_{n_k} \to u \ \text{ in } \ L^\infty_{\text{loc}} \text{ as } k \to \infty \qquad (5.4)$$

where u satisfies

$$u''(r) + \frac{n-1}{r}u'(r) + \lambda g(r)f(u) = 0 \ \text{ for } r > 0 \qquad (5.5)$$

$$u'(0) = 0 \qquad (5.6)$$

$$0 \leq u(r) \leq 1 \ \text{ for } r \geq 0, \ u \in C^2([0,\infty)). \qquad (5.7)$$

The proof is carried out by using Green's function methods, i.e., by writing the o.d.e. as an integral equation of the form

$$r^{n-1}u_k'(r) + \lambda_k \int_0^r t^{n-1}g(t)f(u_k(t))dt = 0 \qquad (5.8)$$

and then by applying the Arzela–Ascoli theorem and the Lebesgue Dominated Convergence theorem. (See [18] for full details.)

Lemma 5.2. *Under the hypotheses of Lemma 5.1, when $u \not\equiv 1$ we have*

$$u_{n_k} \to u \quad in \ L^\infty \quad as \ k \to \infty. \tag{5.9}$$

Proof. By Lemma 3.1 the following estimate holds

$$u_k(x) \le \frac{|u_k|_\infty R^{n-2}}{|x|^{n-2}} \quad |x| \ge R \tag{5.10}$$

and so the result is obvious in the case $u \equiv 0$. It follows that we may suppose the existence of a solution to $(\mathcal{E})_\lambda$. Clearly a similar estimate to (5.10) holds for u. An immediate consequence is that

$$\left| u_{n_k}(x) - u(x) \right| \le \frac{2R^{n-2}}{|x|^{n-2}} \quad |x| \ge R \tag{5.11}$$

and the result follows. q.e.d.

Remark. A similar result to Lemma 5.2 holds when $n = 1, 2$ by using suitable weak supersolutions [16].

Lemma 5.3. *Under the hypotheses of Lemma 5.1, it is impossible to have $u \equiv 1$.*

Proof. Suppose the contrary, that is

$$u_{n_k} \to 1 \quad in \ L^\infty_{\mathrm{loc}} \quad as \ k \to \infty. \tag{5.12}$$

Also by Lemma 3.1 the estimate (5.10) holds. Passing to the limit in (5.10) for $|x| = m > R$ the contradiction follows. q.e.d.

We define

$$\Lambda = \{\lambda \mid \lambda > 0 \text{ such that } (\mathcal{E})_\lambda \text{ has a solution } u_\lambda\}. \tag{5.13}$$

Theorem 5.4. a) Λ *is an open set. Moreover if $\mu \in \Lambda$ then*

$$u_\lambda \to u_\mu \quad in \ L^\infty \quad as \ \lambda \to \mu. \tag{5.14}$$

b) *Suppose $0 < \mu \notin \Lambda$ and there exists $\varepsilon > 0$ such that $(\mu, \mu + \varepsilon) \subset \Lambda$. Then*

$$u_\lambda \to 0 \quad in \quad L^\infty \quad as \quad \lambda \to \mu. \tag{5.15}$$

Proof. a) Suppose $\mu \in \Lambda$. Then it follows from Proposition 2.1 that $A(\mu) \neq \emptyset$. Hence by continuous dependence of solutions of (2.4)–(2.6) in L^∞_{loc} on λ there exists $\varepsilon > 0$ such that

$$\bigcap_{|\lambda - \mu| < \varepsilon} A(\lambda) \neq \emptyset. \tag{5.16}$$

This in conjunction with the existence of weak supersolutions of the form (3.5) establishes that Λ is an open set.

Let p be an element of the set given by (5.16). Therefore

$$p \leq u_\lambda(0) \quad when \quad \lambda \in (\mu - \varepsilon, \mu + \varepsilon) \tag{5.17}$$

which prevents

$$u_{\lambda_k} \to 0 \quad in \quad L^\infty. \tag{5.18}$$

Taking into account Lemma 5.3, Lemma 5.1 gives

$$u_{\lambda_k} \to u \quad in \quad L^\infty \tag{5.19}$$

where u is a solution of $(\mathcal{E})_\mu$. Proposition 2.1 guarantees that $u \equiv u_\mu$.

b) As before Lemma 5.3 establishes convergence as in (5.19). If $u \not\equiv 0$, then Lemma 5.1 gives a solution to $(\mathcal{E})_\mu$. This completes the proof. q.e.d.

To understand the number of bifurcation points from the zero solution we consider the linearized problem

$$u'' + \frac{n-1}{r}u' + \lambda f'(0)g(r)u(r) = 0, \quad r > 0 \tag{5.20}$$

$$u'(0) = 0, \text{ and } 0 < u(r) \leq 1, \quad r \geq 0 \tag{5.21}$$

$$\lim_{r \to \infty} u(r) = 0. \tag{5.22}$$

Theorem 5.5. *If μ, λ are eigenvalues of (5.20)–(5.22) corresponding to positive eigenfunctions u, v respectively, then $\mu = \lambda$ and $u = cv$ for some suitable positive constant c.*

Proof. We may suppose without loss of generality that $\mu \leq \lambda$. On one hand multiplying

$$(r^{n-1}u'(r))' + \mu f'(0)r^{n-1}g(r)u(r) = 0 \tag{5.23}$$

by u and integrating we obtain

$$r^{n-1}u'(r)u(r) - \int_0^r s^{n-1}u'^2(s)ds + \mu f'(0)\int_0^r s^{n-1}g(s)u^2(s)ds = 0. \quad (5.24)$$

On the other hand multiplying

$$(r^{n-1}v'(r))' + \lambda f'(0)r^{n-1}g(r)v(r) = 0 \quad (5.25)$$

by u^2/v and integrating we obtain

$$r^{n-1}\frac{v'(r)}{v(r)}u^2(r) - \int_0^r s^{n-1}v'(s)\left[\frac{2uu'}{v} - \frac{u^2v'}{v^2}\right]ds$$
$$+ \lambda f'(0)\int_0^r s^{n-1}g(s)u^2(s)ds = 0. \quad (5.26)$$

By subtracting (5.24), (5.26) we arrive at

$$r^{n-1}\frac{v'(r)}{v(r)}u^2(r) - r^{n-1}u'(r)u(r) + \int_0^r s^{n-1}\left\{\frac{uv'}{v} - u'\right\}^2 ds$$
$$+ (\lambda - \mu)f'(0)\int_0^r s^{n-1}g(s)u^2(s)ds = 0. \quad (5.27)$$

It can be shown that $\lim_{r\to\infty} r^{n-1}u'(r)u(r) = 0$. Hence it follows from (5.24)
that

$$\int_0^\infty s^{n-1}u^{2\prime}(s)ds = \mu f'(0)\int_0^\infty s^{n-1}g(s)u^2(s)ds < \infty. \quad (5.28)$$

Thus we conclude from (5.26) that

$$\lim_{r\to\infty}\left\{r^{n-1}\frac{v'(r)}{v(r)}u^2(r) + \int_0^r s^{n-1}\left[\frac{uv'}{v} - u'\right]^2 ds\right\} \text{ exists in } \mathbb{R}. \quad (5.29)$$

Therefore either

$$\int_0^\infty s^{n-1}\left[\frac{uv'}{v} - u'\right]^2 ds = \infty \quad (5.30)$$

or

$$\int_0^\infty s^{n-1}\left[\frac{uv'}{v} - u'\right]^2 ds < \infty. \quad (5.31)$$

Suppose (5.30) holds. Then since $u'(r) < 0$, $v'(r) < 0$ for large r, say $r \geq R$ we have from (5.27) that

$$-r^{n-1}\frac{v'(r)}{v(r)} > \frac{1}{2}\int_R^r s^{n-1}\frac{u^2 v'^2}{v^2}ds. \tag{5.32}$$

Set

$$y(r) = \int_R^r s^{n-1}\frac{u^2 v'^2}{v^2}ds, \text{ then } y(r) \to \infty \text{ as } r \to \infty. \tag{5.33}$$

Moreover

$$r^{n-1}u^2(r)y'(r) > \frac{1}{4}y^2(r) \text{ for } r \geq R. \tag{5.34}$$

Therefore

$$\frac{d}{dr}\left[\frac{1}{y(r)} + \int_R^r \frac{ds}{4s^{n-1}u^2(s)}\right] < 0 \text{ for } r \geq R. \tag{5.35}$$

Hence by letting $r \to \infty$, we conclude that

$$\int_R^\infty \frac{ds}{4s^{n-1}u^2(s)} < \infty \tag{5.36}$$

which implies that

$$r_j^{n-2}u^2(r_j) \to \infty, \text{ with } r_j \to \infty \tag{5.37}$$

the contradiction comes from Lemma 3.1.

Therefore we conclude that (5.31) holds, in which case we derive that

$$r_j^n u^2(r_j)\frac{v'^2(r_j)}{v^2(r_j)} \to 0, \text{ for some } r_j \to \infty. \tag{5.38}$$

If we suppose that

$$r_j^{n-1}u^2(r_j)\frac{v'(r_j)}{v(r_j)} \to k \not\equiv 0 \tag{5.39}$$

then from (5.38) we conclude that

$$r_j\frac{v'(r_j)}{v(r_j)} \to 0 \text{ as } j \to \infty. \tag{5.40}$$

But then

$$r_j^{n-1}u^2(r_j)\frac{v'(r_j)}{v(r_j)} = r_j^{n-2}u^2(r_j)r_j\frac{v'(r_j)}{v(r_j)} \tag{5.41}$$

and since we can easily prove as before that

$$r_j^{n-2} u^2(r_j) \quad \text{is bounded} \tag{5.42}$$

we can conclude that

$$r_j^{n-1} u^2(r_j) \frac{v'(r_j)}{v(r_j)} \to 0 \quad \text{as} \quad j \to \infty \tag{5.43}$$

and finally that

$$\int_0^\infty s^{n-1} \left[\frac{uv'}{v} - u' \right]^2 ds + (\lambda - \mu) f'(0) \int_0^\infty s^{n-1} g(s) u^2(s) ds = 0. \tag{5.44}$$

Therefore $\lambda = \mu$ and

$$\frac{d}{dr} \left(\frac{u}{v} \right) = \frac{u'v - uv'}{u^2} = 0 \quad \text{for} \quad r \geq 0. \tag{5.45}$$

Theorem 5.6. *Suppose $0 < \mu \notin \Lambda$ and there exists $\varepsilon > 0$ such that $(\mu, \mu+\varepsilon) \subset \Lambda$. Then μ is an eigenvalue for the linear problem (5.20)–(5.22) and the corresponding eigenfunction is positive.*

Proof. From Theorem 5.4, part b) we conclude that

$$u_\lambda \to 0 \quad \text{in} \quad L^\infty \quad \text{as} \quad \lambda \to \mu. \tag{5.46}$$

Hence by defining

$$v_\lambda = u_\lambda / \|u_\lambda\|_\infty \tag{5.47}$$

it is easily seen that

$$v_\lambda''(r) + \frac{n-1}{r} v_\lambda'(r) + \lambda g(r)[f(u_\lambda)/u_\lambda] v_\lambda = 0 \text{ for } r > 0 \tag{5.48}$$

$$v_\lambda'(0) = 0 \tag{5.49}$$

$$\|v_\lambda\|_\infty = 1, \quad v_\lambda \in C^2([0,\infty)). \tag{5.50}$$

Therefore as in Lemma 5.1,

$$v_\lambda \to \varphi \quad \text{in} \quad L^\infty_{\text{loc}} \quad \text{as} \quad \lambda \to \mu \tag{5.51}$$

with φ a positive solution of (5.20)–(5.22). Moreover by using Lemma 3.1

$$v_\lambda(x) \leq \frac{R^{n-2}}{|x|^{n-2}} \quad |x| \geq R \tag{5.52}$$

and so by (5.51)

$$\varphi(x) \leq \frac{R^{n-2}}{|x|^{n-2}} \quad |x| \geq R. \tag{5.53}$$

Clearly (5.51)–(5.53) implies

$$v_\lambda \to \varphi \quad \text{in} \quad L^\infty \quad \text{as } \lambda \to \mu \tag{5.54}$$

and φ satisfies (5.20)–(5.22).   q.e.d.

6. Global Bifurcation From a Simple Eigenvalue When $n \geq 3$

In this section we discuss the global bifurcation diagram of $(\mathcal{E})$ when $n \geq 3$.

Theorem 6.1. *Suppose $\int_0^\infty r^{n-1} g(r) dr < 0$ (it may be $-\infty$). Then there exists $\lambda_{\mathrm{cr}} > 0$ such that problem $(\mathcal{E})$ has no solutions for $\lambda \in (0, \lambda_{\mathrm{cr}})$. A continuous branch in L^∞ of solutions bifurcates out of the zero solution at λ_{cr}. This branch extends for all $\lambda > \lambda_{\mathrm{cr}}$ and $\|u_\lambda\|_\infty \to 1$ as $\lambda \to \infty$. Moreover λ_{cr} is the unique positive eigenvalue for the linearized problem (5.20)–(5.22).*

Proof. It follows from Proposition 4.1 that Λ is nonempty and by Theorem 5.3 that Λ is an open set. Suppose $(\lambda_{\mathrm{cr}}, b)$ is a connected component of Λ. Clearly $\lambda_{\mathrm{cr}} > 0$ by Theorem 4.4. Also it follows by Theorems 5.4 and 5.6 that λ_{cr} is an eigenvalue of (5.20)–(5.22). Finally Theorem 5.5 establishes the uniqueness of λ_{cr}. Since existence of solutions to $(\mathcal{E})$ for λ large enough comes by Proposition 4.1 we conclude

$$\Lambda = (\lambda_{\mathrm{cr}}, \infty). \tag{6.1}$$

Continuity of the branch follows by Theorem 5.4, part a) and the asymptotics of the branch by Proposition 4.1.   q.e.d.

Remark. If in addition to our assumptions we suppose that g is a radially decreasing function, then we can prove that the branch increases as λ increases, i.e., the function $\lambda \mapsto \|u_\lambda\|_\infty$ is an increasing function for $\lambda \geq \lambda_{\mathrm{cr}}$.

Theorem 6.2. *Suppose that $r^2 |g(r)| \leq c$ for some positive constant c. Then there exists $\lambda_{\mathrm{cr}} > 0$ such that the problem has no solutions for $\lambda \in (0, \lambda_{\mathrm{cr}})$. A continuous branch of solutions bifurcates out of the zero solution at λ_{cr} in L^∞ norm. This branch extends for all $\lambda > \lambda_{\mathrm{cr}}$ and*

$\|u_\lambda\|_\infty \to 1$ *as* $\lambda \to \infty$. *Moreover* λ_{cr} *is a positive eigenvalue for the linear problem* (5.20)–(5.22).

The proof is similar to the proof of Theorem 6.1.

7. Global Bifurcation When $n = 1, 2$ — Concluding Remarks

This section deals with the case $n = 1, 2$. We begin with the case that g satisfies

$$\bar{g} = \int_{\mathbb{R}^n} g\, dx < 0. \tag{7.1}$$

Then in a similar way to the $n \geq 3$ case, and taking into account [5] the following result can be proved.

Theorem 7.1. *Suppose* $n = 1, 2$ *and* $\bar{g} < 0$ *(it may be* $-\infty$*). Then there exists* $\lambda_{\mathrm{cr}} > 0$ *such that problem* $(\mathcal{E})$ *has no solutions for* $\lambda \in (0, \lambda_{\mathrm{cr}})$. *A continuous branch of solutions bifurcates out of the zero solution at* λ_{cr} *in* L^∞ *norm. This branch extends for all* $\lambda > \lambda_{\mathrm{cr}}$ *and* $\|u_\lambda\|_\infty \to 1$ *as* $\lambda \to \infty$. *Moreover* λ_{cr} *is a positive eigenvalue for the linear problem* (5.20)–(5.21).

Remark. Necessary and sufficient conditions for solutions of (1.1)–(1.2) to satisfy (1.3) is given in [5]. Global bifurcation when $n = 1, 2$ of $(\mathcal{E})$ is independent of the asymptotics of solutions at infinity.

In the remainder of this section we consider the case $n = 1, 2$ and $\bar{g} \geq 0$.

It is clear that Hardy's inequality is no longer valid hence Theorem 4.4 does not hold. Moreover every solution of $(\mathcal{E})$ satisfies

Lemma 7.2. *Suppose* u *is a solution of* (1.1)–(1.2), $\bar{g} \geq 0$ *and* $n = 1, 2$. *Then*

$$\lim_{r \to \infty} \frac{r^{n-1} u'(r)}{f(u(r))} = 0. \tag{7.2}$$

Proof. Dividing (1.1) by u and integrating, we obtain

$$\frac{r^{n-1} u'(r)}{u(r)} - \frac{R^{n-1} u'(R)}{u(R)} + \int_R^r s^{n-1} \left\{ \frac{u'(s)}{u(s)} \right\}^2 ds$$
$$+ \lambda \int_R^r s^{n-1} g(s) \frac{f(u(s))}{u(s)} ds = 0. \tag{7.3}$$

Since the second integral converges we conclude

$$\lim_{r\to\infty}\left\{\frac{r^{n-1}u'(r)}{u(r)}+\int_R^r s^{n-1}\left\{\frac{u'(s)}{u(s)}\right\}^2 ds\right\} \tag{7.4}$$

exists in $\mathbb{R}$.

To conclude

$$\lim_{r\to\infty}\frac{r^{n-1}u'(r)}{u(r)}=0 \tag{7.5}$$

we suppose

$$\int_R^\infty s^{n-1}\left\{\frac{u'(s)}{u(s)}\right\}^2 ds=\infty \tag{7.6}$$

and

$$-\frac{r^{n-1}u'(r)}{u(r)}>\frac{1}{2}\int_R^r s^{n-1}\left\{\frac{u'(s)}{u(s)}\right\}^2 ds, \quad r\geq R. \tag{7.7}$$

Set

$$v(r)=\int_R^r s^{n-1}\left\{\frac{u'(s)}{u(s)}\right\}^2 ds. \tag{7.8}$$

Then $\lim_{r\to\infty}v(r)=\infty$ and (7.7) gives

$$v'(r)\geq\frac{1}{4r^{n-1}}v^2(r)\ \text{ for }\ r\geq R. \tag{7.9}$$

Therefore

$$\frac{d}{dr}\left(\frac{1}{v(r)}\right)\leq-\frac{1}{4r^{n-1}}\ \text{ for }\ r\geq R \tag{7.10}$$

and by integrating (7.10)

$$\frac{1}{v(r)}-\frac{1}{v(R)}\leq-\int_R^r\frac{ds}{4s^{n-1}}\ \text{ for }\ r\geq R. \tag{7.11}$$

But

$$\int_R^r\frac{ds}{4s^{n-1}}=\infty\ \text{ when }\ n=1,2 \tag{7.12}$$

and we have arrived at a contradiction.   q.e.d.

Let u be a solution of $(\mathcal{E})$ then

$$\left(r^{n-1}u'(r)\right)'+\lambda r^{n-1}g(r)f(u(r))=0. \tag{7.13}$$

Hence

$$\frac{\left(r^{n-1}u'(r)\right)'}{f(u(r))} + \lambda r^{n-1}g(r) = 0 \tag{7.14}$$

and so

$$r^{n-1}\frac{u'(r)}{f(u(r))} + \int_0^r r^{n-1}\left\{\frac{u'(r)}{f(u(r))}\right\}^2 f'(u(r))dr$$
$$+ \lambda \int_0^r r^{n-1}g(r)dr = 0. \tag{7.15}$$

By Lemma 7.2 we obtain

$$\int_0^\infty r^{n-1}\left\{\frac{u'(r)}{f(u(r))}\right\}^2 f'(u(r))dr + \lambda \int_0^\infty r^{n-1}g(r)dr = 0. \tag{7.16}$$

Therefore is σ is the unique point such that $f'(\sigma) = 0$, then it follows from (7.16) that

$$\|u\|_\infty \geq \sigma. \tag{7.17}$$

This prevents the branch from bifurcating out of the zero solution.

By making one further assumption on g

$$(\text{G}^*) \qquad \begin{cases} \lim_{r\to\infty} r^2 g(r) = -\infty & \text{when } n = 1 \\ \lim_{r\to\infty} r^2(\ln r)^2 g(r) = -\infty & \text{when } n = 2 \end{cases}$$

and taking into account [5] the following can be proved.

Theorem 7.3. *Suppose that $n = 1, 2$, g satisfies G^* and moreover $\bar{g} \geq 0$. Then there exists a continuous branch (in L^∞ norm) of solutions to problem $(\mathcal{E})$ for any $\lambda > 0$ with $\|u_\lambda\|_\infty \to 1$ as $\lambda \to \infty$. If moreover $\bar{g} > 0$ then the branch bifurcates asymptotically from $u \equiv 1$, that is*

$$u_\lambda \to 1 \quad in \quad L^\infty_{\text{loc}} \quad as \quad \lambda \to 0^+.$$

Acknowledgments. The author wish to thank the organizers, Professors N.G. Lloyd, W.-M. Ni, L.A. Peletier, and J. Serrin, for inviting him to participate in the Gregynog conference on Nonlinear Diffusion Equations and their Equilibrium States.

He also thanks the referee for his suggestions that improved the presentation of this work.

Part of this work has been done while the author was visiting Heriot-Watt University, U.K. partially supported by the Royal Society and S.E.R.C.

REFERENCES

[1] C. Bandle, M.A. Pozio and A. Tesei, *Existence and uniqueness of solutions of nonlinear Neumann problems,* Math. Zeitschrift **199**(1988), 257–278.

[2] K.J. Brown, C. Cosner and J. Fleckinger, *Principal eigenvalues for problems with indefinite weight function on $\mathbb{R}^n$,* preprint (1989).

[3] K.J. Brown and P. Hess, *Stability and uniqueness of positive solutions for a semilinear elliptic boundary value problem,* preprint (1989).

[4] K.J. Brown, S.S. Lin and A. Tertikas, *Existence and nonexistence of steady-state solutions for a selection-migration model in population genetics,* J. Math. Biol. **27**(1989), 91–104.

[5] K.J. Brown and A. Tertikas, *On the bifurcation of radially symmetric steady-states solutions arising in population genetics,* to appear in SIAM J. Math. Anal.

[6] P.C. Fife and L.A. Peletier, *Nonlinear diffusion in population genetics,* Arch. Rat. Mech. Anal. **64**(1977), 93–109.

[7] W.H. Fleming, *A selection-migration model in population genetics,* J. Math. Biol. **2**(1975), 219–233.

[8] B. Gidas, W.N. Ni and L. Nirenberg, *Summetry of positive solutions of nonlinear elliptic equations in $\mathbb{R}^n$,* Math. Anal. Appl., part A, Adv. Math. Suppl. Studies **7A**(1981), 369–402.

[9] P. Hess and T. Kato, *On some linear and nonlinear eigenvalue problems with an indefinite weight function,* Comm. Partial Differential Equations 5(1980), 999–1030.

[10] H. Matano, L^∞ *stability of an exponentially decreasing solution of the problem $\Delta u + f(x,u) = 0$ in $\mathbb{R}^n$,* Japan J. Appl. Math. **2**(1985), 85–110.

[11] N. Meyers and J. Serrin, *The exterior Dirichlet problem for second-order elliptic partial differential equations,* J. Math. Mech. **9**(1960), 513–538.

[12] W.M. Ni, *On the elliptic equation $\Delta u + k(x)u^{(n+2)/(n-2)}$ its generalizations and applications in geometry,* Indiana Univ. J. **31**(1982), 493–529.

[13] L.A. Peletier and J. Serrin, *Uniqueness of positive solutions of semilinear equations in $\mathbb{R}^n$,* Arch. Rat. Mech. Anal. **81**(1983), 181–197.

[14] L.A. Peletier and A. Tesei, *Global bifurcation and attractivity of stationary solutions of a degenerate diffusion equation,* Adv. Appl. Math. **7**(1986), 435–454.

[15] M. Slatkin, *Gene flow and selection in a cline,* Genetics **75**(1973), 733–756.

[16] N. Stavrakakis, J. Stratis and A. Tertikas, in preparation.

[17] C.A. Stuart, *Bifurcation in $L^p(\mathbb{R}^n)$ for a semilinear elliptic equation,* Proc. London Math. Soc. **57**(1988), 511–541

[18] A. Tertikas, *Semilinear elliptic equations on $\mathbb{R}^n$,* Ph.D. Thesis, Heriot-Watt University, 1987.

[19] A. Tertikas, *Existence and Uniqueness of solutions for a nonlinear diffusion problem arising in population genetics,* Arch. Rat. Mech. Anal. 4(1988), 289–317.

[20] A. Tertikas, *Uniqueness of solutions for problems arising in population genetics,* in *Differential Equations,* C.M. Dafermos, G. Ladas and G. Papanicolaou, Eds. Lecture Notes in Pure and Applied Mathematics, **118**(1989), 667–672.

[21] A. Tertikas, J.F. Toland, *Graph intersection and uniqueness results for some nonlinear elliptic problems,* to appear in J. Diff. Eqs.

[22] J.F. Toland, *Positive solutions of nonlinear elliptic equations-existence and nonexistence of solutions with radial symmetry in $L^p(\mathbb{R}^n)$,* Trans. Amer. Math. Soc. **282**(1984), 335–354.

Department of Mathematics
University of the Aegean
9 Kanari Street
10671 Athens
GREECE

Present address:

Department of Mathematics
University of Crete
Iraklion – Crete
P.O. Box 1470
GREECE

Conformal Asymptotics of the Isothermal Gas Spheres Equation

LAURENT VERON

1. Introduction

In his famous work on stellar structure [5] Chandrasekhar studied the particular case of an isothermal gas sphere in gravitational equilibrium. Let P be the pressure of the gas, ρ the density and T the absolute temperature; then, according to the second law of thermodynamics, the thermodynamics of a perfect gas in expansion or contraction and Stefan radiation law, the following relation is satisfied

$$(1.1) \qquad P = \left(\frac{k}{\mu H}\right)\rho T + \frac{a}{3}T^4$$

where k is the Boltzmann constant, μ the mean molecular weight, H the mass of the proton and a the Stefan–Boltzmann constant. As T is constant (1.1) reads as follows

$$(1.2) \qquad P = K\rho + D,$$

where the constants K and D are

$$(1.3) \qquad K = \frac{k}{\mu H}T, \qquad D = \frac{a}{3}T^4.$$

The equation of equilibrium between pressure and gravitational attraction is then

$$(1.4) \qquad \operatorname{div}\left(\frac{1}{\rho}\operatorname{grad}P\right) = -4\pi G\rho$$

where G is Newton's constant, and this can be written

$$(1.5) \qquad K\Delta(\operatorname{Ln}\rho) = -4\pi G\rho.$$

If we make the substitution

$$(1.6) \qquad \rho = e^u, \qquad \lambda = \frac{4\pi G}{K}$$

we are led to the study of the following equation

$$(1.7) \qquad -\Delta u = \lambda e^u$$

538   LAURENT VERON

in a 3-dimensional domain, where λ is a positive constant. Radial solutions of (1.7) have been studied for a long time, in particular by Emden [8] and Chandrasekhar [5]; this last author proved in particular that there exists only one radial solution of (1.7) with a positive singularity at 0, and it is the function

$$(1.8) \qquad u_s(x) = \mathrm{Ln}(1/|x|^2) + \mathrm{Ln}(2/\lambda).$$

If we look for singular solutions of (1.7) under the form

$$(1.9) \qquad u(r,\sigma) = \mathrm{Ln}(1/r^2) + \mathrm{Ln}(2/\lambda) + 2\omega(\sigma)$$

where $(r,\sigma) \in \mathbb{R}_*^+ \times S^2$ are the spherical coordinates in $\mathbb{R}^3 \setminus \{0\}$, we find that ω must satisfy

$$(1.10) \qquad \Delta_{S^2}\omega + e^{2\omega} - 1 = 0$$

on S^2 where Δ_{S^2} is the Laplace–Beltrami operator on (S^2, g_0), g_0 being the standard metric. This equation is invariant under the conformal transformations of S^2 and the set G_2 of all continuous (and then analytic) solutions of (1.10) is described in [2], [3].

The set G_2 is the set of functions $\omega = \frac{1}{2}\mathrm{Ln}(\det |d\phi|)$ where ϕ is a conformal transformation of S^2.

The set of conformal transformations of S^2 is isomorphic to the group of homographic transformations of $\mathbb{C}$ with determinant 1 and it has the structure of a 6-dimensional noncompact real Lie group; G_2 can be endowed with a structure of a 3-dimensional noncompact Lie group.

In Section 2 we point out how the set G_2 plays a crucial role to describe the isolated singularities or the asymptotic behaviour of general solutions of (1.7). A key-stone step in this description is the extension of Simon's result [17] concerning the asymptotics of analytic functionals in an infinite cylinder. In Section 3 we study the boundary value problems associated to (1.7) in the complementary of a ball, or in a punctured ball with a prescribed singularity at the center.

The contents of this contribution is the following

-1- Introduction
-2- Asymptotics of $-\Delta u = \lambda e^u$
-3- The boundary value problems
-4- Asymptotics of analytic functionals

Acknowledgments. The results of Sections 2 and 4 are a common work with M.F. Bidaut-Véron and those of Section 3 have been obtained with H. Matano.

2. Asymptotics of $-\Delta u = \lambda e^u$

Throughout this section we assume that u satisfies

$$(2.1) \qquad -\Delta u = \lambda e^u$$

in some 3-*dimensional* euclidean domain where λ is a positive constant. Our first result is that, under a natural growth estimate near a possible isolated singularity we have a complete description of the possible behaviours of u.

Theorem 2.1. *Assume u is a solution of* (2.1) *in* $B_1^* = B_1(0) \setminus \{0\}$ *such that*

$$(2.2) \qquad |x|^2 e^u \in L^\infty_{\mathrm{loc}}(B_1(0)).$$

Then we have the following

(i) *either u can be extended to $B_1(0)$ as a C^∞ solution of* (2.1) *in* $B_1(0)$,

(ii) *or there exist $\gamma < 0$ and $\ell \in \mathbb{R}$ such that*

$$(2.3) \qquad \lim_{x \to 0}(u(x) - \gamma/|x|) = \ell$$

and u satisfies

$$(2.4) \qquad -\Delta u = \lambda e^u + 4\pi\gamma\delta_0$$

in $\mathbf{D}'(B_1(0))$,

(iii) *or there exists $\omega \in G_2$ such that*

$$(2.5) \qquad \lim_{r \to 0}(u(r, \cdot) - \mathrm{Ln}(1/r^2)) = 2\omega(\cdot) + \mathrm{Ln}(2/\lambda)$$

in the C^k-topology of S^2, for any $k \in \mathbb{N}$.

As the complete proof is given in [3] we shall just sketch it.

Lemma 2.1. *Under the hypotheses of Theorem 2.1 there exists $\gamma \leqq 0$ such that*

$$(2.6) \qquad -\Delta u = \lambda e^u + 4\pi\gamma\delta_0$$

in $\mathbf{D}'(B_1(0))$. *There exist also two constants* K_1 *and* $K_2 \geqq 0$ *such that the following estimate holds in* $\bar{B}^*_{1/2} = \bar{B}_{1/2}(0) \setminus \{0\}$

$$(2.7) \qquad \gamma/|x| - K_2 \leqq u(x) \leqq \mathrm{Ln}(1/|x|^2) + K_1.$$

Proof. As (2.2) implies the right-hand side of (2.7) we shall just prove the left-hand side inequality. Defining

$$(2.8) \qquad \tilde{u}(x) = -u(x) + \mathrm{Ln}(1/|x|^2) + K_1$$

then $\tilde{u} \geqq 0$ and

$$(2.9) \qquad \Delta\tilde{u} = \frac{1}{|x|^2}\left(\lambda e^{K_1} e^{-\tilde{u}} - 2\right) = \Phi.$$

As $\Phi \in L^1(B_{1/2}(0))$, $\tilde{u} \in L^1(B_{1/2}(0))$ and there exists $\gamma \leqq 0$ such that

$$(2.10) \qquad \Delta\tilde{u} = \Phi + 4\pi\gamma\delta_0$$

in $\mathbf{D}'(B_{1/2}(0))$ [4]. As u and $\tilde{u}$ belongs to the Marcinkiewicz space $M^3_{\mathrm{loc}}(B_1(0))$ we deduce from maximum principle that the left-hand side inequality (2.7) is valid.

Let $(r, \sigma) \in R^+_* \times S^2$ be the spherical coordinates in $\mathbb{R}^3 \setminus \{0\}$ and $\bar{\phi}(r)$ be the spherical average of any function ϕ, $(r, \sigma) \to \phi(r, \sigma)$, belonging to L^1.

Lemma 2.2. *Under the hypotheses of Theorem 2.1 there exists a constant* $M \geqq 0$ *such that*

$$(2.11) \qquad \left\| u(r, \cdot) - \bar{u}(r) \right\|_{L^\infty(S^2)} \leqq M.$$

Proof. Defining u^* by

$$(2.12) \qquad u^*(t, \sigma) = u(r, \sigma), \qquad t = \mathrm{Ln}(1/r),$$

we see that

$$(2.13) \qquad u^*_{tt} - u^*_t + \Delta_{S^2} u^* + \lambda e^{-2t} e^{u^*} = 0$$

and (2.7) reads as follows

$$(2.14) \qquad \gamma e^t - K_2 \leqq u^*(t, \sigma) \leqq 2t + K_1$$

for $(t,\sigma) \in [a,+\infty) \times S^2$ $(a > 0)$. If we define

$$(2.15) \qquad \psi = u^* - \bar{u}^*, \qquad f = \lambda e^{2t}\left(e^{u^*} - e^{\bar{u}^*}\right),$$

we have

$$(2.16) \qquad \psi_{tt} - \psi_t + \Delta_{S^2}\psi + f = 0.$$

As 2 is the first nonzero eigenvalue of $-\Delta_{S^2}$ we deduce that ψ unique in the class of η, with zero average on S^2, satisfying

$$(2.17) \qquad \lim_{t \to \infty} e^{-2t}\left\|\eta(t,\cdot)\right\|_{L^2(S^2)} = 0.$$

Let $(S(t))_{t \geq 0}$ be the semigroup of contractions of $L^2(S^2)$ generated by $-\left(-\Delta_{S^2} + \frac{1}{4}I\right)^{1/2}$, then we have

$$(2/18)\quad \psi(t+a) = e^{t/2}S(t)\psi(a)$$
$$+ \int_0^\infty e^{s/2}S(s)\int_0^\infty e^{-\tau/2}S(\tau)f(t+\tau-s+a)\,d\tau\,ds$$

as in [18]. As

$$(2.19) \qquad \left\|S(t)\eta\right\|_{L^\infty(S^2)} \leq C\,e^{-3/2t}\left\|\eta\right\|_{L^\infty(S^2)}$$

holds for any $t > 0$ and any η with zero spherical average, we get (2.11).

Lemma 2.3. *Under the hypotheses of Theorem 2.1, we have the following alternative*

(i) *either there exists $\gamma \leq 0$ such that $u(x) - \gamma/|x|$ remains bounded independently of x for $0 < |x| \leq 1/2$,*

(ii) *or $u(x) - \mathrm{Ln}(1/|x|^2)$ remains bounded independently of x for $0 < |x| \leq 1/2$.*

Proof. From concavity there exists $\gamma \in [-\infty, 0]$ such that

$$(2.20) \qquad \lim_{r \to 0} r\,\bar{u}(r) = \gamma,$$

and $\gamma > -\infty$ from Lemma 2.1. If $\gamma < 0$ we have (i). In the case $\gamma = 0$ we claim that

(α) either $\bar{u}(r)$ is bounded near 0,
(β) or $\bar{u}(r) - \mathrm{Ln}(1/r^2)$ is bounded near 0.

In fact (2.7) reads as follows

$$(2.21) \qquad -K_2 \leqq u(x) \leqq \mathrm{Ln}(1/|x|^2) + K_1$$

and (with the notations of Lemma 2.2)

$$(2.22) \qquad \left(e^{-t}\bar{u}_t^*(t)\right)_t < 0.$$

If there exists $t_0 > a$ such that $u_t^*(t_0) \leqq 0$, we deduce that

$$(2.23) \qquad \bar{u}_t^*(t) \leqq e^{t-t_0}\bar{u}_t^*(t_0) \leqq 0$$

holds for $t \geqq t_0$ and we have (α). If such a t_0 does not exist we define

$$(2.24) \qquad v^*(t, \cdot) = u^*(t, \cdot) - 2t, \qquad L = \limsup_{t \to \infty} \bar{v}^*(t).$$

In the case where $L = -\infty$, $v^*(t, \cdot)$ converges to $-\infty$ uniformly on S^2. As $\bar{u}^*$ satisfies

$$(2.25) \qquad \left(e^{-t}\bar{u}_t^*\right)_t + \lambda e^{-t}\overline{e^{v^*}} = 0,$$

we deduce easily that $\bar{u}_t^* \in L^1(a, +\infty)$ which implies (α). In the case where $L > -\infty$ we define

$$(2.26) \qquad \ell = \liminf_{t \to \infty} \bar{v}^*(t).$$

If $\ell > -\infty$ we obtain (β); henceforth we are left with $\ell = -\infty$. Proceeding by contradiction we easily check that there could not exist a sequence $\{t_n\}$ such that $\bar{v}_t^*(t_n) = 0$ and

$$(2.27) \qquad \begin{cases} \bar{v}^*(t_{2n}) \to_{n \to \infty} -\infty, \\ \bar{v}^*(t_{2n+1}) > \bar{v}^*(t_{2n}), \\ \bar{v}^* \text{ is nondecreasing on } [t_{2n}, t_{2n+1}]. \end{cases}$$

Henceforth $\ell > -\infty$, which ends the proof.

Proof of Theorem 2.1. From Lemma 2.3 and elliptic equations regularity theory we are left with the case where

$$(2.28) \qquad |u(r, \sigma) - \mathrm{Ln}(1/r^2)| \leqq M$$

for $(r, \sigma) \in (0, 1/2] \times S^2$. We then define

$$(2.29) \qquad 2v(t, \sigma) = u(r, \sigma) - \mathrm{Ln}(1/r^2) - \mathrm{Ln}(2/\lambda),$$

with $t = \mathrm{Ln}\, r$; v satisfies

$$(2.30) \qquad v_{tt} + v_t + \Delta_{S^2} v + e^{2v} - 1 = 0$$

on $(-\infty, b] \times S^2$. As v is uniformly bounded it is not difficult to see that for any $k \in \mathbb{N}^*$ and any $\ell \in \mathbb{N}$, $\frac{\partial^\ell}{\partial t^\ell} v(t, \cdot)$ is bounded in $C^k(S^2)$ independently of $t \leq b$. Multiplying (2.30) by v_t implies

$$(2.31) \qquad \int_{-\infty}^{b} \int_{S^2} v_t^2 \, d\sigma \, dt < \infty,$$

and from (2.31) we easily deduce that

$$(2.32) \qquad \int_{-\infty}^{b} \int_{S^2} \left| \frac{\partial^\ell}{\partial t^\ell} \nabla^k v \right|^2 d\sigma \, dt < \infty$$

$(\ell \in \mathbb{N}^*)$. Henceforth $v(t, \cdot)$ approaches one of the connected components of the set G_2 in the C^k-topology of S^2 when t tends to $-\infty$. The fact that G_2 is a 3-dimensional manifold gives rise to a huge difficulty which can be overcome by using the analyticity of $u \to e^u$ and Theorem 3.1.

Remark 2.1. It appears that estimate (2.2) is very difficult to obtain. In fact if we compare with the classical Lane–Emden equation in a N-dimensional domain, that is,

$$(2.33) \qquad -\Delta u = u^q;$$

then a priori estimates of positive solutions near an isolated singularity are known only in the case

$$(2.34) \qquad 1 < q < (N+2)/(N-2)$$

([9], [3]). If $N = 3$ the critical exponent is 5.

Remark 2.2. It is easy to check that for any $\phi \in C^0(\partial B_1(0))$ and any $\gamma \leq 0$ there exists a solution of (2.4) in $\mathbf{D}'(B_1(0))$ with value ϕ on $\partial B_1(0)$.

Similarly to Theorem 2.1, one has

Theorem 2.2. *Assume u is a solution of* (2.1) *in $B_1(0)$ such that*

$$(2.35) \qquad |x|^2 e^u \in L^\infty(B_1(0)).$$

Then there exists $\omega \in G_2$ such that

$$(2.36) \qquad \lim_{r \to \infty} (u(r, \cdot) - \mathrm{Ln}(1/r^2)) = 2\omega(\cdot) + \mathrm{Ln}(2/\lambda)$$

in the C^k-topology of S^2, for any $k \in \mathbb{N}$.

If we look now for global solutions of (2.1), in $\mathbb{R}^2 \setminus \{0\}$ we have

Theorem 2.3. *Assume u is a solution of (2.1) in $\mathbb{R}^3 \setminus \{0\}$ such that*

$$(2.37) \qquad |x|^2 e^u \in L^\infty(\mathbb{R}^3).$$

Then we have the following

(i) *either u can be extended to $\mathbb{R}^3$ as a C^∞ solution of (2.1) in $\mathbb{R}^3$,*

(ii) *or there exists $\gamma < 0$ such that (2.3) and (2.4) hold in $\mathbf{D}'(\mathbb{R}^3)$,*

(iii) *or there exists $\omega \in G_2$ such that*

$$(2.38) \qquad u(r, \sigma) = \mathrm{Ln}(1/r^2) + \mathrm{Ln}(2/\lambda) + 2\omega(\sigma)$$

for any $(r, \sigma) \in (0, +\infty) \times S^2$.

Proof. From Theorems 2.1, 2.2, we are left with the case when there exist ω_1 and ω_2 in G_2 such that

$$(2.39) \qquad \lim_{t \to -\infty} v(t, \cdot) = \omega_1(\cdot), \qquad \lim_{t \to +\infty} v(t, \cdot) = \omega_2(\cdot).$$

Defining the energy function ξ on $W^{1,2}(S^2)$ by

$$(2.40) \qquad \xi(\phi) = \int_{S^2} \left\{ \frac{1}{2} |\nabla \phi|^2 - \frac{1}{2} e^{2\phi} + \phi \right\} d\sigma,$$

we have

$$(2.41) \qquad \int_{-\infty}^{\infty} \int_{S^2} v_t^2 \, d\sigma \, dt = \xi(\omega_2) - \xi(\omega_1).$$

It has been noticed in [6] that ξ is constant on G_2 with value -2π. Henceforth v is constant which implies (iii).

3. The Boundary Value Problems

The result that we present here has been obtained in colloration with
H. Matano [13]. We first study the exterior boundary value problem.

Theorem 3.1. *Assume $\lambda > 0$ and $\omega \in G_2$. Then there exists $\delta > 0$
such that for any $\phi \in H^2(S^2)$ satisfying*

$$(3.1) \qquad \|\omega - \phi\|_{H^2(S^2)} \leqq \delta$$

there exists a one-parameter family of solutions u of

$$(3.2) \qquad \begin{cases} -\Delta u = \lambda e^u & in\ \complement B_1(0) \\ u = \phi & on\ \partial B_1(0) \end{cases}$$

such that

$$(3.3) \qquad \lim_{r \to \infty} \left(u(r, \cdot) - \operatorname{Ln}(1/r^2) \right) = \operatorname{Ln}(2/\lambda) + 2\omega(\cdot).$$

The parameter can be identified with $\bar{u}_r(1)$.

The natural way to construct such a u is to look at a perturbation
of the linearized equation following ω: if u is a solution of (2.1) satisfying
(3.3) we write

$$(3.4) \qquad u(r, \sigma) = \operatorname{Ln}(1/r^2) + \operatorname{Ln}(2/\lambda) + 2(\omega(\sigma) + q(t, \sigma))$$

with $t = \operatorname{Ln} r$, and q satisfies

$$(3.5) \qquad q_{tt} + q_t + \Delta_{S^2} q + 2e^{2\omega} q + e^{2\omega}(e^{2q} - 2q - 1) = 0$$

in $\mathbb{R}^+ \times S^2$, with

$$(3.6) \qquad \lim_{t \to \infty} q(t, \cdot) = 0.$$

The linearized equation at $q = 0$ is the following one

$$(3.7) \qquad y_{tt} + y_t + \Delta_{S^2} y + 2e^{2\omega} y = 0$$

and we have the following characterization of $\operatorname{Spec}\left(-(\Delta_{S^2} + 2e^{2\omega})\right)$:

Lemma 3.1. *Let $\{\lambda_j : 1 \leqq j\}$ be the increasing sequence of the eigenvalues of $-(\Delta_{S^2} + 2e^{2\omega})$ in $H^1(S^2)$ and H_j the corresponding eigenspaces.*

Then $\lambda_2 = 0$, $\dim H_2 = 3$ and H_2 is generated by the restriction to S^2 of $x^\ell \circ \phi$, $\ell = 1, 2, 3$ where the x^ℓ are the coordinate functions in $\mathbb{R}^3$ and ϕ is a conformal transformation on S^2 such that

$$(3.8) \qquad \omega = \frac{1}{2} \operatorname{Ln} \det |d\phi|.$$

Proof. If ψ is any function defined on S^2 we have

$$(3.9) \qquad \Delta_{S^2}(\psi \circ \phi) = \det|d\phi^{-1}|(\Delta_{S^2}\psi) \circ \phi,$$

which means

$$(3.10) \qquad \Delta_{S^2}(\psi \circ \phi) = e^{-2\omega}(\Delta_{S^2}\psi) \circ \phi.$$

As the second eigenspace of $-\Delta_{S^2}$ is generated by the restriction to S^2 of (x^1, x^2, x^3), with second eigenvalue 2, we get the result.

Remark 3.1. This result is the consequence of the 3-dimensional Lie group structure on G_2. It is interesting to note that the set G_2 is not degenerated in the sense that its dimension is always equal to the dimension of the kernel of the linearized form at any element of G_2 of the operator characterizing G_2.

It is also interesting to notice that $\lambda_1 < -2$ unless $\omega \equiv 0$.

Lemma 3.2. *For any $\psi \in L^2(S^2)$ and any $\gamma \in \mathbb{R}$ there exists a unique solution y of (3.7) defined on $\mathbb{R}^+ \times S^2$, tending to 0 at infinity and satisfying*

$$(3.11) \qquad y(0, \sigma) = \psi(\sigma); \qquad \bar{y}_t(0) = \gamma.$$

Proof. We project the equation (3.7) onto H_j and we get

$$(3.12) \qquad y_j'' + y_j' - \lambda_j y_j = 0 \qquad (j \geq 1),$$

with associated characteristic equation

$$(3.13) \qquad \rho^2 + \rho - \lambda_j = 0.$$

Henceforth, if the roots of (3.13) are the α_j, β_j, we have

$$(3.14) \qquad \begin{cases} \text{(i)} & \alpha_1 < \beta_1 < 0, \\ \text{(ii)} & \alpha_2 = -1, \ \beta_2 = 0, \\ \text{(iii)} & \alpha_j < 0 < \beta_j \qquad (\forall j \geq 3). \end{cases}$$

If y is a bounded solution of (3.7) tending to 0 at infinity, we get

$$y_1(t) = (\alpha_1 - \beta_1)^{-1}((y_1'(0) - \beta_1 y_1(0))e^{\alpha_1 t}$$
$$- (y_1'(0) - \alpha_1 y_1(0))e^{\beta_1 t})$$
$$y_2(t) = y_2(0)e^{-t}$$
$$y_j(t) = y_j(0)e^{\alpha_j t} \qquad (\forall j \geqq 3).$$

As a consequence the coefficients $y_j(0)$ $(j \geqq 1)$ and $y_1'(0)$ are uniquely determined by (3.11).

Remark 3.2. It is interesting to notice that the stable manifold of 0 in (3.7) is such that for $j \geq 2$ $y_j'(0)$ is a (linear) function of $y_j(0)$. Henceforth the idea for proving Theorem 3.1 is to write (3.5) as a system in the product space $E = H^1(S^2) \times L^2(S^2)$ with

$$v = \begin{pmatrix} q \\ \rho \end{pmatrix}, \ A = \begin{pmatrix} 0 & I \\ -\Delta_{S^2} - 2e^{2\omega}I & -I \end{pmatrix}, \ f(v) = \begin{pmatrix} 0 \\ -(e^{2q} - 2q - 1)e^{2\omega} \end{pmatrix}$$

where $D(A) = H^2(S^2) \times H^1(S^2)$; then (3.5) becomes

$$(3.15) \qquad\qquad v_t = Av + f(v).$$

The eigenvalues of A are the roots of (3.13) and their sign is given in (3.14). If ψ_j is an eigenvector corresponding to the eigenvalue λ_j and Λ_j a root of (3.13), then $v_j = \begin{pmatrix} \psi_j \\ \Lambda_j \psi_j \end{pmatrix}$ is an eigenvector of A with eigenvalue Λ_j. We then define

$$E^0 = \begin{pmatrix} H_2 \\ 0 \end{pmatrix}, E^- = \begin{pmatrix} H_2 \\ -H_2 \end{pmatrix} \oplus \begin{pmatrix} H_1 \\ \beta_1 H_1 \end{pmatrix} \oplus \begin{pmatrix} H_1 \\ \alpha_1 H_1 \end{pmatrix} \bigoplus_{j=3}^{\infty} \begin{pmatrix} H_j \\ \alpha H_j \end{pmatrix}$$

$$E^+ = \bigoplus_{j=3}^{\infty} \begin{pmatrix} H_j \\ \beta_j H_j \end{pmatrix} \quad \text{and we have, algebraically at least,}$$

$$(3.16) \qquad\qquad E = E^0 \oplus E^- \oplus E^+.$$

Let P^+, P^- and P^0 be the projections of E onto E^+, E^-, and E^0, respectively. Then it can be checked that P^+, P^-, and P^0 are continuous; henceforth (3.16) holds topologically. We set

$$(3.17) \qquad\qquad A^+ = AP^+, \qquad A^- = AP^-,$$

and we have

$$(3.18) \qquad \mathrm{Spec}(A^-) \subset \{\lambda \in \mathbb{R}: \lambda \leq \min(-1, \beta_1, \alpha_3)\},$$

$$(3.19) \qquad \mathrm{Spec}(A^+) \subset \{\lambda \in \mathbb{R}: \lambda \geq \beta_3\}.$$

Using Fourier series techniques it is easy to check that

$$(3.20) \qquad (A^+ v, v)_{H^1 \times L^2} \geq \sigma \|v\|^2_{H^1 \times L^2} \qquad (\forall v \in D(A^+))$$

for some $\sigma > 0$ and that $(A^+)^{-1} \in \mathbf{L}(E^+, E^+)$. The same assertions hold for $-A^-$. As a consequence $-A^+$ and A^- are the infinitesimal generators of two analytic exponentially decaying semigroups $(e^{-tA^+})_{t \geq 0}$ and $(e^{tA^-})_{t \geq 0}$. We set

$$(3.21) \qquad H = D((-A^+_{|E^+})^\theta) \oplus D((A^-_{|E^-})^\theta) \oplus E^0 = H^+ \oplus H^- \oplus E^0$$

for some $\theta \in (0, 1)$ (we may define those spaces by Fourier series too). As $D((-A^+_{|E^+})^\theta) \subset L^\infty(S^2) \times H^\theta(S^2)$ (and the same holds for $D((A^-_{|E^-})^\theta)$), f is a C^2 function from H to E satisfying

$$(3.22) \qquad f(0) = f'(0) = 0.$$

If v is a solution of (3.15) with

$$(3.23) \qquad v = v^+ + v^- + v^0 \in E^+ \oplus E^- \oplus E^0,$$

then

$$(3.24) \qquad \begin{cases} v_t^+ = A^+ v_t^+ + P^+ f(v), \\ v_t^- = A^- v_t^- + P^- f(v), \\ v_t^0 = P^0 f(v), \end{cases}$$

and, if $v(t)$ goes to zero "fast enough" when t tends to ∞, we also have

$$(3.25) \qquad v^+(t) = -\int_t^\infty e^{-(\tau - t)A^+} P^+ f(v(\tau)) d\tau,$$

$$(3.26) \qquad v^0(t) = -\int_t^\infty P^0 f(v(\tau)) d\tau,$$

which implies

$$(3.27) \quad v(t) = e^{tA^-} P^- v(t) + \int_0^t e^{(t-\tau)A^-} P^- f(v(\tau)) d\tau$$
$$- \int_t^\infty e^{-(\tau - t)A^+} P^+ f(v(\tau)) d\tau - \int_t^\infty P^0 f(v(\tau)) d\tau.$$

Conversely, a fast enough decaying solution of (3.27) satisfies (3.15). Henceforth the problem is reduced to find this class of "fast enough decaying functions" and to get a solution by fixed point. We start with some estimates: from (3.22) there exists $\eta\colon \mathbb{R}^+ \to \mathbb{R}^+$ such that $\lim_{\rho \to 0} \eta(\rho) = 0$ (we may assume η nondecreasing) and

$$(3.28) \qquad \|f(x) - f(y)\|_E \leqq \eta(\rho)\|x - y\|_H$$

for any $x, y \in H$ such that $\|x\|_H \leqq \rho$, $\|y\|_H \leqq \rho$. We also have

$$(3.29) \qquad \begin{cases} \|e^{tA^-}P^-\|_{\mathbf{L}(E,E)} \leqq M\,e^{-at} \\[4pt] \|e^{-tA^+}P^+\|_{\mathbf{L}(E,E)} \leqq M\,e^{-at} \\[4pt] \|e^{tA^-}P^-\|_{\mathbf{L}(E,H)} \leqq M\,t^{-\theta}\,e^{-at} \\[4pt] \|e^{-tA^+}P^+\|_{\mathbf{L}(E,H)} \leqq M\,t^{-\theta}\,e^{-at} \end{cases}$$

for any $t > 0$, and some $a > 0$ and $M > 0$. For $0 < \alpha < a$ we introduce the class $S(v_0, \delta, \rho)$ where $\delta, \rho > 0$ will be fixed later, $v_0 \in H^-$ with $\|v_0\|_H \leqq \delta$ and

$$(3.30)$$
$$S(v_0, \delta, \rho) = \{\phi \in C(\mathbb{R}^+; H) / P^- \phi(0) = v_0 \text{ and } \|\phi(t)\|_H \leqq \rho e^{-\alpha t} \ (\forall t \geqq 0)\}.$$

If $\phi \in S(v_0, \delta, \rho)$ let T be defined by

$$(3.31)\ (T\phi)(t) = e^{-tA^-}v_0 + \int_0^t e^{(t-\tau)A^-}P^- f(\phi(\tau))d\tau$$

$$- \int_t^\infty e^{-(\tau-t)A^+}P^+ f(\phi(\tau))d\tau - \int_t^\infty P^0 f(\phi(\tau))d\tau.$$

We claim now that for ρ and δ small enough T is a strict contraction from $S(v_0, \delta, \rho)$ into itself. In fact

$$\|(T\phi)(t)\|_H \leq M\,\delta\,e^{-at} + M\,\eta(\rho)\rho \int_0^t \tau^{-\theta}e^{-a\tau}e^{-\alpha(\tau-\tau)}$$

$$+ M\eta(\rho)\rho \int_0^\infty \tau^{-\theta}e^{-a\tau}e^{-\alpha(t+\tau)}d\tau + M\eta(\rho)\rho \int_t^\infty e^{-\alpha\tau}d\tau.$$

Let M_θ be $\max\left(\frac{M}{\alpha}, \int_0^\infty \tau^{-\theta}e^{-(a-\alpha)\tau}d\tau\right)$, then

$$(3.32) \qquad \|(T\phi)(t)\|_H \leqq (M\delta + 3M_\theta\rho\,\eta(\rho))e^{-\alpha t}$$

and we take ρ, δ such that

$$(3.33) \qquad 3M_\theta\eta(\rho) \leqq 1/2, \qquad M\delta \leqq \rho/2.$$

Defining the distance on $S(v_0, \delta, \rho)$ by

$$(3.34) \qquad d(\phi, \psi) = \sup_{t \geq 0}\left(e^{\alpha t}\|\phi(t) - \psi(t)\|_H\right),$$

we easily get the strict contraction property

$$(3.35) \qquad d(T\phi, T\psi) \leq 3M_\theta\, \eta(\rho)\, d(\phi, \psi) \leq \frac{1}{2}d(\phi, \psi),$$

which ends the proof of Theorem 3.1.

Remark 3.3. Using Hale's techniques [10], [11], we can study the regularity of the mapping $v \to v(\cdot, v_0)$ where $v(\cdot, v_0)$ is the unique local fixed point of T in $S(v_0, \delta, \rho)$; it should be easy to check that this mapping is Lipschitz continuous.

Remark 3.4. If we look at the boundary value problem in B_1^* with a prescribed value on ∂B_1 and a prescribed singularity $\omega \in G_2$ at the origin then it is clear that in general this problem is not solvable: for example if u, solution of (2.1) in B_1^*, is constant on ∂B_1 then u must be radial [15] and in fact it must be the restriction to B_1^* of the singular solution (1.8) [5], [14]. However, it is easy to adapt the proof of Theorem 3.1 to get an existence result where H^- is replaced by H^+.

Theorem 3.2. *Assume* $\lambda > 0$ *and* $\omega \in G_2$. *Then there exists* $\delta > 0$ *and four linearly independent continuous linear forms on* $L^2(S^2)$ L_k *(k =* $1, 2, 3, 4$*) such that for any* $\phi \in H^2(S^2)$ *satisfying*

$$(3.36) \qquad \|\omega - \phi\|_{H^2(S^2)} \leq \delta$$

$$(3.37) \qquad L_k(\phi) = 0 \qquad (\forall k = 1, \ldots, k)$$

there exists a solution u of

$$(3.38) \qquad \begin{cases} -\Delta u = \lambda e^u & in\ B_1^*, \\ u = \phi & on\ \partial B_1(0), \\ \lim_{r \to 0}(u(r, \cdot) - \mathrm{Ln}\, 1/r^2) = 2\omega(\cdot) + \mathrm{Ln}(2/\lambda). \end{cases}$$

Remark 3.5. In Theorems 3.1 and 3.2 it is not clear if this simple method provides all the local solutions of (3.2)–(3.3) or (3.38). This can be proved but the proof is much more complicated [13].

4. Asymptotics of Analytic Functionals

We present here an overview of the adaptation of Simon's result [17] concerning the asymptotics of uniformly elliptic and analytic functionals in infinite cylinder. Let (M, g) be a C^∞ Riemannian manifold with or without boundary, TM the tangent bundle and ∇ the covariant derivative on M. Let E be a C^∞ function defined on $M \times \mathbb{R} \times TM$ and

$$(4.1) \qquad \xi(u) = \int_M E(x, u, \nabla u) dv_g$$

where dv_g is the volume element on M. We assume that $E(x, z, p)$ is uniformly convex in the p variable for $p \in T_x M$ and $|z| + |p|$ small enough, thus

$$(4.2) \qquad \frac{d^2}{ds^2} E(x, 0, sp)\Big|_{s=0} \geq C|p|^2.$$

Assume also that E has analytic dependence on $(z, p) \in \mathbb{R} \times T_x M$, uniformly in x for $|z| + |p|$ small enough. Henceforth there exists $\beta \in (0, 1)$ such that, for any $x \in M$,

$$(4.3) \qquad E(x, z + \lambda_1 w, p + \lambda_2 q) = \sum_{|\alpha| \geqq 0} E_\alpha(x, z, w, p, q) \lambda^\alpha$$

where $\lambda^\alpha = \lambda_1^{\alpha_1} \lambda_2^{\alpha_2}$, $(\alpha_1, \alpha_2) \in \mathbb{N}^2$, $\alpha_1 + \alpha_2 = |\alpha|$, $(\lambda_1, \lambda_2) \in \mathbb{R}^2$, and this when $(z, w, p, q) \in \mathbb{R} \times \mathbb{R} \times T_x M \times T_x M$ satisfies

$$(4.4) \qquad \max(|z|, |w|, |p|, |q|) < \beta,$$

and where also

$$(4.5) \qquad \sup_{|\lambda| < 1} \left| \sum_{|\alpha| = j} E_\alpha(x, z, w, p, q) \lambda^\alpha \right| \leqq 1,$$

for any $j \in \mathbb{N}^*$ and (z, w, p, q) satisfying (4.4). We now define the operator $\mathbf{M}$ by

$$(4.6) \qquad \mathbf{M}(u) = -\operatorname{grad} \xi(u)$$

and we assume $\mathbf{M}(0) = 0$. Let $\lambda_1 < \lambda_2 < \lambda_3 < \cdots < \lambda_j < \cdots$ be the eigenvalues of the linearization L of $\mathbf{M}$ at 0 ($\lim_{j \to \infty} \lambda_j = \infty$) and

552 LAURENT VERON

$\phi_1, \phi_2, \ldots, \phi_j, \ldots$, the corresponding eigenfunctions. We then define the operator $\mathbf{L}$ by

$$(4.7) \qquad \mathbf{L}(u) = u_{tt} - u_t + L(u).$$

Henceforth the equation $\mathbf{L}(w) = 0$ has solution of the form

$$(4.8) \qquad w_j = e^{\gamma_j t} \phi_j$$

where

$$(4.9) \qquad \gamma_j^2 - \gamma_j - \gamma_j = 0.$$

We denote

$$(4.10) \qquad \begin{cases} \varepsilon_1 = \min\{\operatorname{Re}\gamma_j : \operatorname{Re}\gamma_j > 0\}, \\ \varepsilon_2 = \min\{-\operatorname{Re}\gamma_j : \operatorname{Re}\gamma_j < 0\}, \\ \varepsilon^* = \min\{\varepsilon_1, \varepsilon_2\}. \end{cases}$$

Let the $L^2(M)$ norm of w be $\|w\|$, dv_g being normalized by $\int_M dv_g = 1$. For any smooth function w on $M \times [0,T)$ we define $|w|_1^*$ and $|w|_2^*$ by

$$|w|_1^*(t) = |w(t,\cdot)|_{C^1(M)} + |w_t(t,\cdot)|_{C^0(M)},$$
$$|w|_2^*(t) = |w(t,\cdot)|_{C^2(M)} + |w_t(t,\cdot)|_{C^1(M)} + |w_{tt}(t,\cdot)|_{C^0(M)}.$$

With those notations we can recall Simon's theorem [17]

Theorem 4.1. *Let ε, T_0, T^* be constants such that*

$$(4.12) \qquad 0 < \varepsilon < \varepsilon^*, \qquad 0 < 2T_0 < T^* \leqq \infty.$$

Then there exist constants $\alpha = \alpha(E) \in (0, 1/2)$, $\tilde{\delta} = \tilde{\delta}(\varepsilon, T_0, E) \in (0, \beta^{1/\alpha})$ such that for any $\delta \in (0, \tilde{\delta})$ the following holds.
  Let u be any smooth solution of

$$(4.13) \qquad u_{tt} - u_t + \mathbf{M}(u) = f$$

in $[0, T^*) \times M$ such that
  (i) u is a δ^α-complete solution in $[0, T^*) \times M$, that is

$$(4.14) \qquad \begin{cases} |u|_2^*(t) < \delta^\alpha & (\forall t \in [0, T^*)) \\ \text{and } T^* = +\infty & \text{or } \limsup_{t \uparrow T^*} |u|_2^*(t) = \delta^\alpha \end{cases}$$

(ii) *u has a $(\varepsilon_1 - \varepsilon, T_0, \varepsilon, \delta)$-bounded growth on $[0, T^*) \times M$, that is: for any $T \in (0, T^* - 2T_0)$*

$$(4.15) \qquad \sup_{T+T_0 \leq t \leq T+2T_0} |u|_2^*(t) \leq e^{(\varepsilon_1 - \varepsilon)T_0} \sup_{T \leq t \leq T+T_0} |u|_2^* t + \delta e^{-\varepsilon T}.$$

Suppose further that

$$(4.16) \qquad |u|_2^*(t) < \delta \qquad (\forall t \in [0, T_0)),$$

$$(4.17) \qquad \xi(u(t)) \geq \xi(0) - \delta \qquad (\forall t \in [0, T^*)),$$

$$(4.18) \qquad |f|_2^*(t) \leq \delta e^{-\varepsilon t} \qquad (\forall t \in [0, T^*)).$$

Then $T^* = \infty$, $\sup_{t \geq 0} |u|_2^*(t) < \delta^\alpha$ and there exists a function $\phi \in C^\infty(M)$ satisfying

$$(4.19) \qquad \mathbf{M}(\phi) = 0$$

such that

$$(4.20) \qquad \lim_{t \to \infty} \left(|u_t(t, \cdot)|_{C^1(M)} + |u(t, \cdot) - \phi|_{C^2(M)} \right) = 0.$$

It is also possible to study the asymptotic behaviour of any smooth solution of (4.13) when t tends to $-\infty$. In that case Simon's result has to be adapted and we have the following

Theorem 4.2. *Theorem 4.1 remains valid if we replace (4.13) by*

$$(4.21) \qquad u_{tt} + u_t + \mathbf{M}(u) = f.$$

In that case (4.17) has to be replaced by

$$(4.22) \qquad |\xi(u(t)) - \xi(0)| \leq \delta \qquad (\forall t \in [0, T^*)).$$

The consequence of Theorems 4.1 and 4.2 is the convergence theorem

Theorem 4.3. *Assume (M, g) is a C^∞ compact manifold, Δ_g the Laplace–Beltrami operator on M, f a real analytic function and $\psi \in C^2(\mathbb{R}^+ \times M)$ a solution of*

$$(4.23) \qquad \psi_{tt} - \varepsilon \psi_t + \Delta_g \psi + f(\psi) = 0$$

in $\mathbb{R}^+ \times M$, where $\varepsilon = \pm 1$. If ψ remains bounded in $\mathbb{R}^+ \times M$ there exists $\omega \in C^\infty(M)$ satisfying

$$(4.24) \qquad \Delta_g \omega + f(\omega) = 0$$

on M such that

$$(4.25) \qquad \lim_{t \to +\infty} |\psi(t, \cdot) - \omega(\cdot)|_{C^k(M)} = 0,$$

for any $k \in \mathbb{N}$.

The proof in the case $\varepsilon = -1$ derives from Theorem 4.2 in the same way as in the case $\varepsilon = 1$ from Theorem 4.1. We shall give an idea of this proof only in this last case.

Step 1. $L^2(M)$ estimates. From Agmon–Douglis–Nirenberg and Schauder estimates we have

$$(4.26) \qquad \left| \frac{\partial^\alpha}{\partial t^\alpha} \nabla^\beta \psi(t, \cdot) \right|_{C^0(M)} \leqq M(\alpha, \beta)$$

for any (α, β) and $t \geqq 1$, and

$$(4.27) \qquad \int_1^\infty \int_M \left| \frac{\partial^\alpha}{\partial t^\alpha} \nabla^\beta \psi \right|^2 dv_g dt < \infty$$

for (α, β) with $\alpha \geqq 1$. Henceforth if $\alpha \geqq 1$, we have

$$(4.28) \qquad \lim_{t \to \infty} \left| \frac{\partial^\alpha}{\partial t^\alpha} \nabla^\beta \psi(t, \cdot) \right|_{C^0(M)} = 0.$$

There exists $\{t_k\}$ tending to $+\infty$ and $\omega \in C^2(M)$ (and in fact $\omega \in C^\infty(M)$) such that

$$(4.29) \qquad \lim_{t_k \to \infty} |\psi(t_k, \cdot) - \omega(\cdot)|_{C^\ell(M)} = 0$$

for any $\ell \in \mathbb{N}$.

Step 2. Reduction to an equation of type (4.13). Let u be $\psi - \omega$, then

$$(4.30) \qquad u_{tt} - u_t + \Delta_g u + g(x, u + \omega) - g(x, \omega) = f.$$

We set

$$(4.31) \quad E(x, z, p) = \frac{1}{2}|p|^2 - (G(x, z + \omega(x)) - G(x, \omega(x)) - zg(x, \omega(x)))$$

where $G(x, r) = \int_0^r g(x, s)ds$, and

$$(4.32) \qquad \mathbf{M}(\phi) = \Delta_g(\phi) + g(x, \phi + \omega) - g(x, \omega).$$

Using the boundedness of ω and Cauchy's majorizations we can check that there exists $\beta \in (0, 1)$ such that E satisfies (4.3). Now we try to apply Theorem 4.1 to

$$(4.33) \qquad u_k(t, \cdot) = u(t + t_k, \cdot) \qquad (t > 0)$$

for k large enough.

Step 3. Energy estimates. From (4.30) we have

$$(4.34) \qquad \frac{1}{2}\|u_t(T)\|^2 - \xi(u(T)) - \frac{1}{2}\|u_t(t)\|^2 + \xi(u(t))$$
$$= \int_t^T \left(\|u_t(\tau)\|^2 + \langle f(\tau), u_t(T)\rangle_{L^2(M)} \right) d\tau,$$

which implies (with (4.18))

$$(4.35) \qquad \frac{1}{2}\|u_t(T)\|^2 - \xi(u(T)) - \frac{1}{2}\|u_t(t)\|^2 + \xi(u(t))$$
$$\geqq \frac{1}{2} \int_t^T \|u_t(\tau)\|^2 d\tau - \frac{1}{2}C_0^2 e^{-2\varepsilon_0 t}$$

for $0 < t < T$. As

$$(4.36) \qquad |\xi(u(t))| \leqq C_1 \|u(t, \cdot)\|_{W^{1,2}(M)},$$

we get

$$(4.37) \qquad \lim_{k \to \infty} \xi(u(t_k)) = 0$$

and

$$(4.38) \qquad \lim_{t \to \infty} \xi(u(t)) = 0.$$

Moreover

$$(4.39) \qquad \|u(\tau_2) - u(\tau_1)\|^2 \leqq (\tau_2 - \tau_1) \int_{\tau_1}^{+\infty} \|u_t(\tau)\|^2 d\tau,$$

which implies

$$(4.40) \quad \|u(\tau_2)\| \leq C_2(1 + \sqrt{\tau_2 - \tau_1}) \, \|u(\tau_1)\|_{W^{1,2}(M)} + C_0 \, e^{-\varepsilon_0 \tau_1} \sqrt{\tau_2 - \tau_1}.$$

Step 4. We claim that for any $\varepsilon \in (0, \varepsilon_0)$, $\delta > 0$, $T_0 > 0$, there exists an integer $\tilde{k} = \tilde{k}(\varepsilon, \delta, \alpha, T_0)$ such that for any $k \geq \tilde{k}$, u_k and f_k satisfy (4.16), (4.17) and (4.18), for any $T^* > 0$, and $\sup_{0 \leq t \leq 2T_0} |u_k(t)|_2^* < \delta^\alpha$.

We first choose $k_2 = k_2(\delta)$ such that $C_0 e^{-\varepsilon_0 t k_2} < \delta$, which implies

$$(4.41) \qquad |f_k|_2^*(t) \leq \delta \, e^{-\delta_0 t} \qquad (\forall t > 0, \ \forall k \geq k_2).$$

From (4.38) we choose $k_3 = k_3(\delta)$ such that

$$(4.42) \qquad \xi(u(t)) \geq -\delta \qquad (\forall t \geq t_k \geq t_{k_3}),$$

which implies (4.17). As we have

$$(4.43) \qquad \|u(t)\| \leq \|u(t_k)\| + \left| \int_{t_k}^{t} \|u_t(\tau)\| d\tau \right|$$

for $t > 0$ we deduce from Step 1 that for any $\lambda > 0$ there exists $k(\lambda) \in \mathbb{N}$ such that

$$(4.44) \qquad \|u(t)\| \leq \lambda(1 + |t - t_k|) \qquad (\forall t > 0, \ \forall k \geq k(\lambda)).$$

From linear elliptic equations theory and the boundedness of u and ψ we deduce that for any $\theta > 1$ we have

$$(4.45) \qquad |u|_{C^{2,1/2}([\theta-1/2, \theta+1/2] \times M)}$$
$$\leq C_3 \big(\|u\|_{L^2([\theta-1, \theta+1] \times M)} + |f|_{C^1([\theta-1, \theta+1] \times M)} \big),$$

where C_3 depends on E and $\|\psi\|_{L^\infty(R^+ \times M)}$ and where $\theta \in (1, +\infty)$ is arbitrary. Henceforth we get

$$(4.46) \qquad \sup_{0 \leq t \leq T} |u_k|_2^*(t) \leq \sqrt{2} \, C_3 \big(\lambda(2 + T) + C_0 \, e^{-\varepsilon_0(t_k - 1)} \big)$$

for $k \geq k(\lambda)$. Choosing $\lambda = \lambda(\varepsilon, \delta, \alpha, T_0)$ small enough there exists $\tilde{k} = \tilde{k}(\varepsilon, \delta, \alpha, T_0)$ such that, for any $k \geq \tilde{k}$, the following holds

$$(4.47) \qquad \sup_{0 \leq t \leq T_0} |u_k|_2^*(t) \leq \delta,$$

$$(4.48) \qquad \sup_{0 \leq t \leq 2T_0} |u_k|_2^*(t) < \delta^\alpha.$$

Step 5. We claim that for any $\varepsilon \in (0, \min(\varepsilon_0, \varepsilon^*))$ there exists $\bar{T}_0 = \bar{T}_0(\varepsilon)$ such that for any $\delta > 0$, u_k has a $(\varepsilon_1 - \varepsilon, \bar{T}, \varepsilon, \delta)$-bounded growth on $[0, T^*) \times M$, for any $k \geq \bar{k}(\varepsilon, \delta)$.

Take $T_0 > 2$ and $T^* > 2T_0$; for any $0 < T < T^* - 2T_0$ any $t \in [T + T_0, T + 2T_0]$ and any $k \in \mathbb{N}$ we have

$$(4.49) \qquad |u_k|_2^* \leq \sqrt{2}\, C_3 \Big(\sup_{t_k+t-1 \leq \tau \leq t_k+t+1} \|u(\tau)\| + C_0\, e^{-\varepsilon_0(t_k+t-1)} \Big).$$

Applying (4.40) with $\tau_2 \in [t_k + t - 1, t_k + t + 1]$ and $\tau_1 = \tau_k + T + T_0/2$, as we have $0 \leq T_2 - \tau_1 \leq 2T_0$, $\tau_1 \geq t_k + T - 3T_0/2$, $\tau_1 - t_k \in [T, T + T_0]$ we get (with $d = \dim M$)

$$(4.50) \qquad |u_k|_2^*(t) \leq C_3\sqrt{2}\,\Big(C_3\sqrt{1+d}\,(1 + \sqrt{2T_0})\, \sup_{T \leq \tau \leq T+T_0} |u_k|_2^*(\tau)$$
$$+ 2C_0\sqrt{2T_0}\, e^{3T_0/2}\, e^{-\varepsilon_0(t+t_k)} \Big).$$

Then we can choose $\bar{T}_0 = \bar{T}_0(\varepsilon)$ and $\bar{k} = \bar{k}(\varepsilon, \delta)$ such that

$$(4.51) \qquad \sup_{T+\bar{T}_0 \leq t \leq T+2\bar{T}_0} |u_k|_2^*(t) \leq e^{(\varepsilon_1-\varepsilon)\bar{T}_0} \sup_{T \leq t \leq T+T_0} |u_k|_2^*(t) + \delta e^{-\varepsilon T}.$$

Step 6. conclusion. Take $\varepsilon \in (0, \min(\varepsilon^*, \varepsilon_0))$ and T_0 as in Step 5; $\tilde{\delta} = \tilde{\delta}(\varepsilon, \bar{T}_0, E)$ be as in Theorem 4.1 and $\delta \in (0, \tilde{\delta})$. From (4.48) we see that, for any $k \geq \tilde{k}(\varepsilon, \delta, \alpha, \bar{T}_0)$, there exists $T^* = T^*(k, \delta, \varepsilon) > 2\bar{T}_0$ such that u_k is a δ^α-complete solution of

$$(4.52) \qquad w_{tt} - w_t + \mathbf{M}(w) = f_k$$

in $[0, T^*) \times M$. Then u_k, f_k satisfy (4.14)–(4.18). Henceforth $T^* = +\infty$ and there exists $\phi \in C^\infty(M)$ such that $\mathbf{M}(\phi) = 0$ and

$$(4.53) \qquad \lim_{t \to +\infty} u_k(t, \cdot) = \phi(\cdot)$$

in the C^2-topology of M. Henceforth $\phi \equiv 0$ and we get (4.25).

Remark 4.1. In [3] we extend Theorem 4.3 to a more general equation where $\Delta_g + g(\cdot, \cdot)$ is replaced by grad $\mathbf{F}(\cdot)$ with

$$(4.54) \qquad \mathbf{F}(w) = \int_M F(x, w, \nabla w)\, dv_g$$

and F satisfies (4.2)–(4.4).

Remark 4.2. We conjecture that the hypothesis of analyticity upon g can be withdrawn. Unfortunately Simon's proof [17] which is based upon Lojasiewicz inequalities [12] no longer works.

REFERENCES

[1] M. Berger, P. Gauduchon, and E. Mazet, *Le Spectre D'une Variété Riemannienne*, Lecture Notes in Math. Vol. 194, Springer-Verlag, 1971.

[2] M.F. Bidaut-Veron and L. Veron, *Groupe conforme de S^2 et propriétés limites des solutions de $-\Delta u = \lambda e^u$*, C.R. Acad. Sci. Paris **308**(1989), 493–498.

[3] M.F. Bidaut-Veron and L. Veron, *Nonlinear elliptic equations on compact Riemannian manifolds and asymptotics of Emden equations*, Inv. Math. (to appear).

[4] H. Brezis and P.L. Lions, *A note on isolated singularities for linear elliptic equations*, Math. Anal. Appl. **7A**(1981), 263–266.

[5] S. Chandrasekhar, *An Introduction to the Study of Stellar Structure*, Dover Publ. Inc., 1967.

[6] S.Y.A. Chang and P.C. Yang, *Prescribing Gaussian curvature on S^2*, Acta Math. **159**(1987), 215–259.

[7] X.Y. Chen, H. Matano, and L. Veron, *Anisotropic singularities of solutions of nonlinear elliptic equations in $\mathbb{R}^2$*, Jl. Funct. Anal. **83**(1989), 50–97.

[8] V.R. Emden, *Gaskugeln*, Teubner, Leipzig, 1897.

[9] B. Gidas and J. Spruck, *Global and local behaviour of positive solutions of nonlinear elliptic equations*, Comm. Pure Appl. Math. **34**(1981), 525–598.

[10] J.C. Hale, *Asymptotic behaviour of dissipative systems*, Math. Surv. Mon., Amer. Math. Soc. (1988).

[11] D. Henry, *Geometric Theory of Semilinear Parabolic Equations*, Lecture Notes in Math. Vol. 840, Springer-Verlag, 1981.

[12] S. Lojasiewicz, *Ensembles semi-analytiques*, I.H.E.S. Notes, 1965.

[13] H. Matano and L. Veron, in preparation.

[14] F. Mignot and J.P. Puel, *Solutions radiales singulières de $-\Delta u = \lambda e^u$*, C.R. Acad. Sci. Paris **307**, Ser. I (1988), 379–382.

[15] W.M. Ni and J. Serrin, *Nonexistence theorems for singular solutions of quasi-linear partial differential equations*, Comm. Pure Appl. Math. **39**(1986), 379–399.

[16] E. Onofri, *On the positivity of the effective action in a theory of random surfaces*, Comm. Math. Phys. **86**(1982), 321–326.

[17] L. Simon, *Asymptotics for a class of nonlinear evolution equations with applications to geometric problems*, Ann. Math. **118**(1983), 525–571.

[18] L. Veron, *Comportement asymptotique des solutions d'équations elliptiques semi-linéaires dans* $\mathbb{R}^N$, Ann. Mat. Pura Appl. **127**(1981), 25–50.

[19] L. Veron, *Singular solutions of some nonlinear elliptic equations*, Nonlinear Anal. **5**(1981), 225–242.

Département de Mathématiques
Faculté des Sciences
Parc de Grandmont
F 37200 Tours, France

Chemical Interfacial Reaction Models with Radial Symmetry

SHOJI YOTSUTANI[1]

Abstract

This article discusses a sort of parabolic-elliptic system with nonlinear boundary conditions, which comes from the chemical interfacial models. The results obtained here are the uniqueness and the existence of the global solutions.

1. Introduction

This is a joint work with Yoshio Yamada (Waseda Univ.).

We consider the following parabolic-elliptic system with nonlinear boundary conditions. Let $Z > 0$ be fixed. Given the initial data $\{u_0(z), v_0(z), w_0(z)\}$, find $(u(r, z), v(r, z), w(r, z), v^*(r, z), w^*(r, z))$ such that

$$\text{(1.1)} \qquad u_{rr} + \frac{1}{r}u_r - a(r)u_z = 0, \quad (r, z) \in (0, r_1) \times (0, Z),$$

$$\text{(1.2)} \qquad v_{rr} + \frac{1}{r}v_r - b(r)v_z = 0, \quad (r, z) \in (r_2, 1) \times (0, Z),$$

$$\text{(1.3)} \qquad w_{rr} + \frac{1}{r}w_r - c(r)w_z = 0, \quad (r, z) \in (r_2, 1) \times (0, Z),$$

$$\text{(1.4)} \qquad u(r, 0) = u_0(r) \geq 0, \quad r \in [0, r_1),$$

$$\text{(1.5)} \qquad v(r, 0) = v_0(r) \geq 0, \quad r \in (r_2, 1),$$

$$\text{(1.6)} \qquad w(r, 0) = w_0(r) \geq 0, \quad r \in (r_2, 1),$$

$$\text{(1.7)} \qquad u_r(0, z) = 0, \quad z \in (0, Z],$$

$$\text{(1.8)} \qquad v_r(1, z) = 0, \quad z \in (0, Z],$$

$$\text{(1.9)} \qquad w_r(1, z) = 0, \quad z \in (0, Z],$$

$$\text{(1.10)} \qquad -u_r(r_1, z) = R(u(r_1, z), v^*(r_1, z), w^*(r_1, z)), \quad z \in (0, Z],$$

$$\text{(1.11)} \qquad v_r^*(r_1, z) = m_1 R(u(r_1, z), v^*(r_1, z), w^*(r_1, z)), \quad z \in (0, Z],$$

$$\text{(1.12)} \qquad w_r^*(r_1, z) = -n_1 R(u(r_1, z), v^*(r_1, z), w^*(r_1, z)), \quad z \in (0, Z],$$

$$\text{(1.13)} \qquad v_r^*(r_2, z) = m_2 v_r(r_2, z), \quad z \in (0, Z],$$

[1]This work is partially supported by grants from the Science Foundation of the Ministry of Education of Japan, No.01540148.

$$(1.14) \qquad w_r^*(r_2, z) = n_2 w_r(r_2, z), \quad z \in (0, Z],$$

$$(1.15) \qquad v^*(r_2, z) = v(r_2, z), \quad z \in (0, Z],$$

$$(1.16) \qquad w^*(r_2, z) = w(r_2, z), \quad z \in (0, Z],$$

$$(1.17) \qquad v_{rr}^* + \frac{1}{r} v_r^* = 0, \quad (r, z) \in (r_1, r_2) \times (0, Z],$$

$$(1.18) \qquad w_{rr}^* + \frac{1}{r} w_r^* = 0, \quad (r, z) \in (r_1, r_2) \times (0, Z].$$

Here $m_1, n_1, m_2, n_2, r_1, r_2$ $(0 < r_1 < r_2 < 1)$ are given positive constants, and $a(r), b(r), c(r)$ and $R(u, v, w)$ are given functions. For simplicity we denote the above initial boundary problem (1.1)-(1.18) by (P).

Concerning $a(r), b(r), c(r)$ and $R(u, v, w)$, we introduce the following assumptions.

$$(A.1) \qquad \begin{cases} a(|x|) \in C^\infty(\{x : |x| \le r_1\}), & a > 0 \quad \text{on} \quad [0, r_1), \\ b(|x|) \in C^\infty(\{x : r_2 \le |x| \le 1\}), & b > 0 \quad \text{on} \quad (r_2, 1), \\ c(|x|) \in C^\infty(\{x : r_2 \le |x| \le 1\}), & c > 0 \quad \text{on} \quad (r_2, 1). \end{cases}$$

(A.2)   There exist non-negative constants k_a, k_b, k_c such that

$$\lim_{\substack{r \to r_1 \\ r \in I_1}} \frac{a(r)}{|r - r_1|^{k_a}} > 0, \ \lim_{\substack{r \to r_2 \\ r \in I_2}} \frac{b(r)}{|r - r_2|^{k_b}} > 0, \ \lim_{\substack{r \to r_2 \\ r \in I_2}} \frac{c(r)}{|r - r_2|^{k_c}} > 0.$$

(R.1)   There exists an open subset U of R^3 and a positive constant δ_R such that

$$U \supset [-\delta_R, \delta_R]^3 \cup [0, \infty)^3, \quad R(u, v, w) \in C^\infty(U),$$

and

$$R_v(u, v, w) \ge 0, \quad R_w(u, v, w) \le 0 \quad \text{for every } (u, v, w) \in [0, \infty)^3,$$

$$R(u, 0, 0) = 0 \text{ for every } u \in [0, \infty),$$

$$R(0, v, 0) = 0 \text{ for every } v \in [0, \infty).$$

(R.2)   The function R has the following properties.

(i) If $R(u, v, 0) = 0$ with $(u, v) \in [0, \infty)^2$, then $u = 0$ or $v = 0$,

(ii) If $R(u, 0, w) = 0$, then $w = 0$.

(R.3)   For any compact subset $K \subset U$, there exists a positive constant c_K such that

$$max\{R(u, v, w)u^-, R(u, v, w)v^-, -R(u, v, w)w^-\}$$
$$\le c_K\{|u^-|^2 + |v^-|^2 + |w^-|^2\}$$

for every $(u, v, w) \in K$, where $u^- = -min\{u, 0\}$, $v^- = -min\{v, 0\}$ and $w^- = min\{w, 0\}$.

(R.4) There exists a positive constant c_R such that

$$max\{-u^{2p-1}R(u, v, w), -v^{2p-1}R(u, v, w), w^{2p-1}R(u, v, w)\}$$
$$\leq c_R\{u^{2p} + v^{2p} + w^{2p}\}$$

for every $(u, v, w) \in [0, \infty)^3$ and $p \in [1, \infty)$.

Our motivation for (P) comes from the chemical interfacial reaction models proposed by Yoshizuka, Kondo and Nakashio [YKN], which is closely related to the models proposed by Kawano, Kusano, Kondo and Nakashio [KKKN] earlier. They analyzed the kinetics of interfacial reaction by comparing the chemical experiments with numerical simulation of their models. After suitable transformation, their typical model is reduced to (P) with

$$(1.19) \qquad \begin{cases} a(r) = k_1(r_1^2 - r^2), \\ b(r) = k_2(1 - r^2 - (1 - r_2^2)(\log r_2)^{-1}\log r), \\ c(r) = k_3(1 - r^2 - (1 - r_2^2)(\log r_2)^{-1}\log r), \\ R(u, v, w) = \dfrac{uv^2 - k_4 w}{(k_5 + k_6 v)^2}, \end{cases}$$

where k_i $(i = 1, \ldots, 6)$ are positve constant determined by chemical substances. In their model, $[0, r_1)$ and $(r_2, 1)$ correspond to the liquid region, (r_1, r_2) corresponds to the region of a hollow fiber, and the boundary $r = r_1$ represents the chemical interface where the following chemical reaction

$$A + 2B \rightleftharpoons AB_2$$

takes place. The unknown functions u, v, w correspond to the concentration of chemical substances A, B and AB_2 in the liquid, and the unknown functions v^*, w^* correspond the concentration of chemical substances B and AB_2 in the hollow fiber region. The function $a(r), b(r)$ and $c(r)$ correspond to the velocity distribution functions of laminar flow in the liquid. $R(u, v, w)$ are related with the kinetics of the chemical reaction at the interface. (See [YKN] and [YY2].) It is easily shown that (A.1), (A.2), (R.1), (R.2), (R.3) and (R.4) are satisfied by following the calculation in [YY2].

There are some results on the system of parabolic equations with nonlinear boundary condition. (See, e.g., [AP], [F], [LSU].) However, because of non-monotone boundary conditions and coupling with elliptic equations, the existence theories do not seem to be useful for deriving the existence of the global solutions.

This problem (P) is closely related to the problem in [YY1] and [YY2]. However this problem is much complicated and difficult. Since parabolic equations (1.1)-(1.3) and elliptic equations (1.17)-(1.18) are coupled, we must understand the meaning of initial condition and compatibility conditions carefully, and need more complicated argument to assure the existence of nonnegative global solutions. Moreover the coefficients $b(r)$ and $c(r)$ in (1.2) and (1.3) vanish at the both boundaries $r = r_2$ and $r = 1$, which derive technical difficulties.

We introduce the following parabolic system to solve the problem (P). Let $Z > 0$ be fixed. Given the initial data $\{u_0(r), v_0(r), w_0(r)\}$, find $(u(r,z), v(r,z), w(r,z), \beta^*(z), \gamma^*(z))$ such that (1.1)-(1.9), and

$$(1.20) \qquad -u_r(r_1, z) = R(u(r_1, z), \beta^*(z), \gamma^*(z)), \quad z \in (0, Z],$$

$$(1.21) \qquad v_r(r_2, z) = \frac{m_1 r_1}{m_2 r_2} R(u(r_1, z), \beta^*(z), \gamma^*(z)), \quad z \in (0, Z],$$

$$(1.22) \qquad w_r(r_2, z) = -\frac{n_1 r_1}{n_2 r_2} R(u(r_1, z), \beta^*(z), \gamma^*(z)), \quad z \in (0, Z],$$

$$(1.23) \qquad v(r_2, z) = \beta^*(z) + q_1 R(u(r_1, z), \beta^*(z), \gamma^*(z)), \quad z \in (0, Z],$$

$$(1.24) \qquad w(r_2, z) = \gamma^*(z) - q_2 R(u(r_1, z), \beta^*(z), \gamma^*(z)), \quad z \in (0, Z],$$

where $q_1 = m_1 r_1 \log (r_2/r_1)$ *and* $q_2 = n_1 r_1 \log (r_2/r_1)$. We denote the above initial boundary value problem (1.1)-(1.9) and (1.20)-(1.24) by (Q). Roughly speaking, since $\beta^*(z)$ and $\gamma^*(z)$ is represented by $u(r_1, z), v(r_2, z)$ and $w(r_2, z)$ in view of (1.23) and (1.24), the right hand side of (1.20), (1.21) and (1.22) are represented by $u(r_1, z)$, $v(r_2, z)$ and $w(r_2, z)$. Therefore (Q) is a parabolic system with the nonlinear boundary condition. We shall show that the problem (P) reduced to the problem (Q) in the next section.

Notation

$$L^2(I) = L^2(I; r\,dr), \qquad H^1(I) = \{u \in L^2(I) : u_r \in L^2(I)\},$$

$$H^1_{\ell oc}((0, Z]) = \bigcup_{\delta > 0} H^1(\delta, Z),$$

$$H^1_{\ell oc}((0, Z]; E) = \bigcup_{\delta > 0} H^1(\delta, Z; E),$$

$$H^2(I) = \{u \in L^2(I) : u_r \in L^2(I), u_{rr} + \frac{1}{r} u_r \in L^2(I)\},$$

$$I_1 = [0, r_1), \qquad I_2^* = (r_1, r_2), \qquad I_2 = (r_2, 1),$$

$$\overline{I_1} = [0, r_1], \qquad \overline{I_2^*} = [r_1, r_2], \qquad \overline{I_2} = [r_2, 1].$$

2. Statement of the results

We shall mainly discuss (P) and (Q) in the framework of L^2-theory. Therefore it is better to define a class of functions in which (P) and (Q) are solved.

Definition 2.1. We say that (u, v, w, v^*, w^*) is a solution of (P) on $[0, Z]$, if it satisfies

(i) (u, v, w, v^*, w^*) belongs to

$$
\begin{aligned}
&C([0, Z]; L^2(I_1) \times L^2(I_2) \times L^2(I_2) \times L^2(I_2^*) \times L^2(I_2^*)) \\
&\cap C((0, Z]; H^2(I_1) \times H^2(I_2) \times H^2(I_2) \times H^2(I_2^*) \times H^2(I_2^*)) \\
&\cap H^1_{\ell oc}((0, Z]; H^2(I_1) \times H^2(I_2) \times H^2(I_2) \times H^2(I_2^*) \times H^2(I_2^*)) \\
&\cap C^\infty(I_1 \times (0, Z]) \times C^\infty(I_2 \times (0, Z]) \times C^\infty(I_2 \times (0, Z]) \\
&\quad \times C^\infty(I_2^* \times (0, Z]) \times C^\infty(I_2^* \times (0, Z]),
\end{aligned}
$$

(ii) u, v, w, v^*, w^* are uniformly bounded on $\overline{I_1} \times (0, Z]$, $\overline{I_2} \times (0, Z]$, $\overline{I_2} \times (0, Z]$, $\overline{I_2^*} \times (0, Z]$, $\overline{I_2^*} \times (0, Z]$, respectively,

(iii) (1.1),(1.2),(1.3),(1.17) and (1.18) are satisfied for every (r, z),

(iv) (1.7)-(1.16) hold z,

(v) (1.4)-(1.6) hold for a.e. r.

Definition 2.2. We say that $(u, v, w, \beta^*, \gamma^*)$ is a solution of (Q) on $[0, Z]$, if it satisfies

(i) (u, v, w) belongs to

$$
\begin{aligned}
&C([0, Z]; L^2(I_1) \times L^2(I_2) \times L^2(I_2)) \\
&\cap C((0, Z]; H^2(I_1) \times H^2(I_2) \times H^2(I_2)) \\
&\cap H^1_{\ell oc}((0, Z]; H^2(I_1) \times H^2(I_2) \times H^2(I_2)) \\
&\cap C^\infty(I_1 \times (0, Z]) \times C^\infty(I_2 \times (0, Z]) \times C^\infty(I_2 \times (0, Z])
\end{aligned}
$$

and $\beta^*, \gamma^* \in H^1_{\ell oc}((0, Z])$,

(ii) $u, v, w, \beta^*, \gamma^*$ are uniformly bounded on $\overline{I_1} \times (0, Z]$, $\overline{I_2} \times (0, Z]$, $\overline{I_2} \times (0, Z]$, $(0, Z]$, $(0, Z]$, respectively,

(iii) (1.1),(1.2),(1.3) are satisfied for every (r, z),

(iv) (1.7)-(1.9),(1.20)-(1.24) hold for every z,

(v) (1.4)-(1.6) hold for a.e. r.

For simplicity we introduce a notion of "nice initial data".

Definition 2.3. We say that a given function $(u_0(r), v_0(r), w_0(r)) \in H^2(I_1) \times H^2(I_2) \times H^2(I_2)$ is a nice initial data if the following conditions are satisfied

(i) $u_0(r) \geq 0$, $v_0(r) \geq 0$, $w_0(r) \geq 0$,

(ii) $u_{0,r}(0) = 0,$ $v_{0,r}(1) = 0,$ $w_{0,r}(1) = 0,$

(iii)   there exist $\beta_0^* \geq 0$ and $\gamma_0^* \geq 0$ such that

$$- u_{0,r}(r_1) = R(u_0(r_1), \beta_0^*, \gamma_0^*),$$

$$v_{0,r}(r_2) = \frac{m_1 r_1}{m_2 r_2} R(u_0(r_1), \beta_0^*, \gamma_0^*)$$

$$w_{0,r}(r_2) = -\frac{n_1 r_1}{n_2 r_2} R(u_0(r_1), \beta_0^*, \gamma_0^*)$$

$$v_0(r_2) = \beta_0^* + q_1 R(u_0(r_1), \beta_0^*, \gamma_0^*),$$

$$w_0(r_2) = \gamma_0^* - q_2 R(u_0(r_1), \beta_0^*, \gamma_0^*),$$

(iv) $\|(u_{0,rr} + \frac{1}{r}u_{0,r})\frac{1}{\sqrt{a}}\|_{L^2(I_1)} < \infty,$ $\|(v_{0,rr} + \frac{1}{r}v_{0,r})\frac{1}{\sqrt{b}}\|_{L^2(I_2)} < \infty,$
$\|(w_{0,rr} + \frac{1}{r}w_{0,r})\frac{1}{\sqrt{c}}\|_{L^2(I_2)} < \infty.$

Remark 2.1. The above conditions are compatibility conditions for the problem (Q), which imply the compatibility condition for the problem (P) by virtue of Proposition 2.1 which we prove later.

Remark 2.2. By Lemma 4.3 in the subsequent section, β_0^* and γ_0^* appeared in the above definition are uniquely determined by $u_0(r_1), v_0(r_2)$ and $w_0(r_2)$.

Our first theorem is concerned with the existence, uniqueness and regularity of solutions of (P) and (Q) for any smooth initial data with the compatibility condition.

Theorem 1.   *Suppose that (A.1),(A.2),(R.1),(R.2),(R.3) and (R.4) hold. Let $\{u_0(r), v_0(r), w_0(r)\}$ be a nice initial data. Then for any $Z > 0$ there exists a unique non-negative solution (u, v, w, v^*, w^*) of (P) (resp. $(u, v, w, \beta^*, \gamma^*)$ of (Q)) on $[0, Z]$ such that*

$$(u, v, w, v^*, w^*) \in C([0, Z]; H^2(I_1) \times H^2(I_2) \times H^2(I_2) \times H^2(I_2^*) \times H^2(I_2^*))$$

(resp. $(u, v, w) \in C([0, Z]; H^2(I_1) \times H^2(I_2) \times H^2(I_2))$, $\beta^*, \gamma^* \in H^1(0, Z)$)

Moreover, *(1.4),(1.5),(1.6) are satisfied for every $r \in [0, r_1]$, $r \in [r_2, 1]$, $r \in [r_2, 1]$,respectively.*

For general initial data, we obtain the following result.

Theorem 2. *Suppose that (A.1),(A.2),(R.1),(R.2),(R.3) and (R.4) hold. Assume that the initial data $\{u_0(r), v_0(r), w_0(r)\}$ satisfies*

$$u_0 \in L^\infty(0, r_1), \qquad u_0 \geq 0 \quad on \quad [0, r_1),$$

$$u_0 \in L^\infty(r_2, 1), \qquad v_0 \geq 0 \quad on \quad (r_2, 1),$$

$$w_0 \in L^\infty(r_2, 1), \qquad w_0 \geq 0 \quad on \quad (r_2, 1).$$

Then for every $Z > 0$ there exists a unique non-negative solution (u, v, w, v^*, w^*) of (P) (resp. $(u, v, w, \beta^*, \gamma^*)$ of (Q)) on $[0, Z]$.

In particular, if $u_0 \in C(\overline{I_1})$, $v_0 \in C(\overline{I_2})$, $w_0 \in C(\overline{I_2})$, then (1.4)-(1.6) are satisfied for every r.

Remark 2.3. In application to the original model [YKN], the initial data are given as constant functions which do not satisfy the compatibility conditions. So, Theorem 2 is useful in application.

We show that the problem (P) is equivalent to the problem (Q).

Proposition 2.1. *Suppose that (u, v, w, v^*, w^*) is a solution of (P), then*

$$(2.1) \qquad v^*(r, z) = v^*(r_1, z) + m_1 r_1 R(u(r_1, z), v^*(r_1, z), w^*(r_1, z)) \log\left(\frac{r}{r_1}\right),$$
$$r \in [r_1, r_2],$$

$$(2.2) \qquad w^*(r, z) = w^*(r_1, z) - n_1 r_1 R(u(r_1, z), v^*(r_1, z), w^*(r_1, z)) \log\left(\frac{r}{r_1}\right),$$
$$r \in [r_1, r_2],$$

and $(u(r, z), v(r, z), w(r, z), \beta^*(z), \gamma^*(z))$ is a solution of (Q), where

$$(2.3) \qquad \beta^*(z) = v^*(r_1, z), \quad \gamma^*(z) = w^*(r_1, z).$$

Conversely, suppose that $(u(r, z), v(r, z), w(r, z), \beta^*(z), \gamma^*(z))$ is a solution of (Q), then $(u(r, z), v(r, z), w(r, z), v^*(r, z), w^*(r, z))$ is a solution of (P), where $v^*(r, z)$ and $w^*(r, z)$ are defined by

$$(2.4) \qquad v^*(r, z) = \beta^*(z) + m_1 r_1 R(u(r_1, z), \beta^*(z), \gamma^*(z)) \log\left(\frac{r}{r_1}\right),$$
$$r \in [r_1, r_2],$$

$$(2.5) \qquad w^*(r, z) = \gamma^*(z) - n_1 r_1 R(u(r_1, z), \beta^*(z), \gamma^*(z)) \log\left(\frac{r}{r_1}\right),$$
$$r \in [r_1, r_2].$$

Proof. Suppose that (u, v, w, v^*, w^*) be a solution of (P). Since v^* and w^* satisfy (1.17) and (1.18), they are reprensented by the form

$$(2.6) \qquad Const. + Const. \log r, \quad r \in [r_1, r_2]$$

for any fixed $z \in (0, Z]$. It follows from (1.11) and (1.12) that (2.1) and (2.2) hold. Substituting $r = r_2$ in (2.1) and (2.2), we obtain (1.23) and (1.24) by

(1.15) and (1.16). We get (1.21) and (1.22) by (1.13),(1.14),(2.1) and (2.2). Thus $(u(r,z),v(r,z),w(r,z),\beta^*(z),\gamma^*(z))$ is a solution of (Q). Converse is easy.   Q.E.D.

In what follows we will mainly consider the problem (Q) insted of (P) since they are equivalent by the above proposition.

3. Outline of the proof

We can prove the existence and the uniqueness of the non-negative global solutions of (Q) as follows.

Step1: Uniqueness of the solutions.

Step2: Existence of a local solution for nice initial data.

Step3: Non-negativity of the local solution obtained in Step 2.

Step4: A priori estimate of L^∞-norm.

Step5: A priori estimate of derivatives.

Step6: Existence of global solutions for nice initial data.

Step7: Existence of global solutions for general initial data.

We explain the idea of the proof of Step 2. We introduce the following Neumann problem which is auxiliary to the problem (Q). Let (f,g,h) be arbitrary given non-negative functions. Find (u,v,w) satisfying (1.1)-(1.9) and

$$-u_r(r_1,z) = R(f(z),\beta^*(z),\gamma^*(z)), \qquad z \in (0,Z],$$

$$v_r(r_2,z) = \frac{m_1 r_1}{m_2 r_2} R(f(z),\beta^*(z),\gamma^*(z)), \qquad z \in (0,Z],$$

$$w_r(r_2,z) = -\frac{n_1 r_1}{n_2 r_2} R(f(z),\beta^*(z),\gamma^*(z)), \qquad z \in (0,Z],$$

where $\beta^*(z)$ and $\gamma^*(z)$ is determined by the relation

$$g(z) = \beta^*(z) + q_1 R(f(z),\beta^*(z),\gamma^*(z)), \quad z \in (0,Z],$$
$$h(z) = \gamma^*(z) - q_2 R(f(z),\beta^*(z),\gamma^*(z)), \quad z \in (0,Z].$$

It is easily seen that $\beta^*(z)$ and $\gamma^*(z)$ are uniquely determined by virtue of the assumption (R.1) and (R.2). (See Lemma 4.3.) We solve the above Neumann problem and denote this solution by $(u(r,z;f,g,h),v(r,z;f,g,h), w(r,z;f,g,h))$. Let us define a complete metric space

$$B_z = \{(f,g,h) \in H^1(0,Z)^3 : f(0) = u_0(0), g(0) = v_0(0), h(0) = w_0(z),$$

$$\int_0^Z |f_z|^2 dz \leq 1, \int_0^Z |g_z|^2 dz \leq 1, \int_0^Z |h_z|^2 dz \leq 1\}.$$

with metric

$$\rho((f_1, g_1, h_1), (f_2, g_2, h_2))$$
$$= \sqrt{\int_0^Z \{|f_{1,z} - f_{2,z}|^2 + |g_{1,z} - g_{2,z}|^2 + |h_{1,z} - h_{2,z}|^2\} dz}.$$

We can show that the mapping

$$(f(z), g(z), h(z)) \mapsto (u(r_1, z; f, g, h), v(r_2, z; f, g, h), w(r_2, z; f, g, h))$$

in B_Z is well-defined and becomes a strictly contraction mapping if Z is sufficiently small.

The most crucial steps of the proofs are Step 3 and Step 4. We apply the Morser's iteration technique in the proof of Step 4. (See,e.g.,[M] and [A]) We omit the detail here. However we shall state the key lemmas of Step 3 and the give the proof of the non-negativity of the local solution obtained in Step 2 in the next section.

4. Non-negativity of solutions

We state the following fundamental inequalities which are slight modification of the well-known Sobolev inequalities.

Lemma 4.1. *Suppose that $p_i(r) \in C(\overline{I_i})$, $p_i > 0$ on I_i $(i = 1, 2)$. If $U \in H^1(I_i)$, then*

$$(4.1) \qquad |U(r_i)| \leq \varepsilon \|U_r\|_{L^2(I_i)} + \|p_i\|_{L^1(I_{i,\varepsilon})}^{-1/2} \|\sqrt{p_i}\, U\|_{L^2(I_i)},$$

$$(4.2) \qquad |U(r_i)| \leq \varepsilon \|U_r\|_{L^2(I_i)} + \|p_i\|_{L^1(I_{i,\varepsilon})}^{-1} \|p_i\, U\|_{L^1(I_i)}$$

for any $\varepsilon \in (0, \varepsilon_i]$, and

$$(4.3) \qquad \|\sqrt{p_i}\, U\|_{L^2(I_i)} \leq C_{I_i, p_i} \{\|U_r\|_{L^2(I_i)} + \|p_i\, U\|_{L^1(I_i)}\},$$

where

$$I_{1,\varepsilon} = [r_1(1 - \varepsilon^2/2), r_1], \qquad I_{2,\varepsilon} = [r_2, r_2(1 + \varepsilon^2)],$$
$$\varepsilon_1 = 1, \qquad \varepsilon_2 = \sqrt{(1 - r_2)/r_2},$$
$$C_{I_1, p_1} = \|\{1 + 2\log(r_1/r)\}\, p_1\|_{L^1(I_1)}^{1/2} + 2\|p_1\|_{L^1(0, 7r_1/8)}^{-1} \|p_1\|_{L^1(I_1)}^{1/2},$$
$$C_{I_2, p_2} = \|\{(1/r_2 - 1)^2 + 2\log(r/r_2)\}\, p_2\|_{L^1(I_2)}^{1/2}$$
$$\qquad + 2\|p_2\|_{L^1(r_2, (r_2+1)/2)}^{-1} \|p_2\|_{L^1(I_2)}^{1/2}.$$

Lemma 4.2. *In addition to the assumptions of Lemma 4.1, suppose that threre exists non-negative constants $k_1, k_2 > 0$ such that*

$$\lim_{\substack{r \to r_i \\ r \in I_i}} \frac{p_i(r)}{|r - r_i|^{k_i}} > 0 \quad (i = 1, 2).$$

If $U \in H^1(I_i)$, then there exists a constant C such that

$$|U(r_i)| \leq \epsilon \|U_r\|_{L^2(I_i)} + C\epsilon^{-2(k_i+1)}\|p_i\, U\|_{L^1(I_i)}$$

for any $\epsilon \in min\{1, \sqrt{(1 - r_2)r_2}/2\}$.

The following simple lemma is fundamental and important.

Lemma 4.3. *Suppose that (R.1) is satisfied. Let $\alpha \geq 0, \beta \geq 0$ and $\gamma \geq 0$ be arbitrary given. Then the following equation*

$$\beta = \beta^* + q_1 R(\alpha, \beta^*, \gamma^*),$$
$$\gamma = \gamma^* - q_2 R(\alpha, \beta^*, \gamma^*)$$

with respect to β^* and γ^* has the unique non-negative solution.

The following lemma is key lemma of the proof of non-negativity of solutions.

Lemma 4.4. *Let $(u, v, w, \beta^*, \gamma^*)$ be a local solutions of (Q). Suppose that*

$$u, v, w, \beta^*, \gamma^* \geq -\delta$$

for some sufficiently small $\delta > 0$. Then there exists a non-negative continuous function $J(z)$ such that

$$R(u(r_1, z), \beta^*(z), \gamma^*(z)) = J(z)R(u(r_1, z), v(r_2, z), w(r_2, z)).$$

Now we shall prove the non-negatity of the local solutions obtained in Step 2.

Proposition 4.1. *Let $(u, v, w, \beta^*, \gamma^*)$ be a local solutions of (Q), and $u, v, w, \beta^*, \gamma^* \geq -\delta$ for some sufficiently small $\delta > 0$. Suppose that*

$$u_0, v_0, w_0 \geq 0,$$

then

$$u, v, w, \beta^*, \gamma^* \geq 0.$$

Proof. It follows from (1.1),(1.20), and Lemma 4.4 that

$$\frac{1}{2}\int_0^{r_1} u^-(r,z)^2 a(r)\,rdr = -\int_0^{r_1} u^- \frac{\partial u}{\partial t} a\,rdr$$

$$= -\int_0^{r_1} u^-\Delta u\,rdr = -\int_0^{r_1} \frac{1}{r}(ru_r)_r u^-\,rdr$$

$$= -r_1 u_r(r_1,z)u^-(r_1,z) - \int_0^{r_1}(u^-{}_r)^2\,rdr$$

$$= r_1 R(u(r_1,z),\beta^*(z),\gamma^*(z))u^-(r_1,z) - \|u^-{}_r\|^2_{L^2(I_1)}$$

$$= r_1 J(z)R(u(r_1,z),v(r_2,z),w(r_2,z))u^-(r_1,z) - \|u^-{}_r\|^2_{L^2(I_1)}.$$

Similarly we have

$$\frac{1}{2}\int_{r_2}^1 v^-(r,z)^2 b(r)\,rdr$$

$$= \frac{m_1 r_1}{m_2}J(z)R(u(r_1,z),v(r_2,z),w(r_2,z))v^-(r_2,z) - \|v^-{}_r\|^2_{L^2(I_2)},$$

and

$$\frac{1}{2}\int_{r_2}^1 w^-(r,z)^2 c(r)\,rdr$$

$$= -\frac{n_1 r_1}{n_2}J(z)R(u(r_1,z),v(r_2,z),w(r_2,z))w^-(r_2,z) - \|w^-{}_r\|^2_{L^2(I_2)},$$

Thus it follows from (R.3) that

$$\frac{1}{2}\frac{d}{dt}\{\int_0^{r_1}|u^-|^2 a\,rdr + \int_{r_2}^1 |v^-|^2 b\,rdr + \int_{r_2}^1 |w^-|^2 c\,rdr\}$$

$$+ \int_0^{r_1}|u^-{}_r|^2\,rdr + \int_{r_2}^1 |v^-{}_r|^2\,rdr + \int_{r_2}^1 |w^-{}_r|^2\,rdr$$

$$\leq C_1\{|u^-(r_1,z)|^2 + |v^-(r_2,z)|^2 + |w^-(r_2,z)|^2\},$$

where C_1 is some positive constant. Therefore by (A.1) and (4.1),

$$\{\int_0^{r_1}|u^-|^2 a\,rdr + \int_{r_2}^1 |v^-|^2 b\,rdr + \int_{r_2}^1 |w^-|^2 c\,rdr\}$$

$$+ \int_0^{r_1}|u^-{}_r|^2\,rdr + \int_{r_2}^1 |v^-{}_r|^2\,rdr + \int_{r_2}^1 |w^-{}_r|^2\,rdr$$

$$\leq C_2\{\int_0^{r_1}|u^-|^2 a\,rdr + \int_{r_2}^1 |v^-|^2 b\,rdr + \int_{r_2}^1 |w^-|^2 c\,rdr\},$$

where C_2 is some positive constant. Integrating the above inequality and applying Gronwall's inequality, we obtain $u^-(r,z) \equiv 0$, because $u^-(r,0) \equiv 0$

in view of the non-negativity of the initial data. Thus we get the non-negativity of u, v, w. Consequently we obtain the non-negativity of β^* and γ^* by (1.23),(1.24) and Lemma 4.3.
$$\text{Q.E.D.}$$

REFERENCES

[A] N.D.Alikakos, *An application of the invariant principle to reactiondiffusion equations*, J. Differential Equations **33** (1979), 201-225.

[AP] D.G.Aronson and L.A.Peletier, *Global stability of symmetric and asymmetric concentration profiles in catalyst particles*, Arch. Rational Mech. Anal. **54** (1974), 175-204.

[F] A.Friedman, *Partial Differential Equations*, Holt-Reinehart-Winston, 1969.

[KKKN] Y.Kawano, K.Kusano, K.Kondo and F.Nakashio, *Extraction rate of acetic acid by long-chain alkylamine in horizontal rectangular channels*, Kagaku Kôgaku Ronbunshu **9** (1983), 44-55.

[LSU] O.A.Ladyženskaya, V.A.Solonikov and N.N.Ural'ceva, *Linear and Quasilinear Equations of Parabolic Type*, Transl. Math. Monogr. 23, Amer. Math. Soc., Providence, R. I., 1968.

[M] J.Moser, *A Harnack inequality for parabolic differential equations*, Comm. Pure Appl. Math. **17** (1964), 101-134.

[YKN] K.Yoshizuka, K.Kondo and F.Nakashio, *Effect of interfacial reaction on rates of extraction and stripping in membrane extraction using a hollow fiber*, J. Chem. Eng. Japan **19** (1986), 312-318.

[YY1] Y.Yamada and S.Yotsutani, *Note on chemical interfacial interfacial reaction models*, Proc. Japan Acad. **62A** (1986), 379-381.

[YY2] Y.Yamada and S.Yotsutani, *A mathematical models on chemical interfacial reactions*, Japan J. Appl. Math. **7** (1990), 369-398.

Faculty of Science and Technology
Department of Applied Mathematics and Informatics
Ryukoku University
Seta, Ohtsu 520-21, Japan

Progress in Nonlinear Differential Equations and Their Applications

Editor
Haim Brezis
Département de Mathématiques
Université P. et M. Curie
4, Place Jussieu
75252 Paris Cedex 05
France
and
Department of Mathematics
Rutgers University
New Brunswick, NJ 08903
U.S.A.

Progress in Nonlinear Differential Equations and Their Applications is a book series that lies at the interface of pure and applied mathematics. Many differential equations are motivated by problems arising in such diversified fields as Mechanics, Physics, Differential Geometry, Engineering, Control Theory, Biology, and Economics. This series is open to both the theoretical and applied aspects, hopefully stimulating a fruitful interaction between the two sides. It will publish monographs, polished notes arising from lectures and seminars, graduate level texts, and proceedings of focused and refereed conferences.

We encourage preparation of manuscripts in some form of TeX for delivery in camera-ready copy, which leads to rapid publication, or in electronic form for interfacing with laser printers or typesetters.

Proposals should be sent directly to the editor or to: Birkhäuser Boston, 675 Massachusetts Avenue, Cambridge, MA 02139